£55.00

The Art of Digital Audio

For Chrissie

The Art of
Digital Audio

Third Edition

John Watkinson

Focal Press

OXFORD   AUCKLAND   BOSTON   JOHANNESBURG   MELBOURNE   NEW DELHI

Focal Press
An imprint of Butterworth-Heinemann
Linacre House, Jordan Hill, Oxford OX2 8DP
225 Wildwood Avenue, Woburn, MA 01801-2041
A division of Reed Educational and Professional Publishing Ltd

A member of the Reed Elsevier plc group

First published 1988
Reprinted 1989, 1991, 1992
Second edition 1994
Reprinted 1994, 1995, 1999, 2000
Third edition 2001

British Library Cataloguing in Publication Data
A catalogue record for this book is available from the British Library

Library of Congress Cataloguing in Publication Data
A catalogue record for this book is available from the Library of Congress

ISBN 0 240 51587 0

Composition by Genesis Typesetting, Rochester, Kent
Printed and bound in Great Britain

FOR EVERY TITLE THAT WE PUBLISH, BUTTERWORTH-HEINEMANN
WILL PAY FOR BTCV TO PLANT AND CARE FOR A TREE.

Contents

Chapter 3 Digital principles 81

Preface

It is over ten years since I wrote the first edition of this book and it seems today that almost everything about the subject has changed except for the laws of physics. As a result there are many things in this edition which have changed beyond recognition, contrasting with some basics which have not changed at all.

Back then I felt that the approach of many technical books limited them to mathematically inclined readers of a certain level, and I resolved to write something where the mathematics were replaced with plain English and where statements were replaced by explanations. I also felt strongly that theories would be much more readily understood if they were illustrated by examples in practical equipment. In short, I wrote a book which I would have wanted to read myself.

Although I was pleased with the way the book turned out, I was not prepared for either the quality or the quantity of the acclaim which it received. To have one's first book commonly referred to as a bible is a great reward for the effort which went into it, and convinced me that the radical approach taken should be maintained in any future edition.

The great success of the first two editions of this book had the effect that I was closely identified with digital audio rather than audio in general. This is partly my own fault because the first edition assumed that the reader would have an existing understanding of analog audio equipment to which the digital knowledge could be added.

Over ten years of teaching digital audio and video technology has shown me that the people who have the most difficulty with digital are those who didn't understand analog in the first place. Today, digital audio is no longer a specialist subject, but has become the norm. It is impossible to function in the audio industry without some knowledge of it.

It follows that this book must change its focus and include more about the general subject of audio so that it is clear what is going to be digitized and how well it needs to be done. The approach of this edition is to begin with a treatment of the human hearing system which defines the necessary quality, and to move on to see how that quality can be reached in the digital domain. Of course, in real systems economic or practical considerations may prevent the goal of transparency being reached, but this is best done with an accurate knowledge of what is being lost. A full chapter is devoted to obtaining the highest quality from a digital audio system.

Another inevitable consequence of the universality of digital audio is that the proportion of those involved who fully grasp the fundamentals must fall. In such a climate pseudoscience can flourish where technically misleading statements are made for the sole purpose of selling equipment. Such statements cause confusion to genuine students and waste the time of their teachers.

The reader may be confident that no such statements appear in this book. On the contrary, this book contains barely any statements at all. Statements are not useful because they don't give confidence. Instead this book contains explanations and arguments from first principles which the reader can follow and gain confidence that what is being said is true. Further confidence can be gained by seeking out the many references quoted. Essentially this is an introductory book, a theory book, an applications book and a reference book all in one.

One of the aspects of digital audio which still fascinates the author is the wide range of disciplines that it embraces. There is still no explanation of the mechanism designers use to combine techologies to make an affordable product. Whilst the theory we use is scientifically based, putting it into practice remains an art, hence the title.

Acknowledgements

I must first thank Denis Mee and Eric Daniel, for it was their invitation to contribute a chapter on digital audio to their definitive work on magnetic recording[1] which gave me the impetus to write a self-contained work.

The Audio Engineering Society provides a continuing forum for discussion of digital audio though its conferences and publications which was essential to my researches. The AES elected me to the status of Fellow in 1991 for my contribution to digital audio and I regard this as the greatest privilege ever extended to me.

Many individuals have found time to discuss complicated subjects and to supply reference material, others have helped by suggesting subjects for inclusion or by reviewing material. I would particularly like to thank the following: Toshi Doi, John Ajimine, David Bush, Roger Lagadec, Kees Schouhamer Immink, Roger Wood, John Mallinson, Tim Shelton, Guy McNally, Tony Griffiths, Steven Harris, Graham Roe, Robin Caine, Steve Lyman, Tom Cavanagh, Joseph Manger, John Vanderkooy and Stanley Lipshitz.

Francis Rumsey of the University of Surrey kindly suggested the title.

John Watkinson
Burghfield Common, England

Reference

1. Mee, C.D. and Daniel, E.D., *Magnetic Recording – Vol III*, New York: McGraw-Hill (1987)

1

Why digital?

1.1 Introduction

The applications of audio technology are numerous, but generally the goal is to reproduce sound at a later time, at another place or both. The consumer needs reasonably affordable equipment which will reproduce recordings or receive transmissions, whereas the record company or broadcaster needs equipment which can manipulate audio signals to produce recordings or programs. In this case flexibility and speed of operation are more important than first cost.

In one sense provided the sound is reproduced to an acceptable standard, the user doesn't care how it is done. The point to be stressed is that it is the service that is needed, not the technology. People don't want technology; instead they want the services it provides. As a result when a new technology comes along it may be adopted if the service is better in some way or if the same service is possible at lower cost or with smaller equipment. Digital audio did just that, irrevocably transforming the face of audio in a very short time for both consumer and professional alike. This is not a history book, but readers interested in the history of digital audio are referred to Chapter 8 of *Magnetic Recording: The first 100 years*.[1]

The first techniques to be used for sound recording, transmission and processing were understandably analog. Some mechanical, electrical or magnetic parameter was caused to vary in the same way that the sound to be recorded had varied the air pressure. The voltage coming from a microphone is an analog of the air pressure (or sometimes velocity), but both vary in the same timescale; the magnetism on a tape or the deflection of a disk groove is an analog of the electrical input signal, but in recorders there is a further analog between time in the input signal and distance along the medium.

In an analog system, information is conveyed by some infinite variation of a continuous parameter such as the voltage on a wire or the strength of flux on a tape. In a recorder, distance along the medium is a further, continuous, analog of time. It does not matter at what point a recording is examined along its length, a value will be found for the recorded signal. That value can itself change with infinite resolution within the physical limits of the system.

Those characteristics are the main weakness of analog signals. Within the allowable bandwidth, *any* waveform is valid. If the speed of the medium is not constant, one valid waveform is changed into another valid waveform; a timebase error cannot be detected in an analog system. In addition, a voltage error simply changes one valid voltage into another; noise cannot be detected in an analog system. We might suspect noise, but how is one to know what proportion of the received voltage is noise and what is the original? If the transfer function of a system is not linear, distortion results, but the distorted waveforms are still valid; an analog system cannot detect distortion. Again we might suspect distortion, but how are we to know how much of the third harmonic energy received is due to the distortion and how much was actually present in the original signal?

It is a characteristic of analog systems that degradations cannot be separated from the original signal, so nothing can be done about them. At the end of a system a signal carries the sum of all degradations introduced in the stages through which it passed. This sets a limit to the number of stages through which a signal can be passed before it is useless. Alternatively, if many stages are envisaged, each piece of equipment must be far better than necessary so that the signal is still acceptable at the end. The equipment will naturally be more expensive.

When setting out to design any audio equipment, it is important to appreciate that the final arbiter is the human hearing system. If the audio signal is reproduced less accurately than our senses, these shortcomings will be audible, whereas if the system is more accurate than our senses, it will *appear* perfect even though it is not. Making the system better still is then a waste of resources. This topic will be explored in more detail in Chapters 2 and 13.

1.2   What is digital audio?

One of the vital concepts to grasp is that digital audio is simply an alternative means of carrying audio information. An ideal digital audio recorder has the same characteristics as an ideal analog recorder: both of

them are totally transparent and reproduce the original applied waveform without error. One need only compare high-quality analog and
digital equipment side by side with the same signals to realize how
transparent modern equipment can be. Needless to say, in the real world
ideal conditions seldom prevail, so analog and digital equipment both fall
short of the ideal. Digital audio simply falls short of the ideal by a smaller
distance than does analog and at lower cost, or, if the designer chooses,
can have the same performance as analog at much lower cost.

Although there are a number of ways in which audio can be
represented digitally, there is one system, known as pulse code
modulation (PCM), which is in virtually universal use. Figure 1.1 shows
how PCM works. Instead of being continuous, the time axis is
represented in a discrete, or stepwise manner. The waveform is not
carried by continuous representation, but by measurement at regular
intervals. This process is called sampling and the frequency with which
samples are taken is called the sampling rate or sampling frequency F_s.
The sampling rate is generally fixed and is thus independent of any
frequency in the signal. If every effort is made to rid the sampling clock
of jitter, or time instability, every sample will be made at an exactly even
time step. Clearly if there is any subsequent timebase error, the instants at
which samples arrive will be changed and the effect can be detected. If
samples arrive at some destination with an irregular timebase, the effect
can be eliminated by storing the samples temporarily in a memory and

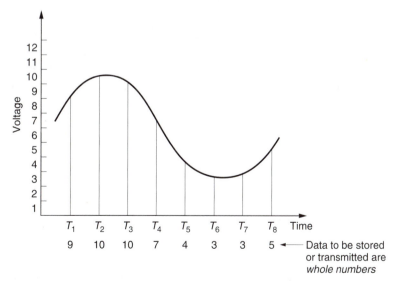

Figure 1.1 In pulse code modulation (PCM) the analog waveform is measured
periodically at the sampling rate. The voltage (represented here by the height) of each
sample is then described by a whole number. The whole numbers are stored or
transmitted rather than the waveform itself.

reading them out using a stable, locally generated clock. This process is called timebase correction and all properly engineered digital audio systems must use it. Clearly timebase error is not reduced; it is totally eliminated. As a result there is little point measuring the wow and flutter of a digital recorder; it doesn't have any. What happens is that the crystal clock in the timebase corrector measures the stability of the flutter meter. It should be stressed that sampling is an analog process. Each sample still varies infinitely as the original waveform did. Sampled analog devices are well known in audio. These are generally implemented with charge-coupled registers and are used for chorus effects in keyboards and for delay in public address systems.

Those who are not familiar with digital audio often worry that sampling takes away something from a signal because it is not taking notice of what happened between the samples. This would be true in a system having infinite bandwidth, but no analog audio signal can have infinite bandwidth. All analog signal sources such as microphones, tape decks, pickup cartridges and so on have a frequency response limit, as indeed do our ears. When a signal has finite bandwidth, the rate at which it can change is limited, and the way in which it changes becomes predictable. When a waveform can only change between samples in one way, the original waveform can be reconstructed from them. A more detailed treatment of the principle will be given in Chapter 4.

Figure 1.1 also shows that each sample is also discrete, or represented in a stepwise manner. The length of the sample, which will be proportional to the voltage of the audio waveform, is represented by a whole number. This process is known as quantizing and results in an approximation, but the size of the error can be controlled until it is negligible. If, for example, we were to measure the height of humans to the nearest metre, virtually all adults would register two metres high and obvious difficulties would result. These are generally overcome by measuring height to the nearest centimetre. Clearly there is no advantage in going further and expressing our height in a whole number of millimetres or even micrometres, although no doubt some Hi-Fi enthusiasts will be able to advance reasons for doing so. The point is that an appropriate resolution can also be found for audio, and a higher figure is not beneficial. The link between audio quality and sample resolution is explored in Chapter 4.

The advantage of using whole numbers is that they are not prone to drift. If a whole number can be carried from one place to another without numerical error, it has not changed at all. By describing audio waveforms numerically, the original information has been expressed in a way which is better able to resist unwanted changes.

Essentially, digital audio carries the original waveform numerically. The number of the sample is an analog of time, and the magnitude of the sample is an analog of the pressure at the microphone. In fact the succession of samples in a digital system is actually *an analog* of the original waveform. This sounds like a contradiction and as a result some authorities prefer the term 'numerical audio' to 'digital audio' and in fact the French word is *numérique*. The term 'digital' is so well established that it is unlikely to change.

As both axes of the digitally represented waveform are discrete, the waveform can accurately be restored from numbers as if it were being drawn on graph paper. If we require greater accuracy, we simply choose paper with smaller squares. Clearly more numbers are then required and each one could change over a larger range.

In simple terms, the audio waveform is conveyed in a digital recorder as if the voltage had been measured at regular intervals with a digital meter and the readings had been written down on a roll of paper. The rate at which the measurements were taken and the accuracy of the meter are the only factors which determine the quality, because once a parameter is expressed as a discrete number, a series of such numbers can be conveyed unchanged. Clearly in this example the handwriting used and the grade of paper have no effect on the information. The quality is determined only by the accuracy of conversion and is independent of the quality of the signal path.

1.3   Why binary?

Humans insist on using numbers expressed to the base of ten, having evolved with that number of digits. Other number bases exist; most people are familiar with the duodecimal system which uses the dozen and the gross. The most minimal system is binary, which has only two digits, 0 and 1. BInary digiTS are universally contracted to bits. These are readily conveyed in switching circuits by an 'on' state and an 'off' state. With only two states, there is little chance of error.

In decimal systems, the digits in a number (counting from the right, or least significant end) represent ones, tens, hundreds and thousands etc. Figure 1.2 shows that in binary, the bits represent one, two, four, eight, sixteen etc. A multi-digit binary number is commonly called a word, and the number of bits in the word is called the wordlength. The right-hand bit is called the least significant bit (LSB) whereas the bit on the left-hand end of the word is called the most significant bit (MSB). Clearly more digits are required in binary than in decimal, but they are more easily handled. A word of eight bits is called a byte, which is a contraction of 'by eight'.

The capacity of memories and storage media is measured in bytes, but to avoid large numbers, kilobytes, megabytes and gigabytes are often

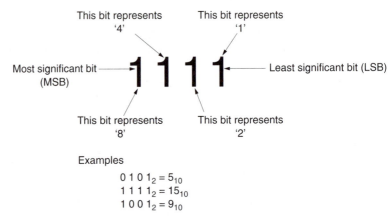

Figure 1.2 In a binary number, the digits represent increasing powers of two from the LSB. Also defined here are MSB and wordlength. When the wordlength is eight bits, the word is a byte. Binary numbers are used as memory addresses, and the range is defined by the address wordlength. Some examples are shown here.

used. As memory addresses are themselves binary numbers, the wordlength limits the address range. The range is found by raising two to the power of the wordlength. Thus a four-bit word has sixteen combinations, and could address a memory having sixteen locations. A ten-bit word has 1024 combinations, which is close to one thousand. In digital terminology, 1K = 1024, so a kilobyte of memory contains 1024 bytes. A megabyte (1 MB) contains 1024 kilobytes and a gigabyte contains 1024 megabytes.

In a digital audio system, the whole number representing the length of the sample is expressed in binary. The signals sent have two states, and change at predetermined times according to some stable clock. Figure 1.3 shows the consequences of this form of transmission. If the binary signal is degraded by noise, this will be rejected by the receiver, which judges the signal solely by whether it is above or below the half-way threshold, a process known as slicing. The signal will be carried in a channel with finite bandwidth, and this limits the slew rate of the signal; an ideally upright edge is made to slope. Noise added to a sloping signal can change the time at which the slicer judges that the level passed through the threshold. This effect is also eliminated when the output of the slicer is reclocked. However many stages the binary signal passes through, the information is unchange except for a delay.

Audio samples which are represented by whole numbers can reliably be carried from one place to another by such a scheme, and if the number is correctly received, there has been no loss of information en route.

There are two ways in which binary signals can be used to carry audio samples and these are shown in Figure 1.4. When each digit of the binary

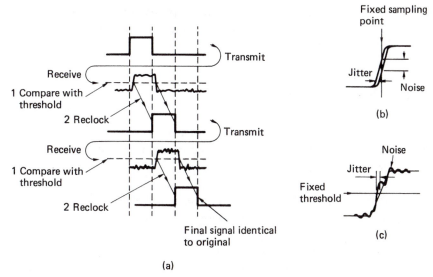

Figure 1.3 (a) A binary signal is compared with a threshold and reclocked on receipt, thus the meaning will be unchanged. (b) Jitter on a signal can appear as noise with respect to fixed timing. (c) Noise on a signal can appear as jitter when compared with a fixed threshold.

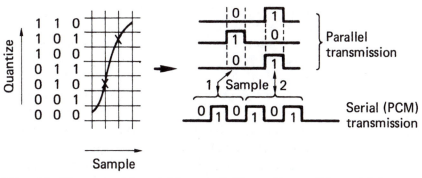

Figure 1.4 When a signal is carried in numerical form, either parallel or serial, the mechanisms of Figure 1.3 ensure that the only degradation is in the conversion processes.

number is carried on a separate wire this is called parallel transmission. The state of the wires changes at the sampling rate. Using multiple wires is cumbersome, particularly where a long wordlength is in use, and a single wire can be used where successive digits from each sample are sent serially. This is the definition of pulse code modulation. Clearly the clock frequency must now be higher than the sampling rate. Whilst the transmission of audio by such a scheme is advantageous in that noise and timebase error have been eliminated, there is a penalty that a single high-quality audio channel requires around one million bits per second. Digital audio came into wide use as soon as such a data rate could be handled economically. Further applications become possible when means to reduce or *compress* the data rate become economic. Chapter 5 considers audio compression.

1.4 Why digital?

There are two main answers to this question, and it is not possible to say which is the most important, as it will depend on one's standpoint.

(a) The quality of reproduction of a well-engineered digital audio system is independent of the medium and depends only on the quality of the conversion processes. If compression is used this can also affect the quality.
(b) The conversion of audio to the digital domain allows tremendous opportunities which were denied to analog signals.

Someone who is only interested in sound quality will judge the former the most relevant. If good-quality convertors can be obtained, all the shortcomings of analog recording can be eliminated to great advantage. One's greatest effort is expended in the design of convertors, whereas those parts of the system which handle data need only be workmanlike. Wow, flutter, particulate noise, print-through, dropouts, modulation noise, HF squashing, azimuth error, and interchannel phase errors are all eliminated.

When a digital recording is copied, the same numbers appear on the copy: it is not a dub, it is a clone. If the copy is indistinguishable from the original, there has been no generation loss. Digital recordings can be copied indefinitely without loss of quality. If you happen to be a sound engineer, this is heaven. If you are a record company executive you take another pill for blood pressure and phone your lawyer to see if you can have it stopped.

In the real world everything has a cost, and one of the greatest strengths of digital technology is low cost. If copying causes no quality loss, recorders do not need to be far better than necessary in order to withstand generation loss. They need only be adequate on the first generation whose quality is then maintained. There is no need for the great size and extravagant tape consumption of professional analog recorders. When the information to be recorded is discrete numbers, they can be packed densely on the medium without quality loss. Should some bits be in error because of noise or dropout, error correction can restore the original value. Digital recordings take up less space than analog recordings for the same or better quality. Tape costs are far less and storage costs are reduced.

Digital circuitry costs less to manufacture. Switching circuitry which handles binary can be integrated more densely than analog circuitry. More functionality can be put in the same chip. Analog circuits are built from a host of different component types which have a variety of shapes and sizes and are costly to assemble and adjust. Digital circuitry uses standardized component outlines and is easier to assemble on automated equipment. Little if any adjustment is needed.

Once audio is in the digital domain, it becomes data, and as such is indistinguishable from any other type of data. Systems and techniques developed in other industries for other purposes can be used for audio. Computer equipment is available at low cost because the volume of production is far greater than that of professional audio equipment. Disk drives and memories developed for computers can be put to use in audio products. A word processor adapted to handle audio samples becomes a workstation. There seems to be little point in waiting for a tape to wind when a disk head can access data in milliseconds. The difficulty of locating the edit point and the irrevocable nature of tape-cut editing are immediately seen as outmoded when the edit point can be located by viewing the audio waveform on a screen or by listening at any speed to audio from a memory. The edit can be simulated or *previewed* and trimmed before it is made permanent.

The merging of digital audio and computation is two-sided. Whilst audio may borrow RAM and hard disk technology from the computer industry, Compact Disc and DAT were borrowed back to create CD-ROM and DDS (digital data storage).

Communications networks developed to handle data can happily carry digital audio over indefinite distances without quality loss. Digital audio broadcasting (DAB) makes use of these techniques to eliminate the interference, fading and multipath reception problems of analog broadcasting. At the same time, more efficient use is made of available bandwidth. In one sense DAB is just conventional radio done with digital transmission. The listener still has to accept what the broadcaster chooses to transmit. In contrast, if the listener uses a data communication channel such as the Internet, any audio program material can in principle be accessed at any time over any distance.

Digital equipment can have self-diagnosis programs built-in. The machine points out its own failures. The days of chasing a signal with an oscilloscope are over. Even if a faulty component in a digital circuit could be located with such a primitive tool, it may be impossible to replace a chip having 60 pins soldered through a six-layer circuit board. The cost of finding the fault may be more than the board is worth. Routine, mindnumbing adjustment of analog circuits to counteract drift is no longer needed. The cost of maintenance falls. A small operation may not need maintenance staff at all; a service contract is sufficient. A larger organization will still need maintenance staff, but they will be fewer in number and their skills will be oriented more to systems than to devices.

As a result of the above, the cost of ownership of digital equipment has for some time now been less than that of analog. Debates about quality are academic; in recording and transmission, analog equipment can no longer compete economically, and it is going out of service as surely as the transistor once replaced the vacuum-tube in electronics and the turbine replaced the piston engine in commercial aviation.

1.5   Some digital audio processes outlined

Whilst digital audio is a large subject, it is not necessarily a difficult one. Every process can be broken down into smaller steps, each of which is relatively easy to follow. The main difficulty with study is to appreciate where the small steps fit in the overall picture. Subsequent chapters of this book will describe the key processes found in digital technology in some detail, whereas this chapter illustrates why these processes are necessary and shows how they are combined in various ways in real equipment. Once the general structure of digital devices is appreciated, the following chapters can be put in perspective.

Figure 1.5(a) shows a minimal digital audio system. This is no more than a point-to-point link which conveys analog audio from one place to another. It consists of a pair of convertors and hardware to serialize and deserialize the samples. There is a need for standardization in serial transmission so that various devices can be connected together. These standards for digital audio interfaces are described in Chapter 7.

Analog audio entering the system is converted in the analog-to-digital convertor (ADC) to samples which are expressed as binary numbers. A typical sample would have a wordlength of sixteen bits. The sample is loaded in parallel into a shift register which is then shifted with a clock running at sixteen times the sampling rate. The data are sent serially to the other end of the line where a slicer rejects noise picked up on the signal. Sliced data are then shifted into a receiving shift register with a bit clock. Once every sixteen bits, the shift register contains a whole sample, and this is read out by the sampling rate clock, or word clock, and sent to the digital-to-analog convertor (DAC), which converts the sample back to an analog voltage.

Following a casual study one might conclude that if the convertors were of transparent quality, the system would be ideal. Unfortunately this is incorrect. As Figure 1.3 showed, noise can change the timing of a sliced signal. Whilst this system rejects noise which threatens to change the numerical value of the samples, it is powerless to prevent noise from

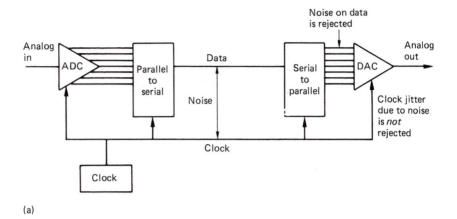

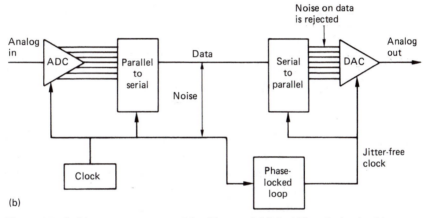

Figure 1.5 In (a) two convertors are joined by a serial link. Although simple, this system is deficient because it has no means to prevent noise on the clock lines causing jitter at the receiver. In (b) a phase-locked loop is incorporated, which filters jitter from the clock.

causing jitter in the receipt of the word clock. Noise on the word clock means that samples are not converted with a regular timebase and the impairment caused can be audible. Stated another way, analog characteristics of the interconnect are not prevented from affecting the reproduced waveform and so the system is not truly digital.

The jitter problem is overcome in Figure 1.5(b) by the inclusion of a phase-locked loop which is an oscillator which synchronizes itself to the *average* frequency of the word clock but which filters out the instantaneous jitter. The operation of a phase-locked loop is analogous to the function of the flywheel on a piston engine. The samples are then fed to the convertor with a regular spacing and the impairment is no longer audible. Chapter 4 shows why the effect occurs and deduces the remarkable clock accuracy needed for accurate conversion.

Whilst this effect is reasonably obvious, it does not guarantee that all convertors take steps to deal with it. Many outboard DACs sold on the consumer market have no phase-locked loop, and one should not be surprised that they can sound worse than the inboard convertor they are supposed to replace. In the absence of timebase correction, the sound quality of an outboard convertor can be affected by such factors as the type of data cable used and the power supply noise of the digital source. Clearly if the sound of a given DAC is affected by cable or source, it is simply not well engineered and should be rejected. Almost by definition a good remote DAC rejects noise and jitter on the digital inputs and its sound is not affected by the digital source or the analog characteristics of the cable.

1.6   The sampler

The system of Figure 1.5 is extended in Figure 1.6 by the addition of some random access memory (RAM). The operation of RAM is described in Chapter 3. What the device does is determined by the way in which the RAM address is controlled. If the RAM address increases by one every time a sample from the ADC is stored in the RAM, a recording can be made for a short period until the RAM is full. The recording can be played back by repeating the address sequence at the same clock rate but reading the memory into the DAC. The result is generally called a sampler. By running the replay clock at various rates, the pitch and duration of the reproduced sound can be altered. At a rate of one million bits per second, a megabyte of memory gives only eight seconds' worth of recording, so clearly samplers will be restricted to a fairly short playing time.

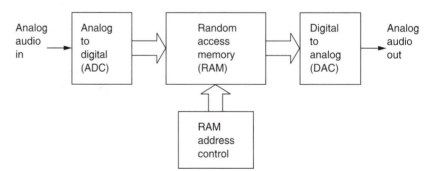

Figure 1.6   In the digital sampler, the recording medium is a random access memory (RAM). Recording time available is short compared with other media, but access to the recording is immediate and flexible as it is controlled by addressing the RAM.

Using compression, the playing time of a RAM-based recorder can be extended. Some telephone answering machines take messages in RAM and eliminate the cassette tape. For pre-determined messages read only memory (ROM) can be used instead as it is non-volatile. Announcements in aircraft, trains and elevators are one application of such devices. RAM-based recorders are now available which can download suitably compressed audio data over the Internet. Having no moving parts these are highly portable.

1.7 The programmable delay

If the RAM is used in a different way, it can be written and read at the same time. The device then becomes an audio delay. Controlling the relationship between the addresses then changes the delay. The addresses are generated by counters which overflow to zero after they have reached a maximum count. As a result the memory space appears to be circular as shown in Figure 1.7. The read and write addresses are driven by a common clock and

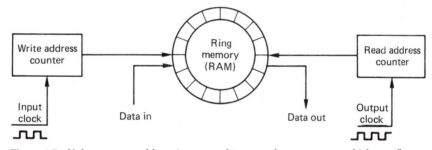

Figure 1.7   If the memory address is arranged to come from a counter which overflows, the memory can be made to appear circular. The write address then rotates endlessly, overwriting previous data once per revolution. The read address can follow the write address by a variable distance (not exceeding one revolution) and so a variable delay takes place between reading and writing.

chase one another around the circle. If the read address follows close behind the write address, the delay is short. If it just stays ahead of the write address, the maximum delay is reached. Programmable delays are useful in TV studios where they allow audio to be aligned with video which has been delayed in various processes. They can also be used in auditoria to align the sound from various loudspeakers.

One of the earliest digital audio products was a delay unit of the type shown here which was used to delay the signal leading to a vinyl disk cutter. The cutter control system could use the input to the delay to obtain advance warning of a loud passage and increase the groove pitch accordingly.

1.8   Time compression

When samples are converted, the ADC must run at a constant clock rate and it outputs an unbroken stream of samples. Time compression allows the sample stream to be broken into blocks for convenient handling.

Figure 1.8 shows an ADC feeding a pair of RAMs. When one is being written by the ADC, the other can be read, and vice versa. As soon as the first RAM is full, the ADC output switched to the input of the other RAM so that there is no loss of samples. The first RAM can then be read at a higher clock rate than the sampling rate. As a result the RAM is read in less time than it took to write it, and the output from the system then pauses until the second RAM is full. The samples are now time compressed. Instead of being an unbroken stream which is difficult to handle, the samples are now arranged in blocks with convenient pauses in between them. In these pauses numerous processes can take place. A rotary head recorder might switch heads; a hard disk might move to another track. On a tape recording, the time compression of the audio samples allows time for synchronizing patterns, subcode and error correction words to be recorded.

In digital audio recorders based on video cassette recorders (VCRs) time compression allows the continuous audio samples to be placed in blocks in the unblanked parts of the video waveform, separated by synchronizing pulses.

Subsequently, any time compression can be reversed by time expansion. Samples are written into a RAM at the incoming clock rate, but read

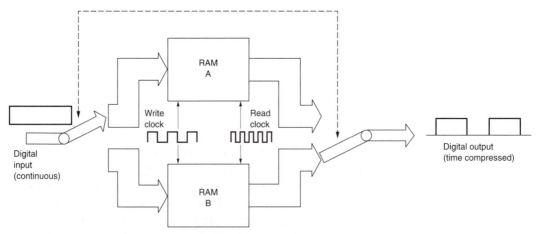

Figure 1.8   In time compression, the unbroken real-time stream of samples from an ADC is broken up into discrete blocks. This is accomplished by the configuration shown here. Samples are written into one RAM at the sampling rate by the write clock. When the first RAM is full, the switches change over, and writing continues into the second RAM whilst the first is read using a higher-frequency clock. The RAM is read faster than it was written and so all the data will be output before the other RAM is full. This opens spaces in the data flow which are used as described in the text.

out at the standard sampling rate. Unless there is a design fault, time compression is totally inaudible. In a recorder, the time expansion stage can be combined with the timebase correction stage so that speed variations in the medium can be eliminated at the same time. The use of time compression is universal in digital audio recording. In general the *instantaneous* data rate at the medium is not the same as the rate at the convertors, although clearly the *average* rate must be the same.

Another application of time compression is to allow more than one channel of audio to be carried in a single channel. If, for example, audio samples are time compressed by a factor of two, it is possible to carry samples from a stereo source in one cable. In digital video recorders both audio and video data are time compressed so that they can share the same heads and tape tracks.

1.9   Synchronization

Transfer of samples between digital audio devices in real time is only possible if both use a common sampling rate and they are synchronized. A digital audio recorder must be able to synchronize to the sampling rate of a digital input in order to record the samples. It is frequently necessary for such a recorder to be able to play back locked to an external sampling rate reference so that it can be connected to, for example, a digital mixer. The process is already common in video systems but now extends to digital audio. Chapter 8 describes a digital audio reference signal (DARS).

Figure 1.9 shows how the external reference locking process works. The timebase expansion is controlled by the external reference which becomes the read clock for the RAM and so determines the rate at which the RAM address changes. In the case of a digital tape deck, the write clock for the RAM would be proportional to the tape speed. If the tape is going too fast, the write address will catch up with the read address in the memory, whereas if the tape is going too slow the read address will catch up with the write address. The tape speed is controlled by subtracting the read address from the write address. The address difference is used to control the tape speed. Thus if the tape speed is too high, the memory will fill faster than it is being emptied, and the address difference will grow larger than normal. This slows down the tape.

Thus in a digital recorder the speed of the medium is constantly changing to keep the data rate correct. Clearly this is inaudible as properly engineered timebase correction totally isolates any instabilities on the medium from the data fed to the convertor.

In multitrack recorders, the various tracks can be synchronized to sample accuracy so that no timing errors can exist between the tracks.

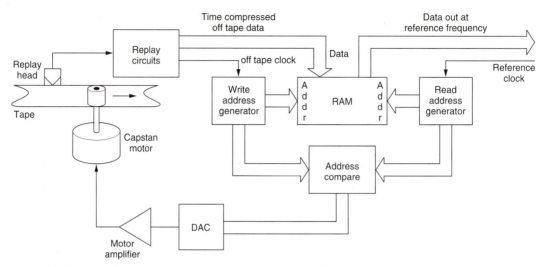

Figure 1.9 In a recorder using time compression, the samples can be returned to a continuous stream using RAM as a timebase corrector (TBC). The long-term data rate has to be the same on the input and output of the TBC or it will lose data. This is accomplished by comparing the read and write addresses and using the difference to control the tape speed. In this way the tape speed will automatically adjust to provide data as fast as the reference clock takes it from the TBC.

Extra transports can be slaved to the first to the same degree of accuracy if more tracks are required. In stereo recorders image shift due to phase errors is eliminated.

In order to replay without a reference, perhaps to provide an analog output, a digital recorder generates a sampling clock locally by means of a crystal oscillator. Provision will be made on professional machines to switch between internal and external references.

1.10   Error correction and concealment

As anyone familiar with analog recording will know, magnetic tape is an imperfect medium. It suffers from noise and dropouts, which in analog recording are audible. In a digital recording of binary data, a bit is either correct or wrong, with no intermediate stage. Small amounts of noise are rejected, but inevitably, infrequent noise impulses cause some individual bits to be in error. Dropouts cause a larger number of bits in one place to be in error. An error of this kind is called a burst error. Whatever the medium and whatever the nature of the mechanism responsible, data are either recovered correctly, or suffer some combination of bit errors and burst errors. In Compact Disc and DVD, random errors can be caused by imperfections in the moulding process, whereas burst errors are due to contamination or scratching of the disc surface.

The audibility of a bit error depends upon which bit of the sample is involved. If the LSB of one sample was in error in a loud passage of music, the effect would be totally masked and no-one could detect it. Conversely, if the MSB of one sample was in error in a quiet passage, no-one could fail to notice the resulting loud transient. Clearly a means is needed to render errors from the medium inaudible. This is the purpose of error correction.

In binary, a bit has only two states. If it is wrong, it is only necessary to reverse the state and it must be right. Thus the correction process is trivial and perfect. The main difficulty is in reliably identifying the bits which are in error. This is done by coding the data by adding redundant bits. Adding redundancy is not confined to digital technology, airliners have several engines and cars have twin braking systems. Clearly the more failures which have to be handled, the more redundancy is needed. If a four-engined airliner is designed to fly normally with one engine failed, three of the engines have enough power to reach cruise speed, and the fourth one is redundant. The amount of redundancy is equal to the amount of failure which can be handled. In the case of the failure of two engines, the plane can still fly, but it must slow down; this is graceful degradation. Clearly the chances of a two-engine failure on the same flight are remote.

In digital audio, the amount of error which can be corrected is proportional to the amount of redundancy, and it will be shown in Chapter 7 that within this limit, the samples are returned to exactly their original value. Consequently *corrected* samples are audibly indistinguishable from the originals. If the amount of error exceeds the amount of redundancy, correction is not possible, and, in order to allow graceful degradation, concealment will be used. Concealment is a process where the value of a missing sample is estimated from those nearby. The estimated sample value is not necessarily exactly the same as the original, and so under some circumstances concealment can be audible, especially if it is frequent. However, in a well-designed system, concealments occur with negligible frequency unless there is an actual fault or problem.

Concealment is made possible by rearranging or shuffling the sample sequence prior to recording. This is shown in Figure 1.10 where odd-numbered samples are separated from even-numbered samples prior to recording. The odd and even sets of samples may be recorded in different places, so that an uncorrectable burst error only affects one set. On replay, the samples are recombined into their natural sequence, and the error is now split up so that it results in every other sample being lost. The waveform is now described half as often, but can still be reproduced with some loss of accuracy. This is better than not being reproduced at all even if it is not perfect. Almost all digital recorders use such an odd/even shuffle for concealment. Clearly if any errors are

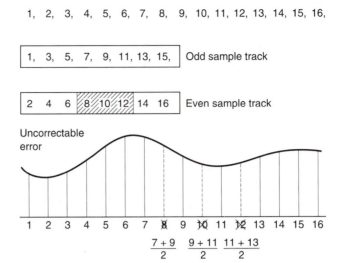

Figure 1.10 In cases where the error correction is inadequate, concealment can be used provided that the samples have been ordered appropriately in the recording. Odd and even samples are recorded in different places as shown here. As a result an uncorrectable error causes incorrect samples to occur singly, between correct samples. In the example shown, sample 8 is incorrect, but samples 7 and 9 are unaffected and an approximation to the value of sample 8 can be had by taking the average value of the two. This interpolated value is substituted for the incorrect value.

fully correctable, the shuffle is a waste of time; it is only needed if correction is not possible.

In high-density recorders, more data are lost in a given sized dropout. Adding redundancy equal to the size of a dropout to every code is inefficient. Figure 1.11 shows that the efficiency of the system can be raised using interleaving. Sequential samples from the ADC are assem-

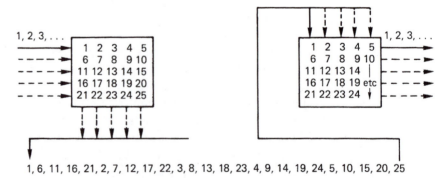

Figure 1.11 In interleaving, samples are recorded out of their normal sequence by taking columns from a memory which was filled in rows. On replay the process must be reversed. This puts the samples back in their regular sequence, but breaks up burst errors into many smaller errors which are more efficiently corrected. Interleaving and de-interleaving cause delay.

bled into codes, but these are not recorded in their natural sequence. A number of sequential codes are assembled along rows in a memory. When the memory is full, it is copied to the medium by reading down columns. On replay, the samples need to be de-interleaved to return them to their natural sequence. This is done by writing samples from tape into a memory in columns, and when it is full, the memory is read in rows. Samples read from the memory are now in their original sequence so there is no effect on the recording. However, if a burst error occurs on the medium, it will damage sequential samples in a vertical direction in the de-interleave memory. When the memory is read, a single large error is broken down into a number of small errors whose size is exactly equal to the correcting power of the codes and the correction is performed with maximum efficiency.

The interleave, de-interleave, time compression and timebase correction processes cause delay and this is evident in the time taken before audio emerges after starting a digital machine. Confidence replay takes place later than the distance between record and replay heads would indicate. In stationary head recorders, confidence replay may be about one tenth of a second behind the input. Synchronous recording requires new techniques to overcome the effect of the delays.

The presence of an error-correction system means that the audio quality is independent of the tape/head quality within limits. There is no point in trying to assess the health of a machine by listening to it, as this will not reveal whether the error rate is normal or within a whisker of failure. The only useful procedure is to monitor the frequency with which errors are being corrected, and to compare it with normal figures. Professional digital audio equipment should have an error rate display.

Some people claim to be able to hear error correction and misguidedly conclude that the above theory is flawed. Not all digital audio machines are properly engineered, however, and if the DAC shares a common power supply with the error- correction logic, a burst of errors will raise the current taken by the logic, which in turn loads the power supply and interferes with the operation of the DAC. The effect is harder to eliminate in small battery-powered machines where space for screening and decoupling components is difficult to find, but it is only a matter of good engineering; there is no flaw in the theory.

1.11   Channel coding

In most recorders used for storing digital information, the medium carries a track which reproduces a single waveform. The audio samples have to be recorded serially, one bit at a time. Some media, such as CD, only have one track, so it must be totally self-contained. Other media, such as digital compact cassette (DCC) have many parallel tracks. At high recording densities, physical tolerances cause phase shifts, or timing errors, between parallel tracks and so it is not possible to read them in parallel. Each track must still be self-contained until the replayed signal has been timebase corrected.

Recording data serially is not as simple as connecting the serial output of a shift register to the head. In digital audio, a common sample value is all zeros, as this corresponds to silence. If a shift register is loaded with all zeros and shifted out serially, the output stays at a constant low level, and no events are recorded on the track. On replay there is nothing to indicate how many zeros were present, or even how fast to move the medium. Clearly serialized raw data cannot be recorded directly, it has to be modulated into a waveform which contains an embedded clock irrespective of the values of the bits in the samples. On replay a circuit called a data separator can lock to the embedded clock and use it to count and separate strings of identical bits.

The process of modulating serial data to make it self-clocking is called channel coding. Channel coding also shapes the spectrum of the serialized waveform to make it more efficient. With a good channel code, more data can be stored on a given medium. Spectrum shaping is used in optical disks to prevent the data from interfering with the focus and tracking servos, and in DAT to allow rerecording without erase heads.

Channel coding is also needed to broadcast digital information where shaping of the spectrum is an obvious requirement to avoid interference with other services. NICAM TV sound, digital video broadcasting (DVB) and digital audio broadcasting (DAB) rely on it.

All the techniques of channel coding are covered in detail in Chapter 6 and digital broadcasting is considered in Chapter 9.

1.12   Compression

The human hearing system comprises not only the physical organs but also processes taking place within the brain. One of purposes of the subconscious processing is to limit the amount of information presented to the conscious mind, to prevent stress and to make everyday life safer and easier. Chapter 2 shows how auditory masking selects only the most important frequencies from the spectrum applied to the ear. Compression takes advantage of this process to reduce the amount of data needed to carry sound of a given subjective quality. The data-reduction process mimics the operation of the hearing mechanism as there is little point in recording information only for the ear to discard it. Compression is explained in detail in Chapter 5.

Compression is essential for services such as DAB and DVB where the bandwidth needed to broadcast regular PCM would be excessive. It can be used to reduce consumption of the medium in consumer recorders such as DCC and MiniDisc. Reduction to around one quarter or one fifth of the PCM data rate with small loss of quality is possible with high-quality compression systems. Greater compression factors inevitably result in quality loss which may be acceptable for certain applications such as communications but not for quality music reproduction.

The output of a compressor is called an elementary stream. This is still binary data, but it is no longer regular PCM, so it cannot be fed to a normal DAC without passing through a matching decoder which provides a conventional PCM output. Compressed data are more sensitive to bit errors than PCM data and concealment is more complex to implement.

There are numerous proprietary compression algorithms, and each needs the appropriate decoder to return to PCM. The combination of a compressor and a decoder is called a codec. The performance of a codec is tested on a single pass, as it would be for use in DAB or in a single-generation recording. The same performance is not necessarily obtained if codecs are cascaded, particularly if they are of different types. If an equalization step is performed on audio which has been through a codec, artifacts may be raised above the masking threshold. As a result, compression may not be suitable for the recording of original material prior to post-production.

1.13   Hard disk recorders

The hard disk stores data on concentric tracks which it accesses by moving the head radially. Clearly while the head is moving it cannot transfer data. Using time compression, a hard disk drive can be made into an audio recorder with the addition of a certain amount of memory.

Figure 1.12 shows the principle. The instantaneous data rate of the disk drive is far in excess of the sampling rate at the convertor, and so a large time-compression factor can be used. The disk drive can read a block of data from disk, and place it in the timebase corrector in a fraction of the real time it represents in the audio waveform. As the timebase corrector read address steadily advances through the memory, the disk drive has time to move the heads to another track before the memory runs out of data. When there is sufficient space in the memory for another block, the drive is commanded to read, and fills up the space. Although the data transfer at the medium is highly discontinuous, the buffer memory provides an unbroken stream of samples to the DAC and so continuous audio is obtained.

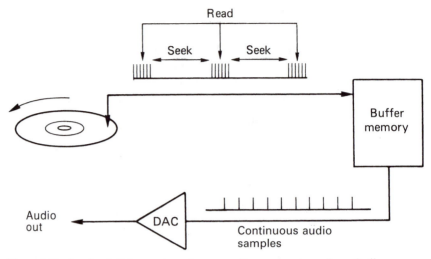

Figure 1.12 In a hard disk recorder, a large-capacity memory is used as a buffer or timebase corrector between the convertors and the disk. The memory allows the convertors to run constantly despite the interruptions in disk transfer caused by the head moving between tracks.

Recording is performed by using the memory to assemble samples until the contents of one disk block is available. This is then transferred to disk at high data rate. The drive can then reposition the head before the next block is available in memory.

An advantage of hard disks is that access to the audio is much quicker than with tape, as all the data are available within the time taken to move the head. This speeds up editing considerably. As hard disks offer so much to digital audio, the entirety of Chapter 10 is devoted to them.

The use of compression allows the recording time of a disk to be extended considerably. This technique is often used in personal computers or organizers to allow them to function as a recorder.

1.14 The PCM adaptor

The PCM adaptor was an early solution to recording the wide bandwidth of PCM audio before high density digital recording developed. The video recorder offered sufficient bandwidth at moderate tape consumption. Whilst they were a breakthrough at the time of their introduction, by modern standards PCM adaptors are crude and obsolescent. Figure 1.13 shows the essential components of a digital audio recorder using this technique. Input analog audio is converted to digital and time compressed to fit into the parts of the video waveform which are not blanked. Time-compressed samples are then odd–even shuffled to allow concealment. Next, redundancy is added and the data

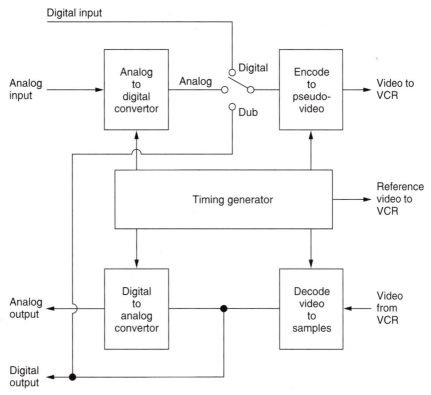

Figure 1.13 Block diagrams of PCM adaptor. Note the dub connection needed for producing a digital copy between two VCRs.

are interleaved for recording. The data are serialized and set on the active line of the video signal as black and white levels shown in Figure 1.14. The video is sent to the recorder, where the analog FM modulator switches between two frequencies representing the black and white levels, a system called frequency shift keying (FSK). This takes the place of the channel coder in a conventional digital recorder.

On replay the FM demodulator of the video recorder acts to return the FSK recording to the black/white video waveform which is sent to the PCM adaptor. The PCM adaptor extracts a clock from the video sync pulses and uses it to separate the serially recorded bits. Error correction is performed after de-interleaving, unless the errors are too great, in which case concealment is used after the de-shuffle. The samples are then returned to the standard sampling rate by the timebase expansion process, which also eliminates any speed variations from the recorder. They can then be converted back to the analog domain.

In order to synchronize playback to a reference and to simplify the circuitry, a whole number of samples is recorded on each unblanked line. The common sampling rate of 44.1 kHz is obtained by recording three

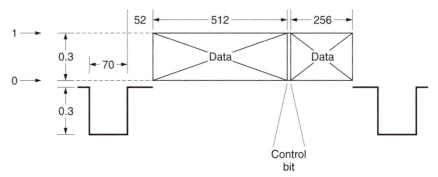

Figure 1.14 Typical line of video from PCM-1610. The control bit conveys the setting of the pre-emphasis switch or the sampling rate depending on position in the frame. The bits are separated using only the timing information in the sync pulses.

samples per line on 245 active lines at 60 Hz. The sampling rate is thus locked to the video sync frequencies and the tape is made to move at the correct speed by sending the video recorder syncs which are generated in the PCM adaptor.

1.15   An open-reel digital recorder

Figure 1.15 shows the block diagram of a machine of this type. Analog inputs are converted to the digital domain by convertors. Clearly there will be one convertor for every audio channel to be recorded. Unlike an

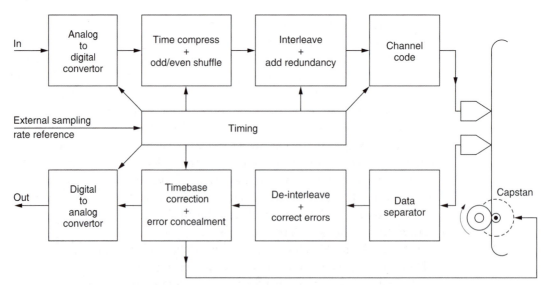

Figure 1.15   Block diagram of one channel of a stationary-head digital audio recorder. See text for details of the function of each block. Note the connection from the timebase corrector to the capstan motor so that the tape is played at such a speed that the TBC memory neither underflows nor overflows.

analog machine, there is not necessarily one tape track per audio channel. In stereo machines the two channels of audio samples may be distributed over a number of tracks each in order to reduce the tape speed and extend the playing time.

The samples from the convertor will be separated into odd and even for concealment purposes, and usually one set of samples wil be delayed with respect to the other before recording. The continous stream of samples from the convertor will be broken into blocks by time compression prior to recording. Time compression allows the insertion of edit gaps, addresses and redundancy into the data stream. An interleaving process is also necessary to reorder the samples prior to recording. As explained above, the subsequent de-interleaving breaks up the effects of burst errors on replay.

The result of the processes so far is still raw data, and these will need to be channel coded before they can be recorded on the medium. On replay a data separator reverses the channel coding to give the original raw data with the addition of some errors. Following de-interleave, the errors are reduced in size and are more readily correctable. The memory required for de-interleave may double as the timebase correction memory, so that variations in the speed of the tape are rendered undetectable. Any errors which are beyond the power of the correction system will be concealed after the odd–even shift is reversed. Following conversion in the DAC an analog output emerges.

On replay a digital recorder works rather differently from an analog recorder, which simply drives the tape at constant speed. In contrast, a digital recorder drives the tape at constant sampling rate. The timebase corrector works by reading samples out to the convertor at constant frequency. This reference frequency comes typically from a crystal oscillator. If the tape goes too fast, the memory will be written faster than it is being read, and will eventually overflow. Conversely, if the tape goes too slow, the memory will become exhausted of data. In order to avoid these problems, the speed of the tape is controlled by the quantity of data in the memory. If the memory is filling up, the tape slows down, if the memory is becoming empty, the tape speeds up. As a result, the tape will be driven at whatever speed is necessary to obtain the correct sampling rate.

1.16   Rotary head digital recorders

The rotary head recorder borrows technology from videorecorders. Rotary heads have a number of advantages which will be detailed in Chapter 9. One of these is extremely high packing density: the number of data bits which can be recorded in a given space. In a digital audio

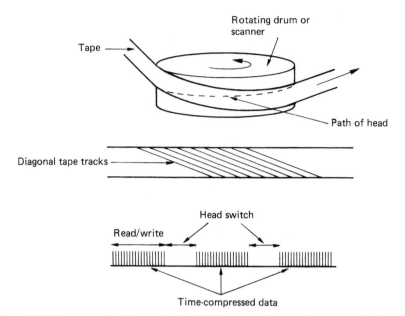

Figure 1.16 In a rotary-head recorder, the helical tape path around a rotating head results in a series of diagonal or slanting tracks across the tape. Time compression is used to create gaps in the recorded data which coincide with the switching between tracks.

recorder packing density directly translates into the playing time available for a given size of the medium.

In a rotary head recorder, the heads are mounted in a revolving drum and the tape is wrapped around the surface of the drum in a helix as can be seen in Figure 1.16. The helical tape path results in the heads traversing the tape in a series of diagonal or slanting tracks. The space between the tracks is controlled not by head design but by the speed of the tape and in modern recorders this space is reduced to zero with corresponding improvement in packing density.

The added complexity of the rotating heads and the circuitry necessary to control them is offset by the improvement in density. These techniques are detailed in Chapter 8. The discontinuous tracks of the rotary head recorder are naturally compatible with time compressed data. As Figure 1.16 illustrates, the audio samples are time compressed into blocks each of which can be contained in one slant track.

In a machine such as DAT (rotary-head digital audio tape) there are two heads mounted on opposite sides of the drum. One rotation of the drum lays down two tracks. Effective concealment can be had by recording odd-numbered samples on one track of the pair and even-numbered samples on the other.

As can be seen from the block diagram shown in Figure 1.17, a rotary head recorder contains the same basic steps as any digital audio recorder.

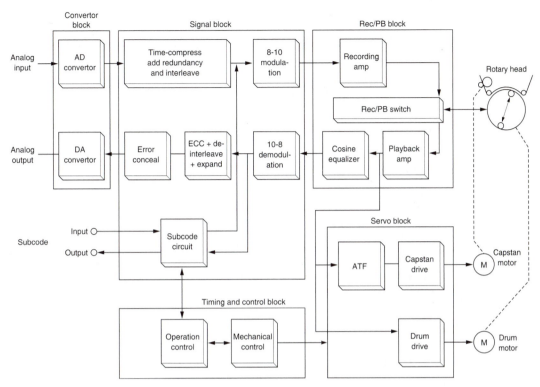

Figure 1.17 Block diagram of DAT

The record side needs ADCs, time compression, the addition of redundancy for error correction, and channel coding. On replay the channel coding is reversed by the data separator, errors are broken up by the de-interleave process and corrected or concealed, and the time compression and any fluctuations from the transport are removed by timebase correction. The corrected, time stable, samples are then fed to the DAC.

1.17   Digital Compact Cassette

Digital Compact Cassette (DCC) is a consumer digital audio recorder using compression. Although the convertors at either end of the machine work with PCM data, these data are not directly recorded, but are reduced to one quarter of their normal rate by processing. This allows a reasonable tape consumption similar to that achieved by a rotary head recorder. In a sense the complexity of the rotary head transport has been exchanged for the electronic complexity of the compression and expansion circuitry.

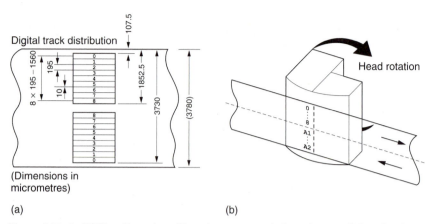

Figure 1.18 In DCC audio and auxiliary data are recorded on nine parallel tracks along each side of the tape as shown in (a). The replay head shown in (b) carries magnetic poles which register with one set of nine tracks. At the end of the tape, the replay head rotates 180° and plays a further nine tracks on the other side of the tape. The replay head also contains a pair of analog audio magnetic circuits which will be swung into place if an analog cassette is to be played.

Figure 1.18 shows that DCC uses stationary heads in a conventional tape transport which can also play analog cassettes. Data are distributed over eight parallel tracks which occupy half the width of the tape. At the end of the tape the head rotates and plays the other eight tracks in reverse. The advantage of the conventional approach with linear tracks is that tape duplication can be carried out at high speed. This makes DCC attractive to record companies.

Owing to the low frequencies recorded, DCC has to use active heads which actually measure the flux on the tape. These magneto-resistive heads are more complex than conventional inductive heads, and have only recently become economic as manufacturing techniques have been developed. DCC is treated in detail in Chapter 9.

As was introduced in section 1.12, compression relies on the phenomenon of auditory masking and this may effectively restrict DCC to being a consumer format. It will be seen from Figure 1.19 that the compression unit adjacent to the input is complemented by the expansion unit or decoder prior to the DAC.

1.18   Digital audio broadcasting

Digital audio broadcasting operates by modulating the transmitter with audio data instead of an analog waveform. Analog modulation works reasonably well for fixed reception sites where a decent directional antenna can be erected at a selected location, but has serious short-

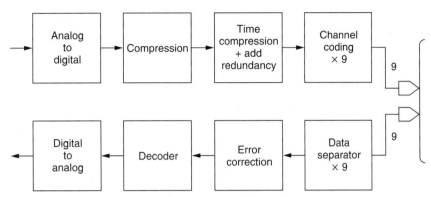

Figure 1.19 In DCC, the PCM data from the convertors are reduced to one-quarter of the original rate prior to distribution over eight tape tracks (plus an auxiliary data track). This allows a slow linear tape speed which can only be read with an MR head. The compression unit is mirrored by the decoder on replay.

comings for mobile reception where there is no control over the location and a large directional antenna is out of the question. The greatest drawback of broadcasting is multipath reception, where the direct signal is received along with delayed echoes from large reflecting bodies such as high-rise buildings. At certain wavelengths the reflection is received antiphase to the direct signal, and cancellation takes place which causes a notch in the received spectrum. In an analog system loss of the signal is inevitable.

In DAB, several digital audio broadcasts are merged into one transmission which is wider than the multipath notches. The data from the different signals are distributed uniformly within the channel so that a notch removes a small part of each channel instead of all of one. Sufficient data are received to allow error correction to re-create the missing values.

A DAB receiver actually receives the entire transmission and the process of 'tuning in' the desired channel is now performed by selecting the appropriate data channel for conversion to analog, making a DAB receiver easier to operate.

DAB resists multipath reception to permit mobile reception and the improvement to reception in car radios is dramatic. The data rate of PCM audio is too great to allow it to be economic for DAB. Compression is essential and this is detailed in Chapter 5.

1.19   Audio in PCs

Whilst the quality digital audio permits in undeniable, the potential of digital audio may turn out to be more important in the long term. Once audio becomes data, there is tremendous freedom to store and process it in computer-related equipment. The restrictions of analog technology are no longer applicable, yet we often needlessly build restrictions into equipment by making a digital replica of an analog system. The analog system evolved to operate within the restrictions imposed by the technology. To take the same system and merely digitize it is to miss the point.

A good example of missing the point was the development of the stereo quarter-inch digital audio tape recorder with open reels. Open-reel tape is sub-optimal for high-density digital recording because it is unprotected from contamination. The recorded wavelengths must be kept reasonably long or the reliability will be poor. Thus the tape consumption of these machines was excessive and more efficient cassette technologies such as DAT proved to have lower purchase cost and running costs as well as being a fraction of the size and weight. The speed and flexibility with which editing could be carried out by hard disk systems took away any remaining advantage. Quarter-inch digital tape found itself trapped between DAT and hard disks and passed into history because it was the wrong approach.

Part of the problem of missed opportunity is that traditionally, professional audio equipment manufacturers have specialized in one area leaving users to assemble systems from several suppliers. Mixer manufacturers may have no expertise in recording. Tape recorder manufacturers may have no knowledge of disk drives.

In contrast, computer companies have always taken a systems view and configure disks, tapes, RAM, processors and communications links as necessary to meet a given requirement. Now that audio is another form of data, this approach is being used to solve audio problems.

Small notebook computers are increasingly available with microphones and audio convertors so that they can act as dictating machines. A personal computer with high-quality audio convertors, compression algorithms and sufficient disk storage becomes an audio recorder. The recording levels and the timer are displayed on screen and soft keys become the rewind, record, etc. controls for the virtual recorder. The recordings can be edited to sample accuracy on disk, with displays of the waveforms in the area of the in and out points on screen. Once edited, the audio data can be sent anywhere in the world using telephone modems and data networks. The PC can be programmed to dial the destination itself at a selected time. At the same time as sending the audio, text files can be sent, along with images from a CCD camera. Without digital technology such a device would be unthinkable.

The market for such devices may well be captured by those with digital backgrounds, but not necessarily in audio. Computer, calculator and other consumer electronics manufacturers have the wider view of the potential of digital techniques.

Digital also blurs the distinction between consumer and professional equipment. In the traditional analog audio world, professional equipment sounded better but cost a lot more than consumer equipment. Now that digital technology is here, the sound quality is determined by the convertors. Once converted, the audio is data. If a bit can only convey whether it is one or zero, how does it know if it is a professional bit or a consumer bit? What is a professional disk drive? The cost of a digital product is a function not of its complexity, but of the volume to be sold. Professional equipment may be forced to use chip sets and transports designed for the volume market because the cost of designing an alternative is prohibitive. A professional machine may be a consumer machine in a stronger box with XLRs instead of phono sockets and PPM level meters. It may be that there will be little room for traditional professional audio manufacturers in the long term.

1.20   Networks

The conventional analog routing structure used in professional installations was simply replicated in the digital domain by the AES/EBU digital audio interface. However, using computer data approaches digital audio routing can also be achieved using networks, interconnecting a number of file servers which store the audio data with workstations from which the recordings can be manipulated. No dedicated audio routeing hardware is required. Chapter 8 considers how data networks operate.

Reference

1. Daniel, E.D., Mee, C.D. and Clark, M.H. (eds), *Magnetic Recording: The first 100 years*, Piscataway: IEEE Press (1999)

2

Some audio principles

2.1 The physics of sound

Sound is simply an airborne version of vibration which is why the two topics are inextricably linked. The air which carries sound is a mixture of gases, mostly nitrogen, some oxygen, a little carbon dioxide and so on. Gases are the highest energy state of matter, for example the application of energy to ice produces water and the application of more energy produces steam. The reason that a gas takes up so much more room than a liquid is that the molecules contain so much energy that they break free from their neighbours and rush around at high speed.

As Figure 2.1(a) shows, the innumerable elastic collisions of these high-speed molecules produce pressure on the walls of any gas container. In fact the distance a molecule can go without a collision, the *mean free path*, is quite short at atmospheric pressure. Consequently gas molecules also collide with each other elastically, so that if left undisturbed, in a container at a constant temperature, every molecule would end up with essentially the same energy and the pressure throughout would be constant and uniform.

Sound disturbs this simple picture. Figure 2.1(b) shows that a solid object which moves *against* gas pressure increases the velocity of the rebounding molecules, whereas in (c) one moving *with* gas pressure reduces that velocity. The average velocity and the displacement of all the molecules in a layer of air near to a moving body is the same as the velocity and displacement of the body. Movement of the body results in a local increase or decrease in pressure of some kind. Thus sound is both a pressure and a velocity disturbance. Integration of the velocity disturbance gives the displacement.

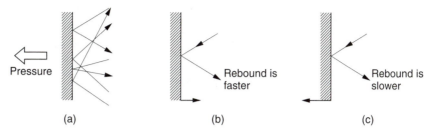

Figure 2.1 (a) The pressure exerted by a gas is due to countless elastic collisions between gas molecules and the walls of the container. (b) If the wall moves against the gas pressure, the rebound velocity increases. (c) Motion with the gas pressure reduces the particle velocity.

Despite the fact that a gas contains endlessly colliding molecules, a small mass or *particle* of gas can have stable characteristics because the molecules leaving are replaced by new ones with identical statistics. As a result acoustics seldom needs to consider the molecular structure of air and the constant motion can be neglected. Thus when particle velocity and displacement is considered in acoustics, this refers to the average values of a large number of molecules. In an undisturbed container of gas the particle velocity and displacement will both be zero everywhere.

When the volume of a fixed mass of gas is reduced, the pressure rises. The gas acts like spring; it is compliant. However, a gas also has mass. Sound travels through air by an interaction between the mass and the compliance. Imagine pushing a mass via a spring. It would not move immediately because the spring would have to be compressed in order to transmit a force. If a second mass is connected to the first by another spring, it would start to move even later. Thus the speed of a disturbance in a mass/spring system depends on the mass and the stiffness.

After the disturbance had propagated the masses would return to their rest position. The mass–spring analogy is helpful for a basic understanding, but is too simple to account for commonly encountered acoustic phenomena such as spherically expanding waves. It must be remembered that the mass and stiffness are distributed throughout the gas in the same way that inductance and capacitance are distributed in a transmission line. Sound travels through air without a net movement of the air.

2.2   The speed of sound

Unlike solids, the elasticity of gas is a complicated process. If a fixed mass of gas is compressed, work has to be done on it. This will generate heat in the gas. If the heat is allowed to escape and the compression does not change the temperature, the process is said to be *isothermal*. However, if

$$\text{Velocity } V = \sqrt{\frac{\gamma R T}{M}}$$

γ = adiabatic constant (1.4 for air)
R = gas constant (8.31 J K^{-1} mole^{-1})
T = absolute temp (K)
M = molecular weight (kg mole^{-1})

Assume air is 21% O_2, 78% N_2, 1% Ar

Molecular weight = 21% × 16 × 2 + 78% × 14 × 2 + 1% × 18 ×1
= 2.87 × 10^{-2} kg mole^{-1}

$$V = \sqrt{\frac{1.4 \times 8.31 \; T}{2.87 \times 10^{-2}}} = 20.1 \sqrt{T}$$

at 20°C T = 293 K V = 20.1 $\sqrt{293}$ = 344 m s^{-1}

Figure 2.2 Calculating the speed of sound from the elasticity of air.

the heat cannot escape the temperature will rise and give a disproportionate increase in pressure. This process is said to be *adiabatic* and the Diesel engine depends upon it. In most audio cases there is insufficient time for much heat transfer and so air is considered to act adiabatically. Figure 2.2 shows how the speed of sound *c* in air can be derived by calculating its elasticity under adiabatic conditions.

If the volume allocated to a given mass of gas is reduced isothermally, the pressure and the density will rise by the same amount so that *c* does not change. If the temperature is raised at constant pressure, the density goes down and so the speed of sound goes up. Gases with lower density than air have a higher speed of sound. Divers who breathe a mixture of oxygen and helium to prevent 'the bends' must accept that the pitch of their voices rises remarkably.

The speed of sound is proportional to the square root of the absolute temperature. On earth, temperature changes with respect to absolute zero (–273°C) also amount to around one per cent except in extremely inhospitable places. The speed of sound experienced by most of us is about 1000 feet per second or 344 metres per second. Temperature falls with altitude in the atmosphere and with it the speed of sound. The local speed of sound is defined as Mach 1. Consequently supersonic aircraft are fitted with Mach meters.

As air acts adiabatically, a propagating sound wave causes cyclic temperature changes. The speed of sound is a function of temperature, yet sound causes a temperature variation. One might expect some effects because of this. Fortunately, sounds which are below the threshold of pain have such a small pressure variation compared with atmospheric pressure that the effect is negligible and air can be assumed to be linear. However, on any occasion where the pressures are higher, this is not a valid assumption. In such cases the positive half cycle significantly increases local temperature and the speed of sound, whereas the negative

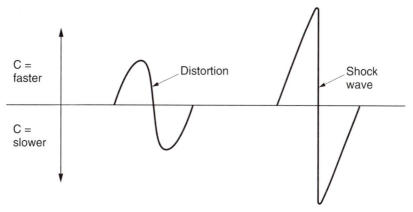

Figure 2.3 At high level, sound distorts itself by increasing the speed of propagation on positive half-cycles. The result is a shock wave.

half cycle reduces temperature and velocity. Figure 2.3 shows that this results in significant distortion of a sine wave, ultimately causing a *shock wave* which can travel faster than the speed of sound until the pressure has dissipated with distance. This effect is responsible for the sharp sound of a handclap.

This behaviour means that the speed of sound changes slightly with frequency. High frequencies travel slightly faster than low because there is less time for heat conduction to take place. Figure 2.4 shows that a

Near source   At a distance

Figure 2.4 In a complex waveform, high frequencies travel slightly faster producing a relative phase change with distance.

complex sound source produces harmonics whose phase relationship with the fundamental advances with the distance the sound propagates. This allows one mechanism (there are others) by which one can judge the distance from a known sound source. Clearly for realistic sound reproduction nothing in the audio chain must distort the phase relationship between frequencies. A system which accurately preserves such relationships is said to be phase linear.

2.3   Wavelength

Sound can be due to a one-off event known as percussion, or a periodic event such as the sinusoidal vibration of a tuning fork. The sound due to percussion is called transient whereas a periodic stimulus produces steady-state sound having a frequency *f*.

Because sound travels at a finite speed, the fixed observer at some distance from the source will experience the disturbance at some later time. In the case of a transient, the observer will detect a single replica of the original as it passes at the speed of sound. In the case of the tuning fork, a periodic sound source, the pressure peaks and dips follow one another away from the source at the speed of sound. For a given rate of vibration of the source, a given peak will have propagated a constant distance before the next peak occurs. This distance is called the wavelength lambda. Figure 2.5 shows that wavelength is defined as the

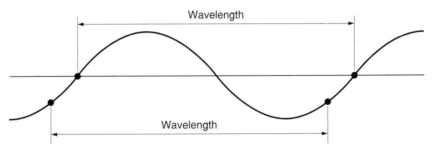

Figure 2.5   Wavelength is defined as the distance between two points at the same place on adjacent cycles. Wavelength is inversely proportional to frequency.

distance between any two identical points on the whole cycle. If the source vibrates faster, successive peaks get closer together and the wavelength gets shorter. Figure 2.5 also shows that the wavelength is inversely proportional to the frequency. It is easy to remember that the wavelength of 1000 Hz is a foot (about 30 cm).

2.4   Periodic and aperiodic signals

Sounds can be divided into these two categories and analysed both in the time domain in which the waveform is considered, or in the frequency domain in which the spectrum is considered. The time and frequency domains are linked by transforms of which the best known is the Fourier tramsform. Transforms will be considered further in Chapter 3.

Figure 2.6(a) shows that a periodic signal is one which repeats after some constant time has elapsed and goes on indefinitely in the time

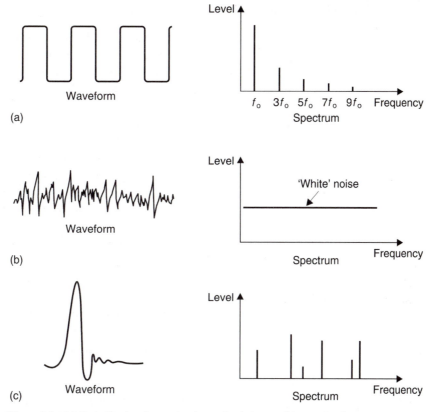

Figure 2.6 (a) Periodic signal repeats after a fixed time and has a simple spectrum consisting of fundamental plus harmonics. (b) Aperiodic signal such as noise does not repeat and has a continuous spectrum. (c) Transient contains an anharmonic spectrum.

domain. In the frequency domain such a signal will be described as having a fundamental frequency and a series of harmonics or partials which are at integer multiples of the fundamental. The timbre of an instrument is determined by the harmonic structure. Where there are no harmonics at all, the simplest possible signal results which has only a single frequency in the spectrum. In the time domain this will be an endless sine wave.

Figure 2.6(b) shows an aperiodic signal known as white noise. The spectrum shows that there is equal level at all frequencies, hence the term 'white' which is analogous to the white light containing all wavelengths. Transients or impulses may also be aperiodic. A spectral analysis of a transient (c) will contain a range of frequencies, but these are not harmonics because they are not integer multiples of the lowest frequency. Generally the narrower an event in the time domain, the broader it will be in the frequency domain and vice versa.

2.5   Sound and the ear

Experiments can tell us that the ear only responds to a certain range of frequencies within a certain range of levels. If sound is defined to fall within those ranges, then its reproduction is easier because it is only necessary to reproduce those levels and frequencies which the ear can detect.

Psychoacoustics can describe how our hearing has finite resolution in both time and frequency domains such that what we perceive is an inexact impression. Some aspects of the original disturbance are inaudible to us and are said to be masked. If our goal is the highest quality, we can design our imperfect equipment so that the shortcomings are masked. Conversely if our goal is economy we can use compression and hope that masking will disguise the inaccuracies it causes.

A study of the finite resolution of the ear shows how some combinations of tones sound pleasurable whereas others are irritating. Music has evolved empirically to emphasize primarily the former. Nevertheless we are still struggling to explain why we enjoy music and why certain sounds can make us happy and others can reduce us to tears. These characteristics must still be present in digitally reproduced sound.

Whatever the audio technology we deal with, there is a common goal of delivering a satisfying experience to the listener. However, some aspects of audio are emotive, some are technical. If we attempt to take an emotive view of a technical problem or vice versa our conclusions will be questionable.

The frequency range of human hearing is extremely wide, covering some ten octaves (an octave is a doubling of pitch or frequency) without interruption. There is hardly any other engineering discipline in which such a wide range is found. For example, in radio different wavebands are used so that the octave span of each is quite small. Whilst video signals have a wide octave span, the signal-to-noise and distortion criteria for video are extremely modest in comparison. Consequently audio is one of the most challenging subjects in engineering. Whilst the octave span required by audio can easily be met in analog or digital electronic equipment, the design of mechanical transducers such as microphones and loudspeakers will always be difficult.

2.6   Hearing

By definition, the sound quality of an audio system can only be assessed by human hearing. Many items of audio equipment can only be designed well with a good knowledge of the human hearing mechanism. The acuity of the human ear is finite but astonishing. It can detect tiny amounts of distortion, and will accept an enormous dynamic range over a wide number of octaves. If the ear detects a different degree of impairment between two audio systems in properly conducted tests, we can say that one of them is superior. Thus quality is completely subjective and can only be checked by listening tests. However, any characteristic of a signal which can be heard can in principle also be measured by a suitable instrument although in general the availability of such instruments lags the requirement. The subjective tests will tell us how sensitive the instrument should be. Then the objective readings from the instrument give an indication of how acceptable a signal is in respect of that characteristic.

The sense we call hearing results from acoustic, mechanical, hydraulic, nervous and mental processes in the ear/brain combination, leading to the term psychoacoustics. It is only possible briefly to introduce the subject here. The interested reader is referred to Moore[1] for an excellent treatment.

Figure 2.7 shows that the structure of the ear is traditionally divided into the outer, middle and inner ears. The outer ear works at low impedance, the inner ear works at high impedance, and the middle ear is an impedance matching device. The visible part of the outer ear is called the pinna which plays a subtle role in determining the direction of arrival of sound at high frequencies. It is too small to have any effect at low frequencies. Incident sound enters the auditory canal or meatus. The pipe-like meatus causes a small resonance at around 4 kHz. Sound vibrates the eardrum or tympanic membrane which seals the outer ear from the middle ear. The inner ear or cochlea works by sound travelling though a fluid. Sound enters the cochlea via a membrane called the oval window.

If airborne sound were to be incident on the oval window directly, the serious impedance mismatch would cause most of the sound to be reflected. The middle ear remedies that mismatch by providing a mechanical advantage. The tympanic membrane is linked to the oval window by three bones known as ossicles which act as a lever system such that a large displacement of the tympanic membrane results in a smaller

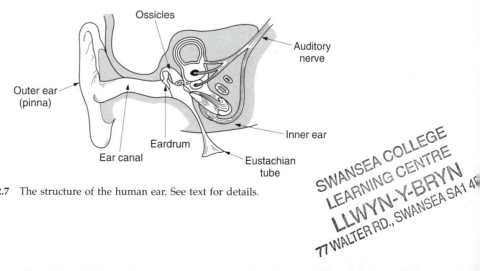

Figure 2.7 The structure of the human ear. See text for details.

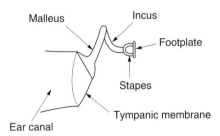

Figure 2.8 The malleus tensions the tympanic membrane into a conical shape. The ossicles provide an impedance-transforming lever system between the tympanic membrane and the oval window.

displacement of the oval window but with greater force. Figure 2.8 shows that the malleus applies a tension to the tympanic membrane rendering it conical in shape. The malleus and the incus are firmly joined together to form a lever. The incus acts upon the stapes through a spherical joint. As the area of the tympanic membrane is greater than that of the oval window, there is a further multiplication of the available force. Consequently small pressures over the large area of the tympanic membrane are converted to high pressures over the small area of the oval window.

The middle ear is normally sealed, but ambient pressure changes will cause static pressure on the tympanic membrane which is painful. The pressure is relieved by the Eustachian tube which opens involuntarily while swallowing. The Eustachian tubes open into the cavities of the head and must normally be closed to avoid one's own speech appearing deafeningly loud.

The ossicles are located by minute muscles which are normally relaxed. However, the middle ear reflex is an involuntary tightening of the *tensor tympani* and *stapedius* muscles which heavily damp the ability of the tympanic membrane and the stapes to transmit sound by about 12 dB at frequencies below 1 kHz. The main function of this reflex is to reduce the audibility of one's own speech. However, loud sounds will also trigger this reflex which takes some 60–120 ms to occur, too late to protect against transients such as gunfire.

2.7   The cochlea

The cochlea, shown in Figure 2.9(a), is a tapering spiral cavity within bony walls which is filled with fluid. The widest part, near the oval window, is called the *base* and the distant end is the *apex*. Figure 2.9(b) shows that the cochlea is divided lengthwise into three volumes by Reissner's membrane and the basilar membrane. The *scala vestibuli* and the *scala tympani* are connected by a small aperture at the apex of the cochlea known as the *helicotrema*. Vibrations from the stapes are

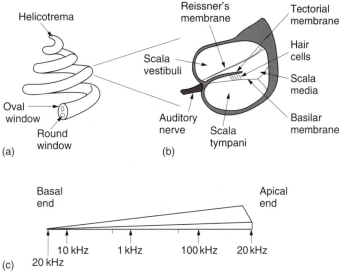

Figure 2.9 (a) The cochlea is a tapering spiral cavity. (b) The cross-section of the cavity is divided by Reissner's membrane and the basilar membrane. (c) The basilar membrane tapers so its resonant frequency changes along its length.

transferred to the oval window and become fluid pressure variations which are relieved by the flexing of the round window. Effectively the basilar membrane is in series with the fluid motion and is driven by it except at very low frequencies where the fluid flows through the helicotrema, bypassing the basilar membrane.

The vibration of the basilar membrane is sensed by the organ of Corti which runs along the centre of the cochlea. The organ of Corti is active in that it contains elements which can generate vibration as well as sense it. These are connected in a regenerative fashion so that the Q factor, or frequency selectivity of the ear, is higher than it would otherwise be. The deflection of hair cells in the organ of Corti triggers nerve firings and these signals are conducted to the brain by the auditory nerve. Some of these signals reflect the time domain, particularly during the transients with which most real sounds begin and also at low frequencies. During continuous sounds, the basilar membrane is also capable of performing frequency analysis.

Figure 2.9(c) shows that the basilar membrane is not uniform, but tapers in width and varies in thickness in the opposite sense to the taper of the cochlea. The part of the basilar membrane which resonates as a result of an applied sound is a function of the frequency. High frequencies cause resonance near to the oval window, whereas low frequencies cause resonances further away. More precisely the distance from the apex where the maximum resonance occurs is a logarithmic function of the frequency. Consequently tones spaced apart in octave steps will excite evenly spaced resonances in the basilar membrane. The prediction of resonance at a particular location on the membrane is called *place theory*. Essentially the basilar membrane is a mechanical frequency analyser. A knowledge of the way it operates is essential to an understanding of musical phenomena such as pitch discrimination, timbre, consonance and dissonance and to auditory phenomena such as critical bands, masking and the precedence effect.

Nerve firings are not a perfect analog of the basilar membrane motion. On continuous tones a nerve firing appears to occur at a constant phase relationship to the basilar vibration, a phenomenon called phase locking, but firings do not necessarily occur on every cycle. At higher frequencies firings are intermittent, yet each is in the same phase relationship.

The resonant behaviour of the basilar membrane is not observed at the lowest audible frequencies below 50 Hz. The pattern of vibration does not appear to change with frequency and it is possible that the frequency is low enough to be measured directly from the rate of nerve firings.

2.8   Mental processes

The nerve impulses are processed in specific areas of the brain which appear to have evolved at different times to provide different types of information. The time domain response works quickly, primarily aiding the direction-sensing mechanism and is older in evolutionary terms. The frequency domain response works more slowly, aiding the determination of pitch and timbre and evolved later, presumably after speech evolved.

The earliest use of hearing was as a survival mechanism to augment vision. The most important aspect of the hearing mechanism was the ability to determine the location of the sound source. Figure 2.10 shows that the brain can examine several possible differences between the signals reaching the two ears. At (a) a phase shift will be apparent. At (b) the distant ear is shaded by the head resulting in a different frequency response compared to the nearer ear. At (c) a transient sound arrives later at the more distant ear. The inter-aural phase, delay and level mechanisms vary in their effectiveness depending on the nature of the sound to be located. At some point a fuzzy logic decision has to be made as to how the information from these different mechanisms will be weighted.

There will be considerable variation with frequency in the phase shift between the ears. At a low frequency such as 30 Hz, the wavelength is around 11.5 metres and so this mechanism must be quite weak at low frequencies. At high frequencies the ear spacing is many wavelengths producing a confusing and complex phase relationship. This suggests a frequency limit of around 1500 Hz which has been confirmed by experiment.

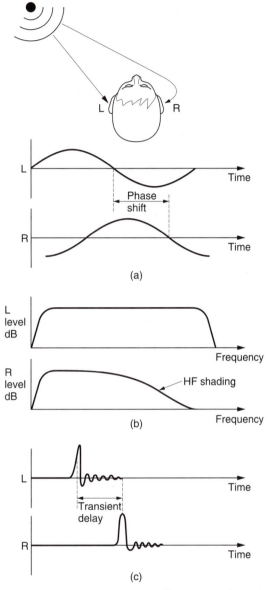

Figure 2.10 Having two spaced ears is cool. (a) Off-centre sounds result in phase difference. (b) Distant ear is shaded by head producing loss of high frequencies. (c) Distant ear detects transient later.

At low and middle frequencies sound will diffract round the head sufficiently well that there will be no significant difference between the level at the two ears. Only at high frequencies does sound become directional enough for the head to shade the distant ear causing what is called an inter-aural intensity difference (IID).

Phase differences are only useful at low frequencies and shading only works at high frequencies. Fortunately real-world sounds are timbral or broadband and often contain transients. Timbral, broadband and transient sounds differ from tones in that they contain many different frequencies.

A transient has an unique aperiodic waveform which, as Figure 2.10(c) shows, suffers no ambiguity in the assessment of inter-aural delay (IAD) between two versions. Note that a one-degree change in sound location causes a IAD of around 10 microseconds. The smallest detectable IAD is a remarkable 6 microseconds. This should be the criterion for spatial reproduction accuracy.

A timbral waveform is periodic at the fundamental frequency but the presence of harmonics means that a greater number of nerve firings can be compared between the two ears. As the statistical deviation of nerve firings with respect to the incoming waveform is about 100 microseconds the only way in which an IAD of 6 microseconds can be resolved is if the timing of many nerve firings is correlated in some way in the brain.

Transient noises produce a one-off pressure step whose source is accurately and instinctively located. Figure 2.11 shows an idealized transient pressure waveform following an acoustic event. Only the initial transient pressure change is required for location. The time of arrival of the transient at the two ears will be different and will locate the source laterally within a processing delay of around a millisecond.

Following the event which generated the transient, the air pressure equalizes. The time taken for this equalization varies and allows the listener to establish the likely size of the sound source. The larger the source, the longer the pressure-equalization time. Only after this does the

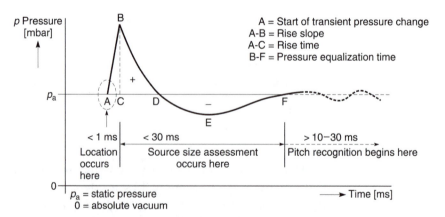

Figure 2.11 Real acoustic event produces a pressure step. Initial step is used for spatial location, equalization time signifies size of source. (Courtesy Manger Schallwandlerbau.)

frequency analysis mechanism tell anything about the pitch and timbre of the sound.

The above results suggest that anything in a sound reproduction system which impairs the reproduction of a transient pressure change will damage localization and the assessment of the pressure-equalization time. Clearly in an audio system which claims to offer any degree of precision, every component must be able to reproduce transients accurately and must have at least a minimum phase characteristic if it cannot be phase linear. In this respect digital audio represents a distinct technical performance advantage although much of this is lost in poor transducer design, especially in loudspeakers.

2.9   Level and loudness

At its best, the ear can detect a sound pressure variation of only 2×10^{-5} Pascals r.m.s. and so this figure is used as the reference against which sound pressure level (SPL) is measured. The sensation of loudness is a logarithmic function of SPL and consequently a logarithmic unit, the deciBel, was adopted for audio measurement. The deciBel is explained in detail in section 2.19.

The dynamic range of the ear exceeds 130 dB, but at the extremes of this range, the ear is either straining to hear or is in pain. Neither of these cases can be described as pleasurable or entertaining, and it is hardly necessary to produce audio of this dynamic range since, among other things, the consumer is unlikely to have anywhere sufficiently quiet to listen to it. On the other hand, extended listening to music whose dynamic range has been excessively compressed is fatiguing.

The frequency response of the ear is not at all uniform and it also changes with SPL. The subjective response to level is called loudness and is measured in *phons*. The phon scale is defined to coincide with the SPL scale at 1 kHz, but at other frequencies the phon scale deviates because it displays the actual SPLs judged by a human subject to be equally loud as a given level at 1 kHz. Figure 2.12 shows the so-called equal loudness contours which were originally measured by Fletcher and Munson and subsequently by Robinson and Dadson. Note the irregularities caused by resonances in the meatus at about 4 kHz and 13 kHz.

Usually, people's ears are at their most sensitive between about 2 kHz and 5 kHz, and although some people can detect 20 kHz at high level, there is much evidence to suggest that most listeners cannot tell if the upper frequency limit of sound is 20 kHz or 16 kHz.[2,3] For a long time it was thought that frequencies below about 40 Hz were unimportant, but it is now clear that reproduction of frequencies down to 20 Hz improves reality and ambience.[4] The generally accepted frequency range for high-

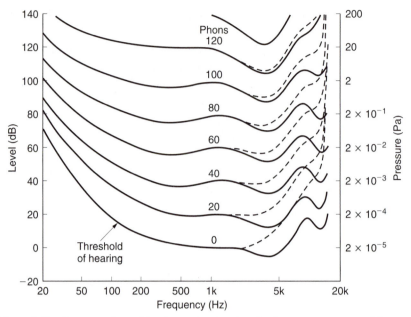

Figure 2.12 Contours of equal loudness showing that the frequency response of the ear is highly level dependent (solid line, age 20; dashed line, age 60).

quality audio is 20 Hz to 20 000 Hz, although for broadcasting an upper limit of 15 000 Hz is often applied.

The most dramatic effect of the curves of Figure 2.12 is that the bass content of reproduced sound is disproportionately reduced as the level is turned down. This would suggest that if a sufficiently powerful yet high-quality reproduction system is available the correct tonal balance when playing a good recording can be obtained simply by setting the volume control to the correct level. This is indeed the case. A further consideration is that many musical instruments as well as the human voice change timbre with level and there is only one level which sounds correct for the timbre.

Audio systems with a more modest specification would have to resort to the use of tone controls to achieve a better tonal balance at lower SPL. A loudness control is one where the tone controls are automatically invoked as the volume is reduced. Although well meant, loudness controls seldom compensate accurately because they must know the original level at which the material was meant to be reproduced as well as the actual level in use. The equalization applied would have to be the difference between the equal loudness curves at the two levels.

There is no standard linking the signal level on a recording with the SPL at the microphone. The SPL resulting from a given signal level leaving a loudness control depends upon the sensitivity of the power amplifier and the loudspeakers and the acoustics of the listening room. Consequently unless set up for a particular installation, loudness controls are doomed to be inaccurate and are eschewed on high-quality equipment.

A further consequence of level-dependent hearing response is that recordings which are mixed at an excessively high level will appear bass light when played back at a normal level. Such recordings are more a product of self-indulgence than professionalism.

Loudness is a subjective reaction and is almost impossible to measure. In addition to the level-dependent frequency response problem, the listener uses the sound not for its own sake but to draw some conclusion about the source. For example, most people hearing a distant motorcycle will describe it as being loud. Clearly at the source, it *is* loud, but the listener has compensated for the distance.

The best that can be done is to make some compensation for the level-dependent response using *weighting curves*. Ideally there should be many, but in practice the A, B and C weightings were chosen where the A curve is based on the 40-phon response. The measured level after such a filter is in units of dBA. The A curve is almost always used because it most nearly relates to the annoyance factor of distant noise sources.

2.10 Frequency discrimination

Figure 2.13 shows an uncoiled basilar membrane with the apex on the left so that the usual logarithmic frequency scale can be applied. The envelope of displacement of the basilar membrane is shown for a single frequency at (a). The vibration of the membrane in sympathy with a single frequency cannot be localized to an infinitely small area, and

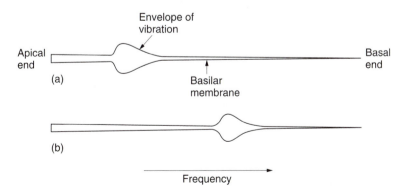

Figure 2.13 The basilar membrane symbolically uncoiled. (a) Single frequency causes the vibration envelope shown. (b) Changing the frequency moves the peak of the envelope.

nearby areas are forced to vibrate at the same frequency with an amplitude that decreases with distance. Note that the envelope is asymmetrical because the membrane is tapering and because of frequency-dependent losses in the propagation of vibrational energy down the cochlea. If the frequency is changed, as in (b), the position of maximum displacement will also change. As the basilar membrane is continuous, the position of maximum displacement is infinitely variable allowing extremely good pitch discrimination of about one twelfth of a semitone which is determined by the spacing of hair cells.

In the presence of a complex spectrum, the finite width of the vibration envelope means that the ear fails to register energy in some bands when there is more energy in a nearby band. Within those areas, other frequencies are mechanically excluded because their amplitude is insufficient to dominate the local vibration of the membrane. Thus the Q factor of the membrane is responsible for the degree of auditory masking, defined as the decreased audibility of one sound in the presence of another.

2.11   Critical bands

The term used in psychoacoustics to describe the finite width of the vibration envelope is *critical bandwidth*. Critical bands were first described by Fletcher.[5] The envelope of basilar vibration is a complicated function. It is clear from the mechanism that the area of the membrane involved will increase as the sound level rises. Figure 2.14 shows the bandwidth as a function of level.

As will be seen in Chapter 3, transform theory teaches that the higher the frequency resolution of a transform, the worse the time accuracy. As the basilar membrane has finite frequency resolution measured in the

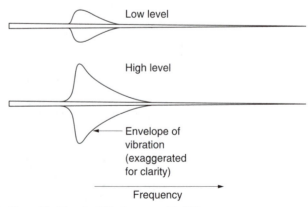

Low level

High level

Envelope of vibration (exaggerated for clarity)

Frequency

Figure 2.14   The critical bandwidth changes with SPL.

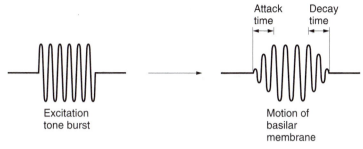

Figure 2.15 Impulse response of the ear showing slow attack and decay due to resonant behaviour.

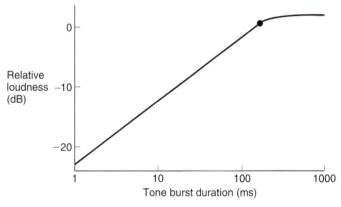

Figure 2.16 Perceived level of tone burst rises with duration as resonance builds up.

width of a critical band, it follows that it must have finite time resolution. This also follows from the fact that the membrane is resonant, taking time to start and stop vibrating in response to a stimulus. There are many examples of this. Figure 2.15 shows the impulse response. Figure 2.16 shows the perceived loudness of a tone burst increases with duration up to about 200 ms due to the finite response time.

The ear has evolved to offer intelligibility in reverberant environments which it does by averaging all received energy over a period of about 30 ms. Reflected sound which arrives within this time is integrated to produce a louder sensation, whereas reflected sound which arrives after that time can be temporally discriminated and is perceived as an echo. Microphones have no such ability, which is why acoustic treatment is often needed in areas where microphones are used.

A further example of the finite time discrimination of the ear is the fact that short interruptions to a continuous tone are difficult to detect. Finite time resolution means that masking can take place even when the masking tone begins after and ceases before the masked sound. This is referred to as forward and backward masking.[6]

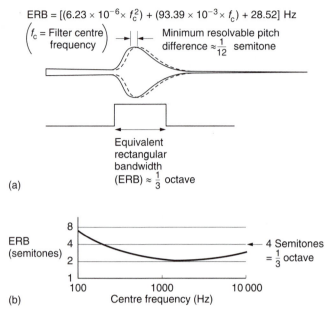

Figure 2.17 Effective rectangular bandwidth of critical band is much wider than the resolution of the pitch discrimination mechanism.

As the vibration envelope is such a complicated shape, Moore and Glasberg have proposed the concept of equivalent rectangular bandwidth to simplify matters. The ERB is the bandwidth of a rectangular filter which passes the same power as a critical band. Figure 2.17(a) shows the expression they have derived linking the ERB with frequency. This is plotted in (b) where it will be seen that one third of an octave is a good approximation. This is about thirty times broader than the pitch discrimination also shown in (b).

Some treatments of human hearing liken the basilar membrane to a bank of fixed filters each of which is the width of a critical band. The frequency response of such a filter can be deduced from the envelope of basilar displacement as has been done in Figure 2.18. The fact that no agreement has been reached on the number of such filters should alert the suspicions of the reader. A third octave filter bank model cannot explain pitch discrimination some thirty times better. The response of the basilar membrane is centred upon the input frequency and no fixed filter can do this. However, the most worrying aspect of the fixed filter model is that according to Figure 2.18(b) a single tone would cause a response in several bands which would be interpreted as several tones. This is at variance with reality. Far from masking higher frequencies, we appear to be creating them!

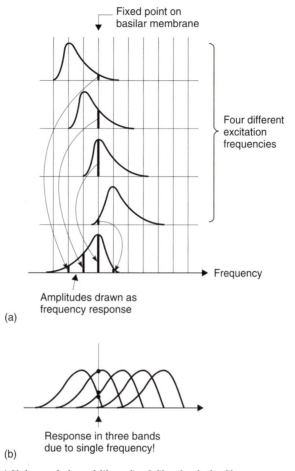

Fixed point on
basilar membrane

Four different
excitation
frequencies

Frequency

Amplitudes drawn as
frequency response

(a)

Response in three bands
due to single frequency!

(b)

Figure 2.18 (a) If the ear behaved like a fixed filter bank the filter response could be derived as shown here. (b) This theory does not hold because a single tone would cause response in several bands.

This author prefers to keep in mind how the basilar membrane is actually vibrating is response to an input spectrum. If a mathematical model of the ear is required, then it has to be described as performing a finite resolution continuous frequency transform.

2.12 Beats

Figure 2.19 shows an electrical signal (a) in which two equal sine waves of nearly the same frequency have been linearly added together. Note that the envelope of the signal varies as the two waves move in and out of phase. Clearly the frequency transform calculated to infinite accuracy is that shown at (b). The two amplitudes are constant and there is no

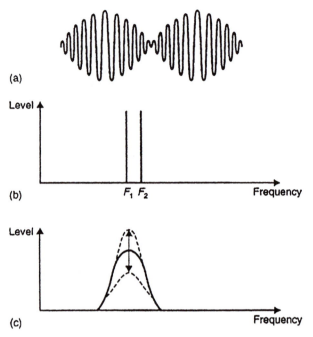

Figure 2.19 (a) Result of adding two sine waves of similar frequency. (b) Spectrum of (a) to infinite accuracy. (c) With finite accuracy only a single frequency is distinguished whose amplitude changes with the envelope of (a) giving rise to beats.

evidence of the envelope modulation. However, such a measurement requires an infinite time. When a shorter time is available, the frequency discrimination of the transform falls and the bands in which energy is detected become broader.

When the frequency discrimination is too wide to distinguish the two tones as in (c), the result is that they are registered as a single tone. The amplitude of the single tone will change from one measurement to the next because the envelope is being measured. The rate at which the envelope amplitude changes is called a *beat* frequency which is not actually present in the input signal. Beats are an artifact of finite frequency resolution transforms. The fact that human hearing produces beats from pairs of tones proves that it has finite resolution.

Measurement of when beats occur allows measurement of critical bandwidth. Figure 2.20 shows the results of human perception of a two-tone signal as the frequency **dF** difference changes. When **dF** is zero, described musically as *unison*, only a single note is heard. As **dF** increases, beats are heard, yet only a single note is perceived. The limited frequency resolution of the basilar membrane has *fused* the two tones together. As **dF** increases further, the sensation of beats ceases at 12–15 Hz and is replaced by a sensation of roughness or *dissonance*. The roughness is due to parts of the basilar membrane being unable to decide the frequency at which to

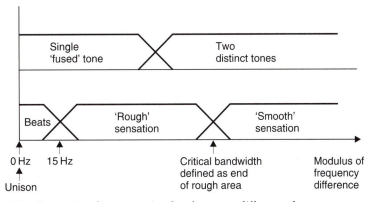

Figure 2.20 Perception of two-tone signal as frequency difference changes.

vibrate. The regenerative effect may well become confused under such conditions. The roughness which persists until **dF** has reached the critical bandwidth beyond which two separate tones will be heard because there are now two discrete basilar resonances. In fact this is the definition of critical bandwidth.

2.13 Music and the ear

The characteristics of the ear, especially critical bandwidth, are responsible for the way music has evolved. Beats are used extensively in music. When tuning a pair of instruments together, a small tuning error will result in beats when both play the same nominal note. In certain pipe organs, pairs of pipes are sounded together with a carefully adjusted pitch error which results in a pleasing tremolo effect.

With certain exceptions, music is intended to be pleasing and so dissonance is avoided. Two notes which sound together in a pleasing manner are described as harmonious or *consonant*. Two sine waves appear consonant if they separated by a critical bandwidth because the roughness of Figure 2.20 is avoided, but real musical instruments produce a series of harmonics in addition to the fundamental.

Figure 2.21 shows the spectrum of a harmonically rich instrument. The fundamental and the first few harmonics are separated by more than a critical band, but from the seventh harmonic more than one harmonic will be in one band and it is possible for dissonance to occur. Musical instruments have evolved to avoid the production of seventh and higher harmonics. Violins and pianos are played or designed to excite the strings at a node of the seventh harmonic to suppress this dissonance.

Harmonic distortion in audio equipment is easily detected even in minute quantities because the first few harmonics fall in non-overlapping

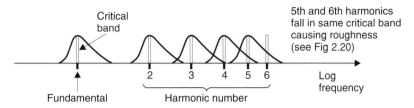

Figure 2.21 Spectrum of a real instrument with respect to critical bandwidth. High harmonics can fall in the same critical band and cause dissonance.

critical bands. The sensitivity of the ear to third harmonic distortion probably deserves more attention in audio equipment than the fidelity of the dynamic range or frequency response. The ear is even more sensitive to anharmonic distortion which can be generated in poor-quality ADCs. This topic will be considered in Chapter 4.

When two harmonically rich notes are sounded together, the harmonics will fall within the same critical band and cause dissonance unless the fundamentals have one of a limited number of simple relationships which makes the harmonics fuse. Clearly an octave relationship is perfect.

Figure 2.22 shows some examples. In (a) two notes with the ratio (interval) 3:2 are considered. The harmonics are either widely separated or fused and the combined result is highly consonant. The interval of 3:2 is known to musicians as a perfect fifth. In (b) the ratio is 4:3. All harmonics are either at least a third of an octave apart or are fused. This relationship is known as a perfect fourth. The degree of dissonance over the range from 1:1 to 2:1 (unison to octave) was investigated by Helmholtz and is shown in (c). Note that the dissonance rises at both ends where the fundamentals are within a critical bandwidth of one another. Dissonances in the centre of the scale are where some harmonics lie within a critical bandwidth of one another. Troughs in the curve indicate areas of consonance. Many of the troughs are not very deep, indicating that the consonance is not perfect. This is because of the effect shown in Figure 2.21 in which high harmonics get closer together with respect to critical bandwidth. When the fundamentals are closer together, the harmonics will become dissonant at a lower frequency, reducing the consonance. Figure 2.22 also shows the musical terms used to describe the consonant intervals.

It is clear from Figure 2.22(c) that the notes of the musical scale have empirically been established to allow the maximum consonance with pairs of notes and chords. Early instruments were tuned to the just diatonic scale in exactly this way. Unfortunately the just diatonic scale does not allow changes of key because the notes are not evenly spaced. A key change is where the frequency of every note in a piece of music is

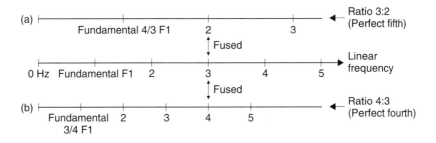

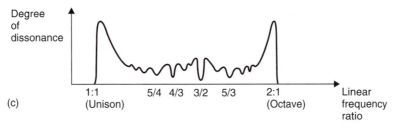

Figure 2.22 (a) Perfect fifth with a frequency ratio of 3:2 is consonant because harmonics are either in different critical bands or are fused. (b) Perfect fourth achieves the same result with 4:3 frequency ratio. (c) Degree of dissonance over range from 1:1 to 2:1.

multiplied by a constant, often to bring the accompaniment within the range of a singer. In continuously tuned instruments such as the violin and the trombone this is easy, but with fretted or keyboard instruments such as a piano there is a problem.

The equal-tempered scale is a compromise between consonance and key changing. The octave is divided into twelve equal intervals called tempered semitones. On a keyboard, seven of the keys are white and produce notes very close to those of the just diatonic scale, and five of the keys are black. Music can be transposed in semitone steps by using the black keys.

Figure 2.23 shows an example of transposition where a scale is played in several keys.

2.14   The sensation of pitch

Frequency is an objective measure whereas *pitch* is the subjective near equivalent. Clearly frequency and level are independent, whereas pitch and level are not. Figure 2.24 shows the relationship between pitch and level. Place theory indicates that the hearing mechanism can sense a single frequency quite accurately as a function of the place or position of maximum basilar vibration. However, most periodic sounds and real musical instruments produce a series of harmonics in addition to the

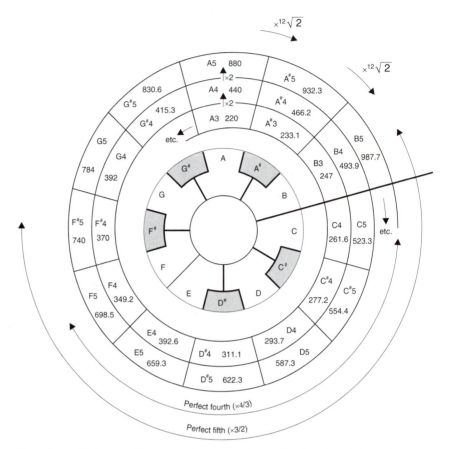

Figure 2.23 With a suitably tempered octave, scales can be played in different keys.

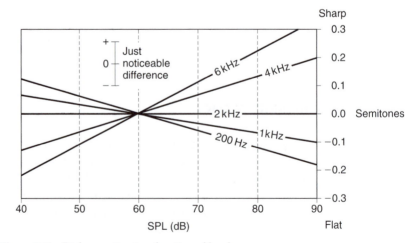

Figure 2.24 Pitch sensation is a function of level.

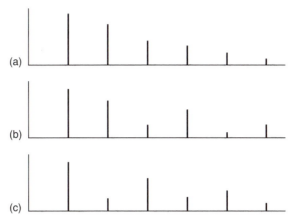

Figure 2.25 (a) Harmonic structure of rich sound. (b) Even harmonic predominance. (c) Odd harmonic predominance. Pitch perception appears independent of harmonic structure.

fundamental. When a harmonically rich sound is present the basilar membrane is excited at spaced locations. Figure 2.25 (a) shows all harmonics, (b) shows even harmonics predominating and (c) shows odd harmonics predominating. It would appear that our hearing is accustomed to hearing harmonics in various amounts and the consequent regular pattern of excitation. It is the overall pattern which contributes to the sensation of pitch even if individual partials vary enormously in relative level.

Experimental signals in which the fundamental has been removed leaving only the harmonics result in unchanged pitch perception. The pattern in the remaining harmonics is enough uniquely to establish the missing fundamental. Imagine the fundamental in (b) to be absent. Neither the second harmonic nor the third can be mistaken for the fundamental because if they were fundamentals a different pattern of harmonics would result. A similar argument can be put forward in the time domain, where the timing of phase-locked nerve firings responding to a harmonic will periodically coincide with the nerve firings of the fundamental. The ear is used to such time patterns and will use them in conjunction with the place patterns to determine the right pitch. At very low frequencies the place of maximum vibration does not move with frequency yet the pitch sensation is still present because the nerve firing frequency is used.

As the fundamental frequency rises it is difficult to obtain a full pattern of harmonics as most of them fall outside the range of hearing. The pitch discrimination ability is impaired and needs longer to operate. Figure 2.26 shows the number of cycles of excitation needed to discriminate pitch as a function of frequency. Clearly at around 5 kHz performance is

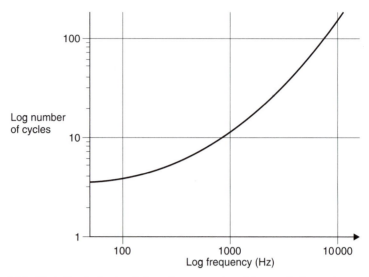

Figure 2.26 Pitch discrimination fails as frequency rises. The graph shows the number of cycles needed to distinguish pitch as a function of frequency.

failing because there are hardly any audible harmonics left. Phase locking also fails at about the same frequency. Musical instruments have evolved accordingly, with the highest notes of virtually all instruments found below 5 kHz.

2.15 Frequency response and linearity

It is a goal in high-quality sound reproduction that the timbre of the original sound shall not be changed by the reproduction process. There are two ways in which timbre can inadvertently be changed, as Figure 2.27 shows. In (a) the spectrum of the original shows a particular relationship between harmonics. This signal is passed through a system (b) which has an unequal response at different frequencies. The result is that the harmonic structure (c) has changed, and with it the timbre. Clearly a fundamental requirement for quality sound reproduction is that the response to all frequencies should be equal.

Frequency response is easily tested using sine waves of constant amplitude at various frequencies as an input and noting the output level for each frequency.

Figure 2.28 shows that another way in which timbre can be changed is by non-linearity. All audio equipment has a transfer function between the input and the output which form the two axes of a graph. Unless the transfer function is exactly straight or *linear*, the output waveform will differ from the input. A non-linear transfer function will cause distortion which changes the distribution of harmonics and changes timbre.

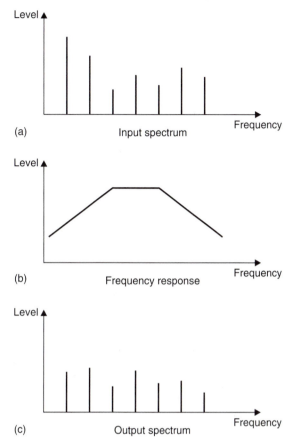

Figure 2.27 Why frequency response matters. Original spectrum at (a) determines timbre of sound. If original signal is passed through a system with deficient frequency response (b), the timbre will be changed (c).

At a real microphone placed before an orchestra a multiplicity of sounds may arrive simultaneously. The microphone diaphragm can only be in one place at a time, so the output waveform must be the sum of all the sounds. An ideal microphone connected by ideal amplification to an ideal loudspeaker will reproduce all of the sounds simultaneously by linear superimposition. However, should there be a lack of linearity anywhere in the system, the sounds will no longer have an independent existence, but will interfere with one another, changing one another's timbre and even creating new sounds which did not previously exist. This is known as *intermodulation*. Figure 2.29 shows that a linear system will pass two sine waves without interference. If there is any non-linearity, the two sine waves will intermodulate to produce sum and difference frequencies which are easily observed in the otherwise pure spectrum.

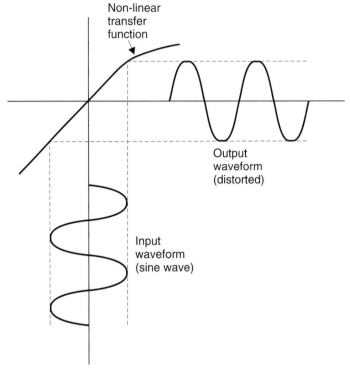

Figure 2.28 Non-linearity of the transfer function creates harmonics by distorting the waveform. Linearity is extremely important in audio equipment.

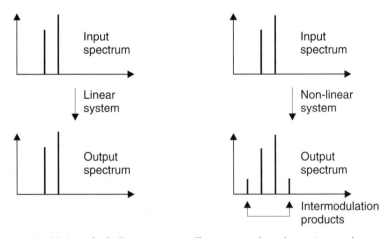

Figure 2.29 (a) A perfectly linear system will pass a number of superimposed waveforms without interference so that the output spectrum does not change. (b) A non-linear system causes inter-modulation where the output spectrum contains sum and difference frequencies in addition to the originals.

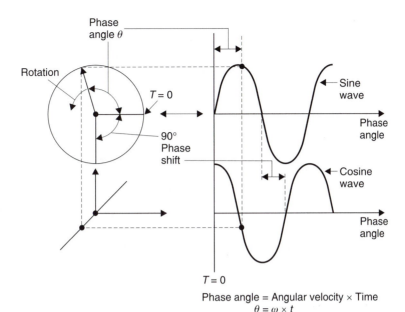

Figure 2.30 A sine wave is one component of a rotation. When a rotation is viewed from two places at right angles, one will see a sine wave and the other will see a cosine wave. The constant *phase shift* between sine and cosine is 90° and should not be confused with the time variant *phase angle* due to the rotation.

2.16 The sine wave

As the sine wave is so useful it will be treated here in detail. Figure 2.30 shows a constant speed rotation viewed along the axis so that the motion is circular. Imagine, however, the view from one side in the plane of the rotation. From a distance only a vertical oscillation will be observed and if the position is plotted against time the resultant waveform will be a sine wave. Geometrically it is possible to calculate the height or displacement because it is the radius multiplied by the sine of the phase angle.

The phase angle is obtained by multiplying the angular velocity w by the time t. Note that the angular velocity is measured in radians per second whereas frequency **f** is measured in rotations per second or Hertz (Hz). As a radian is unit distance at unit radius (about 57°) then there are 2π radians in one rotation. Thus the phase angle at a time **t** is given by $\sin wt$ or $\sin 2\pi ft$.

Imagine a second viewer who is at right angles to the first viewer. He will observe the same waveform, but at a different time. The displacement will be given by the radius multiplied by the cosine of the phase angle. When plotted on the same graph, the two waveforms are *phase-shifted* with respect to one another. In this case the phase-shift is 90° and the two waveforms are said to be *in quadrature*. Incidentally the motions on each side of a steam locomotive are in quadrature so that it can always get started (the term used is quartering). Note that the *phase angle* of a signal is constantly changing with time whereas the *phase-shift* between two signals can be constant. It is important that these two are not confused.

The velocity of a moving component is often more important in audio than the displacement. The vertical component of velocity is obtained by differentiating the displacement. As the displacement is a sine wave, the velocity will be a cosine wave whose amplitude is proportional to frequency. In other words the displacement and velocity are in quadrature with the velocity lagging. This is consistent with the velocity reaching a minimum as the displacement reaches a maximum and vice versa. Figure 2.31 shows the displacement, velocity and acceleration waveforms of a body executing SHM. Note that the acceleration and the displacement are always anti-phase.

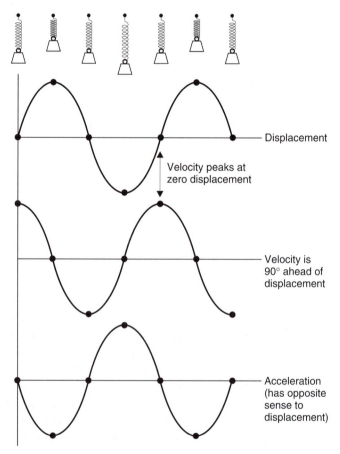

Figure 2.31 The displacement, velocity and acceleration of a body executing simple harmonic motion (SHM).

2.17   Root mean square measurements

Figure 2.32(a) shows that according to Ohm's law, the power dissipated in a resistance is proportional to the square of the applied voltage. This causes no difficulty with direct current (DC), but with alternating signals such as audio it is harder to calculate the power. Consequently a unit of voltage for alternating signals was devised. Figure 2.32(b) shows that the average power delivered during a cycle must be proportional to the mean of the square of the applied voltage. Since power is proportional to the square of applied voltage, the same power would be dissipated by a DC voltage whose value was equal to the square root of the mean of the square of the AC voltage. Thus the Volt rms (root mean square) was specified. An AC signal of a given number of Volts rms will dissipate exactly the same amount of power in a given resistor as the same number of Volts DC.

Figure 2.33(a) shows that for a sine wave the rms voltage is obtained by dividing the peak voltage V_{pk} by the square root of two. However, for a square wave(b) the rms voltage and the peak voltage are the same. Most moving coil AC voltmeters only read correctly on sine waves, whereas many electronic meters incorporate a true rms calculation.

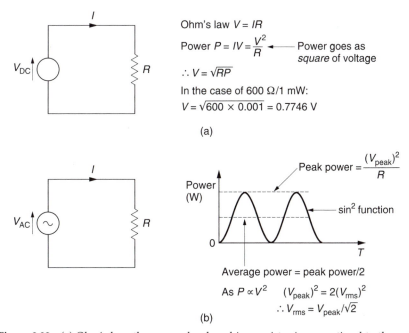

Ohm's law $V = IR$

Power $P = IV = \dfrac{V^2}{R}$ ◄── Power goes as *square* of voltage

$\therefore V = \sqrt{RP}$

In the case of 600 Ω/1 mW:

$V = \sqrt{600 \times 0.001} = 0.7746 \text{ V}$

(a)

Peak power $= \dfrac{(V_{peak})^2}{R}$

Power (W)

sin^2 function

0

T

Average power = peak power/2

As $P \propto V^2$ $(V_{peak})^2 = 2(V_{rms})^2$

$\therefore V_{rms} = V_{peak}/\sqrt{2}$

(b)

Figure 2.32   (a) Ohm's law: the power developed in a resistor is proportional to the square of the voltage. Consequently, 1 mW in 600 Ω requires 0.775 V. With a sinusoidal alternating input (b), the power is a sine squared function which can be averaged over one cycle. A DC voltage which would deliver the same power has a value which is the square root of the mean of the square of the sinusoidal input.

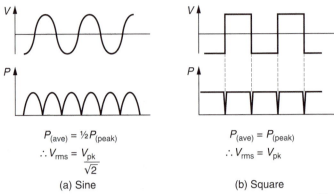

$P_{(ave)} = \frac{1}{2}P_{(peak)}$

$\therefore V_{rms} = \dfrac{V_{pk}}{\sqrt{2}}$

$P_{(ave)} = P_{(peak)}$

$\therefore V_{rms} = V_{pk}$

(a) Sine

(b) Square

Figure 2.33 (a) For a sine wave the conversion factor from peak to rms is $\sqrt{2}$. (b) For a square wave the peak and rms voltage is the same.

On an oscilloscope it is often easier to measure the peak-to-peak voltage which is twice the peak voltage. The rms voltage cannot be measured directly on an oscilloscope since it depends on the waveform although the calculation is simple in the case of a sine wave.

2.18   The deciBel

The first audio signals to be transmitted were on telephone lines. Where the wiring is long compared to the electrical wavelength (not to be confused with the acoustic wavelength) of the signal, a transmission line exists in which the distributed series inductance and the parallel capacitance interact to give the line a characteristic impedance. In telephones this turned out to be about $600\,\Omega$. In transmission lines the best power delivery occurs when the source and the load impedance are the same; this is the process of matching.

It was often required to measure the power in a telephone system, and one milliwatt was chosen as a suitable unit. Thus the reference against which signals could be compared was the dissipation of one milliwatt in $600\,\Omega$. Figure 2.32(a) shows that the dissipation of $1\,mW$ in $600\,\Omega$ will be due to an applied voltage of $0.775\,V$ rms. This voltage became the reference against which all audio levels are compared.

The deciBel is a logarithmic measuring system and has its origins in telephony[7] where the loss in a cable is a logarithmic function of the length. Human hearing also has a logarithmic response with respect to sound pressure level (SPL). In order to relate to the subjective response audio signal level measurements have also to be logarithmic and so the deciBel was adopted for audio.

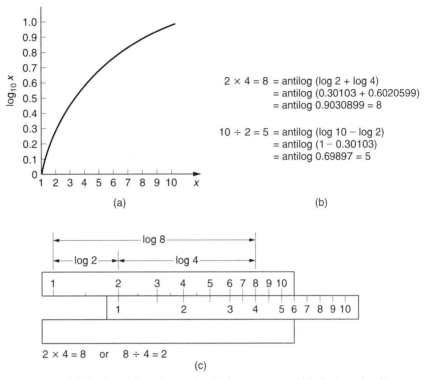

$$2 \times 4 = 8 = \text{antilog (log 2 + log 4)}$$
$$= \text{antilog } (0.30103 + 0.6020599)$$
$$= \text{antilog } 0.9030899 = 8$$

$$10 \div 2 = 5 = \text{antilog (log 10 - log 2)}$$
$$= \text{antilog } (1 - 0.30103)$$
$$= \text{antilog } 0.69897 = 5$$

(a)

(b)

(c)

Figure 2.34 (a) The logarithm of a number is the power to which the base (in this case 10) must be raised to obtain the number. (b) Multiplication is obtained by adding logs, division by subtracting. (c) The slide rule has two logarithmic scales whose length can easily be added or subtracted.

Figure 2.34 shows the principle of the logarithm. To give an example, if it is clear that 10^2 is 100 and 10^3 is 1000, then there must be a power between 2 and 3 to which 10 can be raised to give any value between 100 and 1000. That power is the logarithm to base 10 of the value. e.g. $\log_{10}$ 300 = 2.5 approx. Note that 10^0 is 1.

Logarithms were developed by mathematicians before the availability of calculators or computers to ease calculations such as multiplication, squaring, division and extracting roots. The advantage is that, armed with a set of log tables, multiplication can be performed by adding and division by subtracting. Figure 2.34 shows some examples. It will be clear that squaring a number is performed by adding two identical logs and the same result will be obtained by multiplying the log by 2.

The slide rule is an early calculator which consists of two logarithmically engraved scales in which the length along the scale is proportional to the log of the engraved number. By sliding the moving scale two lengths can easily be added or subtracted and as a result multiplication and division is readily obtained.

$$1 \text{ Bel} = \log_{10} \frac{P_1}{P_2} \qquad 1 \text{ deciBel} = 1/10 \text{ Bel}$$

$$\text{Power ratio (dB)} = 10 \times \log_{10} \frac{P_1}{P_2}$$

(a)

As power α V^2, when using voltages:

$$\text{Power ratio (dB)} = 10 \log \frac{V_1^2}{V_2^2}$$

$$= 10 \times \log \frac{V_1}{V_2} \times 2$$

$$= 20 \log \frac{V_1}{V_2}$$

(b)

Figure 2.35 (a) The Bel is the log of the ratio between two powers, that to be measured and the reference. The Bel is too large so the deciBel is used in practice. (b) As the dB is defined as a power ratio, voltage ratios have to be squared. This is conveniently done by doubling the logs so the ratio is now multiplied by 20.

The logarithmic unit of measurement in telephones was called the Bel after Alexander Graham Bell, the inventor. Figure 2.35(a) shows that the Bel was defined as the log of the *power* ratio between the power to be measured and some reference power. Clearly the reference power must have a level of 0 Bels since $\log_{10} 1$ is 0.

The Bel was found to be an excessively large unit for practical purposes and so it was divided into 10 deciBels, abbreviated dB with a small d and a large B and pronounced deebee. Consequently the number of dB is ten times the log of the power ratio. A device such as an amplifier can have a fixed power gain which is independent of signal level and this can be measured in dB. However, when measuring the power of a signal, it must be appreciated that the dB is a ratio and to quote the number of dBs without stating the reference is about as senseless as describing the height of a mountain as 2000 without specifying whether this is feet or metres. To show that the reference is one milliwatt into $600\,\Omega$, the units will be dB(m). In radio engineering, the dB(W) will be found which is power relative to one watt.

Although the dB(m) is defined as a power ratio, level measurements in audio are often done by measuring the signal voltage using 0.775 V as a reference in a circuit whose impedance is not necessarily 600 Ω. Figure 2.35(b) shows that as the power is proportional to the square of the voltage, the power ratio will be obtained by squaring the voltage ratio. As squaring in logs is performed by doubling, the squared term of the voltages can be replaced by multiplying the log by a factor of two. To give a result in deciBels, the log of the voltage ratio now has to be multiplied by 20.

Whilst 600 Ω matched impedance working is essential for the long distances encountered with telephones, it is quite inappropriate for analog audio wiring in a studio. The wavelength of audio in wires at 20 kHz is 15 km. Studios are built on a smaller scale than this and clearly analog audio cables are *not* transmission lines and their characteristic impedance is not relevant.

In professional analog audio systems impedance matching is not only unnecessary it is also undesirable. Figure 2.36(a) shows that when impedance matching is required the output impedance of a signal source must be artificially raised so that a potential divider is formed with the load. The actual drive voltage must be twice that needed on the cable as the potential divider effect wastes 6 dB of signal level and requires unnecessarily high power supply rail voltages in equipment. A further problem is that cable capacitance can cause an undesirable HF roll-off in conjunction with the high source impedance.

In modern professional analog audio equipment, shown in Figure 2.36(b), the source has the lowest output impedance practicable. This

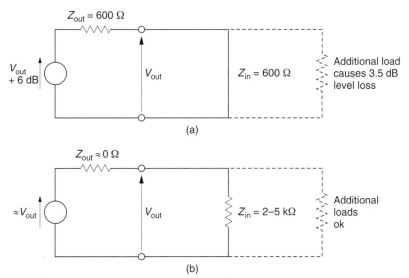

Figure 2.36 (a) Traditional impedance matched source wastes half the signal voltage in the potential divider due to the source impedance and the cable. (b) Modern practice is to use low-output impedance sources with high-impedance loads.

means that any ambient interference is attempting to drive what amounts to a short circuit and can only develop very small voltages. Furthermore shunt capacitance in the cable has very little effect. The destination has a somewhat higher impedance (generally a few kΩ) to avoid excessive currents flowing and to allow several loads to be placed across one driver.

In the absence of a fixed impedance it is meaningless to consider power. Consequently only signal voltages are measured. The reference remains at 0.775 V, but power and impedance are irrelevant. Voltages measured in this way are expressed in dB(u); the most common unit of level in modern analog systems. Most installations boost the signals on interface cables by 4 dB. As the gain of receiving devices is reduced by 4 dB, the result is a useful noise advantage without risking distortion due to the drivers having to produce high voltages.

In order to make the difference between dB(m) and dB(u) clear, consider the lossless matching transformer shown in Figure 2.37. The turns ratio is 2:1 therefore the impedance matching ratio is 4:1. As there is no loss in the transformer, the input power is the same as the output power so that the transformer shows a gain of 0 dB(m). However, the turns ratio of 2:1 provides a voltage gain of 6 dB(u). The doubled output voltage will deliver the same power to the quadrupled load impedance.

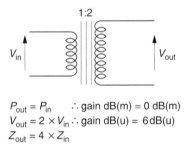

$$P_{out} = P_{in} \quad \therefore \text{gain dB(m)} = 0 \text{ dB(m)}$$
$$V_{out} = 2 \times V_{in} \therefore \text{gain dB(u)} = 6 \text{ dB(u)}$$
$$Z_{out} = 4 \times Z_{in}$$

Figure 2.37 A lossless transformer has no power gain so the level in dB(m) on input and output is the same. However, there is a voltage gain when measurements are made in dB(u).

In a complex system signals may pass through a large number of processes, each of which may have a different gain. If one stays in the linear domain and measures the input level in volts rms, the output level will be obtained by multiplying by the gains of all the stages involved. This is a complex calculation.

The difference between the signal level with and without the presence of a device in a chain is called the *insertion loss* measured in dB. However,

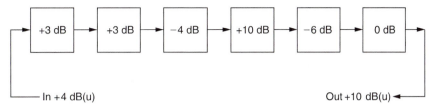

Figure 2.38 In complex systems each stage may have voltage gain measured in dB. By adding all of these gains together and adding to the input level in dB(u), the output level in dB(u) can be obtained.

if the input is measured in dB(u), the output level of the first stage can be obtained by adding the insertion loss in dB. This is shown in Figure 2.38. The output level of the second stage can be obtained by further adding the loss of the second stage in dB and so on. The final result is obtained by adding together all the insertion losses in dB and adding them to the input level in dB(u) to give the output level in dB(u). As the dB is a pure ratio it can multiply anything (by addition of logs) without changing the units. Thus dB(u) of level added to dB of gain are still dB(u).

In acoustic measurements, the sound pressure level (SPL) is measured in deciBels relative to a reference pressure of 2×10^{-5} Pascals (Pa) rms. In order to make the reference clear the units are dB(SPL). In measurements which are intended to convey an impression of subjective loudness, a weighting filter is used prior to the level measurement which reproduces the frequency response of human hearing which is most sensitive in the midrange. The most common standard frequency response is the so-called A-weighting filter, hence the term dB(A) used when a weighted level is being measured. At high or low frequencies, a lower reading will be obtained in dB(A) than in dB(SPL).

2.19   Audio level metering

There are two main reasons for having level meters in audio equipment: to line up or adjust the gain of equipment, and to assess the amplitude of the program material.

Line-up is often done using a 1 kHz sine wave generated at an agreed level such as 0 dB(u). If a receiving device does not display the same level, then its input sensitivity must be adjusted. Tape recorders and other devices which pass signals through are usually lined up so that their input and output levels are identical, i.e. their insertion loss is 0 dB. Line-up is important in large systems because it ensures that inadvertent level changes do not occur.

In measuring the level of a sine wave for the purposes of line-up, the dynamics of the meter are of no consequence, whereas on program material the dynamics matter a great deal. The simplest (and cheapest) level meter is essentially an AC voltmeter with a logarithmic response. As the ear is logarithmic, the deflection of the meter is roughly proportional to the perceived volume, hence the term volume unit (VU) meter.

In audio recording and broadcasting, the worst sin is to overmodulate the tape or the transmitter by allowing a signal of excessive amplitude to pass. Real audio signals are rich in short transients which pass before the sluggish VU meter responds. Consequently the VU meter is also called the virtually useless meter in professional circles.

Broadcasters developed the peak program meter (PPM) which is also logarithmic, but which is designed to respond to peaks as quickly as the ear responds to distortion. Consequently the attack time of the PPM is carefully specified. If a peak is so short that the PPM fails to indicate its true level, the resulting overload will also be so brief that the ear will not hear it. A further feature of the PPM is that the decay time of the meter is very slow, so that any peaks are visible for much longer and the meter is easier to read because the meter movement is less violent.

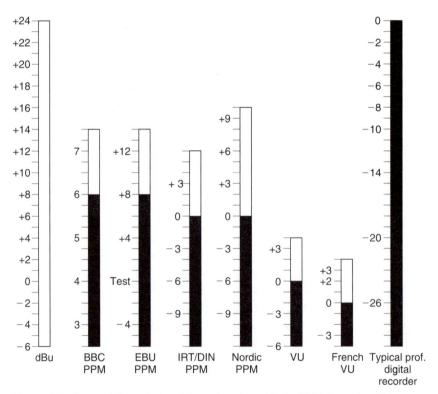

Figure 2.39 Some of the scales used in conjunction with the PPM dynamics. (After Francis Rumsey, with permission.).

The original PPM as developed by the BBC was sparsely calibrated, but other users have adopted the same dynamics and added dB scales, Figure 2.39 shows some of the scales in use.

In broadcasting, the use of level metering and line-up procedures ensures that the level experienced by the listener does not change significantly from program to program. Consequently in a transmission suite, the goal would be to broadcast recordings at a level identical to that which was obtained during production. However, when making a recording prior to any production process, the goal would be to modulate the recording as fully as possible without clipping as this would then give the best signal-to-noise ratio. The level would then be reduced if necessary in the production process.

2.20   Vectors

Often several signals of the same frequency but with differing phases need to be added. When the the two phases are identical, the amplitudes are simply added. When the two phases are 180° apart the amplitudes are subtracted. When there is an arbitrary phase relationship, vector addition is needed. A vector is simply an arrow whose length represents the amplitude and whose direction represents the phase-shift. Figure 2.40

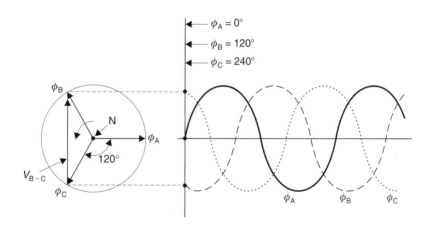

$$V_{B-C} = V_B \sin 120 \; - \; V_C \sin 240 \quad V_B = V_C = V_A$$
$$\therefore \; V_{B-C} = 2 \times V_A \times 0.866 = V_A \times 1.732$$

E.g. if phase voltage is 230 V rms,
    phase-to-phase voltage = 398 V rms

Figure 2.40   Three-phase electricity uses three signals mutually at 120°. Thus the phase-to-phase voltage has to be calculated vectorially from the phase-to-neutral voltage as shown.

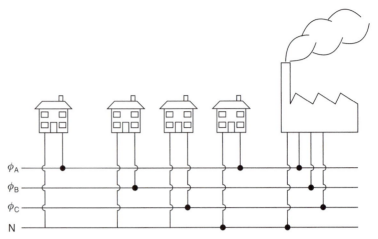

Figure 2.41 The possibility of a phase-to-phase shock is reduced in suburban housing by rotating phases between adjacent buildings. This also balances the loading on the three phases.

shows a vector diagram showing the phase relationship in the common three-phase electricity supply. The length of each vector represents the phase-to-neutral voltage which in many countries is about 230 V rms. As each phase is at 120° from the next, what will the phase-to-phase voltage be? Figure 2.40 shows that the answer can be found geometrically to be about 380 V rms. Consequently whilst a phase-to-neutral shock is not recommended, getting a phase-to-phase shock is recommended even less!

The three-phase electricity supply has the characteristic that although each phase passes through zero power twice a cycle, the total power is constant. This results in less vibration at the generator and in large motors. When a three-phase system is balanced (i.e. there is an equal load on each phase) there is no neutral current. Figure 2.41 shows that most suburban power installations each house only has a single-phase supply for safety. The houses are connected in rotation to balance the load. Business premises such as recording studios and broadcasters will take a three-phase supply which should be reasonably balanced by connecting equal loading to each phase.

2.21   Phase angle and power factor

The power is only obtained by multiplying the voltage by the current when the load is resistive. Only with a resistive load will the voltage and the current be in the same phase. In both electrical and audio power distribution systems, the load may be *reactive* which means that the

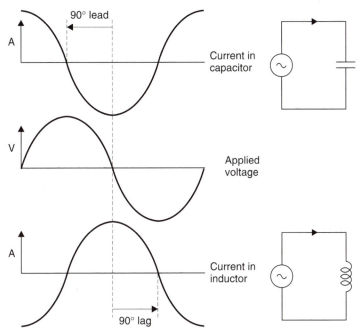

Figure 2.42 Ideal capacitors conduct current with a quadrature phase lead whereas inductors have a quadrature phase lag. In both cases the quadrature makes the product of current and voltage zero so no power is dissipated.

current and voltage waveforms have a relative phase-shift. Mathematicians would describe the load as *complex*.

In a reactive load, the power in Watts **W** is given by multiplying the rms voltage, the rms current and the cosine of the relative phase angle φ. Clearly if the voltage and current are in quadrature there can be no power dissipated because cos φ is zero. Cos φ is called the *power factor*. Figure 2.42 shows that this happens with perfect capacitors and perfect inductors connected to an AC supply. With a perfect capacitor, the current *leads* the voltage by 90°, whereas with a perfect inductor the current *lags* the voltage by 90°.

A power factor significantly less than one is undesirable because it means that larger currents are flowing than are necessary to deliver the power. The losses in distribution are proportional to the current and so a reactive load is an inefficient load. Lightly loaded transformers and induction motors act as inductive loads with a poor power factor. In some industrial installations it is economic to install power factor correction units which are usually capacitor banks that balance the lagging inductive load with a capacitive lead.

As the power factor of a load cannot be anticipated, AC distribution equipment is often rated in Volt-Amps (VA) instead of Watts. With a resistive load, the two are identical, but with a reactive load the power which can be delivered falls. As loudspeakers are almost always reactive, audio amplifiers should be rated in VA. Instead amplifiers are rated in Watts leaving the unfortunate user to find out for himself what reactive load can be driven.

2.22 Audio cabling

Balanced line working was developed for professional analog audio as a means to reject noise. This is particularly important for microphone signals because of the low levels, but is also important for both line level analog and digital signals where interference may be encountered from electrical and radio installations. Figure 2.43 shows how balanced audio should be connected. The receiver subtracts one input from the other which rejects any common mode noise or hum picked up on the wiring. Twisting the wires tightly together ensures that both pick up the same amount of interference.

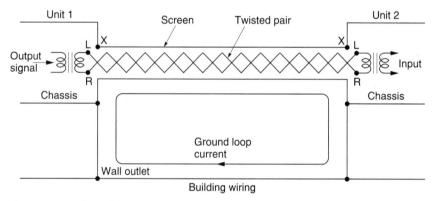

Figure 2.43 Balanced analog audio interface. Note that the braid plays no part in transmitting the audio signal, but bonds the two units together and acts as a screen. Loop currents flowing in the screen are harmless.

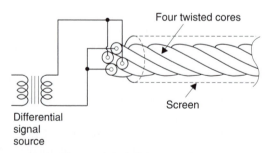

Figure 2.44 In star-quad cabling each leg of the balanced signal is connected to two conductors which are on opposite sides of a four-phase helix. The pickup of interference on the two legs is then as equal as possible so that the differential receiver can reject it.

The star-quad technique is possibly the ultimate interference rejecting cable construction. Figure 2.44 shows that in star-quad cable four conductors are twisted together. Diametrically opposite pairs are connected together at both ends of the cable and used as the two legs of a differential system. The interference pickup on the two legs is rendered as identical as possible by the construction so that it can be perfectly rejected at a well-engineered differential receiver.

The standard connector which has been used for professional audio for many years is the XLR which has three pins. It is easy to remember that pins 1, 2 and 3 connect to eXternal, Live and Return respectively. EXternal is the cable screen, Live is the in-phase leg of the balanced signal and Return is self-explanatory. The metal body shell of the XLR connector should be connected to both the cable screen and pin 1 although cheaper connectors do not provide a tag for the user to make this connection and rely on contact with the chassis socket to ground the shell. Oddly, the male connector (the one with pins) is used for equipment signal outputs,

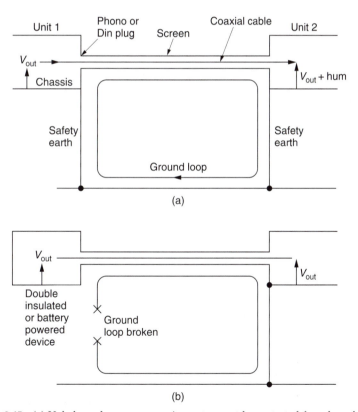

Figure 2.45 (a) Unbalanced consumer equipment cannot be protected from hum loops because the signal return and the screen are the same conductor. (b) With a floating signal source there will be no current in the screen. Source must be double insulated for safety.

whereas the female (the one with receptacles) is used with signal inputs. This is so that when phantom power is used, the live parts are insulated.

When making balanced cables it is important to ensure that the twisted pair is connected identically at both ends. If the two wires are inadvertently interchanged, the cable will still work, but a phase reversal will result, causing problems in analog stereo installations. Digital cables are unaffected by a phase reversal.

In consumer equipment differential working is considered too expensive. Instead single-ended analog signals using coax cable are found using phono, DIN and single-pole jack connectors. Whilst these are acceptable in permanent installations, they will not stand repeated connection and disconnection and become unreliable.

Effective unbalanced transmission of analog or digital signals over long distances is very difficult. When the signal return, the chassis ground and the safety ground are one and the same as in Figure 2.45(a), ground loop currents cannot be rejected. The only solution is to use equipment which is double insulated so that no safety ground is needed. Then each item can be grounded by the coax screen. As Figure 2.45(b) shows, there can then be no ground current as there is no loop. However, unbalanced working also uses higher impedances and lower signal levels and is more prone to interference. For these reasons some better-quality consumer equipment will be found using balanced signals.

2.23   EMC

EMC stands for electromagnetic compatibility which is a way of making electronic equipment more reliable by limiting both the amount of spurious energy radiated and the sensitivity to extraneous radiation. As electronic equipment becomes more common and more of our daily life depends upon its correct operation it becomes important to contain the unwanted effects of interference.

In audio equipment external interference can cause unwanted signals to be superimposed on the wanted audio signal. This is most likely to happen in sensitive stages handling small signals; e.g. microphone preamplifiers and tape replay stages. Interference can enter through any cables, including the power cable, or by radiation. Such stages must be designed from the outset with the idea that radio frequency energy may be present which must be rejected. Whilst one could argue that RF energy from an arcing switch should be suppressed at source one cannot argue that cellular telephones should be banned as they can only operate using RF radiation. When designing from the outset, RF rejection is not too difficult. Putting it in afterwards is often impossible without an uneconomic redesign.

There have been some complaints from the high-end Hi-Fi community that the necessary RF suppression components will impair the sound quality of audio systems but this is nonsense. In fact good EMC design actually improves sound quality because by eliminating common impedances which pick up interference distortion is also reduced.

In balanced signalling the screen does not carry the audio, but serves to extend the screened cabinets of the two pieces of equipment with a metallic tunnel. For this to be effective against RF interference it has to be connected at both ends. This is also essential for electrical safety so that no dangerous potential difference can build up between the units. Figure 2.43 showed that connecting the screen at both ends causes an earth loop with the building ground wiring. Loop currents will circulate as shown but this is not a problem because by shunting loop currents into the screen, they are kept out of the audio wiring.

Some poorly designed equipment routes the X-pin of the XLR via the pcb instead of direct to the equipment frame. As Figure 2.46 shows, this effectively routes loop currents through the circuitry and is prone to interference. This approach does not comply with recent EMC regulations but there is a lot of old equipment still in service which could be put right. A simple track cut and a new chassis-bonded XLR socket is often all that is necessary. Another false economy is the use of plastic XLR shells which cannot provide continuity of screening.

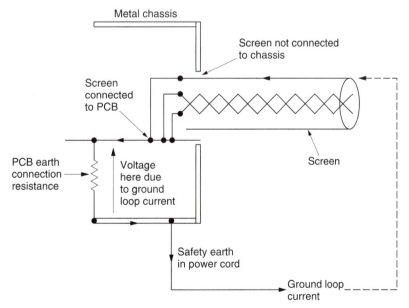

Figure 2.46 Poorly designed product in which screen currents pass to chassis via circuit board. Currents flowing in ground lead will raise voltages which interfere with the audio signal.

Differential working with twisted pairs is designed to reject hum and noise, but it only works properly if both signal legs have identical frequency/impedance characteristics at both ends. The easiest way of achieving this is to use transformers which give much better RF rejection than electronic balancing. Whilst an effective electronic differential receiver can be designed with care, a floating balanced electronic driver cannot compete with a transformer. An advantage of digital audio is that these transformers can be very small and inexpensive, whereas a high-quality analog transformer is very expensive indeed.

Analog audio equipment works at moderate frequencies and seldom has a radiation problem. However, any equipment controlled by a microprocessor or containing digital processing is a potential source of interference and once more steps must be taken in the design stage to ensure that radiation is minimized. It should be borne in mind that poor layout may result in radiation from the digital circuitry actually impairing the performance of analog circuits in the same device. This is a critical issue in convertor design. It is consequently most unlikely that a convertor which could not meet the EMC regulations would meet its audio quality specification.

All AC-powered audio devices contain some kind of power supply which rectifies the AC line to provide DC power. This rectification process is non-linear and can produce harmonics which leave the equipment via the power cable and cause interference elsewhere. Suitable power cord filtering must be provided to limit harmonic generation.

2.24   Electrical safety

Under fault conditions an excess of current can flow and the resulting heat can cause fire. Practical equipment must be fused so that excessive current causes the fuse element to melt, cutting off the supply. In many electronic devices the initial current exceeds the steady current because capacitors need to charge. Safe fusing requires the use of slow-blow fuses which have increased thermal mass. The switch-on surge will not blow the fuse, whereas a steady current of the same value will. Slow blow fuses can be identified by the (T) after the rating, e.g. 3.15 A(T). Blown fuses should only be replaced with items of the same type and rating. Fuses do occasionally blow from old age, but any blown fuse should be regarded as indicating a potential problem. Replacing a fuse with one of a higher rating is the height of folly as no protection against fire is available. When dual-voltage 115/230 equipment is set to a different range a different fuse will often be necessary. In some small power supplies the power taken is small and the use of a fuse is not practicable. Instead a thermal switch is built into the transformer. In the case of overheating this will melt.

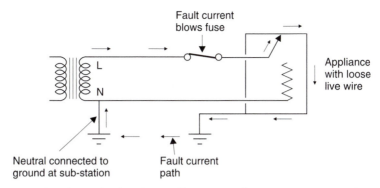

Figure 2.47 For electrical safety the metallic housing of equipment is connected to ground, preventing a dangerous potential existing in the case of a fault.

Generally these switches are designed to work once only after which the transformer must be replaced.

Except for low-voltage battery-powered devices, electrically powered equipment has to be considered a shock hazard to the user and steps must be taken to minimize the hazard. There are two main ways in which this is achieved. Figure 2.47 shows that the equipment is made with a conductive case which is connected to earth via a third conductor. In the event that a fault causes live wiring to contact the case, current is conducted to ground which will blow the fuse. Clearly disconnecting the earth for any purpose could allow the case to become live. The alternative is to construct the equipment in such a way that live wiring physically cannot cause the body to become live. In a double-insulated product all the live components are encased in plastic so that even if a part such as a transformer or motor becomes live it is still insulated from the outside. Double-insulated devices need no safety earth.

Where there is a risk of cables being damaged or cut, earthing and double insulation are of no help because the live conductor in the cable may be exposed. Safety can be enhanced by the use of a residual current breaker (RCB) which detects any imbalance in live and neutral current. An imbalance means that current is flowing somewhere it shouldn't and this results in the breaker cutting off the power.

References

1. Moore, B.C.J., *An Introduction to the Psychology of Hearing*, London: Academic Press (1989)
2. Muraoka, T., Iwahara, M. and Yamada, Y., Examination of audio bandwidth requirements for optimum sound signal transmission. *J. Audio Eng. Soc.*, **29**, 2–9 (1982)
3. Muraoka, T., Yamada, Y. and Yamazaki, M., Sampling frequency considerations in digital audio. *J. Audio Eng. Soc.*, **26**, 252–256 (1978)

4. Fincham, L.R., The subjective importance of uniform group delay at low frequencies. Presented at the 74th Audio Engineering Society Convention (New York, 1983), Preprint 2056(H-1)

5. Fletcher, H., Auditory patterns. *Rev. Modern Physics*, **12**, 47–65 (1940)

6. Carterette, E.C. and Friedman, M.P., *Handbook of Perception*, 305–319. New York: Academic Press (1978)

7. Martin, W.H., Decibel – the new name for the transmission unit. *Bell System Tech. J.*, (January 1929)

3

Digital principles

3.1 Pure binary code

For digital audio use, the prime purpose of binary numbers is to express the values of the samples which represent the original analog sound-velocity or pressure waveform. Figure 3.1 shows some binary numbers and their equivalent in decimal. The radix point has the same significance in binary: symbols to the right of it represent one half, one quarter and so on. Binary is convenient for electronic circuits, which do not get tired, but numbers expressed in binary become very long, and writing them is tedious and error-prone. The octal and hexadecimal notations are both used for writing binary since conversion is so simple. Figure 3.1 also shows that a binary number is split into groups of three or four digits starting at the least significant end, and the groups are individually converted to octal or hexadecimal digits. Since sixteen different symbols are required in hex. the letters A–F are used for the numbers above nine.

There will be a fixed number of bits in a PCM audio sample, and this number determines the size of the quantizing range. In the sixteen-bit samples used in much digital audio equipment, there are 65 536 different numbers. Each number represents a different analog signal voltage, and care must be taken during conversion to ensure that the signal does not go outside the convertor range, or it will be clipped. In Figure 3.2 it will be seen that in a sixteen-bit pure binary system, the number range goes from 0000 hex, which represents the largest negative voltage, through 7FFF hex, which represents the smallest negative voltage, through 8000 hex, which represents the smallest positive voltage, to FFFF hex, which represents the largest positive voltage. Effectively, the zero voltage level

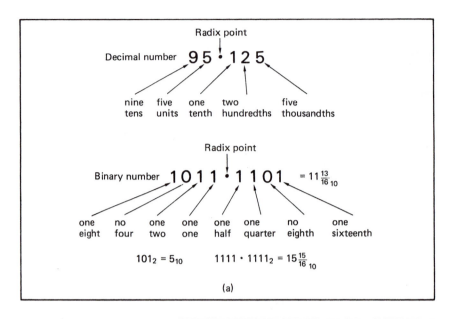

(a)

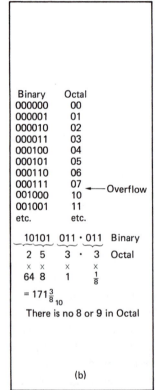

(b)

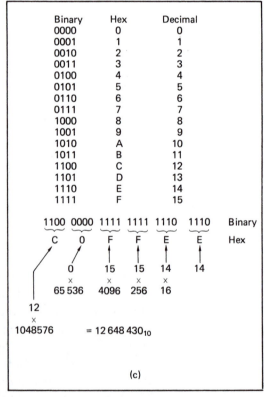

(c)

Figure 3.1 (a) Binary and decimal. (b) In octal, groups of three bits make one symbol 0–7. (c) In hex, groups of four bits make one symbol 0–F. Note how much shorter the number is in hex.

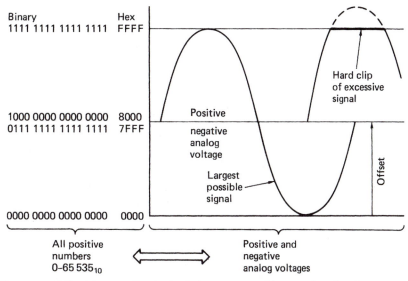

Figure 3.2 Offset binary coding is simple but causes problems in digital audio processing. It is seldom used.

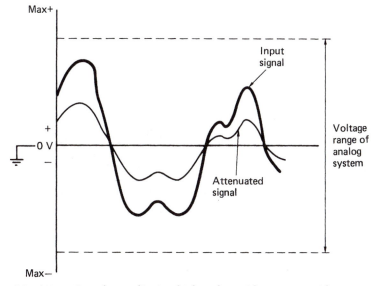

Figure 3.3 Attenuation of an audio signal takes place with respect to midrange.

of the audio has been shifted so that the positive and negative voltages in a real audio signal can be expressed by binary numbers which are only positive. This approach is called offset binary, and is perfectly acceptable where the signal has been digitized only for recording or transmission from one place to another, after which it will be converted directly back to analog. Under these conditions it is not necessary for the quantizing steps to be uniform, provided both ADC and DAC are constructed to the same standard. In practice, it is the requirements of signal processing in the digital domain which make both non-uniform quantizing and offset binary unsuitable.

Figure 3.3 shows that an audio signal voltage is referred to midrange. The level of the signal is measured by how far the waveform deviates from midrange, and attenuation, gain and mixing all take place around that midrange. Audio mixing is achieved by adding sample values from two or more different sources, but unless all the quantizing intervals are of the same size, the sum of two sample values will not represent the sum of the two original analog voltages. Thus sample values which have been obtained by non-uniform quantizing cannot readily be processed.

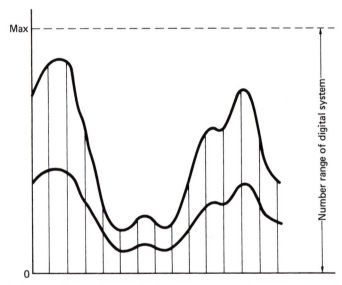

Figure 3.4 The result of an attempted attenuation in pure binary code is an offset. Pure binary cannot be used for digital audio processing.

If two offset binary sample streams are added together in an attempt to perform digital mixing, the result will be that the offsets are also added and this may lead to an overflow. Similarly, if an attempt is made to attenuate by, say, 6.02 dB by dividing all the sample values by two, Figure 3.4 shows that the offset is also divided and the waveform suffers a shifted baseline. The problem with offset binary is that it works with reference to one end of the range. What is needed is a numbering system which operates symmetrically with reference to the centre of the range.

3.2   Two's complement

In the two's complement system, the upper half of the pure binary number range has been redefined to represent negative quantities. If a pure binary counter is constantly incremented and allowed to overflow, it will produce all the numbers in the range permitted by the number of available bits, and these are shown for a four-bit example drawn around the circle in Figure 3.5. As a circle has no real beginning, it is possible to consider it to start wherever it is convenient. In two's complement, the quantizing range represented by the circle of numbers does not start at zero, but starts on the diametrically opposite side of the circle. Zero is midrange, and all numbers with the MSB (most significant bit) set are considered negative. The MSB is thus the equivalent of a sign bit where 1 = minus. Two's complement notation differs from pure binary in that the most significant bit is inverted in order to achieve the half-circle rotation.

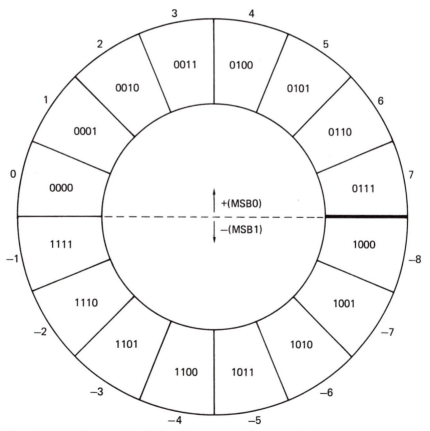

Figure 3.5   In this example of a four-bit two's complement code, the number range is from −8 to +7. Note that the MSB determines polarity.

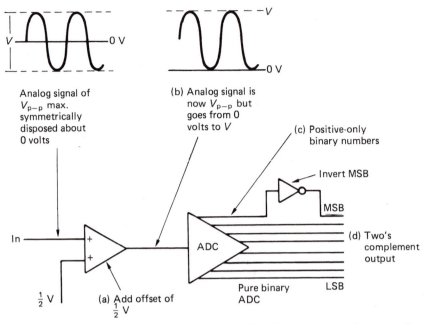

Figure 3.6 A two's complement ADC. At (a) an analog offset voltage equal to one-half the quantizing range is added to the bipolar analog signal in order to make it unipolar as at (b). The ADC produces positive-only numbers at (c), but the MSB is then inverted at (d) to give a two's complement output.

Figure 3.6 shows how a real ADC is configured to produce two's complement output. At (a) an analog offset voltage equal to one half the quantizing range is added to the bipolar analog signal in order to make it unipolar as at (b). The ADC produces positive only numbers at (c) which are proportional to the input voltage. The MSB is then inverted at (d) so that the all-zeros code moves to the centre of the quantizing range. The analog offset is often incorporated into the ADC as is the MSB inversion. Some convertors are designed to be used in either pure binary or two's complement mode. In this case the designer must arrange the appropriate DC conditions at the input. The MSB inversion may be selectable by an external logic level.

The two's complement system allows two sample values to be added, or mixed in audio parlance, and the result will be referred to the system midrange; this is analogous to adding analog signals in an operational amplifier. Figure 3.7 illustrates how adding two's complement samples simulates the audio mixing process. The waveform of input A is depicted by solid black samples, and that of B by samples with a solid outline. The result of mixing is the linear sum of the two waveforms obtained by adding pairs of sample values. The dashed lines depict the output values. Beneath each set of samples is the calculation which will be seen to give

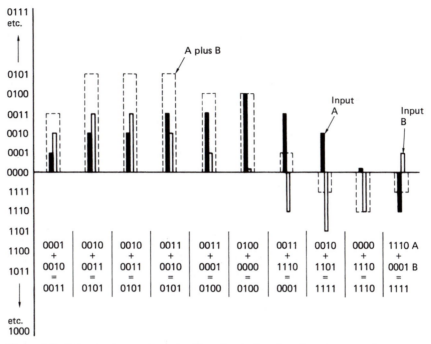

Figure 3.7 Using two's complement arithmetic, single values from two waveforms are added together with respect to midrange to give a correct mixing function.

the correct result. Note that the calculations are pure binary. No special arithmetic is needed to handle two's complement numbers.

Figure 3.8 shows some audio waveforms at various levels with respect to the coding values. Where an audio waveform just fits into the quantizing range without clipping it has a level which is defined as 0 dBFs where Fs indicates *full scale*. Reducing the level by 6.02 dB makes the signal half as large and results in the second bit in the sample

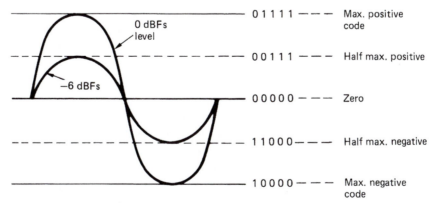

Figure 3.8 0 dBFs is defined as the level of the largest sinusoid which will fit into the quantizing range without clipping.

becoming the same as the sign bit. Reducing the level by a further 6.02 dB to –12 dBFs will make the second and third bits the same as the sign bit and so on. If a signal at –36 dBFs is input to a sixteen-bit system, only ten bits will be active, the remainder will copy the sign bit. For the best performance, analog inputs to digital systems must have sufficient levels to exercise the whole quantizing range.

It is often necessary to phase reverse or invert an audio signal, for example a microphone input to a mixer. The process of inversion in two's complement is simple. All bits of the sample value are inverted to form the one's complement, and one is added. This can be checked by mentally inverting some of the values in Figure 3.5. The inversion is transparent and performing a second inversion gives the original sample values.

Using inversion, signal subtraction can be performed using only adding logic. The inverted input is added to perform a subtraction, just as in the analog domain. This permits a significant saving in hardware complexity, since only carry logic is necessary and no borrow mechanism need be supported.

In summary, two's complement notation is the most appropriate scheme for bipolar signals, and allows simple mixing in conventional binary adders. It is in virtually universal use in digital audio processing, and is accordingly adopted by all the major digital audio interfaces and recording formats.

Two's complement numbers can have a radix point and bits below it just as pure binary numbers can. It should, however, be noted that in two's complement, if a radix point exists, numbers to the right of it are added. For example, 1100.1 is not –4.5, it is –4 + 0.5 = –3.5.

3.3   Introduction to digital processing

However complex a digital process, it can be broken down into smaller stages until finally one finds that there are really only two basic types of element in use. Figure 3.9 shows that the first type is a *logical* element. This produces an output which is a logical function of the input with minimal delay. The second type is a *storage* element which samples the state of the input(s) when clocked and holds or delays that state. The strength of binary logic is that the signal has only two states, and considerable noise and distortion of the binary waveform can be tolerated before the state becomes uncertain. At every logical element, the signal is compared with a threshold, and can thus can pass through any number of stages without being degraded. In addition, the use of a storage element at regular locations throughout logic circuits eliminates time variations or jitter.

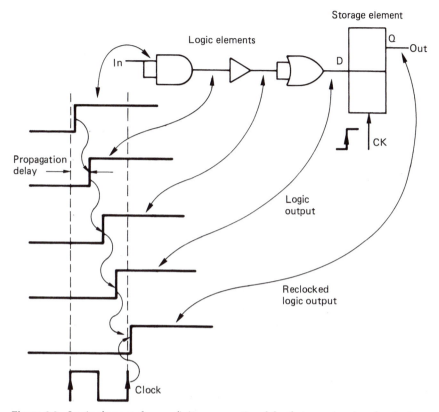

Figure 3.9 Logic elements have a finite propagation delay between input and output and cascading them delays the signal an arbitrary amount. Storage elements sample the input on a clock edge and can return a signal to near coincidence with the system clock. This is known as reclocking. Reclocking eliminates variations in propagation delay in logic elements.

Figure 3.9 shows that if the inputs to a logic element change, the output will not change until the *propagation delay* of the element has elapsed. However, if the output of the logic element forms the input to a storage element, the output of that element will not change until the input is sampled *at the next clock edge*. In this way the signal edge is aligned to the system clock and the propagation delay of the logic becomes irrelevant. The process is known as reclocking.

3.4 Logic elements

The two states of the signal when measured with an oscilloscope are simply two voltages, usually referred to as high and low. The actual voltage levels will depend on the type of logic family in use, and on the supply voltage used. Within logic, these levels are not of much consequence, and it is only necessary to know them when interfacing between different logic families or when driving external devices.

The pure logic designer is not interested at all in the precise signal voltages, only in their meaning. Just as the electrical waveform from a microphone represents sound velocity, so the waveform in a logic circuit represents the truth of some statement. As there are only two states, there can only be *true* or *false* meanings. The true state of the signal can be assigned by the designer to either voltage state. When a high voltage represents a true logic condition and a low voltage represents a false condition, the system is known as *positive logic*, or *high true* logic. This is

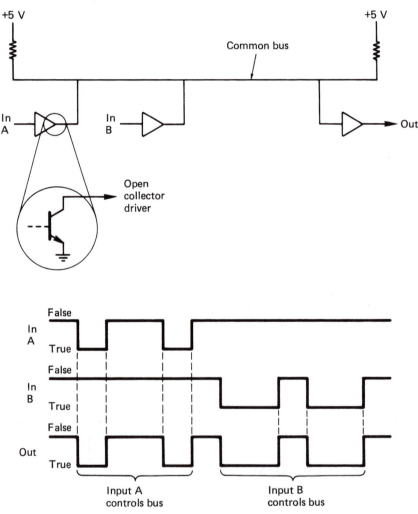

Figure 3.10 Using open-collector drive, several signal sources can share one common bus. If negative logic is used, the bus drivers turn off their output transistors with a false input, allowing another driver to control the bus. This will not happen with positive logic.

the usual system, but sometimes the low voltage represents the true condition and the high voltage represents the false condition. This is known as *negative logic* or *low true* logic. Provided that everyone is aware of the logic convention in use, both work equally well.

Negative logic was found in the TTL (transistor-transistor logic) family, because in this technology it was easier to sink current to ground than to source it from the power supply. Figure 3.10 shows that if it is necessary to connect several logic elements to a common bus so that any one can communicate with any other, an open collector system is used, where high levels are provided by pull-up resistors and the logic elements only pull the common line down. If positive logic were used, when no device was operating the pull-up resistors would cause the common line to take on an absurd true state; whereas if negative logic is used, the common line pulls up to a sensible false condition when there is no device using the bus. Whilst the open collector is a simple way of obtaining a shared bus system, it is limited in frequency of operation due to the time constant of the pull-up resistors charging the bus capacitance. In the so-called tri-state bus systems, there are both active pull-up and pull-down devices connected in the so-called totem-pole output configuration. Both devices can be disabled to a third state, where the output assumes a high impedance, allowing some other driver to determine the bus state.

In logic systems, all logical functions, however complex, can be configured from combinations of a few fundamental logic elements or *gates*. It is not profitable to spend too much time debating which are the truly fundamental ones, since most can be made from combinations of others. Figure 3.11 shows the important simple gates and their derivatives, and introduces the logical expressions to describe them, which can be compared with the truth-table notation. The figure also shows the important fact that when negative logic is used, the OR gate function interchanges with that of the AND gate. Sometimes schematics are drawn to reflect which voltage state represents the true condition. In the so-called intentional logic scheme, a negative logic signal always starts and ends at an inverting 'bubble'. If an AND function is required between two negative logic signals, it will be drawn as an AND symbol with bubbles on all the terminals, even though the component used will be a positive logic OR gate. Opinions vary on the merits of intentional logic.

If numerical quantities need to be conveyed down the two-state signal paths described here, then the only appropriate numbering system is binary, which has only two symbols, 0 and 1. Just as positive or negative logic could be used for the truth of a logical binary signal, it can also be used for a numerical binary signal. Normally, a high voltage level will represent a binary 1 and a low voltage will represent a binary 0, described as a 'high for a one' system. Clearly a 'low for a one' system is just as feasible. Decimal numbers have several columns, each of which repre-

Positive logic name	Boolean expression	Positive logic symbol	Positive logic truth table	Plain English
Inverter or NOT gate	$Q = \overline{A}$	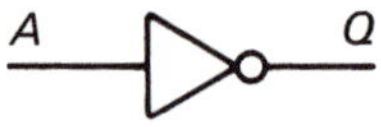	$\begin{array}{c\|c} A & Q \\ \hline 0 & 1 \\ 1 & 0 \end{array}$	Output is opposite of input
AND gate	$Q = A \cdot B$	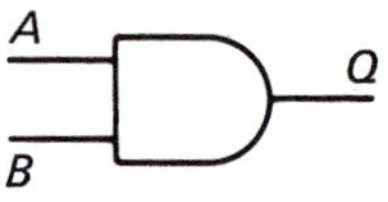	$\begin{array}{cc\|c} A & B & Q \\ \hline 0 & 0 & 0 \\ 0 & 1 & 0 \\ 1 & 0 & 0 \\ 1 & 1 & 1 \end{array}$	Output true when both inputs are true only
NAND (Not AND) gate	$Q = \overline{A \cdot B}$ $= \overline{A} + \overline{B}$	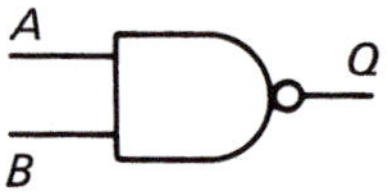	$\begin{array}{cc\|c} A & B & Q \\ \hline 0 & 0 & 1 \\ 0 & 1 & 1 \\ 1 & 0 & 1 \\ 1 & 1 & 0 \end{array}$	Output false when both inputs are true only
OR gate	$Q = A + B$	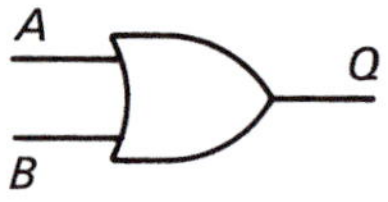	$\begin{array}{cc\|c} A & B & Q \\ \hline 0 & 0 & 0 \\ 0 & 1 & 1 \\ 1 & 0 & 1 \\ 1 & 1 & 1 \end{array}$	Output true if either or both inputs true
NOR (Not OR) gate	$Q = \overline{A + B}$ $= \overline{A} \cdot \overline{B}$	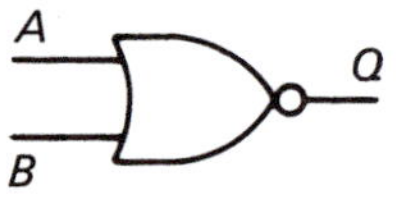	$\begin{array}{cc\|c} A & B & Q \\ \hline 0 & 0 & 1 \\ 0 & 1 & 0 \\ 1 & 0 & 0 \\ 1 & 1 & 0 \end{array}$	Output false if either or both inputs true
Exclusive OR (XOR) gate	$Q = A \oplus B$	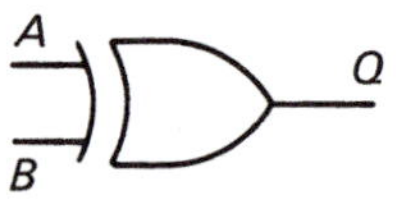	$\begin{array}{cc\|c} A & B & Q \\ \hline 0 & 0 & 0 \\ 0 & 1 & 1 \\ 1 & 0 & 1 \\ 1 & 1 & 0 \end{array}$	Output true if inputs are different

Figure 3.11 The basic logic gates compared.

sents a different power of ten; in binary the column position specifies the power of two.

Several binary digits or bits are needed to express the value of a binary audio sample. These bits can be conveyed at the same time by several signals to form a parallel system, which is most convenient inside equipment because it is fast, or one at a time down a single signal path, which is slower, but convenient for cables between pieces of equipment because the connectors require fewer pins. When a binary system is used to convey numbers in this way, it can be called a digital system.

3.5   Storage elements

The basic memory element in logic circuits is the latch, which is constructed from two gates as shown in Figure 3.12(a), and which can be set or reset. A more useful variant is the D-type latch shown at (b) which remembers the state of the input at the time a separate clock either changes state for an edge-triggered device, or after it goes false for a level-triggered device. D-type latches are commonly available with four or eight latches to the chip. A shift register can be made from a series of latches by connecting the Q output of one latch to the D input of the next and connecting all the clock inputs in parallel. Data are delayed by the number of stages in the register. Shift registers are also useful for converting between serial and parallel data transmissions.

Where large numbers of bits are to be stored, cross-coupled latches are less suitable because they are more complicated to fabricate inside integrated circuits than dynamic memory, and consume more current.

In large random access memories (RAMs), the data bits are stored as the presence or absence of charge in a tiny capacitor as shown in Figure 3.12(c). The capacitor is formed by a metal electrode, insulated by a layer of silicon dioxide from a semiconductor substrate, hence the term MOS (metal oxide semiconductor). The charge will suffer leakage, and the value would become indeterminate after a few milliseconds. Where the delay needed is less than this, decay is of no consequence, as data will be read out before they have had a chance to decay. Where longer delays are necessary, such memories must be refreshed periodically by reading the bit value and writing it back to the same place. Most modern MOS RAM chips have suitable circuitry built in. Large RAMs store thousands of bits, and it is clearly impractical to have a connection to each one. Instead, the desired bit has to be addressed before it can be read or written. The size of the chip package restricts the number of pins available, so that large memories use the same address pins more than once. The bits are arranged internally as rows and columns, and the row address and the column address are specified sequentially on the same pins.

3.6   Binary adding

The binary circuitry necessary for adding two's complement numbers is shown in Figure 3.13. Addition in binary requires two bits to be taken at a time from the same position in each word, starting at the least significant bit. Should both be ones, the output is zero, and there is a *carry-out* generated. Such a circuit is called a half adder, shown in Figure 3.13(a) and is suitable for the least-significant bit of the calculation. All higher stages will require a circuit which can accept a carry input as well

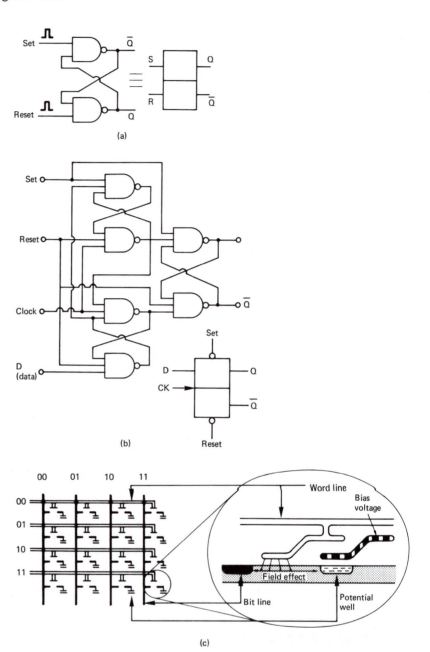

(a)

(b)

(c)

Figure 3.12 Digital semiconductor memory types. In (a), one data bit can be stored in a simple set–reset latch, which has little application because the D-type latch in (b) can store the state of the single data input when the clock occurs. These devices can be implemented with bipolar transistors of FETs, and are called static memories because they can store indefinitely. They consume a lot of power.

In (c), a bit is stored as the charge in a potential well in the substrate of a chip. It is accessed by connecting the bit line with the field effect from the word line. The single well where the two lines cross can then be written or read. These devices are called dynamic RAMs because the charge decays, and they must be read and rewritten (refreshed) periodically.

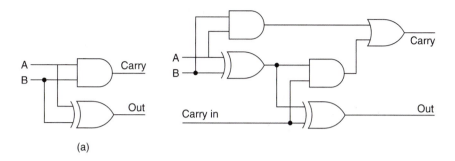

(a)

Data A	Bits B	Carry in	Out	Carry out
0	0	0	0	0
0	0	1	1	0
0	1	0	1	0
0	1	1	0	1
1	0	0	1	0
1	0	1	0	1
1	1	0	0	1
1	1	1	1	1

(b)

A MSB

B MSB

High when
A and B negative
but adder output
positive

High when
A and B positive
but adder output
negative

Adder
MSB

Input A

Two's
complement

Σ

Input B

Out

Maximum
positive
value

Maximum
negative
value

(c)

Figure 3.13 (a) Half adder; (b) full-adder circuit and truth table; (c) comparison of sign bits prevents wraparound on adder overflow by substituting clipping level.

as two data inputs. This is known as a full adder (b). Multibit full adders are available in chip form, and have carry-in and carry-out terminals to allow them to be cascaded to operate on long wordlengths. Such a device is also convenient for inverting a two's complement number, in conjunction with a set of invertors. The adder chip has one set of inputs grounded, and the carry-in permanently held true, such that it adds one to the one's complement number from the invertor.

When mixing by adding sample values, care has to be taken to ensure that if the sum of the two sample values exceeds the number range the result will be clipping rather than wraparound. In two's complement, the action necessary depends on the polarities of the two signals. Clearly if one positive and one negative number are added, the result cannot exceed the number range. If two positive numbers are added, the symptom of positive overflow is that the most significant bit sets, causing an erroneous negative result, whereas a negative overflow results in the most significant bit clearing. The overflow control circuit will be designed to detect these two conditions, and override the adder output. If the MSB of both inputs is zero, the numbers are both positive, thus if the sum has the MSB set, the output is replaced with the maximum positive code (0111. . .). If the MSB of both inputs is set, the numbers are both negative, and if the sum has no MSB set, the output is replaced with the maximum negative code (1000. . .). These conditions can also be connected to warning indicators. Figure 3.13(c) shows this system in hardware. The resultant clipping on overload is sudden, and sometimes a PROM is included which translates values around and beyond maximum to soft-clipped values below or equal to maximum.

A storage element can be combined with an adder to obtain a number of useful functional blocks which will crop up frequently in audio equipment. Figure 3.14(a) shows that a latch is connected in a feedback loop around an adder. The latch contents are added to the input each time it is clocked. The configuration is known as an accumulator in computation because it adds up or accumulates values fed into it. In filtering, it is known as an discrete time integrator. If the input is held at some constant value, the output increases by that amount on each clock. The output is thus a sampled ramp.

Figure 3.14(b) shows that the addition of an invertor allows the difference between successive inputs to be obtained. This is digital differentiation. The output is proportional to the slope of the input.

3.7   The computer

The computer is now a vital part of digital audio systems, being used both for control purposes and to store, access and process audio signals as data. In control, the computer finds applications in database

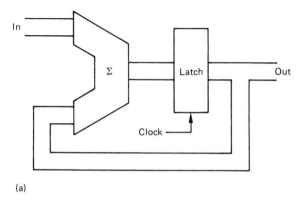

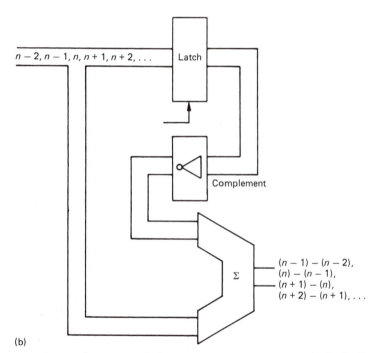

Figure 3.14 Two configurations which are common in processing. In (a) the feedback around the adder adds the previous sum to each input to perform accumulation or digital integration. In (b) an invertor allows the difference between successive inputs to be computed. This is differentiation.

management, automation, editing, and in electromechanical systems such as synchronizers. Some time ago, processing speeds advanced sufficiently to allow computers to manipulate digital audio in real time.

The computer is a programmable device in that its operation is not determined by its construction alone, but instead by a series of *instructions* forming a *program*. The program is supplied to the computer one instruction at a time so that the desired sequence of events takes place.

Programming of this kind has been used for over a century in electromechanical devices, including automated knitting machines and street organs which are programmed by punched cards. However, today's computers differ from these devices in that the program is not fixed, but can be modified by the computer itself. This possibility led to the creation of the term *software* to suggest a contrast to the constancy of hardware.

Computer instructions are binary numbers each of which is interpreted in a specific way. As these instructions don't differ from any other kind of data, they can be stored in RAM. The computer can change its own instructions by accessing the RAM. Most types of RAM are volatile, in that they lose data when power is removed. Clearly if a program is entirely stored in this way, the computer will not be able to recover fom a power failure. The solution is that a very simple starting or *bootstrap* program is stored in non-volatile ROM which will contain instructions which will bring in the main program from a storage system such as a disk drive after power is applied. As programs in ROM cannot be altered, they are sometimes referred to as *firmware* to indicate that they are classified between hardware and software.

Making a computer do useful work requires more than simply a program which performs the required computation. There is also a lot of mundane activity which does not differ significantly from one program to the next. This includes deciding which part of the RAM will be occupied by the program and which by the data, producing commands to the storage disk drive to read the input data from a file and write back the results. It would be very inefficient if all programs had to handle these processes themselves. Consequently the concept of an *operating system* was developed. This manages all the mundane decisions and creates an environment in which useful programs or *applications* can execute.

The ability of the computer to change its own instructions makes it very powerful, but it also makes it vulnerable to abuse. Programs exist which are deliberately written to do damage. These *viruses* are generally attached to plausible messages or data files and enter computers through storage media or communications paths.

There is also the possibility that programs contain logical errors such that in certain combinations of circumstances the wrong result is obtained. If this results in the unwitting modification of an instruction, the next time that instruction is accessed the computer will crash. In consumer-grade software, written for the vast personal computer market, this kind of thing is unfortunately accepted.

For critical applications, software must be *verified*. This is a process which can prove that a program can recover from absolutely every combination of circumstances and keep running properly. This is a non-trivial process, because the number of combinations of states a computer can get into is staggering. As a result most software is unverified.

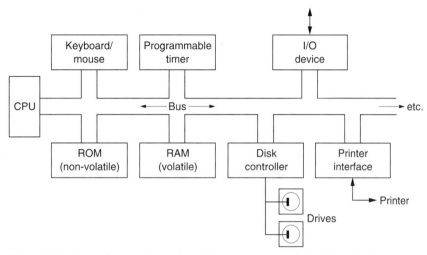

Figure 3.15 A simple computer system. All components are linked by a single data/address/control bus. Although cheap and flexible, such a bus can only make one connection at a time, so it is slow.

It is of the utmost importance that networked computers which can suffer virus infection or computers running unverified software are never used in a life-support or critical application.

Figure 3.15 shows a simple computer system. The various parts are linked by a bus which allows binary numbers to be transferred from one place to another. This will generally use tri-state logic (see section 3.4) so that when one device is sending to another, all other devices present a high impedance to the bus.

The ROM stores the startup program, the RAM stores the operating system, applications programs and the data to be processed. The disk drive stores large quantities of data in a non-volatile form. The RAM only needs to be able to hold part of one program as other parts can be brought from the disk as required. A program executes by *fetching* one instruction at a time from the RAM to the processor along the bus.

The bus also allows keyboard/mouse inputs and outputs to the display and printer. Inputs and outputs are generally abbreviated to I/O. Finally a programmable timer will be present which acts as a kind of alarm clock for the processor.

3.8   The processor

The processor or CPU (central processing unit) is the heart of the system. Figure 3.16 shows the data path of a simple CPU. The CPU has a bus interface which allows it to generate bus addresses and input or output

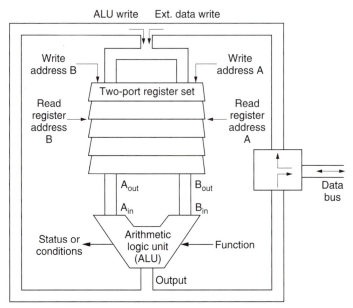

Figure 3.16 The data path of a simple CPU. Under control of an instruction, the ALU will perform some function on a pair of input values from the registers and store or output the result.

data. Sequential instructions are stored in RAM at contiguously increasing locations so that a program can be executed by fetching instructions from a RAM address specified by the program counter (PC) to the instruction register in the CPU. As each instruction is completed, the PC is incremented so that it points to the next instruction. In this way the time taken to execute the instruction can vary.

The processor is notionally divided into data paths and control paths. The CPU contains a number of general-purpose registers or scratchpads which can be used to store partial results in complex calculations. Pairs of these registers can be addressed so that their contents go to the ALU (arithmetic logic unit). This performs various arithmetic (add, subtract, etc.) or logical (AND, OR, etc.) functions on the input data. The output of the ALU may be routed back to a register or output. By reversing this process it is possible to get data into the registers from the RAM. The ALU also outputs the conditions resulting from the calculation, which can control conditional instructions.

Which function the ALU performs and which registers are involved are determined by the instruction currently in the instruction register which is decoded in the control path. One pass through the ALU can be completed in one cycle of the processor's clock. Instructions vary in complexity as do the number of clock cycles needed to complete them. Incoming instructions are decoded and used to access a look-up table which converts them into *microinstructions*, one of which controls the CPU at each clock cycle.

3.9   Interrupts

Ordinarily instructions are executed in the order that they are stored in RAM. However, some instructions direct the processor to jump to a new memory location. If this is a jump to an earlier instruction, the program will enter a loop. The loop must increment a count in a register each time, and contain a conditional instruction called a branch, which allows the processor to jump out of the loop when a predetermined count is reached.

However, it is often required that the sequence of execution should be changeable by some external event. This might be the changing of some value due to a keyboard input. Events of this kind are handled by *interrupts*, which are created by devices needing attention. Figure 3.17 shows that in addition to the PC, the CPU contains another dedicated register called the *stack pointer*. Figure 3.18 shows how this is used. At the end of every instruction the CPU checks to see if an interrupt is asserted on the bus.

If it is, a different set of microinstructions are executed. The PC is incremented as usual, but the next instruction is not executed. Instead, the contents of the PC are stored so that the CPU can resume execution when

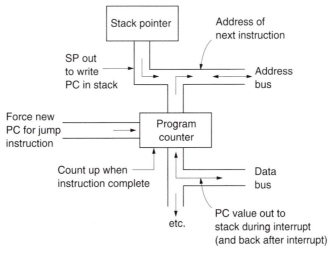

Figure 3.17   Normally the program counter (PC) increments each time an instruction is completed in order to select the next instruction. However, an interrupt may cause the PC state to be stored in the stack area of RAM prior to the PC being forced to the start address of the interrupt subroutine. Afterwards the PC can get its original value back by reading the stack.

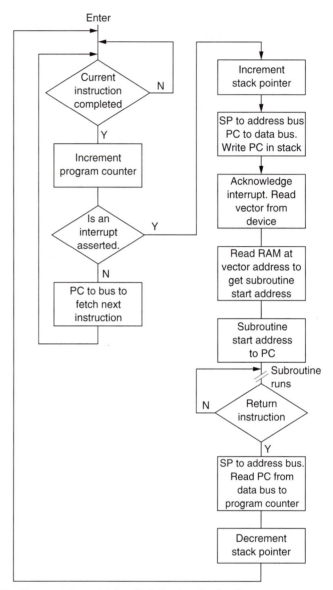

Figure 3.18 How an interrupt is handled. See text for details.

it has handled the current event. The PC state is stored in a reserved area of RAM known as the *stack*, at an address determined by the stack pointer.

Once the PC is stacked, the processor can handle the interrupt. It issues a bus interrupt acknowledge, and the interrupting device replies with an unique code identifying itself. This is known as a *vector* which steers the processor to a RAM address containing a new program counter. This is the RAM address of the first instruction of the *subroutine* which is the program that will handle the interrupt. The CPU loads this address into the PC and begins execution of the subroutine.

At the end of the subroutine there will be a return instruction. This causes the CPU to use the stack pointer as a memory address in order to read the return PC state from the stack. With this value loaded into the PC, the CPU resumes execution where it left off.

The stack exists so that subroutines can themselves be interrupted. If a subroutine is executing when a higher-priority interrupt occurs, the subroutine can be suspended by incrementing the stack pointer and storing the current PC in the next location in the stack.

When the second interrupt has been serviced, the stack pointer allows the PC of the first subroutine to be retrieved. Whenever a stack PC is retrieved, the stack pointer decrements so that it always points to the PC of the next item of unfinished business.

3.10   Programmable timers

Ordinarily processors have no concept of time and simply execute instructions as fast as their clock allows. This is fine for general-purpose processing, but not for time-critical processes such as audio. One way in which the processor can be made time conscious is to use programmable timers. These are devices which reside on the computer bus and which run from a clock. The CPU can set up a timer by loading it with a count. When the count is reached, the timer will interrupt. To give an example, if the count were to be equal to one audio sample period, there would be one interrupt per sample, and this would result in the execution of a subroutine once per sample, provided, of course, that all the instructions could be executed in time.

3.11   Timebase compression and correction

In Chapter 1 it was stated that a strength of digital technology is the ease with which delay can be provided. Accurate control of delay is the essence of timebase correction, necessary whenever the instantaneous time of arrival or rate from a data source does not match the destination. In digital audio, the destination will almost always have perfectly regular timing, namely the sampling rate clock of the final DAC. Timebase correction consists of aligning jittery signals from storage media or transmission channels with that stable reference. In this way, wow and flutter are rendered unmeasurable.

A further function of timebase correction is to reverse the time compression applied prior to recording or transmission. As was shown in section 1.8, digital audio recorders compress data into blocks to facilitate editing and error correction as well as to permit head switching between blocks in rotary-head machines. Owing to the spaces between blocks, data arrive in bursts on replay, but must be fed to the output convertors in an unbroken stream at the sampling rate. The extreme time compression used in DAT to reduce the tape wrap is a further example of the use of the principle (see Chapter 9).

In computer hard-disk drives, which are used in digital audio editing systems, time compression is also used, but a converse problem also arises. Data from the disk blocks arrive at a reasonably constant rate, but cannot necessarily be accepted at a steady rate by the logic because of contention for the use of buses and memory by the different parts of the system. In this case the data must be buffered by a relative of the timebase corrector which is usually referred to as a silo.

Although delay is easily implemented, it is not possible to advance a data stream. Most real machines cause instabilities balanced about the correct timing: the output jitters between too early and too late. Since the information cannot be advanced in the corrector, only delayed, the solution is to run the machine in advance of real time. In this case, correctly timed output signals will need a nominal delay to align them with reference timing. Early output signals will receive more delay, and late output signals will receive less.

Section 2.5 showed the principles of digital storage elements which can be used for delay purposes. The shift-register approach and the RAM approach to delay are very similar, as a shift register can be thought of as a memory whose address increases automatically when clocked. The data rate and the maximum delay determine the capacity of the RAM required. Figure 1.7 showed that the addressing of the RAM is by a counter that overflows endlessly from the end of the memory back to the beginning, giving the memory a ring-like structure. The write address is determined by the incoming data, and the read address is determined by the outgoing data. This means that the RAM has to be able to read and write at the same time. The switching between read and write involves not only a data multiplexer but also an address multiplexer. In general the arbitration between read and write will be done by signals from the stable side of the TBC as Figure 3.19 shows. In the replay case the stable clock will be on the read side. The stable side of the RAM will read a sample when it demands, and the writing will be locked out for that period. The input data cannot be interrupted in many applications, however, so a small buffer silo is installed before the memory, which fills up as the writing is locked out, and empties again as writing is permitted. Alternatively, the memory will be split into blocks as was shown in Chapter 1, such that when one block is reading a different block will be writing and the problem does not arise.

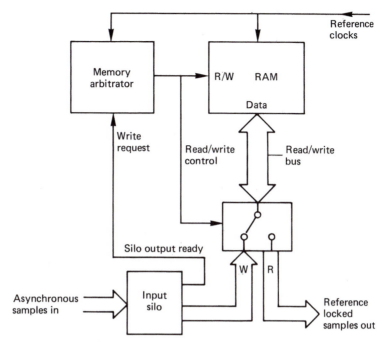

Figure 3.19 In a RAM-based TBC, the RAM is reference synchronous, and an arbitrator decides when it will read and when it will write. During reading, asynchronous input data back up in the input silo, asserting a write request to the arbitrator. Arbitrator will then cause a write cycle between read cycles.

Figure 3.20 shows the operation of a FIFO chip, colloquially known as a silo because the data are tipped in at the top on delivery and drawn off at the bottom when needed. Each stage of the chip has a data register and a small amount of logic, including a data-valid or V bit. If the input register does not contain data, the first V bit will be reset, and this will cause the chip to assert 'input ready'. If data are presented at the input, and clocked into the first stage, the V bit will set, and the 'input ready' signal will become false. However, the logic associated with the next stage sees the V bit set in the top stage, and if its own V bit is clear, it will clock the data into its own register, set its own V bit, and clear the input V bit, causing 'input ready' to reassert, when another word can be fed in. This process then continues as the word moves down the silo, until it arrives at the last register in the chip. The V bit of the last stage becomes the 'output ready' signal, telling subsequent circuitry that there are data to be read. If this word is not read, the next word entered will ripple down to the stage above. Words thus stack up at the bottom of the silo. When a word is read out, an external signal must be provided which resets the bottom V bit. The 'output ready' signal now goes false, and the logic associated with the last stage now sees valid data above, and loads

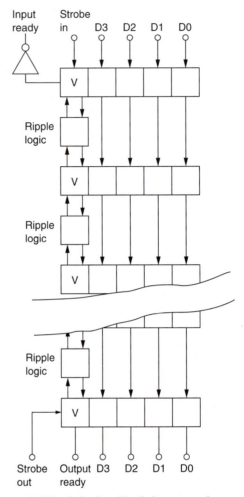

Figure 3.20 Structure of FIFO of silo chip. Ripple logic controls propagation of data down silo.

down the word when it will become ready again. The last register but one will now have no V bit set, and will see data above itself and bring that down. In this way a reset V bit propagates up the chip while the data ripple down, rather like a hole in a semiconductor going the opposite way to the electrons.

3.12   Gain control

When making a digital recording, the gain of the analog input will usually be adjusted so that the quantizing range is fully exercised in order to make a recording of maximum signal-to-noise ratio. During post-

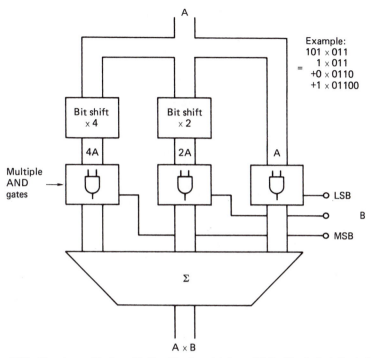

Figure 3.21 Structure of fast multiplier. The input A is multiplied by 1, 2, 4, 8, etc. by bit shifting. The digits of the B input then determine which multiples of A should be added together by enabling AND gates between the shifters and the adder. For long wordlengths, the number of gates required becomes enormous, and the device is best implemented in a chip.

production, the recording may be played back and mixed with other signals, and the desired effect can only be achieved if the level of each can be controlled independently. Gain is controlled in the digital domain by multiplying each sample value by a coefficient. If that coefficient is less than one, attenuation will result; if it is greater than one, amplification can be obtained.

Multiplication in binary circuits is difficult. It can be performed by repeated adding, but this is too slow to be of any use. In fast multiplication, one of the inputs will be simultaneously multiplied by one, two, four, etc., by hard-wired bit shifting. Figure 3.21 shows that the other input bits will determine which of these powers will be added to produce the final sum, and which will be neglected. If multiplying by five, the process is the same as multiplying by four, multiplying by one, and adding the two products. This is achieved by adding the input to itself shifted two places. As the wordlength of such a device increases, the complexity increases exponentially, so this is a natural application for an integrated circuit. It is probably true that digital video would not have been viable without such chips.

3.13   Digital faders and controls

In a digital mixer, the gain coefficients will originate in hand-operated faders, just as in analog. Analog mixers having automated mixdown employ a system similar to the one shown in Figure 3.22. Here, the faders produce a varying voltage and this is converted to a digital code or gain coefficient in an ADC and recorded alongside the audio tracks. On replay the coefficients are converted back to analog voltages which control VCAs (voltage-controlled amplifiers) in series with the analog audio channels. A digital mixer has a similar structure, and the coefficients can be obtained in the same way. However, on replay, the coefficients are not converted back to analog, but remain in the digital domain and control multipliers in the digital audio channels directly. As the coefficients are digital, it is so easy to add automation to a digital mixer that there is not much point in building one without it.

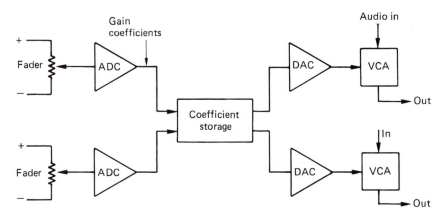

Figure 3.22   The automated mixdown system of an audio console digitizes fader positions for storage and uses the coefficients later to drive VCAs via convertors.

Whilst gain coefficients can be obtained by digitizing the output of an analog fader, it is also possible to obtain coefficients directly in digital faders. Digital faders are a form of displacement transducer in which the mechanical position of the knob is converted directly to a digital code. The position of other controls, such as for equalizers or scrub wheels, will also need to be digitized. Controls can be linear or rotary, and absolute or relative. In an absolute control, the position of the knob determines the output directly. These are inconvenient in automated systems because unless the knob is motorized, the operator does not know the setting the automation system has selected. In a relative control, the knob can be moved to increase or decrease the output, but its absolute position is meaningless. The absolute setting is displayed on a bar LED nearby. In a

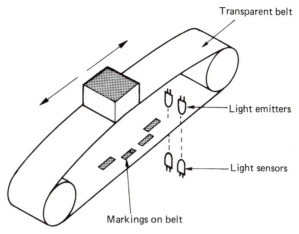

Figure 3.23 An absolute linear fader uses a number of light beams which are interrupted in various combinations according to the position of a grating. A Gray code shown in Figure 3.24 must be used to prevent false codes.

rotary control, the bar LED may take the form of a ring of LEDs around the control. The automation system setting can be seen on the display and no motor is needed. In a relative linear fader, the control may take the form of an endless ridged belt like a caterpillar track. If this is transparent, the bar LED may be seen through it.

Figure 3.23 shows an absolute linear fader. A grating is moved with respect to several light beams, one for each bit of the coefficient required. The interruption of the beams by the grating determines which photocells are illuminated. It is not possible to use a pure binary pattern on the grating because this results in transient false codes due to mechanical tolerances. Figure 3.24 shows some examples of these false codes. For example, on moving the fader from 3 to 4, the MSB goes true slightly before the middle bit goes false. This results in a momentary value of 4 + 2 = 6 between 3 and 4. The solution is to use a code in which only one bit ever changes in going from one value to the next. One such code is the Gray code, which was devised to overcome timing hazards in relay logic but is now used extensively in position encoders.

Gray code can be converted to binary in a suitable PROM or gate array. These are available as industry-standard components.

Figure 3.25 shows a rotary incremental encoder. This produces a sequence of pulses whose number is proportional to the angle through which it has been turned. The rotor carries a radial grating over its entire perimeter. This turns over a second fixed radial grating whose bars are not parallel to those of the first grating. The resultant moiré fringes travel inward or outward depending on the direction of rotation. Two suitably positioned light beams falling on photocells will produce outputs in

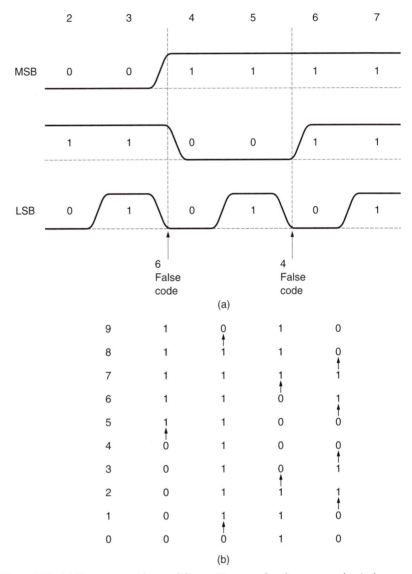

Figure 3.24 (a) Binary cannot be used for position encoders because mechanical tolerances cause false codes to be produced. (b) In Gray code, only one bit (arrowed) changes in between positions, so no false codes can be generated.

quadrature. The relative phase determines the direction and the frequency is proportional to speed. The encoder outputs can be connected to a counter whose contents will increase or decrease according to the direction the rotor is turned. The counter provides the coefficient output and drives the display.

For audio use, a logarithmic characteristic is required in gain control. Linear coefficients can conveniently be converted to logarithmic in a

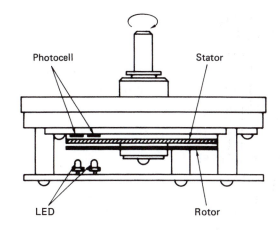

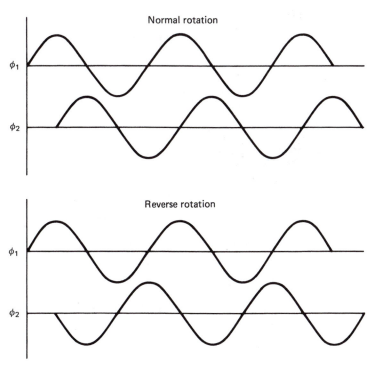

Figure 3.25 The fixed and rotating gratings produce moiré fringes which are detected by two light paths as quadrature sinusoids. The relative phase determines the direction, and the frequency is proportional to speed of rotation.

PROM. The wordlength of the gain coefficients requires some thought as they determine the number of discrete gains available. If the coefficient wordlength is inadequate, the gain control becomes 'steppy' particularly towards the end of a fadeout. This phenomenon is quite noticeable on some low-cost home studio equipment. A compromise between performance and the expense of high-resolution faders is to insert a digital interpolator having a low-pass characteristic between the fader and the gain control stage. This will compute intermediate gains to higher resolution than the coarse fader scale so that the steps cannot be heard. Digital filters used for equalization can also be sensitive to sudden step changes to their control coefficients.[1] Again the solution is to filter the coefficients.

3.14   A digital mixer

The signal path of a simple digital mixer is shown in Figure 3.26. The two inputs are multiplied by their respective coefficients, and added together in two's complement to achieve the mix as was shown in Figure 3.7. Peak limiting will be required as in section 3.6. The sampling rate of the inputs must be exactly the same, and in the same phase, or the circuit will not be able to add on a sample-by-sample basis. If the inputs have come from different sources, they must be synchronized by the same master clock,

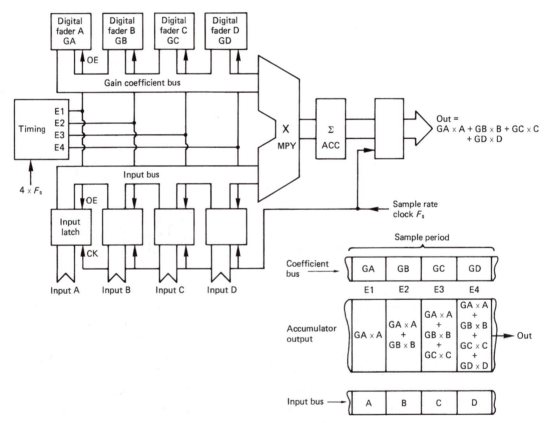

Figure 3.26 One multiplier/accumulator can be time shared between several signals by operating at a multiple of sampling rate. In this example, four multiplications are performed during one sample period.

and/or timebase correction must be provided on the inputs. Synchronization of audio sources follows the principle long established in video in which a reference signal is fed to all devices which then slave or *genlock* to it. This process will be covered in detail in Chapter 8.

Some thought must be given to the wordlength of the system. If a sample is attenuated, it will develop bits which are below the radix point. For example, if an eight-bit sample is attenuated by 24 dB, the sample value will be shifted four places down. Extra bits must be available within the mixer to accommodate this shift. Digital mixers can have an internal wordlength of up to 32 bits. When several attenuated sources are added together to produce the final mix, the result will be a stream of 32-bit or longer samples. As the output will generally need to be of the same format as the input, the wordlength must be shortened. This must be done very carefully, as it is a form of quantizing and will require dithering. The necessary techniques will be treated in Chapter 4.

In practice a digital mixer would not have one multiplier for every input. Multiplier chips are expensive, but can work much faster than the relatively low frequencies used in audio sampling. Figure 3.26 also shows that a more economical system results when a time-shared bus system is used with only one multiplier followed by an accumulator. In one sample period, each of the input samples is fed in turn to the lower input of the multiplier at the same time as the corresponding coefficient is fed to the upper input. The products from the multiplier are accumulated during the sample period, so that at the end of the sample period, the accumulator holds the sum of all the products, which is the digitally mixed sample. The process then repeats for the next sample period. To facilitate the sharing of common circuits by many signals, tri-state logic devices can be used. The outputs of such devices can be wired in parallel, and the state of the parallel connection will be the state of the device whose output is enabled. Clearly only one output can be enabled at a time, and this will be ensured by a sequencer circuit connected to all the device enables. In digital signal processor (DSP) chips, the processes shown above can be simulated in software.

In analog audio mixers, the controls have to be positioned close to the circuitry for performance reasons; thus one control knob is needed for every variable, and the control panel is physically large. Remote control is difficult with such construction. The order in which the signal passes through the various stages of the mixer is determined at the time of design, and any changes are difficult.

In a digital mixer,[2,3] all the filters are controlled by simply changing the coefficients, and remote control is easy. Since control is by digital parameters, it is possible to use assignable controls, such that there need only be one set of filter and equalizer controls, whose setting is conveyed to any channel chosen by the operator.[4] The use of digital processing allows the console to include a video display of the settings. This was seldom attempted in analog desks because the magnetic field from the scan coils tended to break through into the audio circuitry.

Since the audio processing in a digital mixer is by program control, the configuration of the desk can be changed at will by running the programs for the various functions in a different order. The operator can configure the desk to his own requirements by entering symbols on a block diagram on the video display, for example. The configuration and the setting of all the controls can be stored in memory or for a longer term, on disk, and recalled instantly. Such a desk can be in almost constant use, because it can be put back exactly to a known state easily after someone else has used it.

A further advantage of working in the digital domain is that delay can be controlled individually in the audio channels.[5] This allows for the time of arrival of wavefronts at various microphones to be compensated despite their physical position.

Figure 3.27 shows a typical digital mixer installation.[4] The analog microphone inputs are from remote units containing ADCs so that the length of analog cabling can be kept short. The input units communicate with the signal processor using digital fibre-optic links.

The sampling rate of a typical digital audio signal is low compared to the speed at which typical logic gates can operate. It is sensible to

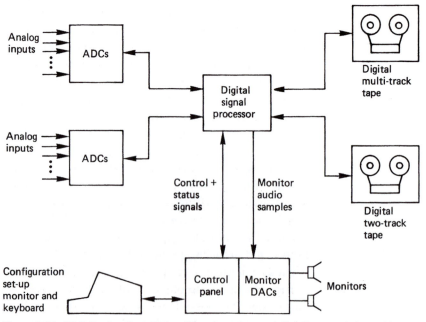

Figure 3.27 Digital mixer installation. The convenience of digital transmission without degradation allows the control panel to be physically remote from the processor.

minimize the quantity of hardware necessary by making each perform many functions in one sampling period. Although general-purpose computers can be programmed to process digital audio, they are not really suitable for the following reasons:

1 The number of arithmetic operations in audio processing, particularly multiplications, is far higher than in data processing.
2 Audio processing is done in real time; data processors do not generally work in real time.
3 The program needed for an audio function generally remains constant for the duration of a session, or changes slowly, whereas a data processor rapidly jumps between many programs.
4 Data processors can suspend a program on receipt of an interrupt; audio processors must work continuously for long periods.
5 Data processors tend to be I/O limited, in that their operating speed is limited by the problems of moving large quantities of data and instructions into the CPU. Audio processors in contrast have a relatively small input and output rate, but compute intensively.

The above is a sufficient case for the development of specialized digital audio signal processors.[6-8] These units are implemented with more internal registers than data processors to facilitate multi-point filter algorithms. The arithmetic unit will be designed to offer high-speed multiply/accumulate using techniques such as pipelining, which allows operations to overlap.[9] The functions of the register set and the arithmetic unit are controlled by a microsequencer.

External control of a DSP will generally be by a smaller processor, often in the operator's console, which passes coefficients to the DSP as the operator moves the controls. In large systems, it is possible for several different consoles to control different sections of the DSP.[10]

3.15   Effects

In addition to equalization and mixing, modern audio production requires numerous effects, and these can be performed in the digital domain by simply mimicking the analog equivalent.

One of the oldest effects is the use of a tape loop to produce an echo, and this can be implemented with memory or, for longer delays, with a disk drive. Figure 3.28(a) shows the basic configuration necessary for echo. If the delay period is dynamically changed from zero to about 10 ms, the result is flanging, where a notch sweeps through the audio spectrum. This was originally done by having two identical analog tapes running, and modifying the capstan speed with hand pressure! A relative

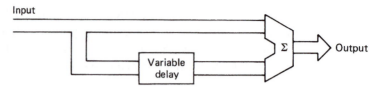

Figure 3.28 (a) A simple configuration to obtain digital echo. The delay would normally be several tens of milliseconds. If the delay is made about 10 ms, the configuration acts as a comb filter, and if the delay is changed dynamically, a notch will sweep the audio spectrum resulting in flanging.

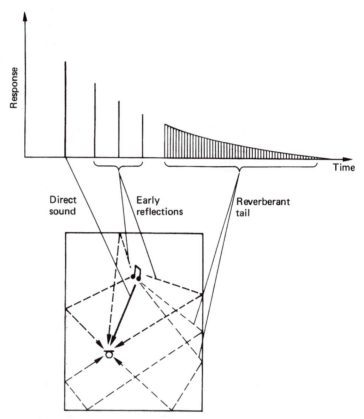

Figure 3.28 (b) In a reverberant room, the signal picked up by a microphone is a mixture of direct sound, early reflections and a highly confused reverberant tail. A digital reverberator will simulate this with various combinations of recursive delay and attenuation.

of echo is reverberation, which is used to simulate ambience on an acoustically dry recording. Figure 3.28(b) shows that reverberation actually consists of a series of distinct early reflections, followed by the reverberation proper, which is due to multiple reflections. The early reflections are simply provided by short delays, but the reverberation is more difficult. A recursive structure is a natural choice for a decaying response, but simple recursion sounds artificial. The problem is that, in a real room, standing waves and interference effects cause large changes in the frequency response at each reflection. The effect can be simulated in a digital reverberator by adding various comb-filter sections which have the required effect on the response.

3.16 The phase-locked loop

All digital audio systems need to be clocked at the appropriate rate in order to function properly. Whilst a clock may be obtained from a fixed frequency oscillator such as a crystal, many operations in audio require *genlocking* or synchronizing the clock to an external source. The phase-locked loop excels at this job, and many others, particularly in connection with recording and transmission.

In phase-locked loops, the oscillator can run at a range of frequencies according to the voltage applied to a control terminal. This is called a voltage-controlled oscillator or VCO. Figure 3.29 shows that the VCO is

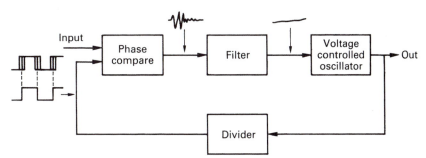

Figure 3.29 A phase-locked loop requires these components as a minimum. The filter in the control voltage serves to reduce clock jitter.

driven by a phase error measured between the output and some reference. The error changes the control voltage in such a way that the error is reduced, so that the output eventually has the same frequency as the reference. A low-pass filter is fitted in the control voltage path to prevent the loop becoming unstable. If a divider is placed between the VCO and the phase comparator, as in the figure, the VCO frequency can be made to be a multiple of the reference. This also has the effect of making the loop more heavily damped, so that it is less likely to change frequency if the input is irregular.

In digital audio, the frequency multiplication of a phase-locked loop is extremely useful. Figure 3.30 shows how the 48 kHz sampling clock is

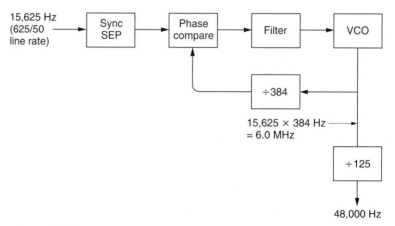

Figure 3.30   Obtaining a 48 kHz sampling clock from the line frequency of 625/50 video using a phase-locked loop.

obtained from the sync pulses of a video reference by such a multiplication process.

Figure 3.31 shows the NLL or numerically locked loop. This is similar to a phase-locked loop, except that the two phases concerned are represented by the state of a binary number. The NLL is useful to generate a remote clock from a master. The state of a clock count in the master is periodically transmitted to the NLL which will recreate the same clock frequency. The technique is used in MPEG transport streams.

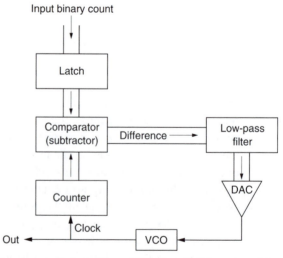

Figure 3.31   The numerically locked loop (NLL) is a digital version of the phase-locked loop.

3.17   Multiplexing principles

Multiplexing is used where several signals are to be transmitted down the same channel. The channel bit rate must be the same as or greater than the sum of the source bit rates. Figure 3.32 shows that when multiplexing is used, the data from each source have to be time compressed. This is done by buffering source data in a memory at the multiplexer. They are written into the memory in real time as they arrive, but will be read from the memory with a clock which has a much higher rate. This means that the readout occurs in a smaller timespan. If, for example, the clock frequency is raised by a factor of ten, the data for a given signal will be transmitted in a tenth of the normal time, leaving time in the multiplex for nine more such signals.

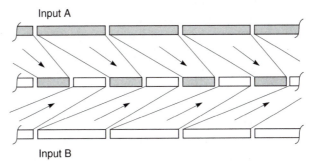

Figure 3.32   Multiplexing requires time compression on each input.

In the demultiplexer another buffer memory will be required. Only the data for the selected signal will be written into this memory at the bit rate of the multiplex. When the memory is read at the correct speed, the data will emerge with its original timebase.

In practice it is essential to have mechanisms to identify the separate signals to prevent them being mixed up and to convey the original signal clock frequency to the demultiplexer. In time-division multiplexing the timebase of the transmission is broken into equal slots, one for each signal. This makes it easy for the demultiplexer, but forces a rigid structure on all the signals such that they must all be locked to one another and have an unchanging bit rate. Packet multiplexing overcomes these limitations.

3.18   Packets

The multiplexer must switch between different time-compressed signals to create the bitstream and this is much easier to organize if each signal

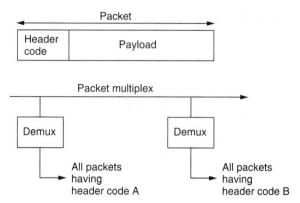

Figure 3.33   Packet multiplexing relles on headers to identify the packets.

is in the form of data packets of constant size. Figure 3.33 shows a packet multiplexing system.

Each packet consists of two components: the header, which identifies the packet, and the payload, which is the data to be transmitted. The header will contain at least an identification code (ID) which is unique for each signal in the multiplex. The demultiplexer checks the ID codes of all incoming packets and discards those which do not have the wanted ID.

In complex systems it is common to have a mechanism to check that packets are not lost or repeated. This is the purpose of the packet continuity count which is carried in the header. For packets carrying the same ID, the count should increase by one from one packet to the next. Upon reaching the maximum binary value, the count overflows and recommences.

3.19   Statistical multiplexing

Packet multiplexing has advantages over time-division multiplexing because it does not set the bit rate of each signal. A demultiplexer simply checks packet IDs and selects all packets with the wanted code. It will do this however frequently such packets arrive. Consequently it is practicable to have variable bit rate signals in a packet multiplex. The multiplexer has to ensure that the total bit rate does not exceed the rate of the channel, but that rate can be allocated arbitrarily between the various signals.

As a practical matter is is usually necessary to keep the bit rate of the multiplex constant. With variable rate inputs this is done by creating null packets which are generally called *stuffing* or *packing*. The headers of these packets contain an unique ID which the demultiplexer does not recognize and so these packets are discarded on arrival.

In an MPEG environment, statistical multiplexing can be extremely useful because it allows for the varying difficulty of real program material. In a multiplex of several television programs, it is unlikely that all the programs will encounter difficult material simultaneously. When one program encounters a detailed scene or frequent cuts which are hard to compress, more data rate can be allocated at the allowable expense of the remaining programs which are handling easy material.

3.20   Filters

Filtering is inseparable from digital audio. Analog or digital filters, and sometimes both, are required in ADCs, DACs, in the data channels of digital recorders and transmission systems and in sampling rate convertors and equalizers. Optical systems used in disk recorders also act as filters.[11] There are many parallels between analog, digital and optical filters, which this section treats as a common subject. The main difference between analog and digital filters is that in the digital domain very complex architectures can be constructed at low cost in LSI and that arithmetic calculations are not subject to component tolerance or drift.

Filtering may modify the frequency response of a system, and/or the phase response. Every combination of frequency and phase response determines the impulse response in the time domain. Figure 3.34 shows that impulse response testing tells a great deal about a filter. In a perfect filter, all frequencies should experience the same time delay. If some groups of frequencies experience a different delay from others, there is a group-delay error. As an impulse contains an infinite spectrum, a filter suffering from group-delay error will separate the different frequencies of an impulse along the time axis.

A pure delay will cause a phase shift proportional to frequency, and a filter with this characteristic is said to be phase-linear. The impulse

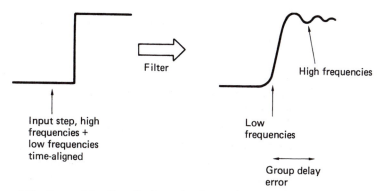

Figure 3.34 Group delay time-displaces signals as a function of frequency.

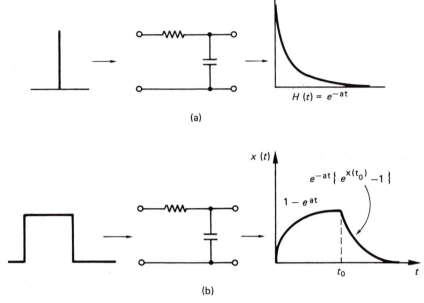

Figure 3.35 (a) The impulse response of a simple *RC* network is an exponential decay. This can be used to calculate the response to a squarewave, as in (b).

response of a phase-linear filter is symmetrical. If a filter suffers from group-delay error it cannot be phase-linear. It is almost impossible to make a perfectly phase-linear analog filter, and many filters have a group-delay equalization stage following them which is often as complex as the filter itself. In the digital domain it is straightforward to make a phase-linear filter, and phase equalization becomes unnecessary.

Because of the sampled nature of the signal, whatever the response at low frequencies may be, all digital channels (and sampled analog channels) act as low-pass filters cutting off at the Nyquist limit, or half the sampling frequency.

Figure 3.35(a) shows a simple RC network and its impulse response. This is the familiar exponential decay due to the capacitor discharging through the resistor (in series with the source impedance which is assumed here to be negligible). The figure also shows the response to a squarewave at (b). These responses can be calculated because the inputs involved are relatively simple. When the input waveform and the impulse response are complex functions, this approach becomes almost impossible.

In any filter, the time domain output waveform represents the convolution of the impulse response with the input waveform. Convolution can be followed by reference to a graphic example in Figure 3.36. Where the impulse response is asymmetrical, the decaying tail occurs *after* the input. As a result it is necessary to reverse the impulse response

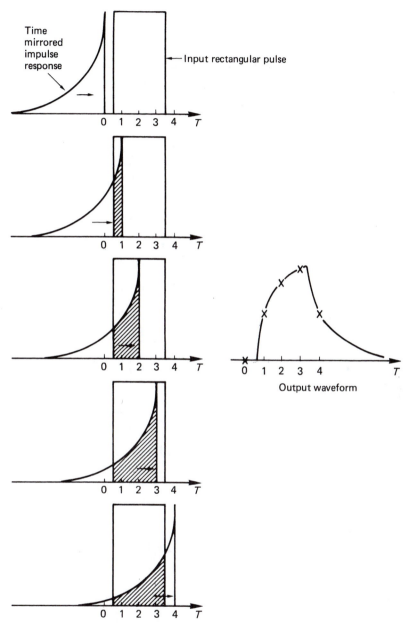

Figure 3.36 In the convolution of two continuous signals (the impulse response with the input), the impulse must be time reversed or mirrored. This is necessary because the impulse will be moved from left to right, and mirroring gives the impulse the correct time-domain response when it is moved past a fixed point. As the impulse response slides continuously through the input waveform, the area where the two overlap determines the instantaneous output amplitude. This is shown for five different times by the crosses on the output waveform.

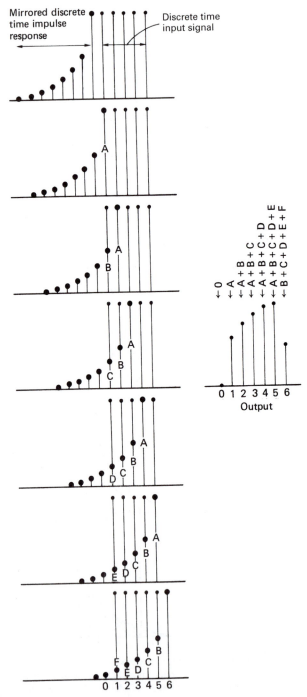

Figure 3.37 In time discrete convolution, the mirrored impulse response is stepped through the input one sample period at a time. At each step, the sum of the cross-products is used to form an output value. As the input in this example is a constant-height pulse, the output is simply proportional to the sum of the coincident impulse response samples. This figure should be compared with Figure 3.36.

in time so that it is mirrored prior to sweeping it through the input waveform. The output voltage is proportional to the shaded area shown where the two impulses overlap.

The same process can be performed in the sampled, or discrete time domain as shown in Figure 3.37. The impulse and the input are now a set of discrete samples which clearly must have the same sample spacing. The impulse response only has value where impulses coincide. Elsewhere it is zero. The impulse response is therefore stepped through the input one sample period at a time. At each step, the area is still proportional to the output, but as the time steps are of uniform width, the area is proportional to the impulse height and so the output is obtained by adding up the lengths of overlap. In mathematical terms, the output samples represent the convolution of the input and the impulse response by summing the coincident cross-products.

As a digital filter works in this way, perhaps it is not a filter at all, but just a mathematical simulation of an analog filter. This approach is quite useful in visualizing what a digital filter does.

3.21   Transforms

Figure 3.38 shows that if a signal with a spectrum or frequency content *a* is passed through a filter with a frequency response *b* the result will be an output spectrum which is simply the product of the two. If the frequency responses are drawn on logarithmic scales (i.e. calibrated in dB) the two can be simply added because the addition of logs is the same as multiplication. Whilst frequency in audio has traditionally meant temporal frequency measured in Hertz, frequency in optics can also be spatial and measured in lines per millimetre (mm^{-1}). Multiplying the spectra of the responses is a much simpler process than convolution.

In order to move to the frequency domain or spectrum from the time domain or waveform, it is necessary to use the Fourier transform, or in sampled systems, the discrete Fourier transform (DFT). Fourier analysis holds that any periodic waveform can be reproduced by adding together an arbitrary number of harmonically related sinusoids of various amplitudes and phases. Figure 3.39 shows how a squarewave can be built up of harmonics. The spectrum can be drawn by plotting the amplitude of the harmonics against frequency. It will be seen that this gives a spectrum which is a decaying wave. It passes through zero at all even multiples of the fundamental. The shape of the spectrum is a $\sin x/x$ curve. If a squarewave has a $\sin x/x$ spectrum, it follows that a filter with a rectangular impulse response will have a $\sin x/x$ spectrum.

A low-pass filter has a rectangular spectrum, and this has a $\sin x/x$ impulse response. These characteristics are known as a transform pair. In

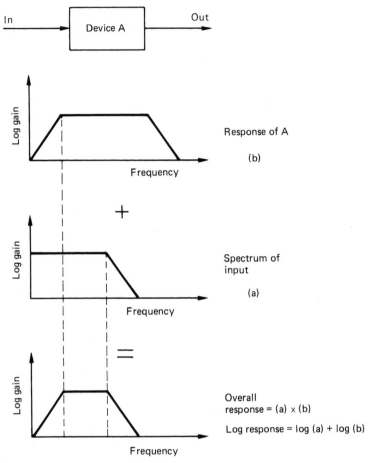

Figure 3.38 In the frequency domain, the response of two series devices is the product of their individual responses at each frequency. On a logarithmic scale the responses are simply added.

transform pairs, if one domain has one shape of the pair, the other domain will have the other shape. Thus a squarewave has a $\sin x/x$ spectrum and a $\sin x/s$ impulse has a square spectrum. Figure 3.40 shows a number of transform pairs. Note the pulse pair. A time domain pulse of infinitely short duration has a flat spectrum. Thus a flat waveform, i.e. DC, has only zero in its spectrum. Interestingly the transform of a Gaussian response in still Gaussian. The impulse response of the optics of a laser disk has a $\sin^2 x/x^2$ function, and this is responsible for the triangular falling frequency response of the pickup.

The spectrum of a pseudo-random sequence is not flat because it has a finite sequence length. The rate at which the sequence repeats is visible in the spectrum. Where pseudo-random sequences are to be used in sample manipulation, i.e. where their effects can be audible, it is essential that the

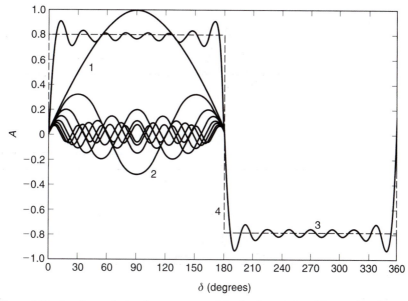

Figure 3.39 Fourier analysis of a squarewave into fundamental and harmonics. *A*, amplitude; δ, phase of fundamental wave in degrees; 1, first harmonic (fundamental); 2 odd harmonics 3–15; 3, sum of harmonics 1–15; 4, ideal squarewave.

sequence length should be long enough to prevent the periodicity being audible.

Figure 3.41 shows that the spectrum of a pseudo-random sequence has a $\sin x/x$ characteristic, with nulls at multiples of the clock frequency. A closer inspection of the spectrum shows that it is not continuous, but takes the form of a comb where the spacing is equal to the repetition rate of the sequence.

3.22   FIR and IIR Filters

Filters can be described in two main classes, as shown in Figure 3.42, according to the nature of the impulse response. Finite-impulse response (FIR) filters are always stable and, as their name suggests, respond to an impulse once, as they have only a forward path. In the temporal domain, the time for which the filter responds to an input is finite, fixed and readily established. The same is therefore true about the distance over which a FIR filter responds in the spatial domain. FIR filters can be made perfectly phase linear if required. Most filters used for sampling rate conversion and oversampling fall into this category.

Infinite-impulse response (IIR) filters respond to an impulse indefinitely and are not necessarily stable, as they have a return path from the

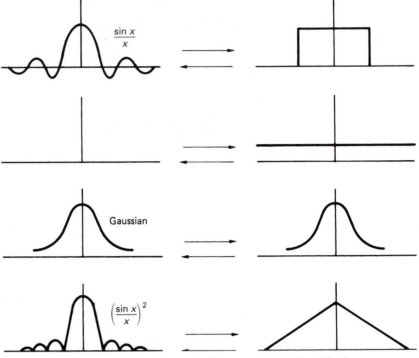

Figure 3.40 The concept of transform pairs illustrates the duality of the frequency (including spatial frequency) and time domains.

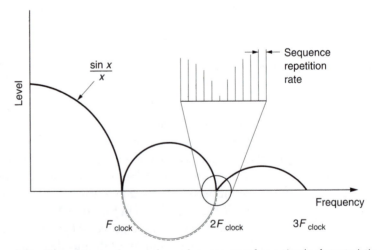

Figure 3.41 The spectrum of a pseudo-random sequence has a $\sin x/x$ characteristic, with nulls at multiples of the clock frequency. The spectrum is not continuous, but resembles a comb where the spacing is equal to the repetition rate of the sequence.

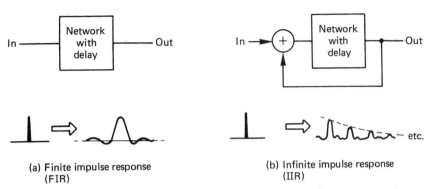

(a) Finite impulse response
(FIR)

(b) Infinite impulse response
(IIR)

Figure 3.42 An FIR filter (a) responds only to an input, whereas the output of an IIR filter (b) continues indefinitely rather like a decaying echo.

output to the input. For this reason they are also called recursive filters. As the impulse response in not symmetrical, IIR filters are not phase linear. In this respect they are similar to analog tone controls.

3.23 FIR filters

A FIR filter works by graphically constructing the impulse response for every input sample. It is first necessary to establish the correct impulse response. Figure 3.43(a) shows an example of a low-pass filter which cuts off at 1/4 of the sampling rate. The impulse response of a perfect low-pass filter is a $\sin x/x$ curve, where the time between the two central zero crossings is the reciprocal of the cut-off frequency. According to the mathematics, the waveform has always existed, and carries on for ever. The peak value of the output coincides with the input impulse. This means that the filter is not causal, because the output has changed before the input is known. Thus in all practical applications it is necessary to truncate the extreme ends of the impulse response, which causes an aperture effect, and to introduce a time delay in the filter equal to half the duration of the truncated impulse in order to make the filter causal.

As an input impulse is shifted through the series of registers in Figure 3.43(b), the impulse response is created, because at each point it is multiplied by a coefficient as in (c). These coefficients are simply the result of sampling and quantizing the desired impulse response. Clearly the sampling rate used to sample the impulse must be the same as the sampling rate for which the filter is being designed. In practice the coefficients are calculated, rather than attempting to sample an actual impulse response. The coefficient wordlength will be a compromise between cost and performance. Because the input sample shifts across the system registers to create the shape of the impulse response, the

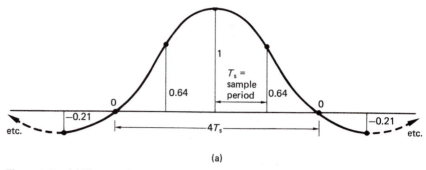

(a)

Figure 3.43 (a) The impulse response of an LPF is a sin$x$/x curve which stretches from $-\infty$ to $+\infty$ in time. The ends of the response must be neglected, and a delay introduced to make the filter causal.

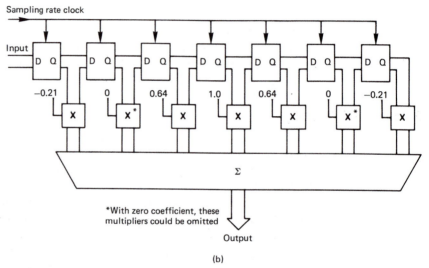

(b)

Figure 3.43 (b) The structure of an FIR LPF. Input samples shift across the register and at each point are multiplied by different coefficients.

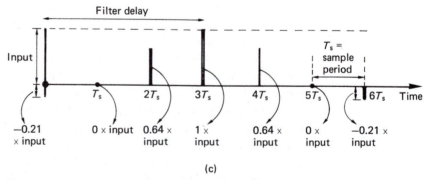

(c)

Figure 3.43 (c) When a single unit sample shifts across the circuit of Figure 3.43(b), the impulse response is created at the output as the impulse is multiplied by each coefficient in turn.

configuration is also known as a transversal filter. In operation with real sample streams, there will be several consecutive sample values in the filter registers at any time in order to convolve the input with the impulse response.

Simply truncating the impulse response causes an abrupt transition from input samples which matter and those which do not. Truncating the filter superimposes a rectangular shape on the time domain impulse response. In the frequency domain the rectangular shape transforms to a $\sin x/x$ characteristic which is superimposed on the desired frequency response as a ripple. One consequence of this is known as Gibb's phenomenon; a tendency for the response to peak just before the cut-off frequency.[12,13] As a result, the length of the impulse which must be considered will depend not only on the frequency response but also on the amount of ripple which can be tolerated. If the relevant period of the impulse is measured in sample periods, the result will be the number of points or multiplications needed in the filter. Figure 3.44 compares the performance of filters with different numbers of points. A high-quality digital audio FIR filter may need in excess of 100 points.

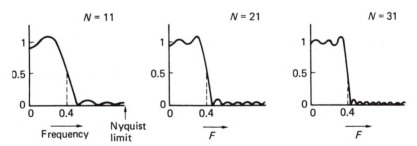

Figure 3.44 The truncation of the impulse in an FIR filter caused by the use of a finite number of points (N) results in ripple in the response. Shown here are three different numbers of points for the same impulse response. The filter is an LPF which rolls off at 0.4 of the fundamental interval. (Courtesy *Philips Technical Review*)

Rather than simply truncate the impulse response in time, it is better to make a smooth transition from samples which do not count to those that do. This can be done by multiplying the coefficients in the filter by a window function which peaks in the centre of the impulse. Figure 3.45 shows some different window functions and their responses. The rectangular window is the case of truncation, and the response is shown at I. A linear reduction in weight from the centre of the window to the edges characterizes the Bartlett window II, which trades ripple for an increase in transition-region width. At III is shown the Hanning window, which is essentially a raised cosine shape. Not shown is the similar Hamming window, which offers a slightly different trade-off between

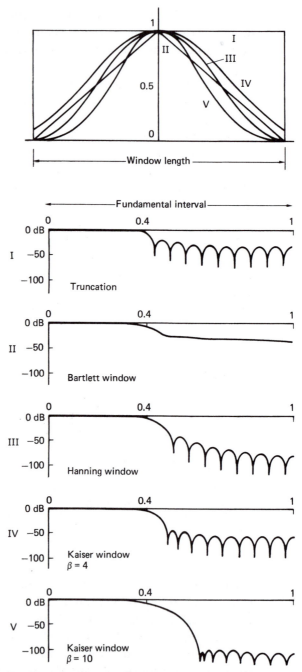

Figure 3.45 The effect of window functions. At top, various window functions are shown in continuous form. Once the number of samples in the window is established, the continuous functions shown here are sampled at the appropriate spacing to obtain window coefficients. These are multiplied by the truncated impulse response coefficients to obtain the actual coefficients used by the filter. The amplitude responses I–V correspond to the window functions illustrated. (Responses courtesy *Philips Technical Review*)

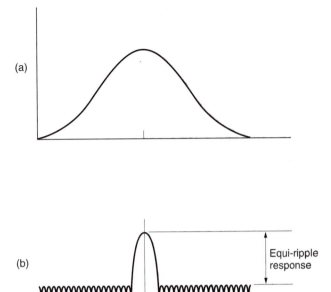

Figure 3.46 The Dolph window shape is shown at (a). The frequency response is at (b). Note the constant height of the response ripples.

ripple and the width of the main lobe. The Blackman window introduces an extra cosine term into the Hamming window at half the period of the main cosine period, reducing Gibb's phenomenon and ripple level, but increasing the width of the transition region. The Kaiser window is a family of windows based on the Bessel function, allowing various trade-offs between ripple ratio and main lobe width. Two of these are shown in IV and V.

The Dolph window[14] shown in Figure 3.46 results in an *equiripple filter* which has the advantage that the attenuation in the stopband never falls below a certain level.

Filter coefficients can be optimized by computer simulation. One of the best-known techniques used is the Remez exchange algorithm, which converges on the optimum coefficients after a number of iterations.

In the example of Figure 3.47, a low-pass FIR filter is shown which is intended to allow downsampling by a factor of two. The key feature is that the stopband must have begun before one half of the output sampling rate. This is most readily achieved using a Hamming window because it was designed empirically to have a flat stopband so that good aliasing attenuation is possible. The width of the transition band determines the number of significant sample periods embraced by the impulse. The Hamming window doubles the width of the transition band. This determines in turn both the number of points in the filter and

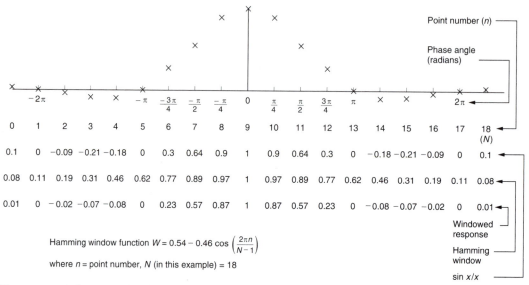

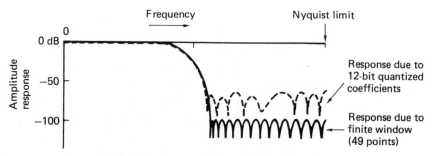

Figure 3.47   A downsampling filter using the Hamming window.

the filter delay. For the purposes of illustration, the number of points is much smaller than would normally be the case in an audio application.

As the impulse is symmetrical, the delay will be half the impulse period. The impulse response is a $\sin x/x$ function, and this has been calculated in the figure. The equation for the Hamming window function is shown with the window values which result. The $\sin x/x$ response is next multiplied by the Hamming window function to give the windowed impulse response shown.

If the coefficients are not quantized finely enough, it will be as if they had been calculated inaccurately, and the performance of the filter will be less than expected. Figure 3.48 shows an example of quantizing coefficients. Conversely, raising the wordlength of the coefficients increases cost.

Figure 3.48   Frequency response of a 49-point transversal filter with infinite precision (solid line) shows ripple due to finite window size. Quantizing coefficients to 12 bits reduces attenuation in the stopband. (Responses courtesy *Philips Technical Review*)

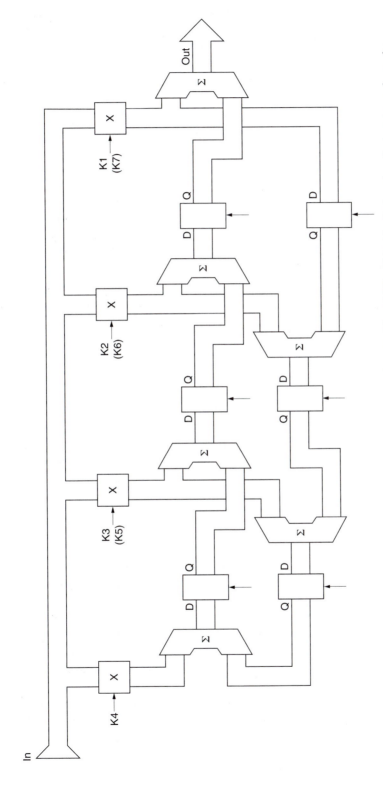

Figure 3.49 A seven-point folded filter for a symmetrical impulse response. In this case K1 and K7 will be identical, and so the input sample can be multiplied once, and the product fed into the output shift system in two different places. The centre coefficient K4 appears once. In an even-numbered filter the centre coefficient would also be used twice.

The FIR structure is inherently phase linear because it is easy to make the impulse response absolutely symmetrical. The individual samples in a digital system do not know in isolation what frequency they represent, and they can only pass through the filter at a rate determined by the clock. Because of this inherent phase-linearity, a FIR filter can be designed for a specific impulse response, and the frequency response will follow.

The frequency response of the filter can be changed at will by changing the coefficients. A programmable filter only requires a series of PROMs to supply the coefficients; the address supplied to the PROMs will select the response. The frequency response of a digital filter will also change if the clock rate is changed, so it is often less ambiguous to specify a frequency of interest in a digital filter in terms of a fraction of the fundamental interval rather than in absolute terms. The configuration shown in Figure 3.43 serves to illustrate the principle. The units used on the diagrams are sample periods and the response is proportional to these periods or spacings, and so it is not necessary to use actual figures.

Where the impulse response is symmetrical, it is often possible to reduce the number of multiplications, because the same product can be used twice, at equal distances before and after the centre of the window. This is known as folding the filter. A folded filter is shown in Figure 3.49.

3.24   Sampling-rate conversion

The topic of sampling-rate conversion will become increasingly important as digital audio equipment becomes more common and attempts are made to create large interconnected systems. Many of the circumstances in which a change of sampling rate is necessary are set out here:

1 To realize the advantages of oversampling converters, an increase in sampling rate is necessary prior to DACs and a reduction in sampling rate is necessary following ADCs. In oversampling the factors by which the rates are changed are very much higher than in other applications.
2 When a digital recording is played back at other than the correct speed to achieve some effect or to correct pitch, the sampling rate of the reproduced signal changes in proportion. If the playback samples are to be fed to a digital mixing console which works at some standard frequency, rate conversion will be necessary.
3 In the past, many different sampling rates were used on recorders which are now becoming obsolete. With sampling-rate conversion, recordings made on such machines can be played back and transferred to more modern formats at standard sampling rates.

4 Different sampling rates exist today for different purposes. Rate conversion allows material to be exchanged freely between rates. For example, master tapes made at 48 kHz on multitrack recorders may be digitally mixed down to two tracks at that frequency, and then converted to 44.1 kHz for Compact Disc or DCC mastering, or to 32 kHz for broadcast use.

5 When digital audio is used in conjunction with film or video, difficulties arise because it is not always possible to synchronize the sampling rate with the frame rate. An example of this is where the digital audio recorder uses its internally generated sampling rate, but also records studio timecode. On playback, the timecode can be made the same as on other units, or the sampling rate can be locked, but not both. Sampling-rate conversion allows a recorder to play back an asynchronous recording locked to timecode.

6 When programs are interchanged over long distances, there is no guarantee that source and destination are using the same timing reference. In this case the sampling rates at both ends of a link will be nominally identical, but drift in reference oscillators will cause the relative sample phase to be arbitrary.

In items 5 and 6 above, the difference of rate between input and output is small, and the process is then referred to as synchronization. This can be simpler than rate conversion, and will be treated in Chapter 8.

Sampling-rate conversion can be effected by returning to the analog domain. A DAC is connected to an ADC. In order to satisfy the requirements of sampling theory, there must be a low-pass filter between the two having a frequency response restricted to one-half of the lower sampling rate. In reality this is seldom done, because all practical machines have anti-aliasing filters at their analog inputs and anti-image filters at their analog outputs. Connecting one machine to another via the analog domain therefore includes one unnecessary filter in the chain. Since analog filters are seldom optimal, degradation may be caused by rate-converting in this way, particularly in the area of phase response, although the introduction of oversampling convertors has lessened the problem.

Analog filters usually have a fixed response, and this is not necessarily the correct one if both input and output rates are to be varied significantly. The increase in noise due to an additional quantizing stage and additional double exposure to clock jitter is not beneficial. Methods of sampling-rate conversion in the digital domain are preferable and will be described here.

There are three basic but related categories of rate conversion, as shown in Figure 3.50. The most straightforward (a) changes the rate by an integer

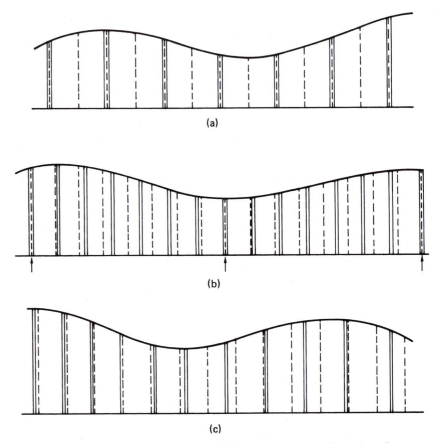

Figure 3.50 Categories of rate conversion. (a) Integer-ratio conversion, where the lower-rate samples are always coincident with those of the higher rate. There are a small number of phases needed. (b) Fractional-ratio conversion, where sample coincidence is periodic. A larger number of phases are required. Example here is conversion from 50.4 kHz to 44.1 kHz (8/7). (c) Variable-ratio conversion, where there is no fixed relationship, and a large number of phases are required.

ratio, up or down. The timing of the system is thus simplified because all samples (input and output) are present on edges of the higher-rate sampling clock. Such a system is generally adopted for oversampling convertors; the exact sampling rate immediately adjacent to the analog domain is not critical, and will be chosen to make the filters easier to implement.

Next in order of difficulty is the category shown at (b) where the rate is changed by the ratio of two small integers. Samples in the input periodically time-align with the output. Many of the early proposals for professional sampling rates were based on simple fractional relationships to 44.1 kHz such as $\frac{8}{7}$ so that this technique could be used. This technique is not suitable for variable-speed replay or for asynchronous operation.

The most complex rate-conversion category is where there is no simple relationship between input and output sampling rates, and indeed they are allowed to vary. This situation, shown at (c), is known as variable-ratio conversion. The time relationship of input and output samples is arbitrary, and independent clocks are necessary. Once it was established that variable-ratio conversion was feasible, the choice of a professional sampling rate became very much easier, because the simple fractional relationships could be abandoned. The conversion fraction between 48 kHz and 44.1 kHz is 160:147, which is indeed not simple.

As the technique of integer-ratio conversion is used almost exclusively for oversampling in digital audio it will be discussed in that context. Sampling-rate reduction by an integer factor is dealt with first.

Figure 3.51(a) shows the spectrum of a typical sampled system where the sampling rate is a little more than twice the analog bandwidth. Attempts to reduce the sampling rate by simply omitting samples, a process known as decimation, will result in aliasing, as shown in (b). Intuitively it is obvious that omitting samples is the same as if the original sampling rate was lower. In order to prevent aliasing, it is necessary to incorporate low-pass filtering into the system where the cut-off frequency reflects the new, lower, sampling rate. An FIR type low-pass filter could be installed, as described earlier in this chapter, immediately prior to the stage where samples are omitted, but this would be wasteful, because for much of its time the FIR filter would be calculating sample values which are to be discarded.

A more effective method is to combine the low-pass filter with the decimator so that the filter only calculates values to be retained in the output sample stream. Figure 3.51(c) shows how this is done. The filter makes one accumulation for every output sample, but that accumulation is the result of multiplying all relevant input samples in the filter window by an appropriate coefficient. The number of points in the filter is determined by the number of *input* samples in the period of the filter window, but the number of multiplications per second is obtained by multiplying that figure by the *output* rate. If the filter is not integrated with the decimator, the number of points has to be multiplied by the input rate. The larger the rate-reduction factor, the more advantageous the decimating filter ought to be, but this is not quite the case, as the greater the reduction in rate, the longer the filter window will need to be to accommodate the broader impulse response.

When the sampling rate is to be increased by an integer factor, additional samples must be created at even spacing between the existing ones. There is no need for the bandwidth of the input samples to be reduced since, if the original sampling rate was adequate, a higher one must also be adequate.

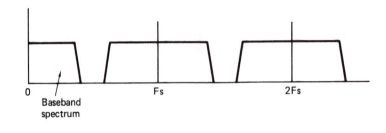

(a)

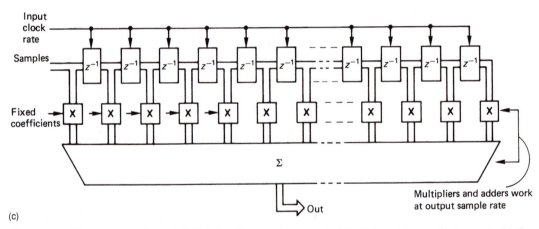

(b)

(c)

Figure 3.51 The spectrum of a typical digital audio sample stream in (a) will be subject to aliasing as in (b) if the baseband width is not reduced by an LPF. In (c) an FIR low-pass filter prevents aliasing. Samples are clocked transversely across the filter at the input rate, but the filter only computes at the output sample rate. Clearly this will only work if the two rates are related by an integer factor.

Figure 3.52 shows that the process of sampling-rate increase can be thought of in two stages. First the correct rate is achieved by inserting samples of zero value at the correct instant, and then the additional samples are given meaningful values by passing the sample stream through a low-pass filter which cuts off at the Nyquist frequency of the original sampling rate. This filter is known as an interpolator, and one of its tasks is to prevent images of the lower input-sampling spectrum from appearing in the extended baseband of the higher-rate output spectrum.

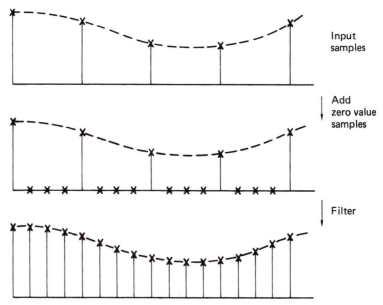

Figure 3.52 In integer-ratio sampling, rate increase can be obtained in two stages. Firstly, zero-value samples are inserted to increase the rate, and then filtering is used to give the extra samples real values. The filter necessary will be an LPF with a response which cuts off at the Nyquist frequency of the input samples.

How do interpolators work? It is important to appreciate that, according to sampling theory, all sampled systems have finite bandwidth. An individual digital sample value is obtained by sampling the instantaneous voltage of the original analog waveform, and because it has zero duration, it must contain an infinite spectrum. However, such a sample can never be heard in that form because of the reconstruction process, which limits the spectrum of the impulse to the Nyquist limit. After reconstruction, one infinitely short digital sample ideally represents a $\sin x/x$ pulse whose central peak width is determined by the response of the reconstruction filter, and whose amplitude is proportional to the sample value. This implies that, in reality, one sample value has meaning over a considerable timespan, rather than just at the sample instant. If this were not true, it would be impossible to build an interpolator.

As in rate reduction, performing the steps separately is inefficient. The bandwidth of the information is unchanged when the sampling rate is increased; therefore the original input samples will pass through the filter unchanged, and it is superfluous to compute them. The combination of the two processes into an interpolating filter minimizes the amount of computation.

As the purpose of the system is purely to increase the sampling rate, the filter must be as transparent as possible, and this implies that a linear-phase configuration is mandatory, suggesting the use of an FIR structure.

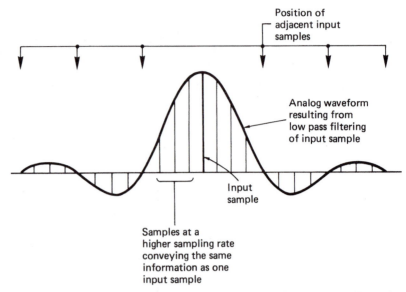

Position of
adjacent input
samples

Analog waveform
resulting from
low pass filtering
of input sample

Input
sample

Samples at a
higher sampling rate
conveying the same
information as one
input sample

Figure 3.53 A single sample results in a $\sin x/x$ waveform after filtering in the analog domain. At a new, higher, sampling rate, the same waveform after filtering will be obtained if the numerous samples of differing size shown here are used. It follows that the value of these new samples can be calculated from the input samples in the digital domain in an FIR filter.

Figure 3.53 shows that the theoretical impulse response of such a filter is a $\sin x/x$ curve which has zero value at the position of adjacent input samples. In practice this impulse cannot be implemented because it is infinite.

The impulse response used will be truncated and windowed as described earlier. To simplify this discussion, assume that a $\sin x/x$ impulse is to be used. The process of interpolation is the same in principle as the reconstruction filtering which takes place in DACs. It will be seen in Chapter 4 that a continuous time analog signal is obtained by summing the analog impulses due to each sample. In a digital interpolating filter, this process is duplicated but in discrete time.[15]

If the sampling rate is to be doubled, new samples must be interpolated exactly halfway between existing samples. The necessary impulse response is shown in Figure 3.54; it can be sampled at the *output* sample period and quantized to form coefficients. If a single input sample is multiplied by each of these coefficients in turn, the impulse response of that sample at the new sampling rate will be obtained. Note that every other coefficient is zero, which confirms that no computation is necessary on the existing samples; they are just transferred to the output. The intermediate sample is computed by adding together the impulse responses of every input sample in the window. The figure shows how this mechanism operates. If the sampling rate is to be increased by a

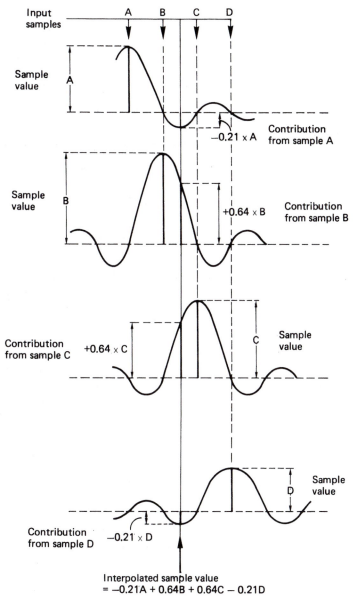

Interpolated sample value
= −0.21A + 0.64B + 0.64C − 0.21D

Figure 3.54 A two times oversampling interpolator. To compute an intermediate sample, the input samples are imagined to be sinx/x impulses, and the contributions from each at the point of interest can be calculated. In practice, rather more samples on either side need to be taken into account.

factor of four, three sample values must be interpolated between existing input samples. Figure 3.55 shows that it is only necessary to sample the impulse response at one-quarter the period of input samples to obtain three sets of coefficients which will be used in turn. In hardware-implemented filters, the input sample which is passed straight to the

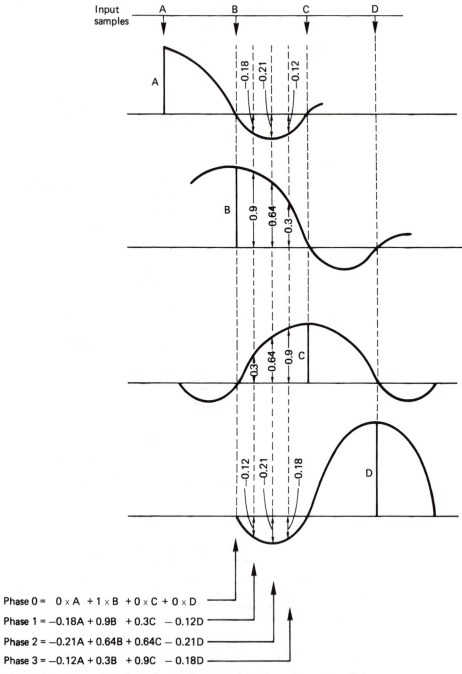

Figure 3.55 In 4× oversampling, for each set of input samples, four phases of coefficients are necessary, each of which produces one of the oversampled values.

output is transferred by using a fourth filter phase where all coefficients are zero except the central one which is unity.

Figure 3.50 showed that when the two sampling rates have a simple fractional relationship *m/n*, there is a periodicity in the relationship between samples in the two streams. It is possible to have a system clock running at the least-common multiple frequency which will divide by different integers to give each sampling rate.[16]

The existence of a common clock frequency means that a fractional-ratio convertor could be made by arranging two integer-ratio convertors in series. This configuration is shown in Figure 3.56(a). The

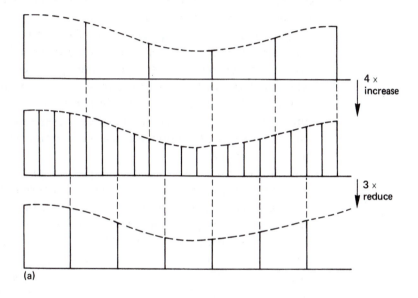

4 ×
increase

3 ×
reduce

(a)

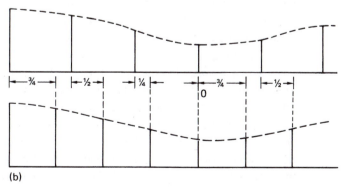

(b)

Figure 3.56 In (a), fractional-ratio conversion of 3/4 in this example is by increasing to 4× input prior to reducing by 3×. The inefficiency due to discarding previously computed values is clear. In (b), efficiency is raised since only needed values will be computed. Note how the interpolation phase changes for each output. Fixed coefficients can no longer be used.

input-sampling rate is multiplied by *m* in an interpolator, and the result is divided by *n* in a decimator. Although this system would work, it would be grossly inefficient, because only one in *n* of the interpolator's outputs would be used. A decimator followed by an interpolator would also offer the correct sampling rate at the output, but the intermediate sampling rate would be so low that the system bandwidth would be quite unacceptable.

As has been seen, a more efficient structure results from combining the processes. The result is exactly the same structure as an integer-ratio interpolator, and requires an FIR filter. The impulse response of the filter is determined by the lower of the two sampling rates, and as before it prevents aliasing when the rate is being reduced, and prevents images when the rate is being increased. The interpolator has sufficient coefficient phases to interpolate *m* output samples for every input sample, but not all of these values are computed; only interpolations which coincide with an output sample are performed. It will be seen in Figure 3.56(b) that input samples shift across the transversal filter at the input sampling rate, but interpolations are performed only at the output sample rate. This is possible because a different filter phase will be used at each interpolation.

In the previous examples, the sample rate of the filter output had a constant relationship to the input, which meant that the two rates had to be phase-locked. This is an undesirable constraint in some applications, including sampling rate convertors used for variable-speed replay. In a variable-ratio convertor, values will exist for the instants at which input samples were made, but it is necessary to compute what the sample values would have been at absolutely any time between available samples. The general concept of the interpolator is the same as for the fractional-ratio convertor, except that an infinite number of filter phases is necessary. Since a realizable filter will have a finite number of phases, it is necessary to study the degradation this causes.

The desired continuous time axis of the interpolator is quantized by the phase spacing, and a sample value needed at a particular time will be replaced by a value for the nearest available filter phase. The number of phases in the filter therefore determines the time accuracy of the interpolation. The effects of calculating a value for the wrong time are identical to sampling with jitter, in that an error occurs proportional to the slope of the signal. The result is program-modulated noise. The higher the noise specification, the greater the desired time accuracy and the larger the number of phases required. The number of phases is equal to the number of sets of coefficients available, and should not be confused with the number of points in the filter, which is equal to the number of coefficients in a set (and the number of multiplications needed to calculate one output value).

In Chapter 4 it will be shown that the sampling jitter accuracy necessary for sixteen-bit working is a few hundred picoseconds. This implies that something like 2^{15} filter phases will be required for adequate performance in a sixteen-bit sampling-rate convertor.[17] The direct provision of so many phases is difficult, since more than a million different coefficients must be stored; so alternative methods have been devised. When several interpolators are cascaded, the number of phases available is the product of the number of phases in each stage. For example, if a filter which could interpolate sample values halfway between existing samples were followed by a filter which could interpolate at one-quarter, one-half and three-quarters the input period, the overall number of phases available would be eight. This is illustrated in Figure 3.57.

For a practical convertor, four filters in series might be needed. To increase the sampling rate, the first two filters interpolate at fixed points between samples input to them, effectively multiplying the input sampling rate by some large factor as well as removing images from the spectrum; the second two work with variable coefficients, like the fractional-ratio convertor described earlier, so that only samples coincident with the output clock are computed. To reduce the sampling rate, the positions of the two pairs of filters are reversed, so that the fixed-response filters perform the anti-aliasing function at the output sampling frequency.

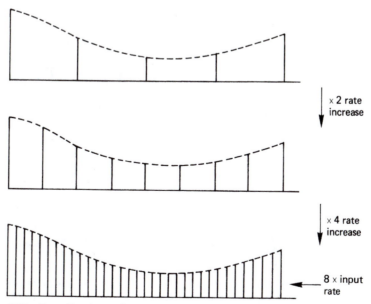

Figure 3.57 Cascading interpolators multiplies the factor of sampling-rate increase of each stage.

As mentioned earlier, the response of a digital filter is always proportional to the sampling rate. When the sampling rate on input or output varies, the phase of the interpolators must change dynamically. The necessary phase must be selected to the stated accuracy, and this implies that the position of the relevant clock edge must be measured in time to the same accuracy. This is not possible because, in real systems, the presence of noise on binary signals of finite-rise time shifts the time where the logical state is considered to have changed. The only way to measure the position of clocks in time without jitter is to filter the measurement digitally, and this can be done with a numerically locked loop. Figure 3.58 shows the essential stages of a variable-ratio convertor of this kind.

When suitable processing speed is available, a digital computer can act as a filter, since each multiplication can be executed serially, and the results accumulated to produce an output sample. For simple filters, the coefficients would be stored in memory, but the number of coefficients needed for rate conversion precludes this. However, it is possible to compute what a set of coefficients should be algorithmically, and this approach permits single-stage conversion.

The two sampling clocks are compared as before, to produce an accurate relative-phase parameter. The lower sampling rate is measured to determine what the impulse response of the filter should be to prevent aliasing or images, and this is fed, along with the phase parameter, to a processor which computes a set of coefficients and multiplies them by a

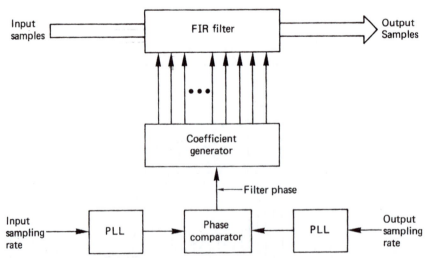

Figure 3.58 (a) In a variable-ratio convertor, the phase relationship of input and output clock edges must be measured to determine the coefficients needed. Jitter on clocks prevents their direct use, and phase-locked loops must be used to average the jitter over many sample clocks.

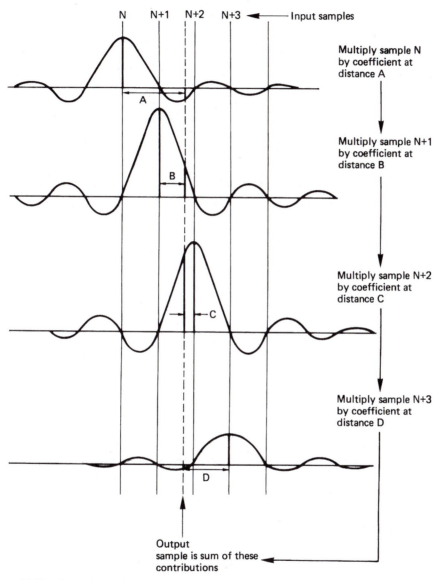

Figure 3.58 (b) The clock relationships in (a) determine the relative phases of output and input samples, which in conjunction with the filter impulse response determine the coefficients necessary.

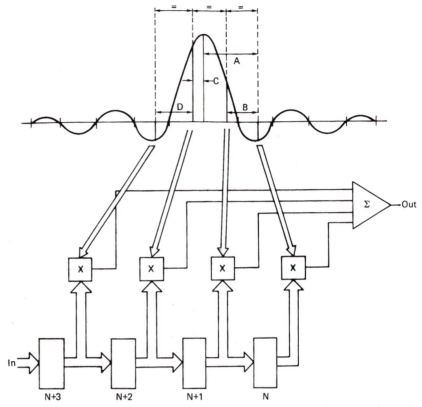

Figure 3.58   (c) The coefficients determined in (b) are fed to the configuration shown (or the equivalent implemented in software) to compute the output sample at the correct interpolated position. Note that actual filter will have many more points than this simple example shows.

window function. These coefficients are then used by the single-filter stage to compute one output sample. The process then repeats for the next output sample.

3.25   IIR filters

Figure 3.59 is a FIR filter which has been adapted in an attempt to simulate an RC network. Because an RC network is causal, i.e. the output cannot appear before the input, the impulse response is asymmetrical, and represents an exponential decay, as shown in Figure 3.59(a). The asymmetry of the impulse response confirms the expected result that this filter will not be phase-linear. The structure of the filter is exactly the same as the earlier examples given in this chapter; only the coefficients have been changed. The simulation of RC networks is common in digital audio

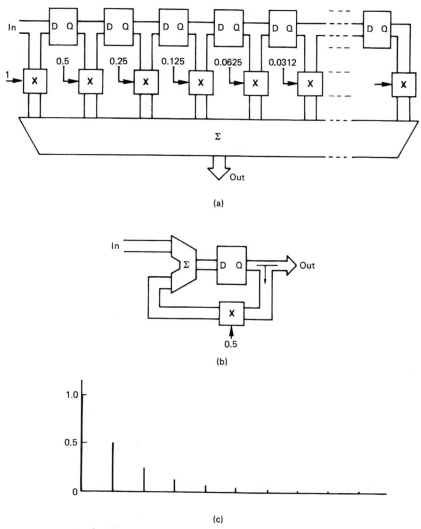

Figure 3.59 In (a) an FIR filter is supplied with exponentially decaying coefficient to simulate an *RC* response. In (b) the configuration of an IIR or recursive filter uses much less hardware (or computation) to give the same response, shown in (c).

for the purposes of equalization or provision of tone controls. A large number of points are required in an FIR filter to create the long exponential decays necessary, and the FIR filter is at a disadvantage here because an exponential decay can be computed as every output sample is a fixed proportion of the previous one.

Figure 3.59(b) shows a much simpler hardware configuration, where the output is returned in attenuated form to the input. The response of this circuit to a single sample is a decaying series of samples, in which the rate of decay is controlled by the gain of the multiplier. If the gain is one,

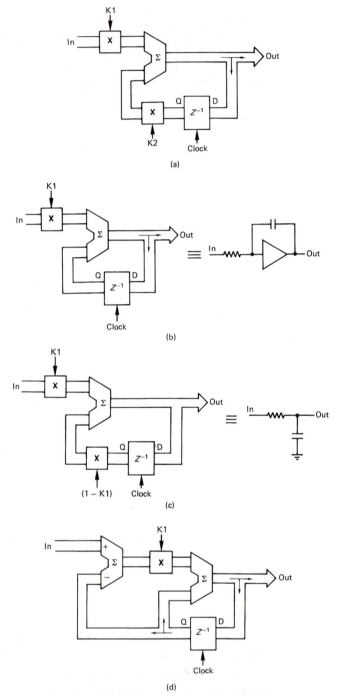

Figure 3.60   (a) First-order lag network IIR filter. Note recursive path through single sample delay latch. (b) The arrangement of (a) becomes an integrator if K2 = 1, since the output is always added to the next sample. (c) When the time coefficients sum to unity, the system behaves as an *RC* lag network. (d) The same performance as in (c) can be realized with only one multiplier by reconfiguring as shown here.

the output can carry on indefinitely. For this reason, the configuration is known as an infinite impulse response (IIR) filter. If the gain of the multiplier is slightly more than one, the output will increase exponentially after a single non-zero input until the end of the number range is reached. Unlike FIR filters, IIR filters are not necessarily stable. FIR filters are easy to understand, but difficult to make in audio applications; IIR filters are easier to make, because less hardware is needed, but they are harder to understand.

One major consideration when recursive techniques are to be used is that the accuracy of the coefficients must be much higher. This is because an impulse response is created by making each output some fraction of the previous one, and a small error in the coefficient becomes a large error after several recursions. This error between what is wanted and what results from using truncated coefficients can often be enough to make the actual filter unstable whereas the theoretical model is not.

By way of introduction to this class of filters, the characteristics of some useful configurations will be discussed. It will be seen that parallels can be drawn with some classical analog circuits.

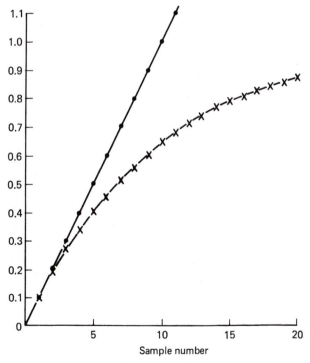

Figure 3.61 The response of the configuration of Figure 3.60 to a unit step. With K2 = 1, the system is an integrator, and the straight line shows the output with K1 = 0.1. With K1 = 0.1 and K2 = 0.9, K1 + K2 = 1 and the exponential response of an *RC* network is simulated.

The terms phase lag and phase lead are used to describe analog circuit characteristics, and they are also applicable to digital circuits. Figure 3.60(a) shows a first-order lag network containing two multipliers, a register to provide one sample period of delay, and an adder. As might be expected, the characteristics of the circuit can be transformed by changing the coefficients. If K2 is greater than unity, the circuit is unstable, as any non-zero input causes the output to increase exponentially. Making K2 equal to unity (Figure 3.60(b)) produces a digital integrator, because the current value in the latch is added to the input to form the next value in the latch. The coefficient K1 determines the time constant in the same way that the RC network does for the analog circuit.

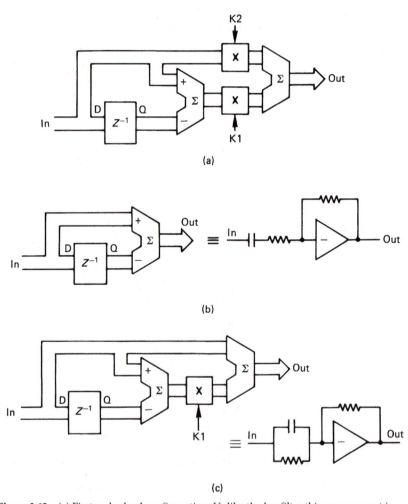

(a)

(b)

(c)

Figure 3.62 (a) First-order lead configuration. Unlike the lag filter this arrangement is always stable, but as before the effect of changing the coefficients is dramatic. (b) When K2 of (a) is made zero, the configuration subtracts successive samples, and thus acts as a differentiator. (c) Setting K2 of (a) to unity gives the high-pass filter response shown here.

Figure 3.60(c) shows the case where K1 + K2 = 1; the response will be the same as an RC lag network. In this case it will be more economical to construct a different configuration shown in (d) having the same characteristics but eliminating one stage of multiplication. The operation of these configurations can be verified by computing their responses to an input step. This is simply done by applying some constant input value, and deducing how the output changes for each applied clock pulse to the register. This has been done for two cases in Figure 3.61 where the linear integrator response and the exponential responses can be seen. It is interesting to experiment with different coefficients to see how the results change.

Figure 3.62(a) shows a first-order lead network using the same basic building blocks. Again, the coefficient values have dramatic power. If K2 is made zero, the circuit simply subtracts the previous sample value from the current one, and so becomes a true differentiator as in (b). K1 determines the time constant. If K2 is made unity, the configuration acts as a high-pass filter as in (c).

3.26 The *z*-transform

Whereas it was possible to design effective FIR filters with relatively simple theory, the IIR filter family are too complicated for that. The z-transform is particularly appropriate for IIR digital filter design because it permits a rapid graphic assessment of the characteristics of a proposed filter. This graphic nature of the z-plane also lends itself to explanation so that an understanding of filter concepts can be obtained without their becoming obscured by mathematics.

Digital filters rely heavily on delaying sample values by one or more sample periods. One tremendous advantage of the z-transform is that a delay which is difficult to handle in time-domain expressions corresponds to a multiplication in the z-domain. This means that the transfer function of a circuit in the z-domain can often be written down by referring to the block diagram, which is why a register causing a sample delay is usually described by z^{-1}.

The circuit configuration of Figure 3.60(c) is repeated in Figure 3.63(a). For simplicity in calculation, the two coefficients have been set to 0.5. The impulse response of this circuit will be found first, followed by the characteristics of the circuit in the z-plane which will immediately give the frequency and phase response.

The impulse response can be found graphically by supplying a series of samples as input which are all zero except for one which has unity value. This has been done in the figure, where it will be seen that once the unity sample has entered, the output *y* will always be one half the previous output, resulting in an exponential decay.

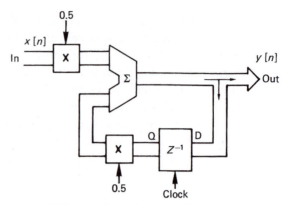

Figure 3.63 (a) A digital filter which simulates an *RC* network. In this example the coefficients are both 0.5.

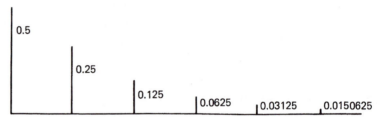

Figure 3.63 (b) The response of (a) to a single unity-value sample. The initial output of 0.5 is due to the input coefficient. Subsequent outputs are always 0.5 of the previous sample, owing to the recursive path through the latch.

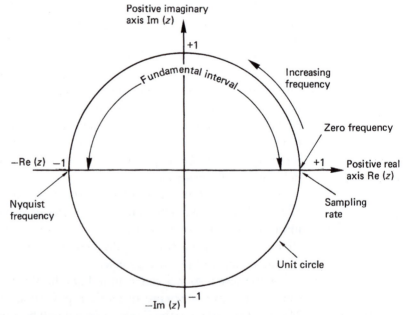

Figure 3.63 (c) The z-plane showing real and imaginary axes. Frequency increases anticlockwise from Re(z) = +1 to the Nyquist limit at Re(z) = −1, returning to Re(z) = +1 at the sampling rate. The origin of aliasing is clear.

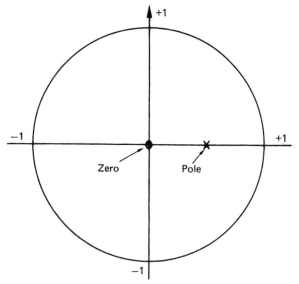

Figure 3.63 (d) The z-plane is used by inserting poles and zeros.

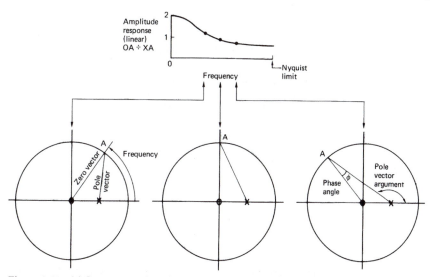

Figure 3.63 (e) Some examples of response of the circuit (a) using the z-plane. The zero vector is divided by the pole vector to obtain the amplitude response, and the phase response is the difference between the arguments of the two vectors.

Where there is a continuous input to the circuit, the output will be one half the previous output y plus one half the current input x. The output then becomes the convolution of the input and the impulse response.

It is possible to express the operation of the circuit mathematically. Time is discrete, and is measured in sample periods, so the time can be expressed by the number of the sample n. Accordingly the input at time

n will be called $x[n]$ and the corresponding output will be called $y[n]$. The previous output is called yn^{-1}. It is then possible to write:

$$y[n] = 0.5\ x[n] + 0.5\ y[n-1]$$

This is called a recurrence relationship, because if it is repeated for all values of n, a convolution results. The relationship can be transformed into an expression in z, referring to Figure 3.63(b):

$$y[n] = y[z],\ x[n] = x[z]$$

$$\text{and} \quad y[n-1] = z^{-1}\ y[z]$$

$$\text{so that} \quad y[z] = 0.5x[z] + 0.5z^{-1}\ y[z] \tag{3.1}$$

As with any system the transfer function is the ratio of the output to the input. Thus:

$$H[z] = \frac{y[z]}{x[z]}$$

Rearranging equation (3.1) gives:

$$y[z] - 0.5z^{-1}y[z] = 0.5x[z]$$

$$y[z](1-0.5z) = 0.5x[z]$$

$$\frac{y[z]}{x[z]} = \frac{0.5}{1-0.5z^{-1}} = \frac{1}{2-z^{-1}} = \frac{z}{2z-1}$$

Thus:

$$H[z] = \frac{z}{2z-1} \tag{3.2}$$

The term in the numerator would make the transfer function zero when it became zero, whereas the terms in the denominator would make the transfer function infinite if they were to become zero. These result in poles.

Poles and zeros are plotted on a z-plane diagram. The basics of the z-plane are shown in Figure 3.63(c). There are two axes at right angles, the real axis $\text{Re}(z)$ and the imaginary axis $\text{Im}(z)$. Upon this plane is drawn a circle of unit radius whose centre is at the intersection of the axes. Frequency increases anticlockwise around this circle from 0 Hz at $\text{Re}(z) = +1$ to the Nyquist limit frequency at $\text{Re}(z) = -1$. Negative frequency increases clockwise around the circle reaching the negative

Nyquist limit at Re(z) = −1. Essentially the circle on the z-plane is produced by taking the graph of a Fourier spectrum and wrapping it round a cylinder. The repeated spectral components at multiples of the sampling frequency in the Fourier domain simply overlap the fundamental interval when rolled up in this way, and so it is an ideal method for displaying the response of a sampled system, since only the response in the fundamental interval is of interest. Figure 3.63(d) shows that the z-plane diagram is used by inserting poles (X) and zeros (0).

In equation (3.2), $H[z]$ can be made zero only if z is zero; therefore a zero is placed on the diagram at $z = 0$. $H[z]$ becomes infinite if z is 0.5, because the denominator becomes zero. A pole (X) is placed on the diagram at Re(z) = 0.5. It will be recalled that 0.5 was the value of the coefficient used in the filter circuit being analysed. The performance of the system can be analysed for any frequency by drawing two vectors. The frequency in question is a fraction of the sampling rate and is expressed as an angle which is the same fraction of 360 degrees. A mark is made on the unit circle at that angle; a pole vector is drawn from the pole to the mark, and the zero vector is drawn from the zero to the mark.

The amplitude response at the chosen frequency is found by dividing the magnitude of the zero vector by the magnitude of the pole vector. A working approximation can be made by taking distances from the diagram with a ruler. The phase response at that frequency can be found by subtracting the argument of the pole vector from the argument of the zero vector, where the argument is the angle between the vector and the positive real z-axis. The phase response is clearly also the angle between the vectors. In Figure 3.63(e) the resulting diagram has been shown for several frequencies, and this has been used to plot the frequency and phase response of the system. This is confirmation of the power of the z-transform, because it has given the results of a Fourier analysis directly from the block diagram of the filter with trivial calculation. There cannot be more zeros than poles in any realizable system, and the poles must remain within the unit circle or instability results.

Figure 3.64(a) shows a slightly different configuration which will be used to support the example of a high-pass filter. High-pass filters with a low cut-off frequency are used extensively in digital audio to remove DC components in sample streams which have arisen due to convertor drift.

Using the same terminology as the previous example:

$$u[n] = x[n] + Ku[n-1]$$

so that in the z-domain:

$$u[z] = x[z] + Kz^{-1} u[z] \tag{3.3}$$

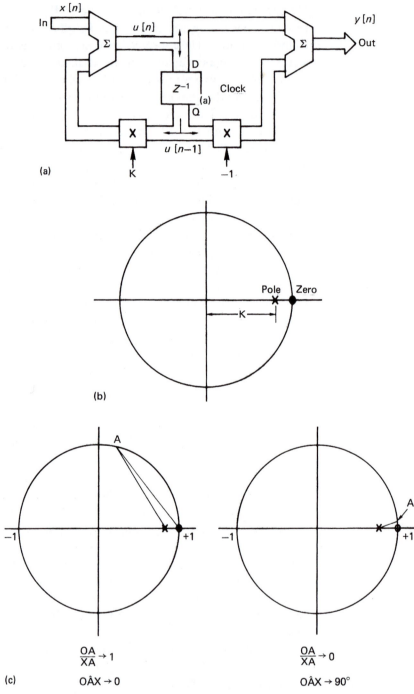

Figure 3.64 (a) Configuration used as a high-pass filter to remove DC offsets in audio samples. (b) The coefficient K determines the position of the pole on the real axis. It would normally be very close to the zero. (c) In the passband, the closeness of the pole and zero means the vectors are almost the same length, so the gain tends to unity and the phase shift is small (left). In the stopband, the gain falls and the phase angle tends to 90°.

Also: $y[n] = u[n] - u[n-1]$

So that in the z-domain:

$$y[z] = u[z] - z^{-1} u[z] = u[z](1-z^{-1})$$

Therefore:

$$\frac{y[z]}{uz} = 1-z^{-1} \tag{3.4}$$

From equation (3.3) $u[z] - Kz^{-1} u[z] = x[z]$

Therefore:

$$u[z](1-Kz^{-1}) = x[z]$$

and

$$\frac{u[z]}{x[z]} = \frac{1}{1-Kz^{-1}} \tag{3.5}$$

From equations (3.4) and (3.5):

$$\frac{y[z]}{x[z]} = \frac{y[z]}{u[z]} \times \frac{u[z]}{x[z]} = \frac{1-z}{1-Kz^{-1}}$$

Since the numerator determines the position of the pole, this will be at $z = 1$, and the zero will be at $z = K$ because this makes the denominator go to zero. Figure 3.64(b) shows that if the pole is put close to the zero by making K almost unity, the filter will only attenuate very low frequencies. At high frequencies, the ratio of the length of the pole vector to the zero vector will be almost unity, whereas at very low frequencies the ratio falls steeply becoming zero at DC. The phase characteristics can also be established. At high frequencies the pole and zero vectors will be almost parallel, so phase shift will be minimal. It is only in the area of the zero that the phase will change.

3.27 Bandpass filters

The low- and high-pass cases have been examined, and it has been seen that they can be realized with simple first-order filters. Bandpass circuits will now be discussed; these will generally involve higher-order configurations. Bandpass filters are used extensively for presence filters and their more complex relative the graphic equalizer, and are essentially filters which are tuned to respond to a certain band of frequencies more than others.

Figure 3.65(a) shows a bandpass filter, which is essentially a lead filter and a lag filter combined. The coefficients have been made simple for clarity; in fact two of them are set to zero. The adder now sums three terms, so the recurrence relationship is given by:

$$y[n] = x[n] - x[n-2] - 0.25y[n-2]$$

Since in the z-transform $x[n-1] = z^{-1}x[n]$,

$$y[z] = x[z] - z^{-2}x[z] - 0.25z^{-2}y[z]$$

$$H[z] = \frac{y[z]}{x[z]} = \frac{1-z^{-2}}{1 + 0.25z^{-2}} = \frac{z^2 - 1}{z + 0.25^2}$$

As before, the position of poles and zeros is determined by finding what values of z will cause the denominator or the numerator to become zero. This can be done by factorizing the terms. In the denominator the roots will be complex:

$$\frac{z^2 - 1}{z^2 + 0.25} = \frac{(z+1)(z-1)}{(z+j0.5)(z-j0.5)}$$

Figure 3.65(b) illustrates that the poles have come off the real axis of the z-plane, and then appear as a complex conjugate pair, so that the diagram is essentially symmetrical about the real axis. There are also two zeros, at Re+1 and Re–1.

With more poles and zeros, the graphical method of determining the frequency response becomes more complicated. The procedure is to multiply together the lengths of all the zero vectors, and divide by the product of the lengths of all the pole vectors. The process is shown in Figure 3.65(c). The frequency response shows an indistinct peak at half the Nyquist frequency. This is because the poles are some distance from the unit radius.

A more pronounced peak can be obtained by placing the poles closer to unit radius. In contrast to the previous examples, which have accepted a particular configuration and predicted what it will do, this example will decide what response is to be obtained and then compute the coefficients needed. Figure 3.66(a) shows the z-plane diagram. The resonant frequency is one third of the Nyquist limit, or 60 degrees around the circle, and the poles have been placed at a radius of 0.9 to give a peakier response.

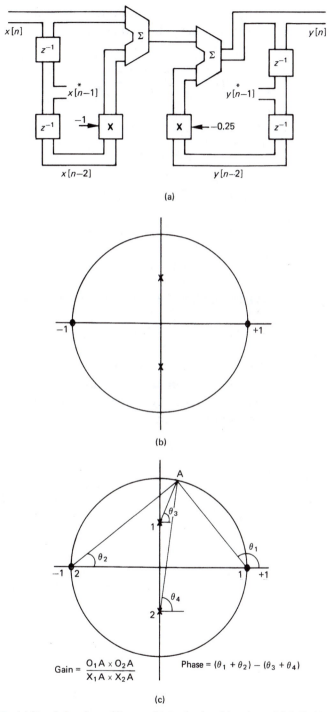

Figure 3.65 (a) Simple bandpass filter combining lead and lag stages. Note that two terms marked have zero coefficients, reducing the complexity of implementation. (b) Bandpass filters are characterized by poles which are away from the real axis. (c) With multiple poles and zeros, the computation of gain and phase is a little more complicated.

It is possible to write down the transfer function directly from the position of the poles and zeros:

$$H[z] = \frac{(z + 1)(z - 1)}{(z - re^{j\theta T})(z - re^{-j\theta T})}$$

$$= \frac{z^2 - 1}{z^2 - 2rz\cos\omega T + r^2} = \frac{y[z]}{x[z]}$$

$$y[z](z^2 - 2rz\cos\omega T + r^2) = x[z](z^2 - 1)$$

$$y[z]z^2 = x[z]z^2 - x[z] + y[z]2rz\cos\omega T - y[z]r^2$$

$$y[z] = x[z] - x[z]z^{-2} + y[z]2rz^{-1}\cos\theta T - y[z]r^2z^{-2}$$

As $\cos 60 = 0.5$, the recurrence relationship can be written:

$$y[n] = x[n] - x[n-2] + 0.9y[n-1] - 0.81y[n-2]$$

The configuration necessary can now be seen in Figure 3.66(b), along with the necessary coefficients. Since the transfer function is the ratio of two quadratic expressions, this arrangement is often referred to as a biquadratic section.

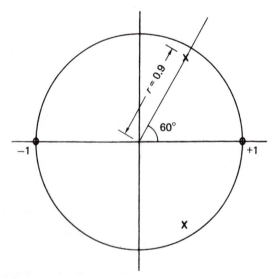

Figure 3.66   (a) The peak frequency chosen here is one-sixth of the sample rate, which requires the poles to be on radii at ±60 from the real axis. The radius of 0.9 places the poles close to the unit cycle, resulting in a pronounced peak.

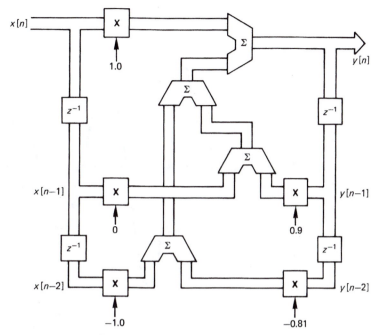

Figure 3.66 (b) The biquadratic configuration shown here implements the recurrence relationship derived in the text: $y[n] = x[n] - x[n-2] + 0.9y[n-1] - 0.81y[n-2]$.

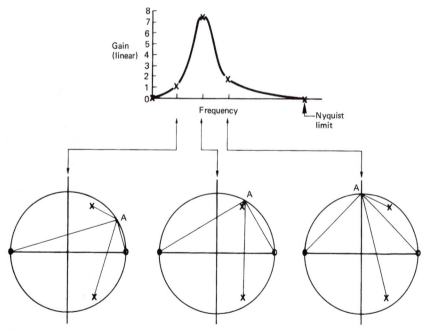

Figure 3.66 (c) Calculating the frequency response of the filter of (b). The method of Figure 3.65(c) is used. Note how the pole vector becomes very short as the resonance is reached. As this is in the denominator, the response is large. The instability resulting from a pole on or outside the unit circle is thus clearly demonstrated.

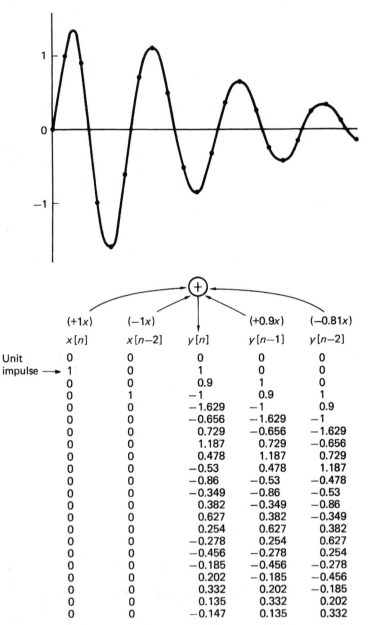

	(+1x)	(−1x)		(+0.9x)	(−0.81x)
	$x[n]$	$x[n-2]$	$y[n]$	$y[n-1]$	$y[n-2]$
Unit	0	0	0	0	0
impulse →	1	0	1	0	0
	0	0	0.9	1	0
	0	1	−1	0.9	1
	0	0	−1.629	−1	0.9
	0	0	−0.656	−1.629	−1
	0	0	0.729	−0.656	−1.629
	0	0	1.187	0.729	−0.656
	0	0	0.478	1.187	0.729
	0	0	−0.53	0.478	1.187
	0	0	−0.86	−0.53	−0.478
	0	0	−0.349	−0.86	−0.53
	0	0	0.382	−0.349	−0.86
	0	0	0.627	0.382	−0.349
	0	0	0.254	0.627	0.382
	0	0	−0.278	0.254	0.627
	0	0	−0.456	−0.278	0.254
	0	0	−0.185	−0.456	−0.278
	0	0	0.202	−0.185	−0.456
	0	0	0.332	0.202	−0.185
	0	0	0.135	0.332	0.202
	0	0	−0.147	0.135	0.332

Figure 3.66 (d) The response of the filter of (b) to a unit impulse has been tabulated by using the recurrence relation. The filter rings with a period of six samples, as expected.

As with the previous example, the frequency response is computed by multiplying the vector lengths, and it will be seen in (c) to have the desired response. The impulse response has been computed for a single non-zero input sample in Figure 3.66(d), and this will be seen to ring with a period of six samples as expected.

3.28 Higher-order filters: cascading

In the example above, the calculations were performed with precision, and the result was as desired. In practice, the coefficients will be represented by a finite wordlength, which means that some inaccuracy will be unavoidable. Owing to the recursion of previous calculations, IIR filters are sensitive to the accuracy of the coefficients, and the higher the order of the filter, the more sensitivity will be shown. In the worst case, a stable filter with a pole near the unit circle may become unstable if the coefficients are represented to less than the required accuracy. Whilst it is possible to design high-order digital filters with a response fixed for a given application, programmable filters of the type required for audio are seldom attempted above the second order to avoid undue coefficient sensitivity. The same response as for a higher-order filter can be obtained by cascading second-order filter sections. For certain applications, such as graphic equalizers, filter sections might be used in parallel.

A further issue which demands attention is the effect of truncation of the wordlength within the data path of the filter. Truncation of coefficients causes only a fixed change in the filter performance which can be calculated. The same cannot be said for data-path truncation. When a sample is multiplied by a coefficient, the necessary wordlength increases dramatically. In a recursive filter the output of one multiplication becomes the input to the next, and so on, making the theoretical wordlength required infinite. By definition, the required wordlength is not available within a realizable filter. Some low-order bits of the product will be lost, which causes noise or distortion depending on the input signal. A series cascade will produce more noise of this kind than a parallel implementation.[18]

In some cases, truncation can cause oscillation. Consider a recursive decay following a large input impulse. Successive output samples become smaller and smaller, but if truncation takes place, the sample may be coarsely quantized as it becomes small, and at some point will not be the correct proportion of the previous sample. In an extreme case, the decay may reverse, and the output samples will grow in magnitude until they are great enough to be represented accurately in the truncation, when they will again decay. The filter is then locked in an endless loop known as a limit cycle.[19] It is a form of instability which cannot exist on large signals because the larger a signal becomes, the smaller the effect of a given truncation. It can be prevented by the injection of digital dither at an appropriate point in the data path. The randomizing effect of the dither destroys the deterministic effect of truncation and prevents the occurrence of the limit cycle.

In the opposite case from truncation due to losing low-order sample bits, products are also subject to overflow if sufficient high-order bits are not available after a multiplication.[20] A simple overflow results in a wraparound of the sample value, and is most disturbing as well as being a possible source of large-amplitude limit cycles. The clipping or saturating adders described in Chapter 3 find an application in digital filters, since the output clips or limits instead of wrapping, and limit cycles are prevented. In order to balance the requirements of truncation and saturation, the output of one stage may be shifted one or more binary places before entering the next stage. This process is known as scaling, since shifting down in binary divides by powers of two; it can be used to prevent overflow. Conversely if the coefficients in use dictate that the high-order bits of a given multiplier output will never be exercised, a shift up may be used to reduce the effects of truncation.

The configuration shown in Figure 3.66 is not the only way of implementing a two-pole two-zero filter. Figures 3.67 shows some alternatives. Starting with the direct form 2 filter, the delays are exchanged with the adders to give a structure which is sometimes referred to as the canonic form.

Filters can also be transposed to yield a different structure with the same transfer function. Transposing is done by reversing signal flow in all the branches, the delays and the multipliers, and replacing nodes in the flow with adders and vice versa. Coefficients and delay lengths are unchanged. The transposed configurations are shown in Figure 3.67(b) and (c). The transposed direct form 2 filter has advantages for audio use,[21] since it has less tendency to problems with overflow, and can be made so that the dominant truncation takes place at one node, which eases the avoidance of limit cycles.

3.29 Pole/zero positions

Since the coefficients of a digital filter are all binary numbers of finite wordlength, it follows that there must be a finite number of positions of poles and zeros in the z-plane. For audio use, the frequencies at which the greatest control is required are usually small compared to the sampling frequency. In presence filters, the requirement for a sharp peak places the poles near to the unit circle, and the low frequencies used emphasize the area adjacent to unity on the real axis. For maximum flexibility of

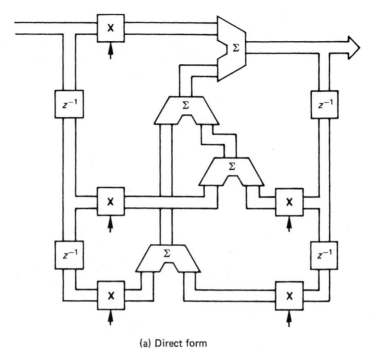

(a) Direct form

Figure 3.67 The same filter can be realized in the direct and canonical forms, (a) and (b), and each can be transposed, (c) and (d).

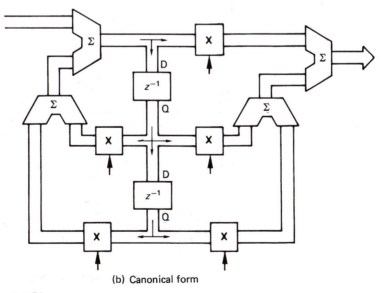

(b) Canonical form

Figure 3.67(b)

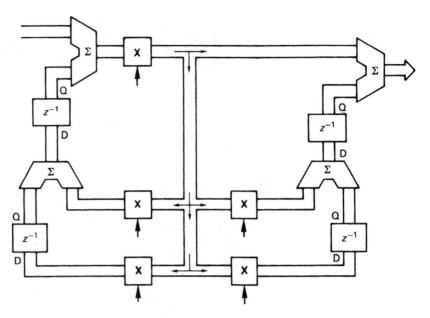

(c) Transposed direct form

Figure 3.67(c)

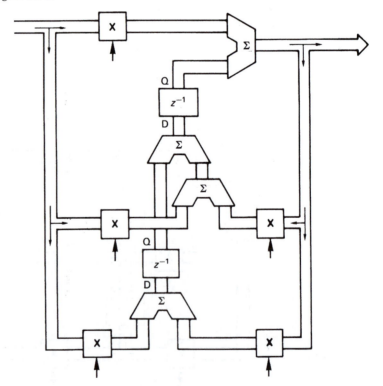

(d) Transposed canonical form

Figure 3.67(d)

response, a large number of pole and zero positions are needed in this area for given coefficient wordlengths. The structure of the filter has a great bearing on the pole and zero positions available.

Canonic structures result in highly non-uniform distributions of available pole positions, and the direct form is little better.[22] Figure 3.68 shows a comparison of direct form and coupled form, where the latter has a uniform pole distribution. The transfer function of the arrangement of Figure 3.68(a) is given by:

$$H[z] = \frac{1}{1 + az^{-1} + bz^{-2}}$$

Reference to section 3.27 will show that

$$a = -2r\cos\theta T \text{ and } b = r^2$$

When poles are near to unit radius, r is nearly 1, and so also is coefficient b. When ωT is small, a will be nearly -2. In this case a more accurate representation of the coefficients can be had by expressing them as the difference between the wanted values and 1 and -2 respectively. Thus b becomes $1-b'$ and a becomes $2-a'$. Since a' and b' are small, representing them with a reasonable wordlength means that the low-order bit represents a much smaller quantizing step with respect to unity. This is the approach of the Agarwal–Burrus filter structure.[23]

Figure 3.69 shows the equivalent of Figure 3.68(a) where the use of difference coefficients can be seen. The multiplication by -1 requires only complementing, and by -2 requires a single-bit shift in addition.

An alternative method of providing accurate coefficients is to transform the z-plane with a horizontal shift, as in Figure 3.70(a). A new z' plane is defined, where the origin is at unity on the real z-axis:

$$z' = z - 1 \text{ and } z'^{-1} = \frac{1}{z-1} = \frac{z^{-1}}{1 - z-1}$$

Figure 3.70(b) shows that the above expression is used in the realization of a z' stage, which will be seen to be a digital integrator. The general expression for a biquadratic section is given in (c). $z - 1$ is replaced by $1/(z' + 1)$ throughout, and the expression is multiplied out. It will be seen that the z-plane coefficients gathered in brackets can be replaced by z'-plane coefficients. When poles and zeros are required near $\text{Re}[z] = 1$, it will be found that all the z' coefficients are small, allowing accurate multiplication with short-coefficient wordlengths. Figure 3.70(d) shows a filter constructed in this way. These configurations demonstrate their accuracy with minimal truncation noise and limit cycles.

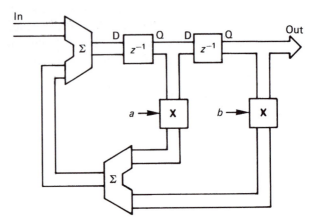

Figure 3.68 (a) A second-order, direct-form digital filter, with coefficients quantized to three bits (not including the sign bit). This means that each coefficient can have only eight values. As the pole radius is the square root of the coefficient b the pole distribution is non-uniform, and poles are not available in the area near to Re[z] = 1 which is of interest in audio.

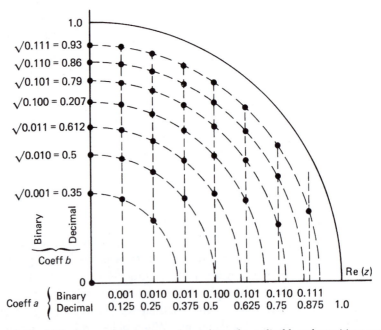

Figure 3.68 (b) In the coupled structure shown here, the realizable pole positions are now on a uniform grid, with the advantage of more pole positions near to Re[z] = 1, but with the penalty that more processing is required.

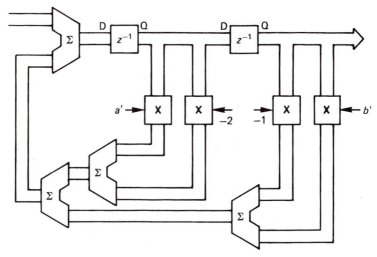

Figure 3.69 In the Agarwal–Burrus filter structure, advantage is taken of the fact that the coefficients are nearly 1 and 2 when the poles are close to unit radius. The coefficients actually used are the difference between these round numbers and the desired value. This configuration should be compared with that of Figure 3.68(a).

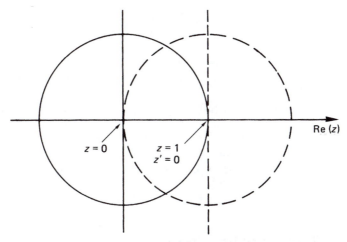

Figure 3.70 (a) By defining a new z'-plane whose origin is at $Re[z] = 1$, small coefficients in z' will correspond to poles near to $z = 1$.

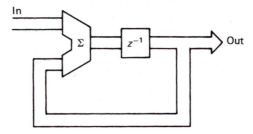

Figure 3.70 (b) The above configuration implements a z'^{-1} transfer function. It is in fact a digital integrator.

$$H[z] = \frac{a_0 + a_1 z^{-1} + a_2 z^{-2}}{1 + b_1 z^{-1} + b_2 z^{-2}}$$

Multiplying by z^2 throughout:

$$H[z] = \frac{a_0 z^2 + a_1 z + a_2}{z^2 + b_1 z + b_2}$$

Since $z = z' + 1$, then $z^2 = z'^2 = z'^2 + 2z' + 1$. Therefore:

$$H[z] = \frac{a_0(z'^2 + 2z' + 1) + a_1(z' + 1) + a_2}{z'^2 + 2z' + 1 + b_1(z' + 1) + b_2}$$

$$= \frac{a_0 z'^2 + 2a_0 z' + a_0 + a_1 z' + a_1 + a_2}{z'^2 + 2z' + 1 + b_1 z' + b_1 + b_2}$$

$$= \frac{a_0 + 2a_0 z'^{-1} + a_0 z'^{-2} + a_1 z^{-1} + a_1 z'^{-2} + a_2 z'^{-2}}{1 + 2z'^{-1} + z'^{-2} + b_2 z'^{-1} + b_1 z'^{-2} + b_2 z'^{-2}}$$

$$= \frac{a_0 + (2a_0 + a_1)z'^{-1} + (a_0 + a_1 + a_2) z'^{-2}}{1 + (2 + b_1)z'^{-1} + (1 + b_1 + b_2) z'^{-2}}$$

$$= \frac{a'_0 + a'_1 z'^{-1} + a'_2 z'^{-2}}{1 + b'_1 z'^{-1} + b'_{22} z'^{-2}}$$

where $\quad a' = a_0 \quad a'_1 = 2a_0 + a_1 \quad a'_2 = a_0 + a_1 + a_2$

$$b'_1 = 2 + b_1 \quad b'_2 = 1 + b_1 + b_2$$

Figure 3.70 (c) Conversion to the z'-plane requires the coefficients in z to be combined as shown here.

3.30 The Fourier transform

Figure 3.39 showed that if the amplitude and phase of each frequency component is known, linearly adding the resultant components in an inverse transform results in the original waveform. In digital systems the waveform is expressed as a number of discrete samples. As a result the Fourier transform analyses the signal into an equal number of discrete frequencies. This is known as a discrete Fourier transform or DFT in which the number of frequency coefficients is equal to the number of input samples. The fast Fourier transform is no more than an efficient way of computing the DFT.[24] As was seen in the previous section, practical systems must use windowing to create short-term transforms.

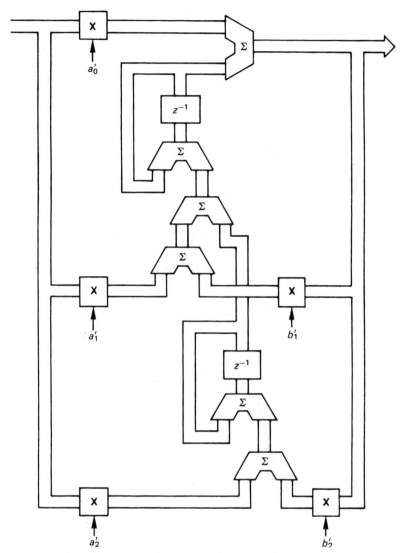

Figure 3.70 (d) Implementation of a z'-plane filter. The z'^{-1} sections can be seen at centre. As the coefficients are all small, scaling (not shown) must be used.

It will be evident from Figure 3.71 that the knowledge of the phase of the frequency component is vital, as changing the phase of any component will seriously alter the reconstructed waveform. Thus the DFT must accurately analyse the phase of the signal components.

There are a number of ways of expressing phase. Figure 3.72 shows a point which is rotating about a fixed axis at constant speed. Looked at from the side, the point oscillates up and down at constant frequency. The waveform of that motion is a sine wave, and that is what we would see if the rotating point were to translate along its axis whilst we continued to look from the side.

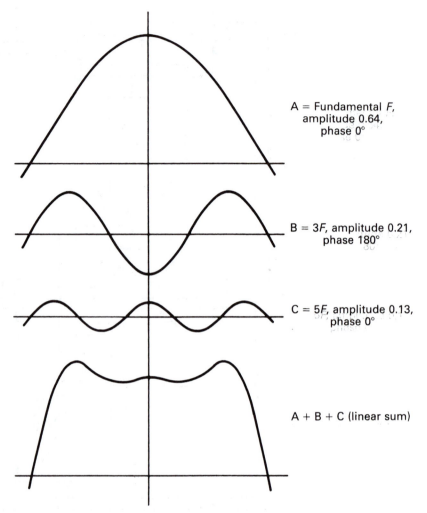

A = Fundamental *F*, amplitude 0.64, phase 0°

B = 3*F*, amplitude 0.21, phase 180°

C = 5*F*, amplitude 0.13, phase 0°

A + B + C (linear sum)

Figure 3.71 Fourier analysis allows the synthesis of any waveform by the addition of discrete frequencies of appropriate amplitude and phase.

One way of defining the phase of a waveform is to specify the angle through which the point has rotated at time zero ($T = 0$). If a second point is made to revolve at 90 degrees to the first, it would produce a cosine wave when translated. It is possible to produce a waveform having arbitrary phase by adding together the sine and cosine wave in various proportions and polarities. For example, adding the sine and cosine waves in equal proportion results in a waveform lagging the sine wave by 45 degrees.

Figure 3.72 shows that the proportions necessary are respectively the sine and the cosine of the phase angle. Thus the two methods of describing phase can be readily interchanged.

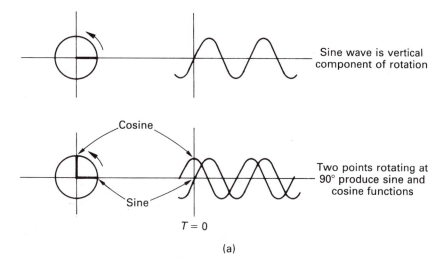

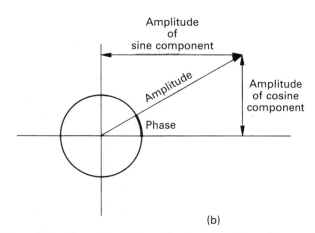

Figure 3.72 The origin of sine and cosine waves is to take a particular viewpoint of a rotation. Any phase can be synthesized by adding proportions of sine and cosine waves.

The discrete Fourier transform spectrum-analyses a string of samples by searching separately for each discrete target frequency. It does this by multiplying the input waveform by a sine wave, known as the basis function, having the target frequency and adding up or integrating the products. Figure 3.73(a) shows that multiplying by basis functions gives a non-zero integral when the input frequency is the same, whereas (b) shows that with a different input frequency (in fact all other different frequencies) the integral is zero showing that no component of the target frequency exists. Thus from a real waveform containing many frequencies all frequencies except the target frequency are excluded. The magnitude of the integral is proportional to the amplitude of the target component.

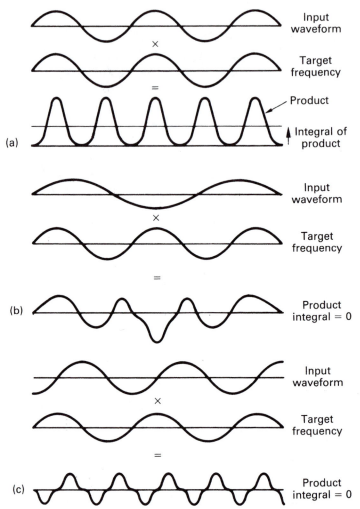

Figure 3.73 The input waveform is multiplied by the target frequency and the result is averaged or integrated. At (a) the target frequency is present and a large integral results. With another input frequency the integral is zero as at (b). The correct frequency will also result in a zero integral shown at (c) if it is at 90° to the phase of the search frequency. This is overcome by making two searches in quadrature.

Figure 3.73(c) shows that the target frequency will not be detected if it is phase shifted 90 degrees as the product of quadrature waveforms is always zero. Thus the discrete Fourier transform must make a further search for the target frequency using a cosine basis function. It follows from the arguments above that the relative proportions of the sine and cosine integrals reveals the phase of the input component. Thus each discrete frequency in the spectrum must be the result of a pair of quadrature searches.

Searching for one frequency at a time as above will result in a DFT, but only after considerable computation. However, a lot of the calculations are repeated many times over in different searches. The fast Fourier transform gives the same result with less computation by logically gathering together all the places where the same calculation is needed and making the calculation once.

The amount of computation can be reduced by performing the sine and cosine component searches together. Another saving is obtained by

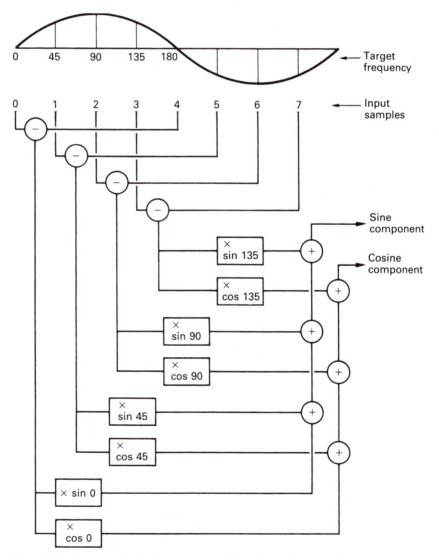

Figure 3.74 An example of a filtering search. Pairs of samples are subtracted and multiplied by sampled sine and cosine waves. The products are added to give the sine and cosine components of the search frequency.

noting that every 180 degrees the sine and cosine have the same
magnitude but are simply inverted in sign. Instead of performing four
multiplications on two samples 180 degrees apart and adding the pairs of
products it is more economical to subtract the sample values and
multiply twice, once by a sine value and once by a cosine value.

The first coefficient is the arithmetic mean which is the sum of all the
sample values in the block divided by the number of samples. Figure 3.74
shows how the search for the lowest frequency in a block is performed.
Pairs of samples are subtracted as shown, and each difference is then
multiplied by the sine and the cosine of the search frequency. The process
shifts one sample period, and a new sample pair are subtracted and
multiplied by new sine and cosine factors. This is repeated until all the

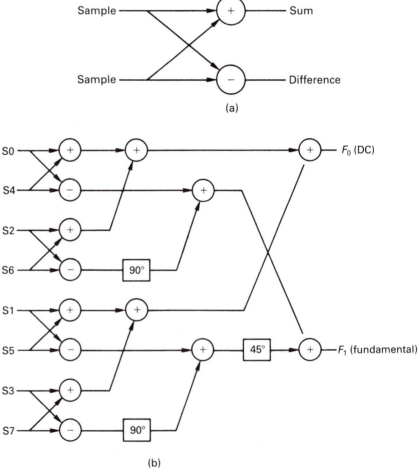

Figure 3.75 The basic element of an FFT is known as a butterfly as at (a) because of the
shape of the signal paths in a sum and difference system. The use of butterflies to
compute the first two coefficients is shown in (b).

sample pairs have been multiplied. The sine and cosine products are then added to give the value of the sine and cosine coefficients respectively.

It is possible to combine the calculation of the DC component which requires the sum of samples and the calculation of the fundamental which requires sample differences by combining stages shown in Figure 3.75(a) which take a pair of samples and add and subtract them. Such a stage is called a butterfly because of the shape of the schematic. Figure 3.75(b) shows how the first two components are calculated. The phase rotation boxes attribute the input to the sine or cosine component outputs according to the phase angle. As shown the box labelled 90 degrees attributes nothing to the sine output, but unity gain to the cosine output. The 45 degree box attributes the input equally to both components.

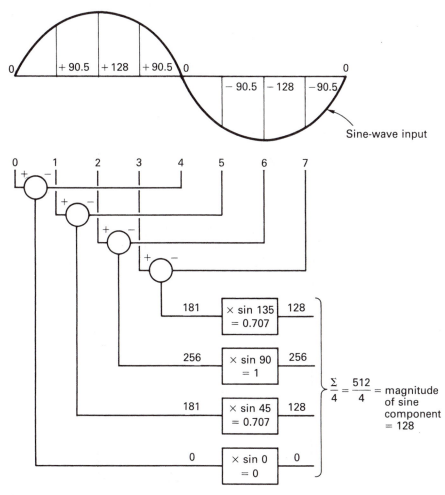

Figure 3.75 (c) An actual calculation of a sine coefficient. This should be compared with the result shown in (d).

Figure 3.75(c) shows a numerical example. If a sinewave input is considered where zero degrees coincides with the first sample, this will produce a zero sine coefficient and non-zero cosine coefficient. Figure 3.75(d) shows the same input waveform shifted by 90 degrees. Note how the coefficients change over.

Figure 3.75(e) shows how the next frequency coefficient is computed. Note that exactly the same first-stage butterfly outputs are used, reducing the computation needed.

A similar process may be followed to obtain the sine and cosine coefficients of the remaining frequencies. The full FFT diagram for eight samples is shown in Figure 3.76(a). The spectrum this calculates is shown in (b). Note that only half of the coefficients are useful in a real band-limited system because the remaining coefficients represent frequencies above one half of the sampling rate.

In short-time Fourier transforms (STFTs) the overlapping input sample blocks must be multiplied by window functions. The principle is the

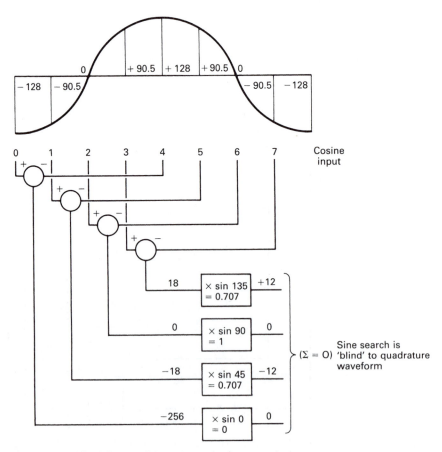

Figure 3.75 (d) With a quadrature input the frequency is not seen.

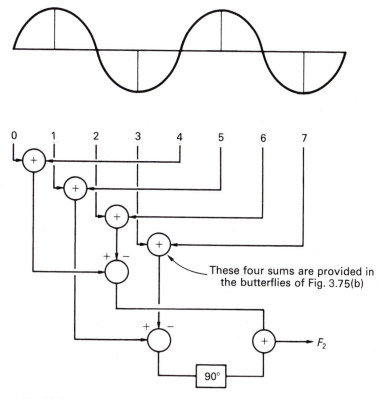

Figure 3.75 (e) The butterflies used for the first coefficients form the basis of the computation of the next coefficient.

same as for the application in FIR filters shown in section 3.23. Figure 3.77 shows that multiplying the search frequency by the window has exactly the same result except that this need be done only once and much computation is saved. Thus in the STFT the basis function is a windowed sine or cosine wave.

The FFT is used extensively in such applications as phase correlation, where the accuracy with which the phase of signal components can be analysed is essential. It also forms the foundation of the discrete cosine transform.

3.31 The discrete cosine transform (DCT)

The DCT is a special case of a discrete Fourier transform in which the sine components of the coefficients have been eliminated leaving a single number. This is actually quite easy. Figure 3.78(a) shows a block of input samples to a transform process. By repeating the samples in a

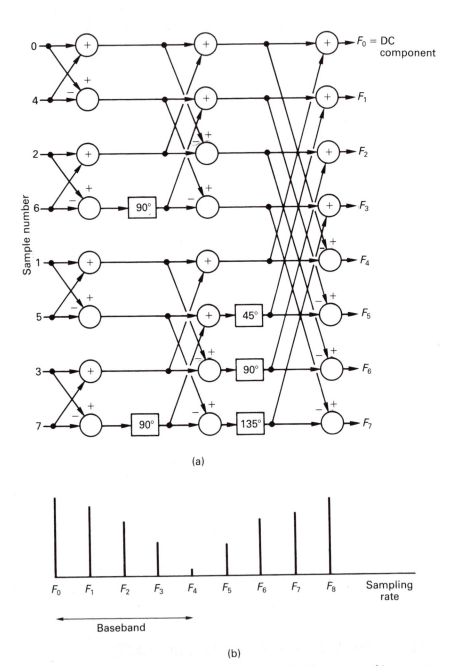

(a)

(b)

Figure 3.76 At (a) is the full butterfly diagram for an FFT. The spectrum this computes is shown at (b).

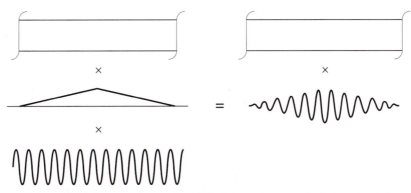

Figure 3.77 Multiplication of a windowed block by a sine wave basis function is the same as multiplying the raw data by a windowed basis function but requires less multiplication as the basis function is constant and can be pre-computed.

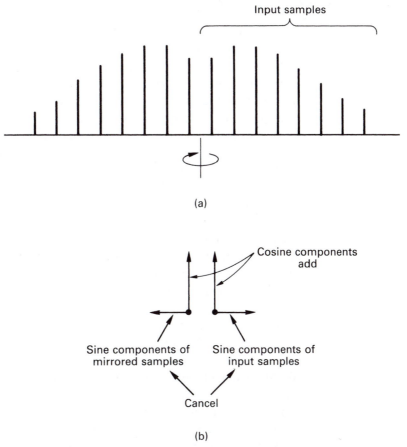

Figure 3.78 The DCT is obtained by mirroring the input block as shown at (a) prior to an FFT. The mirroring cancels out the sine components as at (b), leaving only cosine coefficients.

time-reversed order and performing a discrete Fourier transform on the double-length sample set a DCT is obtained. The effect of mirroring the input waveform is to turn it into an even function whose sine coefficients are all zero. The result can be understood by considering the effect of individually transforming the input block and the reversed block.

Figure 3.78(b) shows that the phase of all the components of one block are in the opposite sense to those in the other. This means that when the components are added to give the transform of the double length block all the sine components cancel out, leaving only the cosine coefficients, hence the name of the transform.[25] In practice the sine component calculation is eliminated. Another advantage is that doubling the block length by mirroring doubles the frequency resolution, so that twice as many useful coefficients are produced. In fact a DCT produces as many useful coefficients as input samples.

3.32   The wavelet transform

The wavelet transform was not discovered by any one individual, but has evolved via a number of similar ideas and was only given a strong mathematical foundation relatively recently.[26–29] The wavelet transform is similar to the Fourier transform in that it has basis functions of various frequencies which are multiplied by the input waveform to identify the frequencies it contains. However, the Fourier transform is based on periodic signals and endless basis functions and requires windowing. The wavelet transform is fundamentally windowed, as the basis functions employed are not endless sine waves, but are finite on the time axis; hence the name. Wavelet transforms do not use a fixed window, but instead the window period is inversely proportional to the frequency being analysed. As a result a useful combination of time and frequency resolutions is obtained. High frequencies corresponding to transients in audio or edges in video are transformed with short basis functions and therefore are accurately located. Low frequencies are transformed with long basis functions which have good frequency resolution.

Figure 3.79 shows that that a set of wavelets or basis functions can be obtained simply by scaling (stretching or shrinking) a single wavelet on the time axis. Each wavelet contains the same number of cycles such that as the frequency reduces, the wavelet gets longer. Thus the frequency discrimination of the wavelet transform is a constant fraction of the signal frequency. In a filter bank such a characteristic would be described as 'constant Q'. Figure 3.80 shows the division of the frequency domain by a wavelet transform is logarithmic whereas in the Fourier transform the division is uniform. The logarithmic coverage is effectively dividing the frequency domain into octaves and as such parallels the frequency discrimination of human hearing.

Fourier transform   Wavelet transform

Figure 3.79 Unlike discrete Fourier transforms, wavelet basis functions are scaled so that they contain the same number of cycles irrespective of frequency. As a result their frequency discrimination ability is a constant proportion of the centre frequency.

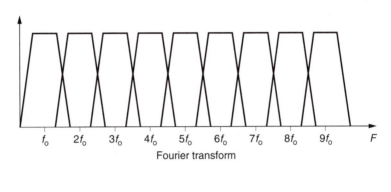

Fourier transform

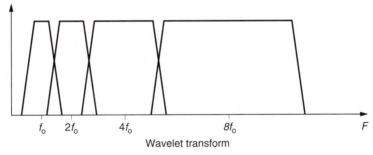

Wavelet transform

Figure 3.80 Wavelet transforms divide the frequency domain into octaves instead of the equal bands of the Fourier transform.

As it is relatively recent, the wavelet transform has yet to be widely used although it shows great promise. It has been successfully used in audio and in commercially available non-linear video editors and in other fields such as radiology and geology.

3.33   Modulo-*n* arithmetic

Conventional arithmetic which is in everyday use relates to the real world of counting actual objects, and to obtain correct answers the concepts of borrow and carry are necessary in the calculations.

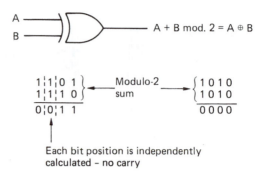

Each bit position is independently
calculated – no carry

Figure 3.81 In modulo-2 calculations, there can be no carry or borrow operations and conventional addition and subtraction become identical. The XOR gate is a modulo-2 adder.

There is an alternative type of arithmetic which has no borrow or carry which is known as modulo arithmetic. In modulo-n no number can exceed n. If it does, n or whole multiples of n are subtracted until it does not. Thus 25 modulo-16 is 9 and 12 modulo-5 is 2. The count shown in Figure 3.81 is from a four-bit device which overflows when it reaches 1111 because the carry-out is ignored. If a number of clock pulses m are applied from the zero state, the state of the counter will be given by m mod.16. Thus modulo arithmetic is appropriate to systems in which there is a fixed wordlength and this means that the range of values the system can have is restricted by that wordlength. A number range which is restricted in this way is called a finite field.

Modulo-2 is a numbering scheme which is used frequently in digital processes. Figure 3.81 also shows that in modulo-2 the conventional addition and subtraction are replaced by the XOR function such that: A + B Mod.2 = A XOR B. When multi-bit values are added Mod.2, each column is computed quite independently of any other. This makes Mod.2 circuitry very fast in operation as it is not necessary to wait for the carries from lower-order bits to ripple up to the high-order bits.

Modulo-2 arithmetic is not the same as conventional arithmetic and takes some getting used to. For example, adding something to itself in Mod.2 always gives the answer zero.

3.34 The Galois field

Figure 3.82 shows a simple circuit consisting of three D-type latches which are clocked simultaneously. They are connected in series to form a shift register. At (a) a feedback connection has ben taken from the output to the input and the result is a ring counter where the bits contained will recirculate endlessly. At (b) one XOR gate is added so that the output is

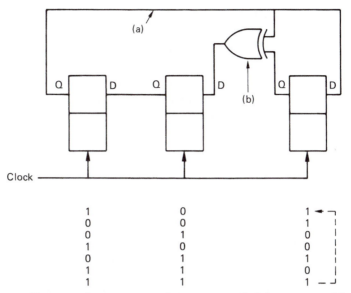

Figure 3.82 The circuit shown is a twisted-ring counter which has an unusual feedback arrangement. Clocking the counter causes it to pass through a series of non-sequential values. See text for details.

fed back to more than one stage. The result is known as a twisted-ring counter and it has some interesting properties. Whenever the circuit is clocked, the left-hand bit moves to the right-hand latch, the centre bit moves to the left-hand latch and the centre latch becomes the XOR of the two outer latches. The figure shows that whatever the starting condition of the three bits in the latches, the same state will always be reached again after seven clocks, except if zero is used.

The states of the latches form an endless ring of non-sequential numbers called a Galois field after the French mathematical prodigy Evariste Galois who discovered them. The states of the circuit form a maximum-length sequence because there are as many states as are permitted by the wordlength. As the states of the sequence have many of the characteristics of random numbers, yet are repeatable, the result can also be called a pseudo-random sequence (prs). As the all-zeros case is disallowed, the length of a maximum length sequence generated by a register of m bits cannot exceed (2^m-1) states. The Galois field, however includes the zero term. It is useful to explore the bizarre mathematics of Galois fields which use modulo-2 arithmetic. Familiarity with such manipulations is helpful when studying the error correction, particularly the Reed–Solomon codes used in recorders and treated in Chapter 6. They will also be found in processes which require pseudo-random numbers such as digital dither, treated in section 3.14, and randomized channel codes used in, for example, NICAM 728 and discussed in Chapters 6 and 8.

The circuit of Figure 3.82 can be considered as a counter and the four points shown will then be representing different powers of 2 from the MSB on the left to the LSB on the right. The feedback connection from the MSB to the other stages means that whenever the MSB becomes 1, two other powers are also forced to one so that the code of 1011 is generated.

Each state of the circuit can be described by combinations of powers of x, such as

$$x^2 = 100,$$

$$x = 010,$$

$$x^2 + x = 110, \text{ etc.}$$

The fact that three bits have the same state because they are connected together is represented by the Mod.2 equation:

$$x^3 + x + 1 = 0$$

Let $x = a$, which is a primitive element. Now

$$a^3 + a + 1 = 0 \tag{1}$$

In modulo 2

$$a + a = a^2 + a^2 = 0$$

$$a = x = 010$$

$$a^2 = x^2 = 100$$

$$a^3 = a + 1 \quad = 011 \text{ from (1)}$$

$$a^4 = a \times a^3 = a(a + 1) = a^2 + a = 110$$

$$a^5 = a^2 + a + 1 = 111$$

$$a^6 = a \times a^5 = a(a^2 + a + 1)$$

$$= a^3 + a^2 + a = a + 1 + a^2 + a$$

$$= a^2 + 1 = 101$$

$$a^7 = a(a^2 + 1) = a^3 + a$$

$$= a + 1 + a = 1 = 001$$

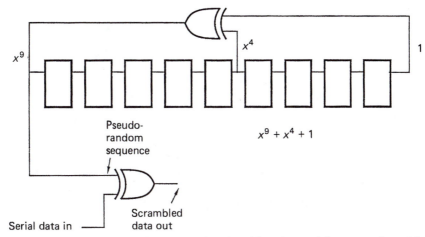

Figure 3.83 In NICAM, randomizing is done by adding the serial data to a polynomial generated by the circuit shown here. The receiver needs an identical system which is synchronous with that of the transmitter.

In this way it can be seen that the complete set of elements of the Galois field can be expressed by successive powers of the primitive element. Note that the twisted-ring circuit of Figure 3.82 simply raises *a* to higher and higher powers as it is clocked; thus the seemingly complex multibit changes caused by a single clock of the register become simple to calculate using the correct primitive and the appropriate power.

The numbers produced by the twisted-ring counter are not random; they are completely predictable if the equation is known. However, the sequences produced are sufficiently similar to random numbers that in many cases they will be useful. They are thus referred to as pseudo-random sequences. The feedback connection is chosen such that the expression it implements will not factorize. Otherwise a maximum-length sequence could not be generated because the circuit might sequence around one or other of the factors depending on the initial condition. A useful analogy is to compare the operation of a pair of meshed gears. If the gears have a number of teeth which is relatively prime, many revolutions are necessary to make the same pair of teeth touch again. If the number of teeth have a common multiple, far fewer turns are needed.

Figure 3.83 shows the pseudo-random sequence generator used in NICAM 728. Its purpose is to break up the transmitted spectrum so that the sound carrier does not cause patterning on the TV picture. The sequence length of the circuit shown is 511 because the expression will not factorize. Further details of NICAM 728 can be found in Chapter 8.

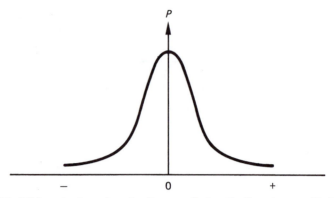

Figure 3.84 White noise in analog circuits generally has the Gaussian amplitude distribution.

3.35   Noise and probability

Probability is a useful concept when dealing with processes which are not completely predictable. Thermal noise in electronic components is random, and although under given conditions the noise power in a system may be constant, this value only determines the heat that would be developed in a resistive load. In digital systems, it is the instantaneous voltage of noise which is of interest, since it is a form of interference which could alter the state of a binary signal if it were large enough. Unfortunately the instantaneous voltage cannot be predicted; indeed if it could the interference could not be called noise. Noise can only be quantified statistically, by measuring or predicting the likelihood of a given noise amplitude.

Figure 3.84 shows a graph relating the probability of occurrence to the amplitude of noise. The noise amplitude increases away from the origin along the horizontal axis, and for any amplitude of interest, the probability of that noise amplitude occurring can be read from the curve. The shape of the curve is known as a Gaussian distribution, which crops up whenever the overall effect of a large number of independent phenomena is considered. Thermal noise is due to the contributions from countless molecules in the component concerned. Magnetic recording depends on superimposing some average magnetism on vast numbers of magnetic particles.

If it were possible to isolate an individual noise-generating microcosm of a tape or a head on the molecular scale, the noise it could generate would have physical limits because of the finite energy present. The noise distribution might then be rectangular as shown in Figure 3.85(a), where all amplitudes below the physical limit are equally likely. The output of a twisted-ring counter such as that in Figure 3.82 can have a uniform

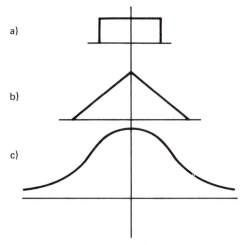

Figure 3.85 At (a) is a rectangular probability; all values are equally likely but between physical limits. At (b) is the sum of two rectangular probabilities, which is triangular, and at (c) is the Gaussian curve which is the sum of an infinite number of rectangular probabilities.

probability. Each value occurs once per sequence. The outputs are positive only but do not include zero, but every value from 1 up to 2^{n-1} is then equally likely.

The output of a prs generator can be made into the two's complement form by inverting the MSB. This has the effect of exchanging the missing all zeros value for a missing fully negative value as can be seen by considering the number ring in Figure 3.5. In this example, inverting the MSB causes the code of 1000 representing −8 to become 0000. The result is a four-bit prs generating uniform probability from −7 to +7 as shown in Figure 3.85(a).

If the combined effect of two of these uniform probability processes is considered, clearly the maximum amplitude is now doubled, because the two effects can add, but provided the two effects are uncorrelated, they can also subtract, so the probability is no longer rectangular, but becomes triangular as in Figure 3.85(b). The probability falls to zero at peak amplitude because the chances of two independent mechanisms reaching their peak value with the same polarity at the same time are understandably small.

If the number of mechanisms summed together is now allowed to increase without limit, the result is the Gaussian curve shown in Figure 3.85(c), where it will be seen that the curve has no amplitude limit, because it is just possible that all mechanisms will simultaneously reach their peak value together, although the chances of this happening are incredibly remote. Thus the Gaussian curve is the overall probability of a large number of uncorrelated uniform processes.

References

1. Spreadbury, D., Harris, N. and Lidbetter, P. So you think performance is cracked using standard floating point DSPs? *Proc. 10th. Int. AES Conf.*, 105–110 (1991)
2. Richards, J.W., Digital audio mixing. *The Radio and Electron. Eng.*, **53**, 257–264 (1983)
3. Richards, J.W. and Craven, I., An experimental 'all digital' studio mixing desk. *J. Audio Eng. Soc.*, **30**, 117–126 (1982)
4. Jones, M.H., Processing systems for the digital audio studio. In *Digital Audio*, edited by B. Blesser, B. Locanthi and T.G. Stockham Jr, pp. 221–225, New York: Audio Engineering Society (1982)
5. Lidbetter, P.S., A digital delay processor and its applications. Presented at the 82nd Audio Engineering Society Convention (London, 1987), Preprint 2474(K-4)
6. McNally, G.J., COPAS A high speed real time digital audio processor. *BBC Research Dept Report*, RD 1979/26
7. McNally, G.W., Digital audio: COPAS-2, a modular digital audio signal processor for use in a mixing desk. *BBC Research Dept Report*, RD 1982/13
8. Vandenbulcke, C. *et al.*, An integrated digital audio signal processor. Presented at the 77th Audio Engineering Society Convention (Hamburg, 1985), Preprint 2181(B-7)
9. Moorer, J.A., The audio signal processor: the next step in digital audio. In *Digital Audio*, edited by B. Blesser, B. Locanthi and T.G. Stockham Jr, pp. 205–215, New York: Audio Engineering Society (1982)
10. Gourlaoen, R. and Delacroix, P., The digital sound mixing desk: architecture and integration in the future all-digital studio. Presented at the 80th Audio Engineering Society Convention (Montreux, 1986), Preprint 2327(D-1)
11. Ray, S.F., *Applied Photographic Optics*. Oxford: Focal Press (1988) (Ch. 17)
12. van den Enden, A.W.M. and Verhoeckx, N.A.M., Digital signal processing: theoretical background. *Philips Tech. Rev.*, **42**, 110–144, (1985)
13. McClellan, J.H., Parks, T.W. and Rabiner, L.R., A computer program for designing optimum FIR linear-phase digital filters. *IEEE Trans. Audio and Electroacoustics*, **AU-21**, 506–526 (1973)
14. Dolph, C.L., A current distribution for broadside arrays which optimises the relationship between beam width and side lobe level. *Proc. IRE*, **34**, 335–348 (1946)
15. Crochiere, R.E. and Rabiner, L.R., Interpolation and decimation of digital signals – a tutorial review. *Proc. IEEE*, **69**, 300–331 (1981)
16. Rabiner, L.R., Digital techniques for changing the sampling rate of a signal. In *Digital Audio*, edited by B. Blesser, B. Locanthi and T.G. Stockham Jr, pp. 79–89, New York: Audio Engineering Society (1982)
17. Lagadec, R., Digital sampling frequency conversion. In *Digital Audio*, edited by B. Blesser, B. Locanthi and T.G. Stockham Jr, pp. 90–96, New York: Audio Engineering Society (1982)
18. Jackson, L.B., Roundoff noise analysis for fixed-point digital filters realized in cascade or parallel form. *IEEE Trans. Audio and Electroacoustics*, **AU–18**, 107–122 (1970)
19. Parker, S.R., Limit cycles and correlated noise in digital filters. In *Digital Signal Processing*, Western Periodicals Co., 177–179 (1979)
20. Claasen, T.A.C.M., Mecklenbrauker, W.F.G. and Peek, J.B.H., Effects of quantizing and overflow in recursive digital filters. *IEEE Trans. ASSP*, **24**, 517–529 (1976)
21. McNally, G.J., Digital audio: recursive digital filtering for high-quality audio signals. *BBC Res. Dept Report*, RD 1981/10
22. Rabiner, L.R. and Gold, B., *Theory and Application of Digital Signal Processing*, New Jersey: Prentice Hall (1975)
23. Agarwal, R.C. and Burrus, C.S., New recursive digital filter structures having very low sensitivity and roundoff noise. *IEEE Trans. Circuits. Syst.*, **CAS–22**, 921–927 (1975)
24. Kraniauskas, P., *Transforms in Signals and Systems*, Wokingham: Addison-Wesley (1992)
25. Ahmed, N., Natarajan, T. and Rao, K., Discrete Cosine Transform, *IEEE Trans. Computers*, **C-23** 90–93 (1974)

26. Goupillaud, P., Grossman, A. and Morlet, J., Cycle-Octave and related transforms in seismic signal analysis. *Geoexploration*, **23**, 85–102, Elsevier Science (1984/5)
27. Daubechies, I., The wavelet transform, time–frequency localisation and signal analysis. *IEEE Trans. Info. Theory*, **36**, No.5, 961–1005 (1990)
28. Rioul, O. and Vetterli, M., Wavelets and signal processing. *IEEE Signal Process. Mag.*, 14–38 (Oct. 1991)
29. Strang, G. and Nguyen, T., *Wavelets and Filter Banks*, Wellesly, MA: Wellesley-Cambridge Press (1996)

4

Conversion

Chapter 1 introduced the fundamental characteristic of digital audio which is that the quality is independent of the storage or transmission medium and is determined instead by the accuracy of conversion between the analog and digital domains. This chapter will examine in detail the theory and practice of this critical aspect of digital audio.

4.1   Introduction to conversion

Any analog audio source can be characterized by a given useful bandwidth and signal-to-noise ratio. If a well-engineered digital channel having a wider bandwidth and a greater signal-to-noise ratio is put in series with such a source, it is only necessary to set the levels correctly and the analog signal is then subject to no loss of information whatsoever. The digital clipping level is above the largest analog signal, the digital noise floor is below the inherent noise in the signal and the low- and high-frequency response of the digital channel extends beyond the frequencies in the analog signal.

The digital channel is a 'wider window' than the analog signal needs and its extremities cannot be explored by that signal. As a result there is no test known which can reliably tell whether or not the digital system was or was not present, unless, of course, it is deficient in some quantifiable way.

The wider-window effect is obvious on certain Compact Discs which are made from analog master tapes. The CD player faithfully reproduces the tape hiss, dropouts and HF squashing of the analog master, which render the entire CD mastering and reproduction system transparent by comparison.

On the other hand, if an analog source can be found which has a wider window than the digital system, then the digital system will be evident due to either the reduction in bandwidth or the reduction in dynamic range. No analog recorder comes into this category, but certain high-quality capacitor microphones can slightly outperform many digital audio systems in dynamic range and considerably outperform the frequency range of human hearing.

The sound conveyed through a digital system travels as a stream of bits. Because the bits are discrete, it is easy to quantify the flow, just by counting the number per second. It is much harder to quantify the amount of information in an analog signal (from a microphone, for example) but if this were done using the same units, it would be possible to decide just what bit rate was necessary to convey that signal without loss of information, i.e. to make the window just wide enough. If a signal can be conveyed without loss of information, and without picking up any unwanted signals on the way, it will have been transmitted perfectly.

The connection between analog signals and information capacity was made by Shannon, in one of the most significant papers in the history of this technology,[1] and those parts which are important for this subject are repeated here. The principles are straightforward, and offer an immediate insight into the relative performances and potentials of different modulation methods, including digitizing.

Figure 4.1 shows an analog signal with a certain amount of super-imposed noise, as is the case for all real audio signals. Noise is defined as

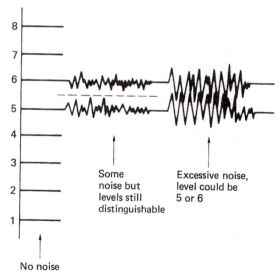

Figure 4.1 To receive eight different levels in a signal unambiguously, the peak-to-peak noise must be less than the difference in level. Signal-to-noise ratio must be at least 8:1 or 18 dB to convey eight levels. This can also be conveyed by three bits ($2^3 = 8$). For 16 levels, SNR would have to be 24 dB, which would be conveyed by four bits.

a random superimposed signal which is not correlated with the wanted signal. To avoid pitfalls in digital audio, this definition must be adhered to with what initially seems like pedantry. The noise is random, and so the actual voltage of the wanted signal is uncertain; it could be anywhere in the range of the noise amplitude. If the signal amplitude is, for the sake of argument, sixteen times the noise amplitude, it would only be possible to convey sixteen different signal levels unambiguously, because the levels have to be sufficiently different that noise will not make one look like another. It is possible to convey sixteen different levels in all combinations of four data bits, and so the connection between the analog and quantized domains is established.

The choice of sampling rate (the rate at which the signal voltage must be examined to convey the information in a changing signal) is important in any system; if it is too low, the signal will be degraded, and if it is too high, the number of samples to be recorded will rise unnecessarily, as will the cost of the system. Here it will be established just what sampling rate is necessary in a given situation, initially in theory, then taking into account practical restrictions. By multiplying the number of bits needed to express the signal voltage by the rate at which the process must be updated, the bit rate of the digital data stream resulting from a particular analog signal can be determined.

There are a number of ways in which an audio waveform can be digitally represented, but the most useful and therefore common is pulse code modulation or PCM which was introduced in Chapter 1. The input is a continuous-time, continuous-voltage waveform, and this is converted into a discrete-time, discrete-voltage format by a combination of sampling and quantizing. As these two processes are orthogonal (a 64-dollar word for at right angles to one another) they are totally independent and can be performed in either order. Figure 4.2(a) shows an analog sampler preceding a quantizer, whereas (b) shows an asynchronous quantizer preceding a digital sampler. Ideally, both will give the same results; in practice each has different advantages and suffers from different deficiencies. Both approaches will be found in real equipment.

The independence of sampling and quantizing allows each to be discussed quite separately in some detail, prior to combining the processes for a full understanding of conversion.

4.2   Sampling and aliasing

Sampling is no more than periodic measurement, and it will be shown here that there is no theoretical need for sampling to be audible. Practical equipment may, of course be less than ideal, but, given good engineering practice, the ideal may be approached quite closely.

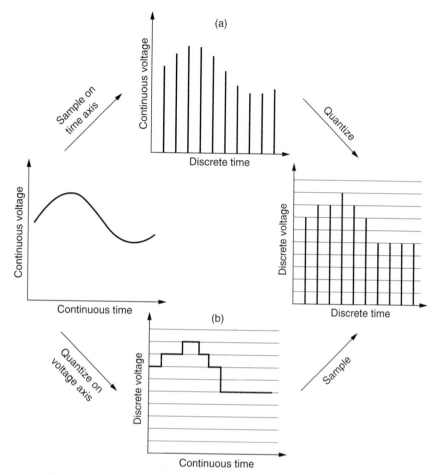

Figure 4.2 Since sampling and quantizing are orthogonal, the order in which they are performed is not important. In (a) sampling is performed first and the samples are quantized. This is common in audio convertors. In (b) the analog input is quantized into an asynchronous binary code. Sampling takes place when this code is latched on sampling clock edges. This approach is universal in video convertors.

Audio sampling must be regular, because the process of timebase correction prior to conversion back to analog assumes a regular original process as was shown in Chapter 1. The sampling process originates with a pulse train which is shown in Figure 4.3(a) to be of constant amplitude and period. The audio waveform amplitude-modulates the pulse train in much the same way as the carrier is modulated in an AM radio transmitter. One must be careful to avoid over-modulating the pulse train as shown in (b) and this is achieved by applying a DC offset to the analog waveform so that silence corresponds to a level half-way up the pulses as at (c). Clipping due to any excessive input level will then be symmetrical.

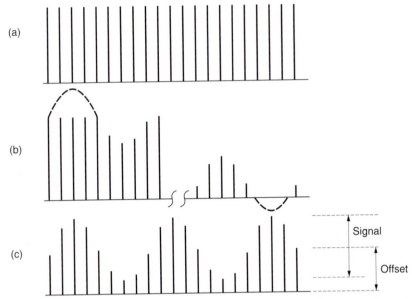

Figure 4.3 The sampling process requires a constant-amplitude pulse train as shown in (a). This is amplitude modulated by the waveform to be sampled. If the input waveform has excessive amplitude or incorrect level, the pulse train clips as shown in (b). For an audio waveform, the greatest signal level is possible when an offset of half the pulse amplitude is used to centre the waveform as shown in (c).

In the same way that AM radio produces sidebands or images above and below the carrier, sampling also produces sidebands although the carrier is now a pulse train and has an infinite series of harmonics as shown in Figure 4.4(a). The sidebands repeat above and below each harmonic of the sampling rate as shown in (b).

The sampled signal can be returned to the continuous-time domain simply by passing it into a low-pass filter. This filter has a frequency response which prevents the images from passing, and only the baseband signal emerges, completely unchanged. If considered in the frequency domain, this filter can be called an anti-image filter; if considered in the time domain it can be called a reconstruction filter.

If an input is supplied having an excessive bandwidth for the sampling rate in use, the sidebands will overlap (Figure 4.4(c)) and the result is aliasing, where certain output frequencies are not the same as their input frequencies but instead become difference frequencies (d)). It will be seen from Figure 4.4 that aliasing does not occur when the input frequency is equal to or less than half the sampling rate, and this derives the most fundamental rule of sampling, which is that the sampling rate must be at least twice the highest input frequency. Sampling theory is usually attributed to Shannon[2] who applied it to information theory at around

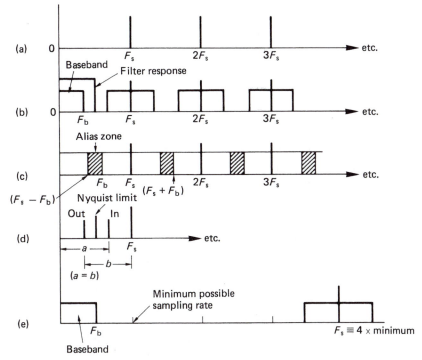

Figure 4.4 (a) Spectrum of sampling pulses. (b) Spectrum of samples. (c) Aliasing due to sideband overlap. (d) Beat-frequency production. (e) 4× oversampling.

the same time as Kotelnikov in Russia. These applications were pre-dated by Whittaker. Despite that it is often referred to as Nyquist's theorem.

Whilst aliasing has been described above in the frequency domain, it can be described equally well in the time domain. In Figure 4.5(a) the sampling rate is obviously adequate to describe the waveform, but at (b) it is inadequate and aliasing has occurred.

Aliasing is commonly seen on television and in the cinema, owing to the relatively low frame rates used. With a frame rate of 24 Hz, a film camera will alias on any object changing at more than 12 Hz. Such objects

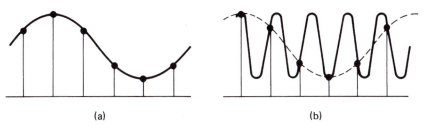

Figure 4.5 In (a) the sampling is adequate to reconstruct the original signal. In (b) the sampling rate is inadequate, and reconstruction produces the wrong waveform (dashed). Aliasing has taken place.

include the spokes of stagecoach wheels. When the spoke-passing frequency reaches 24 Hz the wheels appear to stop. Aliasing does, however, have useful applications, including the stroboscope, which makes rotating machinery appear stationary, the sampling oscilloscope, which can display periodic waveforms of much greater frequency than the sweep speed of the tube normally allows and the spectrum analyser.

One often has no control over the spectrum of input signals and in practice it is necessary also to have a low-pass filter at the input to prevent aliasing. This anti-aliasing filter prevents frequencies of more than half the sampling rate from reaching the sampling stage.

4.3   Reconstruction

If ideal low-pass anti-aliasing and anti-image filters are assumed, having a vertical cut-off slope at half the sampling rate, an ideal spectrum shown in Figure 4.6(a) is obtained. It was shown in Chapter 2 that the impulse response of a phase linear ideal low-pass filter is a $\sin x/x$ waveform in the time domain, and this is repeated in (b). Such a waveform passes through zero volts periodically. If the cut-off frequency of the filter is one-half of the sampling rate, the impulse passes through zero *at the sites of all other samples*. It can be seen from Figure 4.6(c) that at the output of such a filter, the voltage at the centre of a sample is due to that sample alone, since the value of *all* other samples is zero at that instant. In other words the continuous time output waveform must join up the tops of the input samples. In between the sample instants, the output of the filter is the sum of the contributions from many impulses, and the waveform smoothly joins the tops of the samples. If the time domain is being considered, the anti-image filter of the frequency domain can equally well be called the reconstruction filter. It is a consequence of the band-limiting of the original anti-aliasing filter that the filtered analog waveform could only travel between the sample points in one way. As the reconstruction filter has the same frequency response, the reconstructed output waveform must be identical to the original band-limited waveform prior to sampling. It follows that sampling need not be audible. The reservations expressed by some journalists about 'hearing the gaps between the samples' clearly have no foundation whatsoever. A rigorous mathematical proof of reconstruction can be found in Betts.[3]

The ideal filter with a vertical 'brick-wall' cut-off slope is difficult to implement. As the slope tends to vertical, the delay caused by the filter goes to infinity: the quality is marvellous but you don't live to hear it. In practice, a filter with a finite slope has to be accepted as shown in Figure 4.7. The cut-off slope begins at the edge of the required band, and

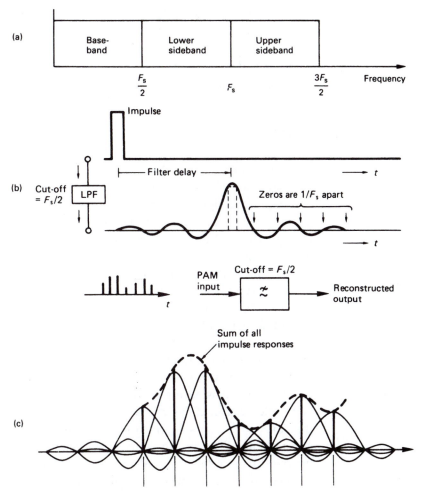

Figure 4.6 If ideal 'brick wall' filters are assumed, the efficient spectrum of (a) results. An ideal low-pass filter has an impulse response shown in (b). The impulse passes through zero at intervals equal to the sampling period. When convolved with a pulse train at the sampling rate, as shown in (c), the voltage at each sample instant is due to that sample alone as the impulses from all other samples pass through zero there.

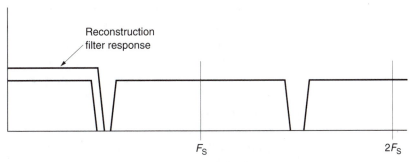

Figure 4.7 As filters with finite slope are needed in practical systems, the sampling rate is raised slightly beyond twice the highest frequency in the baseband.

consequently the sampling rate has to be raised a little to drive aliasing products to an acceptably low level. There is no absolute factor by which the sampling rate must be raised; it depends upon the filters which are available and the level of aliasing products which are acceptable. The latter will depend upon the wordlength to which the signal will be quantized.

4.4   Filter design

The discussion so far has assumed that perfect anti-aliasing and reconstruction filters are used. Perfect filters are not available, of course, and because designers must use devices with finite slope and rejection, aliasing can still occur. It is not easy to specify anti-aliasing filters, particularly the amount of stopband rejection needed. The amount of aliasing resulting would depend on, among other things, the amount of out-of-band energy in the input signal. Very little is known about the energy in typical source material outside the audible range. As a further complication, an out-of-band signal will be attenuated by the response of the anti-aliasing filter to that frequency, but the residual signal will then alias, and the reconstruction filter will reject it according to its attenuation at the new frequency to which it has aliased. To take the opposite extreme, if a microphone were used which had no response at all above the audio band, no anti-aliasing filter would be needed.

It could be argued that the reconstruction filter is unnecessary, since all the images are outside the range of human hearing. However, the slightest non-linearity in subsequent stages would result in gross intermodulation distortion. Most transistorized audio power amplifiers become grossly non-linear when fed with signals far beyond the audio band. It is this non-linearity which enables amplifiers to demodulate strong radio transmissions. The simple solution is to curtail the response of power amplifiers somewhat beyond the audio band so that they become immune to passing taxis and refrigerator thermostats. This is seldom done in Hi-Fi amplifiers because of the mistaken belief that response far beyond the audio band is needed for high fidelity. The truth of the belief is academic as all known recorded or broadcast music sources, whether analog or digital, are band-limited. As a result there is nothing to which a power amplifier of excess bandwidth can respond except RF interference and inadequately suppressed images from digital sources. The possibility of damage to tweeters and beating with the bias systems of analog tape recorders must also be considered.

Consequently a reconstruction filter is a practical requirement. It would, however, be acceptable to bypass one of the filters involved in a copy from one digital machine to another via the analog domain, although a digital transfer is, of course, to be preferred.

Every signal which has been through the digital domain has passed through both an anti-aliasing filter and a reconstruction filter. These filters must be carefully designed in order to prevent artifacts, particularly those due to lack of phase linearity as they may be audible.[4–6] The nature of the filters used has a great bearing on the subjective quality of the system. Entire books have been written about analog filters, so they will only be treated briefly here.

Figures 4.8 and 4.9 show the terminology used to describe the common elliptic low-pass filter. These filters are popular because they can be realized with fewer components than other filters of similar response. It is a characteristic of these elliptic filters that there are ripples in the passband and stopband. Lagadec and Stockham[7] found that filters with passband ripple cause dispersion: the output signal is smeared in time and, on toneburst signals, pre-echoes can be detected. In much equipment the anti-aliasing filter and the reconstruction filter will have the same specification, so that the passband ripple is doubled with a corresponding increase in dispersion. Sometimes slightly different filters are used to reduce the effect.

It is difficult to produce an analog filter with low distortion. Passive filters using inductors suffer non-linearity at high levels due to the B/H curve of the cores. It seems a shame to go to such great lengths to remove the non-linearity of magnetic tape from a recording using digital techniques only to pass the signal through magnetic inductors in the filters. Active filters can simulate inductors which are linear using op-amp techniques, but they tend to suffer non-linearity at high frequencies where the falling open-loop gain reduces the effect of feedback. Active filters can also contribute noise, but this is not necessarily a bad thing in controlled amounts, since it can act as a dither source.

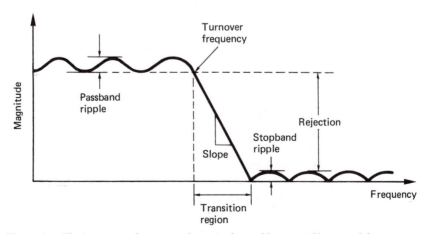

Figure 4.8 The important features and terminology of low-pass filters used for anti-aliasing and reconstruction.

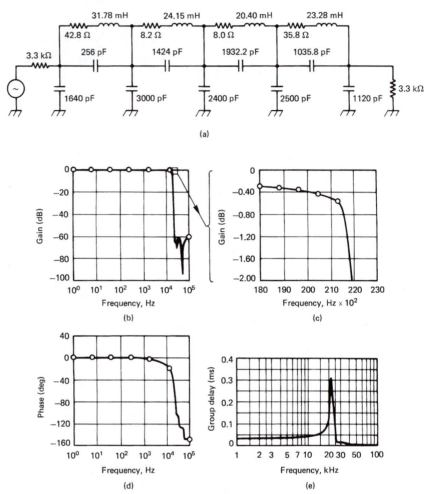

Figure 4.9 (a) Circuit of typical nine-pole elliptic passive filter with frequency response in (b) shown magnified in the region of cut-off in (c). Note phase response in (d) beginning to change at only 1 kHz, and group delay in (e), which require compensation for quality applications. Note that in the presence of out-of-band signals, aliasing might only be 60 dB down. A 13-pole filter manages in excess of 80 dB, but phase response is worse.

It is instructive to examine the phase response of such filters. Since a sharp cut-off is generally achieved by cascading many filter sections which cut at a similar frequency, the phase responses of these sections will accumulate. The phase may start to leave linearity at only a few kiloHertz, and near the cut-off frequency the phase may have completed several revolutions. As stated, these phase errors can be audible and phase equalization is necessary. An advantage of linear phase filters is that ringing is minimized, and there is less possibility of clipping on transients.

It is possible to construct a ripple-free phase-linear filter with the required stopband rejection,[8,9] but it is expensive in terms of design effort and component complexity, and it might drift out of specification as components age. The money may be better spent in avoiding the need for such a filter. Much effort can be saved in analog filter design by using oversampling. Strictly, oversampling means no more than that a higher sampling rate is used than is required by sampling theory. In the loose sense an 'oversampling convertor' generally implies that some combination of high sampling rate and various other techniques has been applied. Oversampling is treated in depth in a later section of this chapter. The audible superiority and economy of oversampling convertors has led them to be almost universal. Accordingly the treatment of oversampling in this volume is more prominent than that of filter design.

4.5   Choice of sampling rate

Sampling theory is only the beginning of the process which must be followed to arrive at a suitable sampling rate. The finite slope of realizable filters will compel designers to raise the sampling rate. For consumer products, the lower the sampling rate, the better, since the cost of the medium is directly proportional to the sampling rate: thus sampling rates near to twice 20 kHz are to be expected. For professional products, there is a need to operate at variable speed for pitch correction. When the speed of a digital recorder is reduced, the offtape sampling rate falls, and Figure 4.10 shows that with a minimal sampling rate the first image frequency can become low enough to pass the reconstruction filter. If the sampling frequency is raised without changing the response of the filters, the speed can be reduced without this problem. It follows that variable-speed recorders, generally those with stationary heads, must use a higher sampling rate.

In the early days of digital audio research, the necessary bandwidth of about 1 megabit per second per audio channel was difficult to store. Disk drives had the bandwidth but not the capacity for long recording time, so attention turned to video recorders. In Chapter 9 it will be seen that these were adapted to store audio samples by creating a pseudo-video waveform which could convey binary as black and white levels. The sampling rate of such a system is constrained to relate simply to the field rate and field structure of the television standard used, so that an integer number of samples can be stored on each usable TV line in the field. Such a recording can be made on a monochrome recorder, and these recordings are made in two standards, 525 lines at 60 Hz and 625 lines at 50 Hz. Thus it is possible to find a frequency which is a common multiple of the two and also suitable for use as a sampling rate.

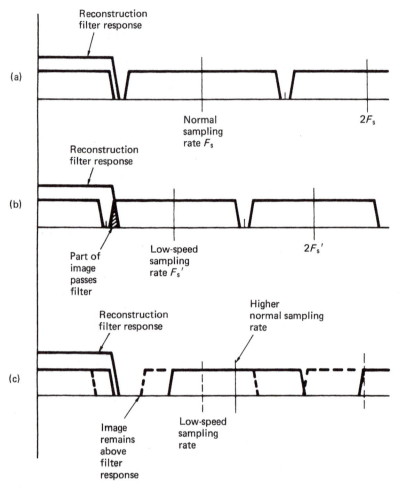

Figure 4.10 At normal speed, the reconstruction filter correctly prevents images entering the baseband, as at (a). When speed is reduced, the sampling rate falls, and a fixed filter will allow part of the lower sideband of the sampling frequency to pass. If the sampling rate of the machine is raised, but the filter characteristic remains the same, the problem can be avoided, as at (c).

The allowable sampling rates in a pseudo-video system can be deduced by multiplying the field rate by the number of active lines in a field (blanked lines cannot be used) and again by the number of samples in a line. By careful choice of parameters it is possible to use either 525/60 or 625/50 video with a sampling rate of 44.1 kHz.

In 60 Hz video, there are 35 blanked lines, leaving 490 lines per frame, or 245 lines per field for samples. If three samples are stored per line, the sampling rate becomes

$$60 \times 245 \times 3 = 44.1 \, \text{kHz}$$

In 50 Hz video, there are 37 lines of blanking, leaving 588 active lines per frame, or 294 per field, so the same sampling rate is given by

$$50.00 \times 294 \times 3 = 44.1\,\text{kHz}.$$

The sampling rate of 44.1 kHz came to be that of the Compact Disc. Even though CD has no video circuitry, the equipment originally used to make CD masters was video based and determines the sampling rate.

For landlines to FM stereo broadcast transmitters having a 15 kHz audio bandwidth, the sampling rate of 32 kHz is more than adequate, and has been in use for some time in the United Kingdom and Japan. This frequency is also in use in the NICAM 728 stereo TV sound system and in DAB. It is also used for the Sony NT format mini-cassette. The professional sampling rate of 48 kHz was proposed as having a simple relationship to 32 kHz, being far enough above 40 kHz for variable-speed operation.

Although in a perfect world the adoption of a single sampling rate might have had virtues, for practical and economic reasons digital audio now has essentially three rates to support: 32 kHz for broadcast, 44.1 kHz for CD and its mastering equipment, and 48 kHz for 'professional' use.[10] In fact the use of 48 kHz is not as common as its title would indicate. The runaway success of CD has meant that much equipment is run at 44.1 kHz to suit CD. With the advent of digital filters, which can track the sampling rate, a higher sampling rate is no longer necessary for pitch changing. 48 kHz is extensively used in television where it can be synchronized to both line standards relatively easily. The currently available DVTR formats offer only 48 kHz audio sampling. A number of formats can operate at more than one sampling rate. Both DAT and DASH formats are specified for all three rates, although not all available hardware implements every possibility. Most hard disk recorders will operate at a range of rates.

Recently there have been proposals calling for dramatically increased audio sampling rates. These are misguided and will not be considered further here. The subject will, however, be treated in Chapter 13.

4.6   Sample and hold

In practice many analog to digital convertors require a finite time to operate, and instantaneous samples must be extended by a device called a sample-and-hold or, more accurately, a track-hold circuit.

The simplest possible track-hold circuit is shown in Figure 4.11(a). When the switch is closed, the output will follow the input. When the switch is opened, the capacitor holds the signal voltage which existed at

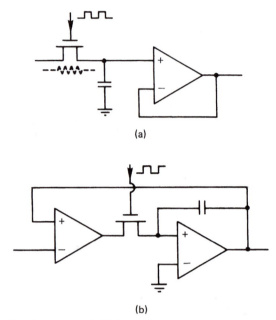

(a)

(b)

Figure 4.11 (a) The simple track-hold circuit shown has poor frequency response as the resistance of the FET causes a rolloff in conjunction with the capacitor. In (b) the resistance of the FET is now inside a feedback loop and will be eliminated, provided the left-hand op-amp never runs out of gain or swing.

the instant of opening. This simple arrangement has a number of shortcomings, particularly the time constant of the on-resistance of the switch with the capacitor, which extends the settling time. The effect can be alleviated by putting the switch in a feedback loop as shown in Figure 4.11(b). The buffer amplifiers must meet a stringent specification, because they need bandwidth well in excess of audio frequencies to ensure that operation is always feedback controlled between holding periods. When the switch is opened, the slightest change in input voltage causes the input buffer to saturate, and it must be able to rapidly recover from this condition when the switch next closes. The feedback minimizes the effect of the on-resistance of the switch, but the off-resistance must be high to prevent the input signal affecting the held voltage. The leakage current of the integrator must be low to prevent droop which is the term given to an unwanted slow change in the held voltage.

Figure 4.12 shows the various events during a track-hold sequence and catalogs the various potential sources of inaccuracy. A further phenomenon which is not shown in Figure 4.12 is that of dielectric relaxation. When a capacitor is discharged rapidly by connecting a low resistance path across its terminals, not all the charge is removed. After the discharge circuit is disconnected, the capacitor voltage may rise again slightly as charge which was trapped in the high-resistivity dielectric

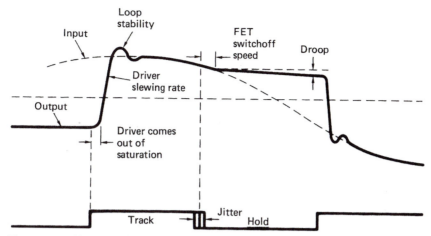

Figure 4.12 Characteristics of the feedback track-hold circuit of Figure 4.11(b) showing major sources of error.

slowly leaks back to the electrodes. In track-hold circuits dielectric relaxation can cause the value of one sample to be affected by the previous one. Some dielectrics display less relaxation than others. Mica capacitors, traditionally regarded as being of high quality, actually display substantially worse relaxation characteristics than many other types. Polypropylene and teflon are significantly better.

The track-hold circuit is extremely difficult to design because of the accuracy demanded by audio applications. In particular it is very difficult to meet the droop specification for much more than sixteen-bit applications. Greater accuracy has been reported by modelling the effect of dielectric relaxation and applying an inverse correction signal.[11]

When a performance limitation such as the track-hold stage is found, it is better to find an alternative approach. It will be seen later in this chapter that more advanced conversion techniques allow the track-hold circuit and its shortcomings to be eliminated.

4.7 Sampling clock jitter

The instants at which samples are taken in an ADC and the instants at which DACs make conversions must be evenly spaced, otherwise unwanted signals can be added to the audio. Figure 4.13 shows the effect of sampling clock jitter on a sloping waveform. Samples are taken at the wrong times. When these samples have passed through a system, the timebase correction stage prior to the DAC will remove the jitter, and the result is shown at (b). The magnitude of the unwanted signal is proportional to the slope of the audio waveform and so the amount of

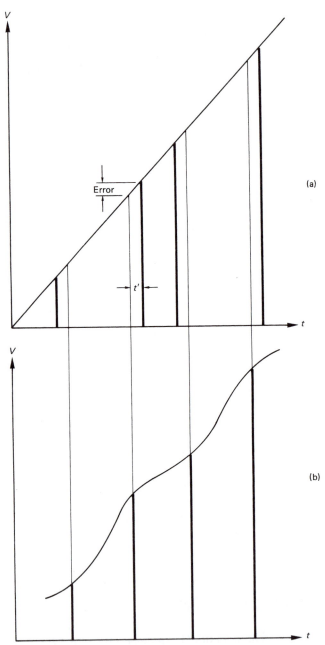

Figure 4.13 The effect of sampling timing jitter on noise, and calculation of the required accuracy for a sixteen-bit system. (a) Ramp sampled with jitter has error proportional to slope. (b) When jitter is removed by later circuits, error appears as noise added to samples. For a sixteen-bit system there are $2^{16}Q$, and the maximum slope at 20 kHz will be $20\,000\,\pi \times 2^{16}\,Q$ per second. If jitter is to be neglected, the noise must be less than $\frac{1}{2}Q$, thus timing accuracy t' multiplied by maximum slope $= \frac{1}{2}Q$ or $20\,000\,\pi \times 2^{16}Qt' = \frac{1}{2}Q$

$$\therefore 2' = \frac{1}{2 \times 20\,000\, \times\pi \times 2^{16}} = 121 \text{ ps}$$

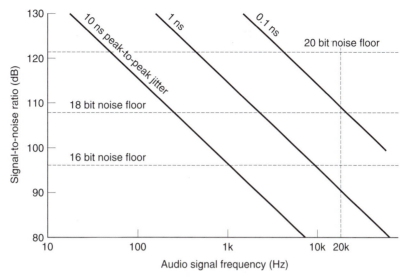

Figure 4.14 Effects of sample clock jitter on signal-to-noise ratio at different frequencies, compared with theoretical noise floors of systems with different resolutions. (After W. T. Shelton, with permission)

jitter which can be tolerated falls at 6 dB per octave. As the resolution of the system is increased by the use of longer sample wordlength, tolerance to jitter is further reduced. The nature of the unwanted signal depends on the spectrum of the jitter. If the jitter is random, the effect is noise-like and relatively benign unless the amplitude is excessive. Figure 4.14 shows the effect of differing amounts of random jitter with respect to the noise floor of various wordlengths. Note that even small amounts of jitter can degrade a twenty-bit convertor to the performance of a good sixteen-bit unit. There is thus no point in upgrading to higher-resolution convertors if the clock stability of the system is insufficient to allow their performance to be realized.

Clock jitter is not necessarily random. Figure 4.15 shows that one source of clock jitter is crosstalk or interference on the clock signal. A balanced clock line will be more immune to such crosstalk, but the consumer electrical digital audio interface is unbalanced and prone to external interference. The unwanted additional signal changes the time at which the sloping clock signal appears to cross the threshold voltage of the clock receiver. This is simply the same phenomenon as that of Figure 4.13 but in reverse. The threshold itself may be changed by ripple on the clock receiver power supply. There is no reason why these effects should be random; they may be periodic and potentially audible.[12,13]

The allowable jitter is measured in picoseconds, as shown in Figure 4.13 and clearly steps must be taken to eliminate it by design. Convertor clocks must be generated from clean power supplies which are well

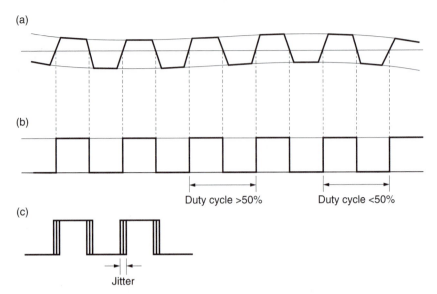

Figure 4.15 Crosstalk in transmission can result in unwanted signals being added to the clock waveform. It can be seen here that a low-frequency interference signal affects the slicing of the clock and causes a periodic jitter.

decoupled from the power used by the logic because a convertor clock must have a signal-to-noise ratio of the same order as that of the audio. Otherwise noise on the clock causes jitter which in turn causes noise in the audio.

Power supply ripple from conventional 50/60 Hz transformer rectifiers is difficult to eliminate, but these supplies are giving way to switched mode power supplies on grounds of cost and efficiency. If the switched mode power supply is locked to the sampling clock, the power supply ripple is sampled at its own frequency and appears to be DC. Clock jitter is thus avoided and samples are taken in between switching transients. This approach is used in some digital multi-track recorders where the amount of logic and power required is considerable. In variable-speed operation the power supply switching speed varies along with the capstan speed and the sampling rate.

If an external clock source is used, it cannot be used directly, but must be fed through a well-designed, well-damped phase-locked loop which will filter out the jitter. The operation of a phase-locked loop was described in Chapter 2. The phase-locked loop must be built to a higher accuracy standard than in most applications. Noise reaching the frequency control element will cause the very jitter the device is meant to eliminate. Some designs use a crystal oscillator whose natural frequency can be shifted slightly by a varicap diode. The high Q of the crystal produces a cleaner clock. Unfortunately this high Q also means that the frequency swing which can be achieved is quite small. It is sufficient for locking to a single standard sampling rate reference, but not for locking to a range of sampling rates or for variable-speed operation. In this case a conventional varicap VCO is required. Some machines can switch between a crystal VCO and a wideband VCO depending on the sampling rate accuracy. As will be seen in Chapter 8, the AES/EBU interface has provision for conveying sampling rate accuracy in the channel status data and this could be used to select the appropriate oscillator. Some machines which need to operate at variable speed but with the highest quality use a double-phase-locked loop arrangement where the residual jitter in the first loop is further reduced by the second. The external clock signal is sometimes fed into the clean circuitry using an optical coupler to improve isolation.

Although it has been documented for many years, attention to control of clock jitter is not as great in actual hardware as it might be. It accounts for much of the slight audible differences between convertors reproducing the same data. A well-engineered convertor should substantially reject jitter on an external clock and should sound the same when reproducing the same data irrespective of the source of the data. A remote convertor which sounds different when reproducing, for example, the same Compact Disc via the digital outputs of a variety of CD players is simply not well engineered and should be rejected. Similarly if the effect of changing the type of digital cable feeding the convertor can be heard, the unit is a dud. Unfortunately many consumer external DACs fall into this category, as the steps outlined above have not been taken. Some consumer external DACs, however, have RAM timebase correction which has a large enough correction range that the convertor can run from a local fixed frequency crystal. The incoming clock does no more than control the memory write cycles. Any incoming jitter is rejected totally.

Many portable digital machines have compromised jitter performance because their small size and weight constraints make the provision of adequate screening, decoupling and phase-locked loop circuits difficult.

4.8   Aperture effect

The reconstruction process of Figure 4.6 only operates exactly as shown if the impulses are of negligible duration. In many DACs this is not the case, and many keep the analog output constant for a substantial part of the sample period or even until a different sample value is input. This produces a waveform which is more like a staircase than a pulse train. The case where the pulses have been extended in width to become equal to the sample period is known as a zero-order-hold system and has a 100 per cent aperture ratio. Note that the aperture effect is not apparent in a

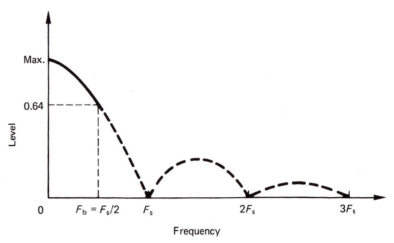

Figure 4.16 Frequency response with 100 per cent aperture has nulls at multiples of sampling rate. Area of interest is up to half sampling rate.

track-hold system; the holding period is only for the convenience of the quantizer which then outputs a value corresponding to the input voltage at the instant hold mode was entered.

It was shown in Chapter 3 that whereas pulses of negligible width have a uniform spectrum, which is flat within the audio band, pulses of 100 per cent aperture ratio have a $\sin x/x$ spectrum which is shown in Figure 4.16. The frequency response falls to a null at the sampling rate, and as a result is about 4 dB down at the edge of the audio band. If the pulse width is stable, the reduction of high frequencies is constant and predictable, and an appropriate equalization circuit can render the overall response flat once more. An alternative is to use resampling which is shown in Figure 4.17. Resampling passes the zero-order-hold waveform through a further synchronous sampling stage which consists of an analog switch which closes briefly in the centre of each sample period. The output of the switch will be pulses which are narrower than the original. If, for example, the aperture ratio is reduced to 50 per cent of the sample period, the first frequency response null is now at twice the sampling rate, and the loss at the edge of the audio band is reduced. As the figure shows, the frequency response becomes flatter as the aperture ratio falls. The process should not be carried too far, as with very small aperture ratios there is little energy in the pulses and noise can be a problem. A practical limit is around 12.5 per cent where the frequency response is virtually ideal.

The term resampling will also be found in descriptions of sampling rate convertors, where it refers to the process of finding samples at new locations to describe the original waveform. The context usually makes it clear which meaning is intended.

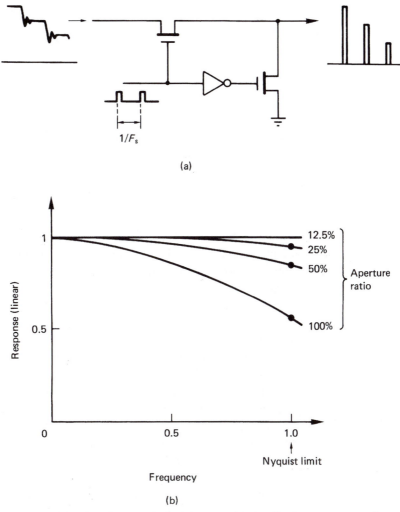

(a)

(b)

Figure 4.17 (a) Resampling circuit eliminates transients and reduces aperture ratio. (b) Response of various aperture ratios.

4.9 Quantizing

Quantizing is the process of expressing some infinitely variable quantity by discrete or stepped values. Quantizing turns up in a remarkable number of everyday guises. Figure 4.18 shows that an inclined ramp enables infinitely variable height to be achieved, whereas a step-ladder allows only discrete heights to be had. A step-ladder quantizes height. When accountants round off sums of money to the nearest pound or dollar they are quantizing. Time passes continuously, but the display on a digital clock changes suddenly every minute because the clock is quantizing time.

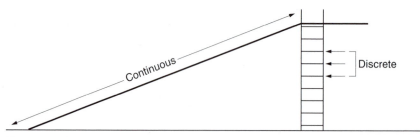

Figure 4.18 An analog parameter is continuous whereas a quantized parameter is restricted to certain values. Here the sloping side of a ramp can be used to obtain any height whereas a ladder only allows discrete heights.

In audio the values to be quantized are infinitely variable voltages from an analog source. Strict quantizing is a process which operates in the voltage domain only. For the purpose of studying the quantizing of a single sample, time is assumed to stand still. This is achieved in practice either by the use of a track-hold circuit or the adoption of a quantizer technology which operates before the sampling stage.

Figure 4.19(a) shows that the process of quantizing divides the voltage range up into quantizing intervals Q, also referred to as steps S. In applications such as telephony these may advantageously be of differing size, but for digital audio the quantizing intervals are made as identical as possible. If this is done, the binary numbers which result are truly proportional to the original analog voltage, and the digital equivalents of mixing and gain changing can be performed by adding and multiplying sample values. If the quantizing intervals are unequal this cannot be done. When all quantizing intervals are the same, the term uniform quantizing is used. The term linear quantizing will be found, but this is, like military intelligence, a contradiction in terms.

The term LSB (least significant bit) will also be found in place of quantizing interval in some treatments, but this is a poor term because quantizing works in the voltage domain. A bit is not a unit of voltage and can have only two values. In studying quantizing, voltages within a quantizing interval will be discussed, but there is no such thing as a fraction of a bit.

Whatever the exact voltage of the input signal, the quantizer will locate the quantizing interval in which it lies. In what may be considered a separate step, the quantizing interval is then allocated a code value which is typically some form of binary number. The information sent is the number of the quantizing interval in which the input voltage lies. Whereabouts that voltage lies within the interval is not conveyed, and this mechanism puts a limit on the accuracy of the quantizer. When the number of the quantizing interval is converted back to the analog domain, it will result in a voltage at the centre of the quantizing interval

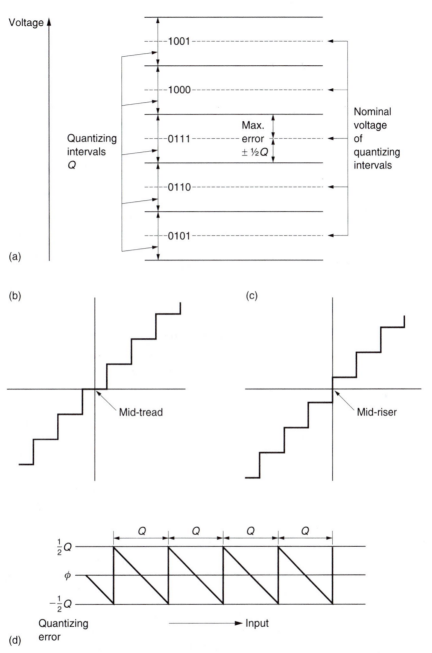

Figure 4.19 Quantizing assigns discrete numbers to variable voltages. All voltages within the same quantizing interval are assigned the same number which causes a DAC to produce the voltage at the centre of the intervals shown by the dashed lines in (a). This is the characteristic of the mid-tread quantizer shown in (b). An alternative system is the mid-riser system shown in (c). Here 0 volts analog falls between two codes and there is no code for zero. Such quantizing cannot be used prior to signal processing because the number is no longer proportional to the voltage. Quantizing error cannot exceed $\pm\frac{1}{2}Q$ as shown in (d).

as this minimizes the magnitude of the error between input and output. The number range is limited by the wordlength of the binary numbers used. In a sixteen-bit system, 65 536 different quantizing intervals exist, although the ones at the extreme ends of the range have no outer boundary.

4.10   Quantizing error

It is possible to draw a transfer function for such an ideal quantizer followed by an ideal DAC, and this is also shown in Figure 4.19. A transfer function is simply a graph of the output with respect to the input. In audio, when the term linearity is used, this generally means the straightness of the transfer function. Linearity is a goal in audio, yet it will be seen that an ideal quantizer is anything but linear.

Figure 4.19(b) shows the transfer function is somewhat like a staircase, and zero volts analog, corresponding to all zeros digital or muting, is half-way up a quantizing interval, or on the centre of a tread. This is the so-called mid-tread quantizer which is universally used in audio. Figure 4.19(c) shows the alternative mid-riser transfer function which causes difficulty in audio because it does not have a code value at muting level and as a result the numerical code value is not proportional to the analog signal voltage.

Quantizing causes a voltage error in the audio sample which is given by the difference between the actual staircase transfer function and the ideal straight line. This is shown in Figure 4.19(d) to be a sawtooth-like function which is periodic in Q. The amplitude cannot exceed $\pm \frac{1}{2}Q$ peak-to-peak unless the input is so large that clipping occurs.

Quantizing error can also be studied in the time domain where it is better to avoid complicating matters with the aperture effect of the DAC. For this reason it is assumed here that output samples are of negligible duration. Then impulses from the DAC can be compared with the original analog waveform and the difference will be impulses representing the quantizing error waveform. This has been done in Figure 4.20. The horizontal lines in the drawing are the boundaries between the quantizing intervals, and the curve is the input waveform. The vertical bars are the quantized samples which reach to the centre of the quantizing interval. The quantizing error waveform shown at (b) can be thought of as an unwanted signal which the quantizing process adds to the perfect original. If a very small input signal remains within one quantizing interval, the quantizing error *is* the signal.

As the transfer function is non-linear, ideal quantizing can cause distortion. As a result practical digital audio devices deliberately use non-ideal quantizers to achieve linearity. The quantizing error of an ideal

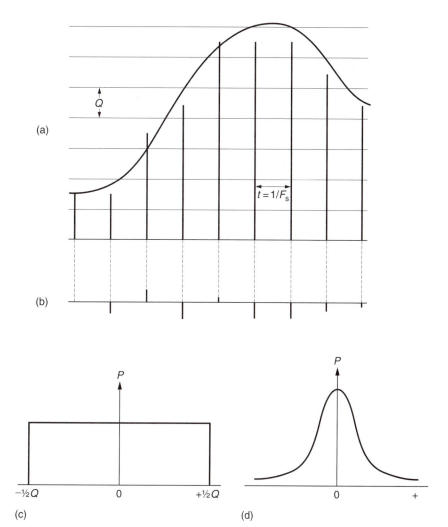

Figure 4.20 At (a) an arbitrary signal is represented to finite accuracy by PAM needles whose peaks are at the centre of the quantizing intervals. The errors caused can be thought of as an unwanted signal (b) added to the original. In (c) the amplitude of a quantizing error needle will be from $-\frac{1}{2}Q$ to $+\frac{1}{2}Q$ with equal probability. Note, however, that white noise in analog circuits generally has Gaussian amplitude distribution, shown in (d).

quantizer is a complex function, and it has been researched in great depth.[14–16] It is not intended to go into such depth here. The characteristics of an ideal quantizer will be pursued only far enough to convince the reader that such a device cannot be used in quality audio applications.

As the magnitude of the quantizing error is limited, its effect can be minimized by making the signal larger. This will require more quantizing intervals and more bits to express them. The number of quantizing intervals multiplied by their size gives the quantizing range of the convertor. A signal outside the range will be clipped. Provided that clipping is avoided, the larger the signal, the less will be the effect of the quantizing error.

Where the input signal exercises the whole quantizing range and has a complex waveform (such as from orchestral music), successive samples will have widely varying numerical values and the quantizing error on a given sample will be independent of that on others. In this case the size of the quantizing error will be distributed with equal probability between the limits. Figure 4.20(c) shows the resultant uniform probability density. In this case the unwanted signal added by quantizing is an additive broadband noise uncorrelated with the signal, and it is appropriate in this case to call it quantizing noise. This is not quite the same as thermal noise which has a Gaussian probability shown in Figure 4.20(d) (see Chapter 3 for a treatment of probability). The difference is of no consequence as in the large signal case the noise is masked by the signal. Under these conditions, a meaningful signal-to-noise ratio can be calculated as follows:

In a system using n-bit words. there will be 2^n quantizing intervals. The largest sinusoid which can fit without clipping will have this peak-to-peak amplitude. The peak amplitude will be half as great, i.e. $2^{n-1}Q$ and the rms amplitude will be this value divided by $\sqrt{2}$.

The quantizing error has an amplitude of $\frac{1}{2}Q$ peak which is the equivalent of $Q/\sqrt{12}$ rms. The signal-to-noise ratio for the large signal case is then given by:

$$20 \log_{10} \frac{\sqrt{12} \times 2^{n-1}}{\sqrt{2}} \text{ dB}$$

$$= 20 \log_{10} (\sqrt{6} \times 2^{n-1}) \text{ dB}$$

$$= 20 \log \left(2^n \times \frac{\sqrt{6}}{2}\right) \text{ dB}$$

$$= 20n \log 2 + 20 \log \frac{\sqrt{6}}{2} \text{ dB}$$

$$= 6.02n + 1.76 \text{ dB} \tag{4.1}$$

By way of example, a sixteen-bit system will offer around 98.1 dB SNR.

Whilst the above result is true for a large complex input waveform, treatments which then assume that quantizing error is *always* noise give results which are at variance with reality. The expression above is only valid if the probability density of the quantizing error is uniform. Unfortunately at low levels, and particularly with pure or simple waveforms, this is simply not the case.

At low audio levels, quantizing error ceases to be random, and becomes a function of the input waveform and the quantizing structure as Figure 4.20 showed. Once an unwanted signal becomes a deterministic function of the wanted signal, it has to be classed as a distortion rather than a noise. Distortion can also be predicted from the non-linearity, or staircase nature, of the transfer function. With a large signal, there are so many steps involved that we must stand well back, and a staircase with 65 000 steps appears to be a slope. With a small signal there are few steps and they can no longer be ignored.

The non-linearity of the transfer function results in distortion, which produces harmonics. Unfortunately these harmonics are generated *after* the anti-aliasing filter, and so any which exceed half the sampling rate will alias. Figure 4.21 shows how this results in anharmonic distortion within the audio band. These anharmonics result in spurious tones known as birdsinging. When the sampling rate is a multiple of the input frequency the result is harmonic distortion. This is shown in Figure 4.22. Where more than one frequency is present in the input, intermodulation distortion occurs, which is known as granulation.

As the input signal is further reduced in level, it may remain within one quantizing interval. The output will be silent because the signal is now the quantizing error. In this condition, low-frequency signals such as air-conditioning rumble can shift the input in and out of a quantizing interval so that the quantizing distortion comes and goes, resulting in noise modulation.

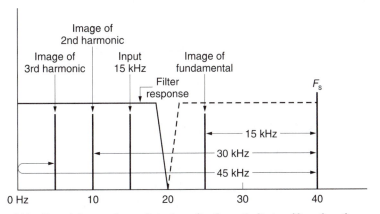

Figure 4.21 Quantizing produces distortion *after* the anti-aliasing filter; thus the distortion products will fold back to produce anharmonics in the audio band. Here the fundamental of 15 kHz produces second and third harmonic distortion at 30 and 45 kHz. This results in aliased products at 40 − 30 = 10 kHz and 40 − 45 = (−)5 kHz.

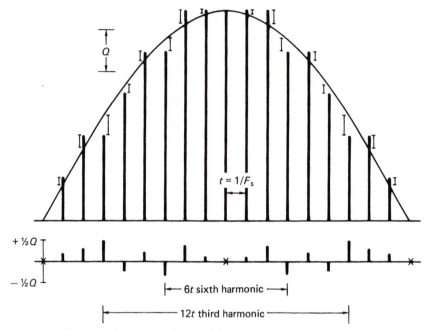

Figure 4.22 Mathematically derived quantizing error waveform for sine wave sampled at a multiple of itself. The numerous autocorrelations between quantizing errors show that there are harmonics of the signal in the error, and that the error is not random, but deterministic.

Needless to say, any one of the above effects would preclude the use of an ideal quantizer for high-quality work. There is little point in studying the adverse effects further as they should be and can be eliminated completely in practical equipment by the use of dither. The importance of correctly dithering a quantizer cannot be emphasized enough, since failure to dither irrevocably distorts the converted signal: there can be no process which will subsequently remove that distortion.

The signal-to-noise ratio derived above has no relevance to practical audio applications as it will be modified by the dither and by any noise shaping used.

4.11   Introduction to dither

At high signal levels, quantizing error is effectively noise. As the audio level falls, the quantizing error of an ideal quantizer becomes more strongly correlated with the signal and the result is distortion. If the quantizing error can be decorrelated from the input in some way, the system can remain linear but noisy. Dither performs the job of decorrelation by making the action of the quantizer unpredictable and gives the system a noise floor like an analog system.

The first documented use of dither was by Roberts[17] in picture coding. In this system, pseudo-random noise (see Chapter 3) with rectangular probability and a peak-to-peak amplitude of Q was added to the input signal prior to quantizing, but was subtracted after reconversion to analog. This is known as subtractive dither and was investigated by Schuchman[18] and much later by Sherwood.[19] Subtractive dither has the advantages that the dither amplitude is non-critical, the noise has full statistical independence from the signal[15] and has the same level as the quantizing error in the large signal undithered case.[20] Unfortunately, it suffers from practical drawbacks, since the original noise waveform must accompany the samples or must be synchronously recreated at the DAC. This is virtually impossible in a system where the audio may have been edited or where its level has been changed by processing, as the noise needs to remain synchronous and be processed in the same way. All practical digital audio systems use non-subtractive dither where the dither signal is added prior to quantization and no attempt is made to remove it at the DAC.[21] The introduction of dither prior to a conventional quantizer inevitably causes a slight reduction in the signal-to-noise ratio

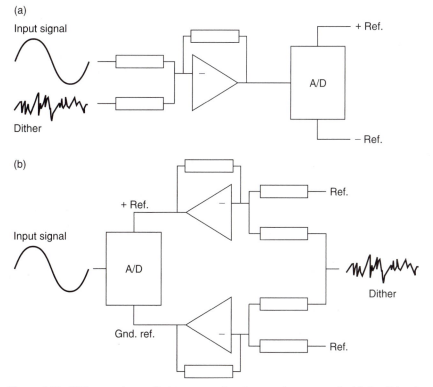

Figure 4.23 Dither can be applied to a quantizer in one of two ways. In (a) the dither is linearly added to the analog input signal, whereas in (b) it is added to the reference voltages of the quantizer.

attainable, but this reduction is a small price to pay for the elimination of non-linearities. The technique of noise shaping in conjunction with dither will be seen to overcome this restriction and produce performance in excess of the subtractive dither example above.

The ideal (noiseless) quantizer of Figure 4.19 has fixed quantizing intervals and must always produce the same quantizing error from the same signal. In Figure 4.23 it can be seen that an ideal quantizer can be dithered by linearly adding a controlled level of noise either to the input signal or to the reference voltage which is used to derive the quantizing intervals. There are several ways of considering how dither works, all of which are equally valid.

The addition of dither means that successive samples effectively find the quantizing intervals in different places on the voltage scale. The quantizing error becomes a function of the dither, rather than a predictable function of the input signal. The quantizing error is not eliminated, but the subjectively unacceptable distortion is converted into a broadband noise which is more benign to the ear.

Some alternative ways of looking at dither are shown in Figure 4.24. Consider the situation where a low-level input signal is changing slowly within a quantizing interval. Without dither, the same numerical code is output for a number of sample periods, and the variations within the interval are lost. Dither has the effect of forcing the quantizer to switch between two or more states. The higher the voltage of the input signal within a given interval, the more probable it becomes that the output code will take on the next higher value. The lower the input voltage within the interval, the more probable it is that the output code will take the next lower value. The dither has resulted in a form of duty cycle modulation, and the resolution of the system has been extended indefinitely instead of being limited by the size of the steps.

Dither can also be understood by considering what it does to the transfer function of the quantizer. This is normally a perfect staircase, but in the presence of dither it is smeared horizontally until with a certain amplitude the average transfer function becomes straight.

In an extension of the application of dither, Blesser[22] has suggested digitally generated dither which is converted to the analog domain and added to the input signal prior to quantizing. That same digital dither is then subtracted from the digital quantizer output. The effect is that the transfer function of the quantizer is smeared diagonally (Figure 4.25). The significance of this diagonal smearing is that the amplitude of the dither is not critical. However much dither is employed, the noise amplitude will remain the same. If dither of several quantizing intervals is used, it has the effect of making all the quantizing intervals in an imperfect convertor appear to have the same size.

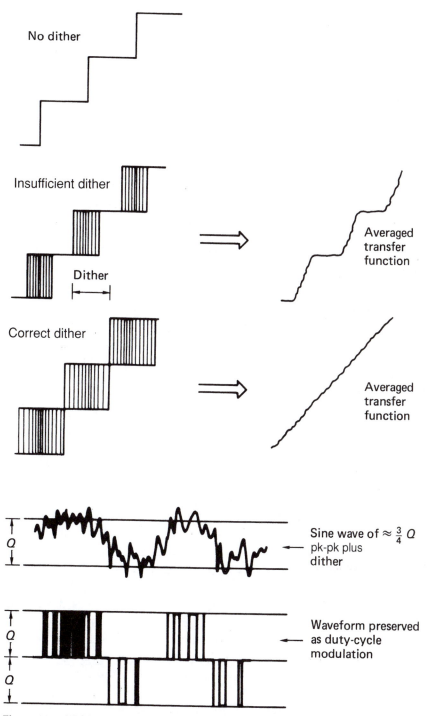

Figure 4.24 Wideband dither of the appropriate level linearizes the transfer function to produce noise instead of distortion. This can be confirmed by spectral analysis. In the voltage domain, dither causes frequent switching between codes and preserves resolution in the duty cycle of the switching.

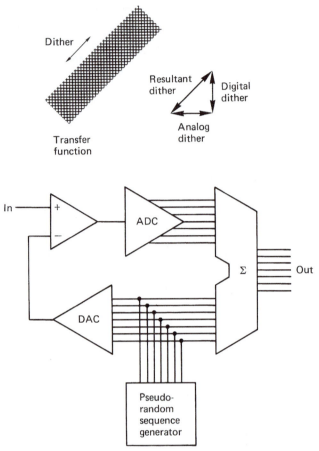

Figure 4.25 In this dither system, the dither added in the analog domain shifts the transfer function horizontally, but the same dither is subtracted in the digital domain, which shifts the transfer function vertically. The result is that the quantizer staircase is smeared diagonally as shown top left. There is thus no limit to dither amplitude, and excess dither can be used to improve differential linearity of the convertor.

4.12   Requantizing and digital dither

The advanced ADC technology which is detailed later in this chapter allows as much as 24-bit resolution to be obtained, with perhaps more in the future. The situation then arises that an existing sixteen-bit device such as a digital recorder needs to be connected to the output of an ADC with greater wordlength. The words need to be shortened in some way.

Chapter 3 showed that when a sample value is attenuated, the extra low-order bits which come into existence below the radix point preserve the resolution of the signal and the dither in the least significant bit(s) which linearizes the system. The same word extension will occur in any process involving multiplication, such as digital filtering. It will subsequently be necessary to shorten the wordlength. Clearly the high-order bits cannot be discarded in two's complement as this would cause clipping of positive half-cycles and a level shift on negative half-cycles due to the loss of the sign bit. Low-order bits must be removed instead. Even if the original conversion was correctly dithered, the random element in the low-order bits will now be some way below the end of the intended word. If the word is simply truncated by discarding the unwanted low-order bits or rounded to the nearest integer the linearizing effect of the original dither will be lost.

Shortening the wordlength of a sample reduces the number of quantizing intervals available without changing the signal amplitude. As Figure 4.26 shows, the quantizing intervals become larger and the original signal is *requantized* with the new interval structure. This will introduce requantizing distortion having the same characteristics as quantizing distortion in an ADC. It then is obvious that when shortening the wordlength of a twenty-bit convertor to sixteen bits, the four low-order bits must be removed in a way that displays the same overall quantizing structure as if the original convertor had been only of sixteen-bit wordlength. It will be seen from Figure 4.26 that truncation cannot be used because it does not meet the above requirement but results in signal-dependent offsets because it always rounds in the same direction. Proper

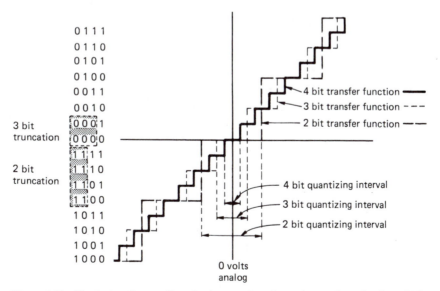

Figure 4.26 Shortening the wordlength of a sample reduces the number of codes which can describe the voltage of the waveform. This makes the quantizing steps bigger, hence the term requantizing. It can be seen that simple truncation or omission of the bits does not give analogous behaviour. Rounding is necessary to give the same result as if the larger steps had been used in the original conversion.

numerical rounding is essential in audio applications. Rounding in two's complement is a little more complex than in pure binary as can be seen in Chapter 3.

Requantizing by numerical rounding accurately simulates analog quantizing to the new interval size. Unfortunately the twenty-bit convertor will have a dither amplitude appropriate to quantizing intervals one sixteenth the size of a sixteen-bit unit and the result will be highly non-linear.

In practice, the wordlength of samples must be shortened in such a way that the requantizing error is converted to noise rather than distortion. One technique which meets this requirement is to use digital dithering[23] prior to rounding. This is directly equivalent to the analog dithering in an ADC. It will be shown later in this chapter that in more complex systems noise shaping can be used in requantizing just as well as it can in quantizing.

Digital dither is a pseudo-random sequence of numbers. If it is required to simulate the analog dither signal of Figures 4.23 and 24, then it is obvious that the noise must be bipolar so that it can have an average voltage of zero. Two's complement coding must be used for the dither values as it is for the audio samples.

Figure 4.27 shows a simple digital dithering system (i.e. one without noise shaping) for shortening sample wordlength. The output of a two's complement pseudo-random sequence generator (see Chapter 3) of appropriate wordlength is added to input samples prior to rounding. The most significant of the bits to be discarded is examined in order to

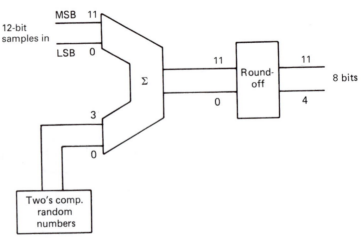

Figure 4.27 In a simple digital dithering system, two's complement values from a random number generator are added to low-order bits of the input. The dithered values are then rounded up or down according to the value of the bits to be removed. The dither linearizes the requantizing.

determine whether the bits to be removed sum to more or less than half a quantizing interval. The dithered sample is either rounded down, i.e. the unwanted bits are simply discarded, or rounded up, i.e. the unwanted bits are discarded but one is added to the value of the new short word. The rounding process is no longer deterministic because of the added dither which provides a linearizing random component.

If this process is compared with that of Figure 4.23 it will be seen that the principles of analog and digital dither are identical; the processes simply take place in different domains using two's complement numbers which are rounded or voltages which are quantized as appropriate. In fact quantization of an analog dithered waveform is identical to the hypothetical case of rounding after bipolar digital dither where the number of bits to be removed is infinite, and remains identical for practical purposes when as few as eight bits are to be removed. Analog dither may actually be generated from bipolar digital dither (which is no more than random numbers with certain properties) using a DAC.

4.13   Dither techniques

The intention here is to treat the processes of analog and digital dither as identical except where differences need to be noted. The characteristics of the noise used are rather important for optimal performance, although many sub-optimal but nevertheless effective systems are in use. The main parameters of interest are the peak-to-peak amplitude, the amplitude probability distribution function (pdf) and the spectral content.

The most comprehensive ongoing study of non-subtractive dither has been that of Vanderkooy and Lipshitz.[21,23,24] and the treatment here is based largely upon their work.

4.13.1   Rectangular pdf dither

Chapter 3 showed that the simplest form of dither (and therefore the easiest to generate) is a single sequence of random numbers which have uniform or rectangular probability. The amplitude of the dither is critical. Figure 4.28(a) shows the time-averaged transfer function of one quantizing interval in the presence of various amplitudes of rectangular dither. The linearity is perfect at an amplitude of $1Q$ peak-to-peak and then deteriorates for larger or smaller amplitudes. The same will be true of all levels which are an integer multiple of Q. Thus there is no freedom in the choice of amplitude.

With the use of such dither, the quantizing noise is not constant. Figure 4.28(b) shows that when the analog input is exactly centred in a

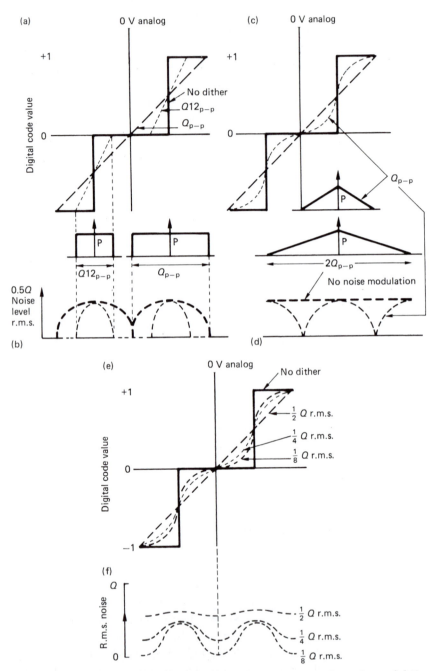

Figure 4.28 (a) Use of rectangular probability dither can linearize, but noise modulation (b) results. Triangular pdf dither (c) linearizes, but noise modulation is eliminated as at (d). Gaussian dither (e) can also be used, almost eliminating noise modulation at (f).

quantizing interval (such that there is no quantizing error) the dither has no effect and the output code is steady. There is no switching between codes and thus no noise. On the other hand, when the analog input is exactly at a riser or boundary between intervals, there is the greatest switching between codes and the greatest noise is produced. Mathematically speaking, the first moment, or mean error is zero but the second moment, which in this case is equal to the variance, is not constant. From an engineering standpoint, the system is linear but suffers noise modulation: the noise floor rises and falls with the signal content and this is audible in the presence of low-frequency signals.

The dither adds an average noise amplitude of $Q/\sqrt{12}$ rms to the quantizing noise of the same level. In order to find the resultant noise level it is necessary to add the powers as the signals are uncorrelated. The total power is given by:

$$2 \times \frac{Q^2}{12} = \frac{Q^2}{6}$$

and the rms voltage is $Q/\sqrt{6}$. Another way of looking at the situation is to consider that the noise power doubles and so the rms noise voltage has increased by 3 dB in comparison with the undithered case. Thus for an n-bit wordlength, using the same derivation as expression (4.1) above, the signal to noise ratio for Q pk-pk rectangular dither will be given by:

$$6.02n - 1.24 \, \text{dB} \tag{4.2}$$

Unlike the undithered case, this is a true signal-to-noise ratio and linearity is maintained at all signal levels. By way of example, for a sixteen-bit system 95.1 dB SNR is achieved. The 3 dB loss compared to the undithered case is a small price to pay for linearity.

4.13.2 Triangular pdf dither

The noise modulation due to the use of rectangular-probability dither is undesirable. It comes about because the process is too simple. The undithered quantizing error is signal dependent and the dither represents a single uniform-probability random process. This is only capable of decorrelating the quantizing error to the extent that its mean value is zero, rendering the system linear. The signal dependence is not eliminated, but is displaced to the next statistical moment. This is the variance and the result is noise modulation. If a further uniform-probability random process is introduced into the system, the signal dependence is displaced to the next moment and the second moment or variance becomes constant.

Adding together two statistically independent rectangular probability functions produces a triangular probability function. A signal having this characteristic can be used as the dither source.

Figure 4.28(c) shows the averaged transfer function for a number of dither amplitudes. Linearity is reached with a peak-to-peak amplitude of $2Q$ and at this level there is no noise modulation. The lack of noise modulation is another way of stating that the noise is constant. The triangular pdf of the dither matches the triangular shape of the quantizing error function.

The dither adds two noise signals with an amplitude of $Q/\sqrt{12}$ rms to the quantizing noise of the same level. In order to find the resultant noise level it is necessary to add the powers as the signals are uncorrelated. The total power is given by:

$$3 \times \frac{Q^2}{12} = \frac{Q^2}{4}$$

and the rms voltage is $Q/\sqrt{4}$. Another way of looking at the situation is to consider that the noise power is increased by 50 per cent in comparison to the rectangular dithered case and so the rms noise voltage has increased by 1.76 dB. Thus for an n-bit wordlength, using the same derivation as expressions (4.1) and (4.2) above, the signal to noise ratio for Q peak-to-peak rectangular dither will be given by:

$$6.02n - 3 \text{ dB} \tag{4.3}$$

Continuing the use of a sixteen-bit example, a SNR of 93.3 dB is available which is 4.8 dB worse than the SNR of an undithered quantizer in the large-signal case. It is a small price to pay for perfect linearity and an unchanging noise floor.

4.13.3 Gaussian pdf dither

Adding more uniform probability sources to the dither makes the overall probability function progressively more like the Gaussian distribution of analog noise. Figure 4.28(d) shows the averaged transfer function of a quantizer with various levels of Gaussian dither applied. Linearity is reached with $\frac{1}{2}Q$ rms and at this level noise modulation is negligible. The total noise power is given by:

$$\frac{Q^2}{4} + \frac{Q^2}{12} = \frac{3 \times Q^2}{12} + \frac{Q^2}{12} = \frac{Q^2}{3}$$

and so the noise level will be $Q\sqrt{3}$ rms. The noise level of an undithered quantizer in the large signal case is $Q\sqrt{12}$ and so the noise is higher by a factor of:

$$\frac{Q}{\sqrt{3}} \times \frac{\sqrt{12}}{Q} = \frac{Q}{\sqrt{3}} \times \frac{2\sqrt{3}}{Q} = 2 = 6.02 \text{ dB} \tag{4.4}$$

Thus the SNR is given by $6.02(n-1) + 1.76$ dB. A sixteen-bit system with correct Gaussian dither has a SNR of 92.1 dB.

This is inferior to the figure in expression (4.3) by 1.1 dB. In digital dither applications, triangular probability dither of $2Q$ peak-to-peak is optimum because it gives the best possible combination of nil distortion, freedom from noise modulation and SNR. Using dither with more than two rectangular processes added is detrimental. Whilst this result is also true for analog dither, it is not practicable to apply it to a real ADC as all real analog signals contain thermal noise which is Gaussian. If triangular dither is used on a signal containing Gaussian noise, the results derived above are not obtained. ADCs should therefore use Gaussian dither of $Q/2$ rms and the performance will be given by expression (4.4).

It should be stressed that all the results in this section are for conventional quantizing and requantizing. The use of techniques such as oversampling and/or noise shaping require an elaboration of the theory in order to give meaningful SNR figures.

4.14   Basic digital-to-analog conversion

This direction of conversion will be discussed first, since ADCs often use embedded DACs in feedback loops.

The purpose of a digital-to-analog convertor is to take numerical values and reproduce the continuous waveform that they represent. Figure 4.29 shows the major elements of a conventional conversion subsystem, i.e. one in which oversampling is not employed. The jitter in the clock needs to be removed with a VCO or VCXO. Sample values are buffered in a latch and fed to the convertor element which operates on each cycle of the clean clock. The output is then a voltage proportional to the number for at least a part of the sample period. A resampling stage may be found next, in order to remove switching transients, reduce the aperture ratio or allow the use of a convertor which takes a substantial part of the sample

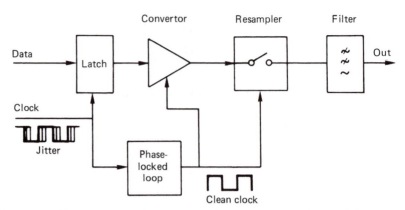

Figure 4.29 The components of a conventional convertor. A jitter-free clock drives the voltage conversion, whose output may be resampled prior to reconstruction.

period to operate. The resampled waveform is then presented to a reconstruction filter which rejects frequencies above the audio band. This section is primarily concerned with the implementation of the convertor element. There are two main ways of obtaining an analog signal from PCM data. One is to control binary-weighted currents and sum them; the other is to control the length of time a fixed current flows into an integrator. The two methods are contrasted in Figure 4.30. They appear simple, but are of no use for audio in these forms because of practical limitations. In Figure 4.30(c), the binary code is about to have a major overflow, and all the low-order currents are flowing. In Figure 4.30(d), the binary input has increased by one, and only the most significant current flows. This current must equal the sum of all the others plus one. The accuracy must be such that the step size is within the required limits. In this simple four-bit example, if the step size needs to be a rather casual 10 per cent accurate, the necessary accuracy is only one part in 160, but for a sixteen-bit system it would become one part in 655 360, or about 2 ppm. This degree of accuracy is almost impossible to achieve, let alone maintain in the presence of ageing and temperature change.

The integrator-type convertor in this four-bit example is shown in Figure 4.30(e); it requires a clock for the counter which allows it to count up to the maximum in less than one sample period. This will be more than sixteen times the sampling rate. However, in a sixteen-bit system, the clock rate would need to be 65 536 times the sampling rate, or about 3 GHz. Whilst there may be a market for a CD player which can defrost a chicken, clearly some refinements are necessary to allow either of these convertor types to be used in audio applications.

One method of producing currents of high relative accuracy is *dynamic element matching*.[25,26] Figure 4.31 shows a current source feeding a pair of nominally equal resistors. The two will not be the same owing to

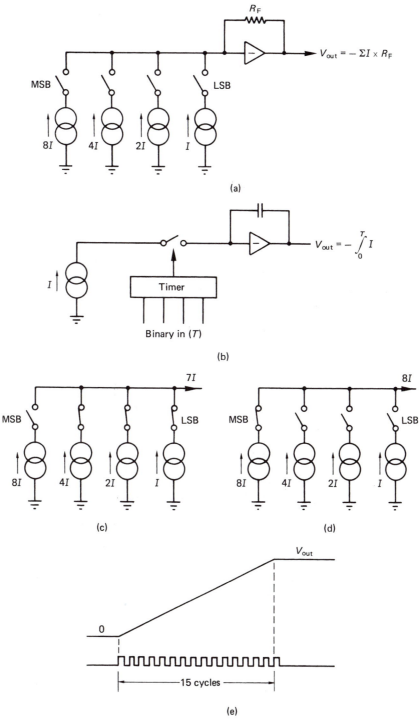

Figure 4.30 Elementary conversion: (a) weighted current DAC; (b) timed integrator DAC; (c) current flow with 0111 input; (d) current flow with 1000 input; (e) integrator ramps up for 15 cycles of clock for input 1111.

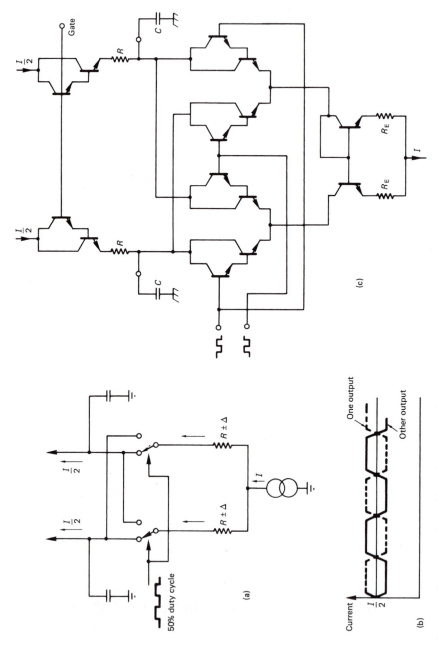

Figure 4.31 Dynamic element matching. (a) Each resistor spends half its time in each current path. (b) Average current of both paths will be identical if duty cycle is accurately 50 per cent. (c) Typical monolithic implementation. Note clock frequency is arbitrary.

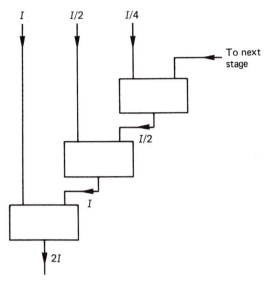

Figure 4.32 Cascading the current dividers of Figure 4.31 produces a binary-weighted series of currents.

manufacturing tolerances and drift, and thus the current is only approximately divided between them. A pair of change-over switches places each resistor in series with each output. The average current in each output will then be identical, provided that the duty cycle of the switches is exactly 50 per cent. This is readily achieved in a divide-by-two circuit. The accuracy criterion has been transferred from the resistors to the time domain in which accuracy is more readily achieved. Current averaging is performed by a pair of capacitors which do not need to be of any special quality. By cascading these divide-by-two stages, a binary-weighted series of currents can be obtained, as in Figure 4.32. In practice, a reduction in the number of stages can be obtained by using a more complex switching arrangement. This generates currents of ratio 1:1:2 by dividing the current into four paths and feeding two of them to one output, as shown in Figure 4.33. A major advantage of this approach is that no trimming is needed in manufacture, making it attractive for mass production. Freedom from drift is a further advantage.

To prevent interaction between the stages in weighted-current convertors, the currents must be switched to ground or into the virtual earth by change-over switches. The on-resistance of these switches is a source of error, particularly the MSB, which passes most current. A solution in monolithic convertors is to fabricate switches whose area is proportional to the weighted current, so that the voltage drops of all the switches are the same. The error can then be removed with a suitable offset. The layout of such a device is dominated by the MSB switch since, by definition, it is as big as all the others put together.

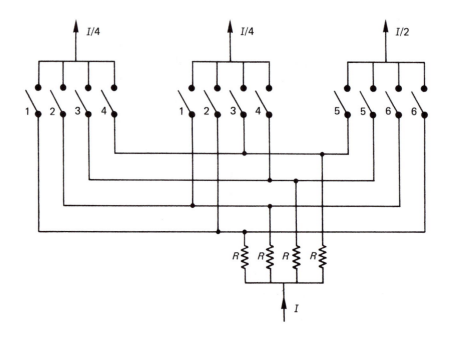

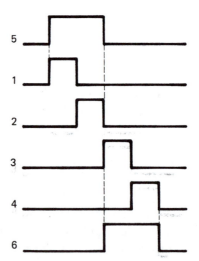

Figure 4.33 More complex dynamic element-matching system. Four drive signals (1, 2, 3, 4) of 25 per cent duty cycle close switches of corresponding number. Two signals (5, 6) have 50 per cent duty cycle, resulting in two current shares going to right-hand output. Division is thus into 1:1:2.

The practical approach to the integrator convertor is shown in Figures 4.34 and 4.35 where two current sources whose ratio is 256:1 are used; the larger is timed by the high byte of the sample and the smaller is timed by the low byte. The necessary clock frequency is reduced by a factor of 256.

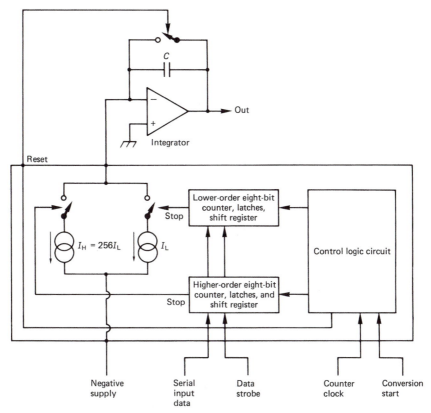

Figure 4.34 Simplified diagram of Sony CX-20017. The high-order and low-order current sources (I_H and I_L) and associated timing circuits can be seen. The necessary integrator is external.

Any inaccuracy in the current ratio will cause one quantizing step in every 256 to be of the wrong size as shown in Figure 4.36, but current tracking is easier to achieve in a monolithic device. The integrator capacitor must have low dielectric leakage and relaxation, and the operational amplifier must have low bias current as this will have the same effect as leakage.

The output of the integrator will remain constant once the current sources are turned off, and the resampling switch will be closed during the voltage plateau to produce the pulse amplitude modulated output. Clearly this device cannot produce a zero-order-hold output without an additional sample-hold stage, so it is naturally complemented by resampling. Once the output pulse has been gated to the reconstruction filter, the capacitor is discharged with a further switch in preparation for the next conversion. The conversion count must take place in rather less than one sample period to permit the resampling and discharge phases. A clock frequency of about 20 MHz is adequate for a sixteen-bit 48 kHz

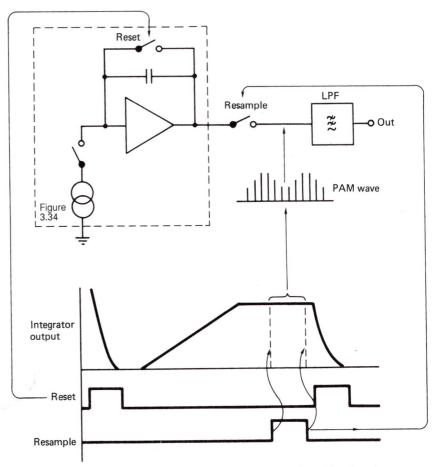

Figure 4.35 In an integrator convertor, the output level is only stable when the ramp finishes. An analog switch is necessary to isolate the ramp from subsequent circuits. The switch can also be used to produce a PAM (pulse amplitude modulated) signal which has a flatter frequency response than a zero-order-hold (staircase) signal.

unit, which permits the ramp to complete in 12.8 μs, leaving 8 μs for resampling and reset.

4.15 Basic analog-to-digital conversion

A conventional analog-to-digital subsystem is shown in Figure 4.37. Following the anti-aliasing filter there will be a sampling process. Many of the ADCs described here will need a finite time to operate, whereas an instantaneous sample must be taken from the input. The solution is to use a track-hold circuit, which was described in section 4.7. Following sampling the sample voltage is quantized. The number of the quantized level is then converted to a binary code, typically two's complement. This

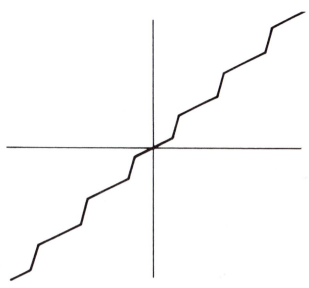

Figure 4.36 Imprecise tracking in a dual-slope convertor results in the transfer function shown here.

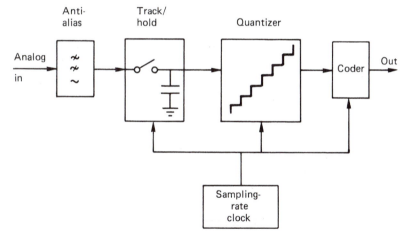

Figure 4.37 A conventional analog-to-digital subsystem. Following the anti-aliasing filter there will be a sampling process, which may include a track-hold circuit. Following quantizing, the number of the quantized level is then converted to a binary code, typically two's complement.

section is concerned primarily with the implementation of the quantizing step.

The general principle of a quantizer is that different quantized voltages are compared with the unknown analog input until the closest quantized voltage is found. The code corresponding to this becomes the output. The comparisons can be made in turn with the minimal amount of hardware, or simultaneously.

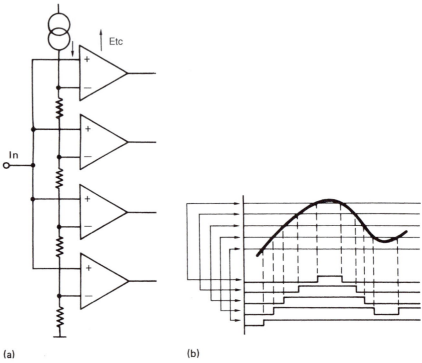

(a)   (b)

Figure 4.38 The flash convertor. In (a) each quantizing interval has its own comparator, resulting in waveforms of (b). A priority encoder is necessary to convert the comparator outputs to a binary code. Shown in (c) is a typical eight-bit flash convertor primarily intended for video applications. (Courtesy TRW)

The flash convertor is probably the simplest technique available for PCM and DPCM conversion. The principle is shown in Figure 4.38. The threshold voltage of every quantizing interval is provided by a resistor chain which is fed by a reference voltage. This reference voltage can be varied to determine the sensitivity of the input. There is one voltage comparator connected to every reference voltage, and the other input of all of these is connected to the analog input. A comparator can be considered to be a one-bit ADC. The input voltage determines how many of the comparators will have a true output. As one comparator is necessary for each quantizing interval, then, for example, in an eight-bit system there will be 255 binary comparator outputs, and it is necessary to use a priority encoder to convert these to a binary code. Note that the quantizing stage is asynchronous; comparators change state as and when the variations in the input waveform result in a reference voltage being crossed. Sampling takes place when the comparator outputs are clocked into a subsequent latch. This is an example of quantizing before sampling as was illustrated in Figure 4.2. Although the device is simple in principle, it contains a lot of circuitry and can only be practicably implemented on

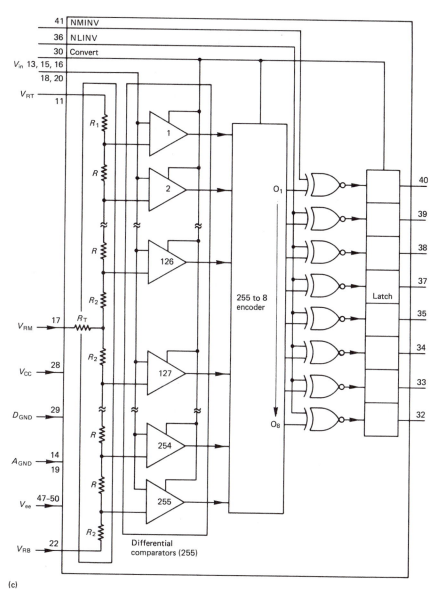

(c)

Figure 4.38 (c) *Note*: RT goes to junction of R2s.

a chip. A sixteen-bit device would need a ridiculous 65 535 comparators, and thus these convertors are not practicable for direct audio conversion, although they will be used to advantage in the DPCM and oversampling convertors described later in this chapter. The analog signal has to drive a lot of inputs which results in a significant parallel capacitance, and a low-impedance driver is essential to avoid restricting the slewing rate of the input. The extreme speed of a flash convertor is a distinct advantage in oversampling. Because computation of all bits is performed simultaneously, no track-hold circuit is required, and droop is eliminated. Figure 4.38 shows a flash convertor chip. Note the resistor ladder and the comparators followed by the priority encoder. The MSB can be selectively inverted so that the device can be used either in offset binary or two's complement mode.

Reduction in component complexity can be achieved by quantizing serially. The most primitive method of generating different quantized voltages is to connect a counter to a DAC as in Figure 4.39. The resulting staircase voltage is compared with the input and used to stop the clock to the counter when the DAC output has just exceeded the input. This method is painfully slow, and is not used, as a much faster method exists which is only slightly more complex. Using successive approximation, each bit is tested in turn, starting with the MSB. If the input is greater than half-range, the MSB will be retained and used as a base to test the next bit, which will be retained if the input exceeds three-quarters range and so on. The number of decisions is equal to the number of bits in the word,

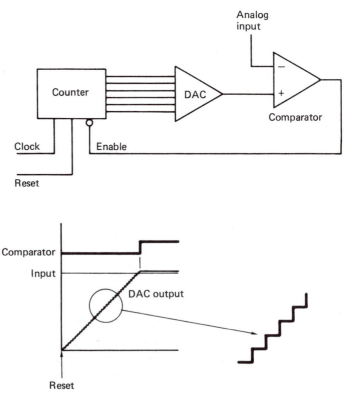

Figure 4.39 Simple ramp ADC compares output of DAC with input. Count is stopped when DAC output just exceeds input. This method, although potentially accurate, is much too slow for digital audio.

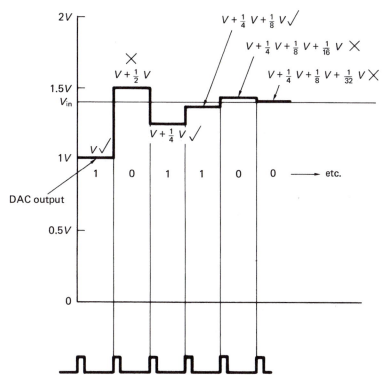

Figure 4.40 Successive approximation tests each bit in turn, starting with the most significant. The DAC output is compared with the input. If the DAC output is below the input (✓) the bit is made 1; if the DAC output is above the input (✗) the bit is made zero.

in contrast to the number of quantizing intervals which was the case in the previous example. A drawback of the successive approximation convertor is that the least significant bits are computed last, when droop is at its worst. Figures 4.40 and 4.41 show that droop can cause a successive approximation convertor to make a significant error under certain circumstances.

Analog-to-digital conversion can also be performed using the dual-current-source type DAC principle in a feedback system; the major difference is that the two current sources must work sequentially rather than concurrently. Figure 4.42 shows a sixteen-bit application in which the capacitor of the track-hold circuit is also used as the ramp integrator. The system operates as follows. When the track-hold FET switches off, the capacitor C will be holding the sample voltage. Two currents of ratio 128:1 are capable of discharging the capacitor. Owing to this ratio, the smaller current will be used to determine the seven least significant bits, and the larger current will determine the nine most significant bits. The currents are provided by current sources of ratio 127:1. When both run

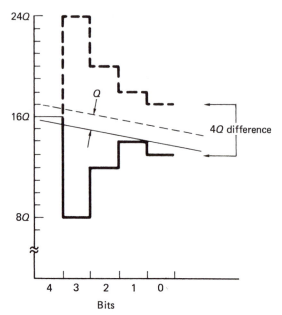

Figure 4.41 Two drooping track-hold signals (solid and dashed lines) which differ by one quantizing interval Q are shown here to result in conversions which are $4Q$ apart. Thus droop can destroy the monotonicity of a convertor. Low-level signals (near the midrange of the number system) are especially vulnerable.

together, the current produced is 128 times that from the smaller source alone. This approach means that the current can be changed simply by turning off the larger source, rather than by attempting a change-over.

With both current sources enabled, the high-order counter counts up until the capacitor voltage has fallen below the reference of $-128Q$ supplied to comparator 1. At the next clock edge, the larger current source is turned off. Waiting for the next clock edge is important, because it ensures that the larger source can only run for entire clock periods, which will discharge the integrator by integer multiples of $128Q$. The integrator voltage will overshoot the $128Q$ reference, and the remaining voltage on the integrator will be less than $128Q$ and will be measured by counting the number of clocks for which the smaller current source runs before the integrator voltage reaches zero. This process is termed residual expansion. The break in the slope of the integrator voltage gives rise to the alternative title of gear-change convertor. Following ramping to ground in the conversion process, the track-hold circuit must settle in time for the next conversion. In this sixteen-bit example, the high-order conversion needs a maximum count of 512, and the low order needs 128: a total of 640. Allowing 25 per cent of the sample period for the track-hold circuit to operate, a 48 kHz convertor would need to be clocked at some 40 MHz. This is rather faster than the clock needed for the DAC using the same technology.

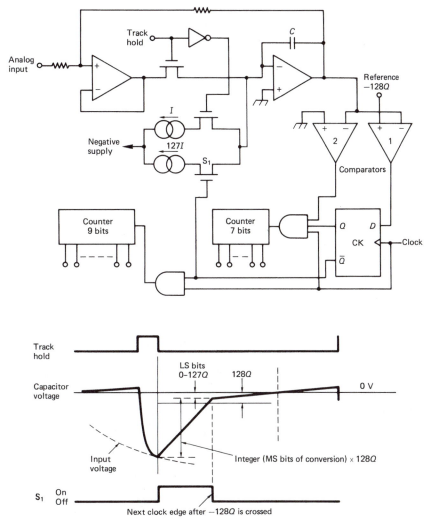

Figure 4.42 Dual-ramp ADC using track-hold capacitor as integrator.

4.16   Alternative convertors

Although PCM audio is universal because of the ease with which it can be recorded and processed numerically, there are several alternative related methods of converting an analog waveform to a bitstream. The output of these convertor types is not Nyquist rate PCM, but this can be obtained from them by appropriate digital processing. In advanced conversion systems it is possible to adopt an alternative convertor technique specifically to take advantage of a particular characteristic. The output is then digitally converted to Nyquist rate PCM in order to obtain the advantages of both.

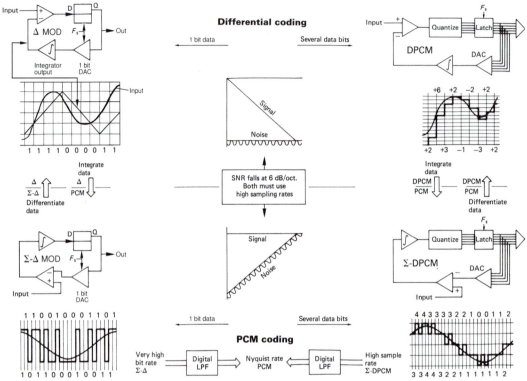

Figure 4.43 The four main alternatives to simple PCM conversion are compared here. Delta modulation is a one-bit case of differential PCM, and conveys the slope of the signal. The digital output of both can be integrated to give PCM. Σ–Δ (sigma–delta) is a one-bit case of Σ-DPCM. The application of integrator before differentiator makes the output true PCM, but tilts the noise floor; hence these can be referred to as 'noise-shaping' convertors.

Conventional PCM has already been introduced. In PCM, the amplitude of the signal only depends on the number range of the quantizer, and is independent of the frequency of the input. Similarly, the amplitude of the unwanted signals introduced by the quantizing process is also largely independent of input frequency.

Figure 4.43 introduces the alternative convertor structures. The top half of the diagram shows convertors which are differential. In differential coding the value of the output code represents the difference between the current sample voltage and that of the previous sample. The lower half of the diagram shows convertors which are PCM. In addition, the left side of the diagram shows single-bit convertors, whereas the right side shows multi-bit convertors.

In differential pulse code modulation (DPCM), shown at top right, the difference between the previous absolute sample value and the current one is quantized into a multi-bit binary code. It is possible to produce a DPCM signal from a PCM signal simply by subtracting successive samples; this is digital differentiation. Similarly the reverse process is possible by using an accumulator or digital integrator (see Chapter 2) to compute sample values from the differences received. The problem with this approach is that it is very easy to lose the baseline of the signal if it commences at some arbitrary time. A digital high-pass filter can be used to prevent unwanted offsets.

Differential convertors do not have an absolute amplitude limit. Instead there is a limit to the maximum rate at which the input signal voltage can change. They are said to be slew rate limited, and thus the permissible signal amplitude falls at 6 dB per octave. As the quantizing steps are still uniform, the quantizing error amplitude has the same limits as PCM. As input frequency rises, ultimately the signal amplitude available will fall down to it.

If DPCM is taken to the extreme case where only a binary output signal is available then the process is described as delta modulation (top-left in Figure 4.43). The meaning of the binary output signal is that the current analog input is above or below the accumulation of all previous bits. The characteristics of the system show the same trends as DPCM, except that there is severe limiting of the rate of change of the input signal. A DPCM decoder must accumulate all the difference bits to provide a PCM output for conversion to analog, but with a one-bit signal the function of the accumulator can be performed by an analog integrator.

If an integrator is placed in the input to a delta modulator, the integrator's amplitude response loss of 6 dB per octave parallels the convertor's amplitude limit of 6 dB per octave; thus the system amplitude limit becomes independent of frequency. This integration is responsible for the term sigma-delta modulation, since in mathematics sigma is used to denote summation. The input integrator can be combined with the integrator already present in a delta-modulator by a slight rearrangement of the components (bottom-left in Figure 4.43). The transmitted signal is now the amplitude of the input, not the slope; thus the receiving integrator can be dispensed with, and all that is necessary to after the DAC is an LPF to smooth the bits. The removal of the integration stage at the decoder now means that the quantizing error amplitude rises at 6 dB per octave, ultimately meeting the level of the wanted signal.

The principle of using an input integrator can also be applied to a true DPCM system and the result should perhaps be called sigma DPCM (bottom-right in Figure 4.43). The dynamic range improvement over delta–sigma modulation is 6 dB for every extra bit in the code. Because the level of the quantizing error signal rises at 6 dB per octave in both delta–sigma modulation and sigma DPCM, these systems are sometimes referred to as 'noise-shaping' convertors, although the word 'noise' must be used with some caution. The output of a sigma DPCM system is again PCM, and a DAC will be needed to receive it, because it is a binary code.

As the differential group of systems suffer from a wanted signal that converges with the unwanted signal as frequency rises, they must all use very high sampling rates.[27] It is possible to convert from sigma DPCM to conventional PCM by reducing the sampling rate digitally. When the sampling rate is reduced in this way, the reduction of bandwidth excludes a disproportionate amount of noise because the noise shaping concentrated it at frequencies beyond the audio band. The use of noise shaping and oversampling is the key to the high resolution obtained in advanced convertors.

4.17   Oversampling

Oversampling means using a sampling rate which is greater (generally substantially greater) than the Nyquist rate. Neither sampling theory nor quantizing theory *require* oversampling to be used to obtain a given signal quality, but Nyquist rate conversion places extremely high demands on component accuracy when a convertor is implemented. Oversampling allows a given signal quality to be reached without requiring very close tolerance, and therefore expensive, components. Although it can be used alone, the advantages of oversampling are better realized when it is used in conjunction with noise shaping. Thus in practice the two processes are generally used together and the terms are often seen used in the loose sense as if they were synonymous. For a detailed and quantitative analysis of oversampling having exhaustive references the serious reader is referred to Hauser.[28]

In section 4.4, where dynamic element matching was described, it was seen that component accuracy was traded for accuracy in the time domain. Oversampling is another example of the same principle.

Figure 4.44 shows the main advantages of oversampling. At (a) it will be seen that the use of a sampling rate considerably above the Nyquist rate allows the anti-aliasing and reconstruction filters to be realized with a much more gentle cut-off slope. There is then less likeliehood of phase linearity and ripple problems in the audio passband.

Figure 4.44(b) shows that information in an analog signal is two-dimensional and can be depicted as an area which is the product of bandwidth and the linearly expressed signal-to-noise ratio. The figure also shows that the same amount of information can be conveyed down a channel with a SNR of half as much (6 dB less) if the bandwidth used is doubled, with 12 dB less SNR if bandwidth is quadrupled, and so on, provided that the modulation scheme used is perfect.

The information in an analog signal can be conveyed using some analog modulation scheme in any combination of bandwidth and SNR which yields the appropriate channel capacity. If bandwidth is replaced

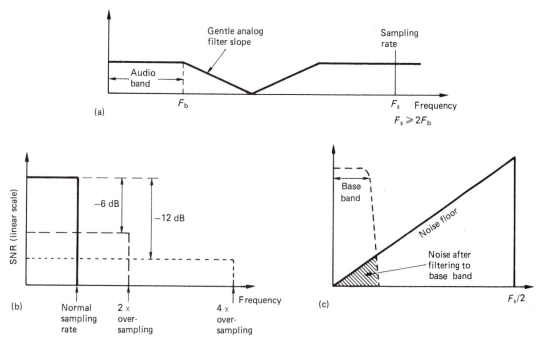

Figure 4.44 Oversampling has a number of advantages. In (a) it allows the slope of analog filters to be relaxed. In (b) it allows the resolution of convertors to be extended. In (c) *a noise-shaped* convertor allows a disproportionate improvement in resolution.

by sampling rate and SNR is replaced by a function of wordlength, the same must be true for a digital signal as it is no more than a numerical analog. Thus raising the sampling rate potentially allows the wordlength of each sample to be reduced without information loss.

Oversampling permits the use of a convertor element of shorter wordlength, making it possible to use a flash convertor. The flash convertor is capable of working at very high frequency and so large oversampling factors are easily realized. The flash convertor needs no track-hold system as it works instantaneously. The drawbacks of track-hold set out in section 4.6 are thus eliminated. If the sigma-DPCM convertor structure of Figure 4.43 is realized with a flash convertor element, it can be used with a high oversampling factor. Figure 4.44(c) shows that this class of convertor has a rising noise floor. If the highly oversampled output is fed to a digital low-pass filter which has the same frequency response as an analog anti-aliasing filter used for Nyquist rate sampling, the result is a disproportionate reduction in noise because the majority of the noise was outside the audio band. A high-resolution convertor can be obtained using this technology without requiring unattainable component tolerances.

Information theory predicts that if an audio signal is spread over a much wider bandwidth by, for example, the use of an FM broadcast

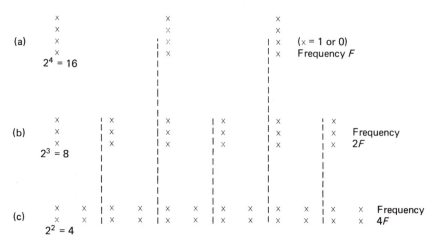

Figure 4.45 Information rate can be held constant when frequency doubles by removing one-bit from each word. In all cases here it is 16*F*. Note bit rate of (c) is double that of (a). Data storage in oversampled form is inefficient.

						Digital audio sample values
					0000 0	0000
					0001 1	
					0010 2	
					0011 3	
					0100 4	
					0101 5	
					0110 6	
					0111 7	
				000 do	1000 8	
				001 re	1001 9	
				010 mi	1010 A	
				011 fa	1011 B	
			00 = Spring	100 so	1100 C	
			01 = Summer	101 la	1101 D	
		0 = No	10 = Autumn	110 te	1110 E	
		1 = Yes	11 = Winter	111 do	1111 F	FFFF
No of bits		1	2	3	4	16
Information per word		2	4	8	16	65 536
Information per bit		2	2	≈3	4	4096

Figure 4.46 The amount of information per bit increases disproportionately as wordlength increases. It is always more efficient to use the longest words possible at the lowest word rate. It will be evident that sixteen-bit PCM is 2048 times as efficient as delta modulation. Oversampled data are also inefficient for storage.

transmitter, the SNR of the demodulated signal can be higher than that of the channel it passes through, and this is also the case in digital systems. The concept is illustrated in Figure 4.45. At (a) four-bit samples are delivered at sampling rate *F*. As four bits have sixteen combinations, the information rate is 16 F. At (b) the same information rate is obtained with three-bit samples by raising the sampling rate to 2 F and at (c) two-bit samples having four combinations require to be delivered at a rate of 4 F. Whilst the information rate has been maintained, it will be noticed that the bit-rate of (c) is twice that of (a). The reason for this is shown in Figure 4.46. A single binary digit can only have two states; thus it can only convey two pieces of information, perhaps 'yes' or 'no'. Two binary digits together can have four states, and can thus convey four pieces of information, perhaps 'spring summer autumn or winter', which is two pieces of information per bit. Three binary digits grouped together can have eight combinations, and convey eight pieces of information, perhaps 'doh re mi fah so lah te or doh', which is nearly three pieces of information per digit. Clearly the further this principle is taken, the greater the benefit. In a sixteen-bit system, each bit is worth 4K pieces of information. It is always more efficient, in information-capacity terms, to use the combinations of long binary words than to send single bits for every piece of information. The greatest efficiency is reached when the longest words are sent at the slowest rate which must be the Nyquist rate. This is one reason why PCM recording is more common than delta modulation, despite the simplicity of implementation of the latter type of convertor. PCM simply makes more efficient use of the capacity of the binary channel.

As a result, oversampling is confined to convertor technology where it gives specific advantages in implementation. The storage or transmission system will usually employ PCM, where the sampling rate is a little more than twice the audio bandwidth. Figure 4.47 shows a digital audio tape recorder such as DAT using oversampling convertors. The ADC runs at n times the Nyquist rate, but once in the digital domain the rate needs to be

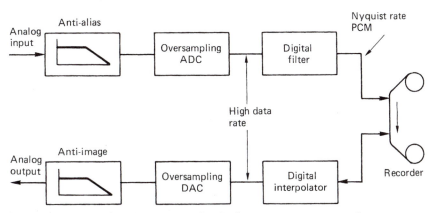

Figure 4.47 A recorder using oversampling in the convertors overcomes the shortcomings of analog anti-aliasing and reconstruction filters and the convertor elements are easier to construct; the recording is made with Nyquist rate PCM which minimizes tape consumption.

reduced in a type of digital filter called a *decimator*. The output of this is conventional Nyquist rate PCM, according to the tape format, which is then recorded. On replay the sampling rate is raised once more in a further type of digital filter called an *interpolator*. The system now has the best of both worlds: using oversampling in the convertors overcomes the shortcomings of analog anti-aliasing and reconstruction filters and the wordlength of the convertor elements is reduced making them easier to construct; the recording is made with Nyquist rate PCM which minimizes tape consumption. Digital filters have the characteristic that their frequency response is proportional to the sampling rate. If a digital recorder is played at a reduced speed, the response of the digital filter will reduce automatically and prevent images passing the reconstruction process.

Oversampling is a method of overcoming practical implementation problems by replacing a single critical element or bottleneck by a number of elements whose overall performance is what counts. As Hauser[28] properly observed, oversampling tends to overlap the operations which are quite distinct in a conventional convertor. In earlier sections of this chapter, the vital subjects of filtering, sampling, quantizing and dither have been treated almost independently. Figure 4.48(a) shows that it is possible to construct an ADC of predictable performance by taking a suitable anti-aliasing filter, a sampler, a dither source and a quantizer and assembling them like building bricks. The bricks are effectively in series and so the performance of each stage can only limit the overall

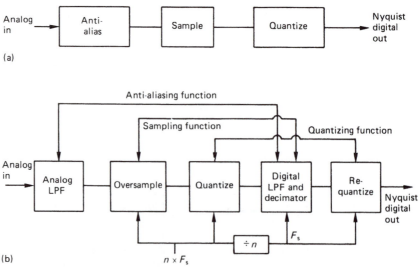

(a)

(b)

Figure 4.48 A conventional ADC performs each step in an identifiable location as in (a). With oversampling, many of the steps are distributed as shown in (b).

performance. In contrast, Figure 4.48(b) shows that with oversampling the overlap of operations allows different processes to augment one another allowing a synergy which is absent in the conventional approach.

If the oversampling factor is n, the analog input must be bandwidth limited to $n.F_s/2$ by the analog anti-aliasing filter. This unit need only have flat frequency response and phase linearity within the audio band. Analog dither of an amplitude compatible with the quantizing interval size is added prior to sampling at $n.F_s/2$ and quantizing.

Next, the anti-aliasing function is completed in the digital domain by a low-pass filter which cuts off at $F_s/2$. Using an appropriate architecture this filter can be absolutely phase linear and implemented to arbitrary accuracy. Such filters are discussed in Chapter 3. The filter can be considered to be the demodulator of Figure 4.44 where the SNR improves as the bandwidth is reduced. The wordlength can be expected to increase. As Chapter 3 illustrated, the multiplications taking place within the filter extend the wordlength considerably more than the bandwidth reduction alone would indicate. The analog filter serves only to prevent aliasing into the audio band at the oversampling rate; the audio spectrum is determined with greater precision by the digital filter.

With the audio information spectrum now Nyquist limited, the sampling process is completed when the rate is reduced in the decimator. One sample in n is retained.

The excess wordlength extension due to the anti-aliasing filter arithmetic must then be removed. Digital dither is added, completing the dither process, and the quantizing process is completed by requantizing

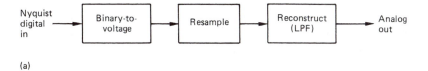

(a)

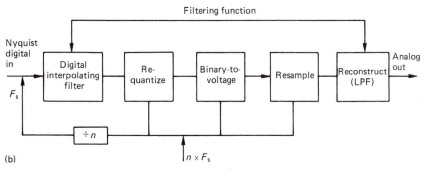

(b)

Figure 4.49 A conventional DAC in (a) is compared with the oversampling implementation in (b).

the dithered samples to the appropriate wordlength which will be greater than the wordlength of the first quantizer. Alternatively noise shaping may be employed.

Figure 4.49(a) shows the building-brick approach of a conventional DAC. The Nyquist rate samples are converted to analog voltages and then a steep-cut analog low-pass filter is needed to reject the sidebands of the sampled spectrum.

Figure 4.49(b) shows the oversampling approach. The sampling rate is raised in an interpolator which contains a low-pass filter which restricts the baseband spectrum to the audio bandwidth shown. A large frequency gap now exists between the baseband and the lower sideband. The multiplications in the interpolator extend the wordlength considerably and this must be reduced within the capacity of the DAC element by the addition of digital dither prior to requantizing. Again noise shaping may be used as an alternative.

4.18   Oversampling without noise shaping

If an oversampling convertor is considered which makes no attempt to shape the noise spectrum, it will be clear that if it contains a perfect quantizer, no amount of oversampling will increase the resolution of the system, since a perfect quantizer is blind to all changes of input within one quantizing interval, and looking more often is of no help. It was shown earlier that the use of dither would linearize a quantizer, so that input changes much smaller than the quantizing interval would be reflected in the output and this remains true for this class of convertor.

Figure 4.50 shows the example of a white-noise-dithered quantizer, oversampled by a factor of four. Since dither is correctly employed, it is valid to speak of the unwanted signal as noise. The noise power extends over the whole baseband up to the Nyquist limit. If the basebandwidth is reduced by the oversampling factor of four back to the bandwidth of the original analog input, the noise bandwidth will also be reduced by a factor of four, and the noise power will be one-quarter of that produced at the quantizer. One-quarter noise power implies one-half the noise voltage, so the SNR of this example has been increased by 6 dB, the equivalent of one extra bit in the quantizer. Information theory predicts that an oversampling factor of four would allow an extension by two bits. This method is suboptimal in that very large oversampling factors would be needed to obtain useful resolution extension, but it would still realize some advantages, particularly the elimination of the steep-cut analog filter.

The division of the noise by a larger factor is the only route left open, since all the other parameters are fixed by the signal bandwidth required.

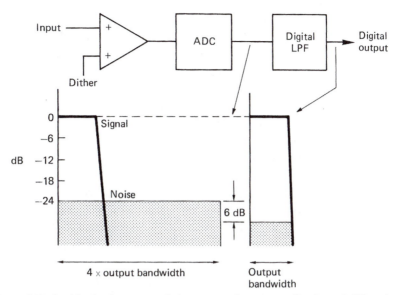

Figure 4.50 In this simple oversampled convertor, 4× oversampling is used. When the convertor output is low-pass filtered, the noise power is reduced to one-quarter, which in voltage terms is 6 dB. This is a suboptimal method and is not used.

The reduction of noise power resulting from a reduction in bandwidth is only proportional if the noise is white, i.e. it has uniform power spectral density (PSD). If the noise from the quantizer is made spectrally non-uniform, the oversampling factor will no longer be the factor by which the noise power is reduced. The goal is to concentrate noise power at high frequencies, so that after low-pass filtering in the digital domain down to the audio input bandwidth, the noise power will be reduced by more than the oversampling factor.

4.19   Noise shaping

Noise shaping dates from the work of Cutler[29] in the 1950s. It is a feedback technique applicable to quantizers and requantizers in which the quantizing process of the current sample is modified in some way by the quantizing error of the previous sample.

When used with requantizing, noise shaping is an entirely digital process which is used, for example, following word extension due to the arithmetic in digital mixers or filters in order to return to the required wordlength. It will be found in this form in oversampling DACs. When used with quantizing, part of the noise-shaping circuitry will be analog. As the feedback loop is placed around an ADC it must contain a DAC. When used in convertors, noise shaping is primarily an implementation technology. It allows processes which are conveniently available in integrated circuits to be put to use in audio conversion. Once integrated circuits can be employed, complexity ceases to be a drawback and low-cost mass production is possible.

It has been stressed throughout this chapter that a series of numerical values or samples is just another analog of an audio waveform. Chapter 3 showed that all analog processes such as mixing, attenuation or integration all have exact numerical parallels. It has been demonstrated that digitally dithered requantizing is no more than a digital simulation of analog quantizing. It should be no surprise that in this section noise shaping will be treated in the same way. Noise shaping can be performed by manipulating analog voltages or numbers representing them or both. If the reader is content to make a conceptual switch between the two, many obstacles to understanding fall, not just in this topic, but in digital audio in general.

The term noise shaping is idiomatic and in some respects unsatisfactory because not all devices which are called noise shapers produce true noise. The caution which was given when treating quantizing error as noise is also relevant in this context. Whilst 'quantizing-error-spectrum shaping' is a bit of a mouthful, it is useful to keep in mind that noise shaping means just that in order to avoid some pitfalls. Some noise-shaper architectures do not produce a signal decorrelated quantizing error and need to be dithered.

Figure 4.51(a) shows a requantizer using a simple form of noise shaping. The low-order bits which are lost in requantizing are the quantizing error. If the value of these bits is added to the next sample before it is requantized, the quantizing error will be reduced. The process is somewhat like the use of negative feedback in an operational amplifier except that it is not instantaneous, but encounters a one sample delay. With a constant input, the mean or average quantizing error will be brought to zero over a number of samples, achieving one of the goals of additive dither. The more rapidly the input changes, the greater the effect of the delay and the less effective the error feedback will be. Figure 4.51(b) shows the equivalent circuit seen by the quantizing error, which is created at the requantizer and subtracted from itself one sample period later. As a result the quantizing error spectrum is not uniform, but has the shape of a raised sinewave shown at (c), hence the term noise shaping. The noise is very small at DC and rises with frequency, peaking at the Nyquist frequency at a level determined by the size of the quantizing step. If used with oversampling, the noise peak can be moved outside the audio band.

Figure 4.52 shows a simple example in which two low-order bits need to be removed from each sample. The accumulated error is controlled by using the bits which were neglected in the truncation, and adding them

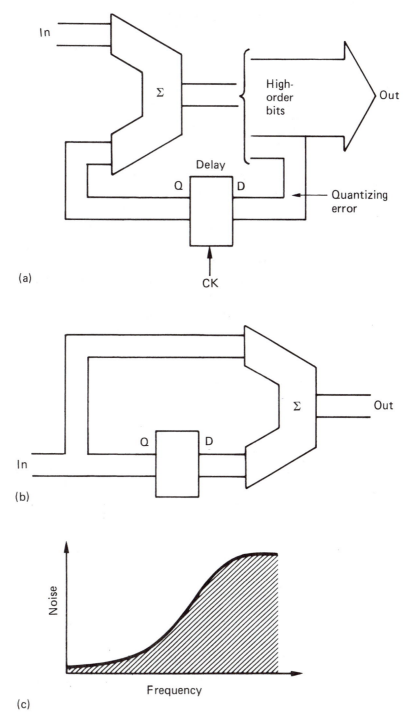

(a)

(b)

(c)

Figure 4.51 (a) A simple requantizer which feeds back the quantizing error to reduce the error of subsequent samples. The one-sample delay causes the quantizing error to see the equivalent circuit shown in (b) which results in a sinusoidal quantizing error spectrum shown in (c).

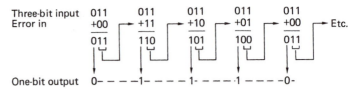

Figure 4.52 By adding the error caused by truncation to the next value, the resolution of the lost bits is maintained in the duty cycle of the output. Here, truncation of 011 by 2 bits would give continuous zeros, but the system repeats 0111, 0111, which, after filtering, will produce a level of three-quarters of a bit.

to the next sample. In this example, with a steady input, the roundoff mechanism will produce an output of 01110111 ... If this is low-pass filtered, the three ones and one zero result in a level of three-quarters of a quantizing interval, which is precisely the level which would have been obtained by direct conversion of the full digital input. Thus the resolution is maintained even though two bits have been removed.

The noise-shaping technique was used in the first-generation Philips CD players which oversampled by a factor of four. Starting with sixteen-bit PCM from the disc, the 4× oversampling will in theory permit the use of an ideal fourteen-bit convertor, but only if the wordlength is reduced optimally. The oversampling DAC system used is shown in Figure 4.53.[30] The interpolator arithmetic extends the wordlength to 28 bits, and this is reduced to 14 bits using the error feedback loop of Figure 4.51. The noise floor rises slightly towards the edge of the audio band, but remains below the noise level of a conventional sixteen-bit DAC which is shown for comparison.

The fourteen-bit samples then drive a DAC using dynamic element matching. The aperture effect in the DAC is used as part of the reconstruction filter response, in conjunction with a third-order Bessel

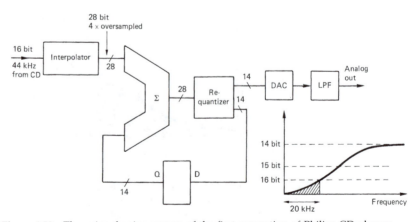

Figure 4.53 The noise-shaping system of the first generation of Philips CD players.

filter which has a response 3 dB down at 30 kHz. Equalization of the aperture effect within the audio passband is achieved by giving the digital filter which produces the oversampled data a rising response. The use of a digital interpolator as part of the reconstruction filter results in extremely good phase linearity.

Noise shaping can also be used without oversampling. In this case the noise cannot be pushed outside the audio band. Instead the noise floor is shaped or weighted to complement the unequal spectral sensitivity of the ear to noise.[20,31,32] Unless we wish to violate Shannon's theory, this psychoacoustically optimal noise shaping can only reduce the noise power at certain frequencies by increasing it at others. Thus the average log PSD over the audio band remains the same, although it may be raised slightly by noise induced by imperfect processing.

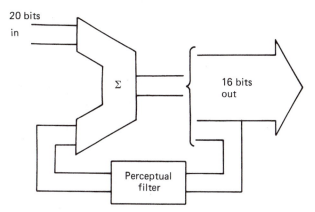

Figure 4.54 Perceptual filtering in a requantizer gives a subjectively improved SNR.

Figure 4.54 shows noise shaping applied to a digitally dithered requantizer. Such a device might be used when, for example, making a CD master from a twenty-bit recording format. The input to the dithered requantizer is subtracted from the output to give the error due to requantizing. This error is filtered (and inevitably delayed) before being subtracted from the system input. The filter is not designed to be the exact inverse of the perceptual weighting curve because this would cause extreme noise levels at the ends of the band. Instead the perceptual curve is levelled off[33] such that it cannot fall more than e.g. 40 dB below the peak.

Psychoacoustically optimal noise shaping can offer nearly three bits of increased dynamic range when compared with optimal spectrally flat dither. Enhanced Compact Discs recorded using these techniques are now available.

4.20 Noise-shaping ADCs

The sigma DPCM convertor introduced in Figure 4.43 has a natural application here and is shown in more detail in Figure 4.55. The current digital sample from the quantizer is converted back to analog in the embedded DAC. The DAC output differs from the ADC input by the quantizing error. The DAC output is subtracted from the analog input to produce an error which is integrated to drive the quantizer in such a way that the error is reduced. With a constant input voltage the average error will be zero because the loop gain is infinite at DC. If the average error is zero, the mean or average of the DAC outputs must be equal to the analog input. The instantaneous output will deviate from the average in what is called an idling pattern. The presence of the integrator in the error feedback loop makes the loop gain fall with rising frequency. With the feedback falling at 6 dB per octave, the noise floor will rise at the same rate.

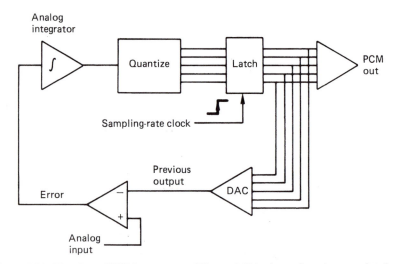

Figure 4.55 The sigma DPCM convertor of Figure 4.43 is shown here in more detail.

Figure 4.56 shows a simple oversampling system using a sigma-DPCM convertor and an oversampling factor of only four. The sampling spectrum shows that the noise is concentrated at frequencies outside the audio part of the oversampling baseband. Since the scale used here means that noise power is represented by the area under the graph, the area left under the graph after the filter shows the noise-power reduction. Using the relative areas of similar triangles shows that the reduction has been by a factor of sixteen. The corresponding noise-voltage reduction would be a factor of four, or 12 dB, which corresponds to an additional two bits in

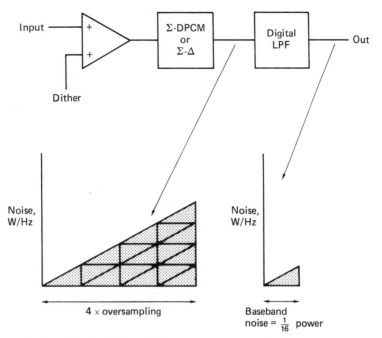

Figure 4.56 In a sigma-DPCM or $\Sigma-\Delta$ convertor, noise amplitude increases by 6 dB/octave, noise power by 12dB/octave. In this 4× oversampling convertor, the digital filter reduces bandwidth by four, but noise power is reduced by a factor of 16. Noise voltage falls by a factor of four or 12 dB.

wordlength. These bits will be available in the wordlength extension which takes place in the decimating filter. Owing to the rise of 6 dB per octave in the PSD of the noise, the SNR will be 3 dB worse at the edge of the audio band.

One way in which the operation of the system can be understood is to consider that the coarse DAC in the loop defines fixed points in the audio transfer function. The time averaging which takes place in the decimator then allows the transfer function to be interpolated between the fixed points. True signal-independent noise of sufficient amplitude will allow this to be done to infinite resolution, but by making the noise primarily outside the audio band the resolution is maintained but the audio band signal-to-noise ratio can be extended. A first-order noise shaping ADC of the kind shown can produce signal-dependent quantizing error and requires analog dither. However, this can be outside the audio band and so need not reduce the SNR achieved.

A greater improvement in dynamic range can be obtained if the integrator is supplanted to realize a higher-order filter.[34] The filter is in the feedback loop and so the noise will have the opposite response to the filter and will therefore rise more steeply to allow a greater SNR enhancement after decimation. Figure 4.57 shows the theoretical SNR

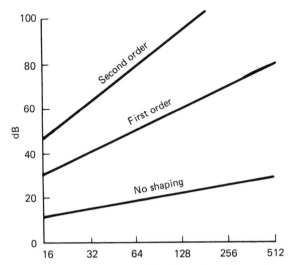

Figure 4.57 The enhancement of SNR possible with various filter orders and oversampling factors in noise-shaping convertors.

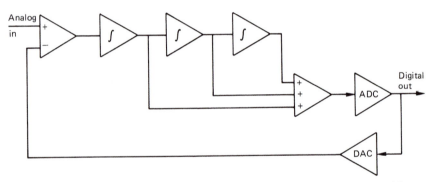

Figure 4.58 Stabilizing the loop filter in a noise-shaping convertor can be assisted by the incorporation of feedforward paths as shown here.

enhancement possible for various loop filter orders and oversampling factors. A further advantage of high-order loop filters is that the quantizing noise can be decorrelated from the signal, making dither unnecessary. High-order loop filters were at one time thought to be impossible to stabilize, but this is no longer the case, although care is necessary. One technique which may be used is to include some feedforward paths as shown in Figure 4.58.

An ADC with high-order noise shaping was disclosed by Adams[35] and a simplified diagram is shown in Figure 4.59. The comparator outputs of the 128 times oversampled four-bit flash ADC are directly fed to the DAC which consists of fifteen equal resistors fed by CMOS switches. As with all feedback loops, the transfer characteristic cannot be more accurate than the feedback, and in this case the feedback accuracy is determined

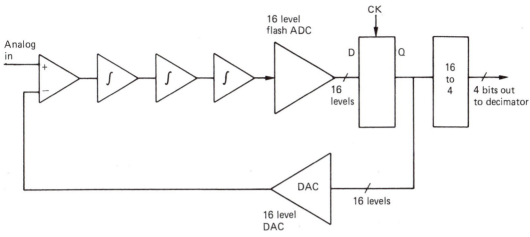

Figure 4.59 An example of a high-order noise-shaping ADC. See text for details.

by the precision of the DAC.[36] Driving the DAC directly from the ADC comparators is more accurate because each input has equal weighting. The stringent MSB tolerance of the conventional binary weighted DAC is then avoided. The comparators also drive a 16 to 4 priority encoder to provide the four-bit PCM output to the decimator. The DAC output is subtracted from the analog input at the integrator. The integrator is followed by a pair of conventional analog operational amplifiers having frequency-dependent feedback and a passive network which gives the loop a fourth-order response overall. The noise floor is thus shaped to rise at 24 dB per octave beyond the audio band. The time constants of the loop filter are optimized to minimize the amplitude of the idling pattern as this is an indicator of the loop stability. The four-bit PCM output is low-pass filtered and decimated to the Nyquist frequency. The high oversampling factor and high-order noise shaping extend the dynamic range of the four-bit flash ADC to 108 dB at the output.

4.21   A one-bit DAC

It might be thought that the waveform from a one-bit DAC is simply the same as the digital input waveform. In practice this is not the case. The input signal is a logic signal which need only be above or below a threshold for its binary value to be correctly received. It may have a variety of waveform distortions and a duty cycle offset. The area under the pulses can vary enormously. In the DAC output the amplitude needs to be extremely accurate. A one-bit DAC uses only the binary information from the input, but reclocks to produce accurate timing and uses a

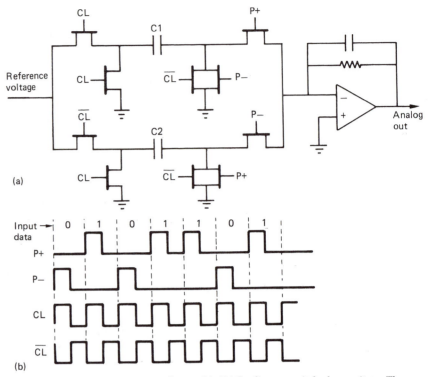

Figure 4.60 In (a) the operation of a one-bit DAC relies on switched capacitors. The switching waveforms are shown in (b).

reference voltage to produce accurate levels. The area of pulses produced is then constant. One-bit DACs will be found in noise-shaping ADCs as well as in the more obvious application of producing analog audio.

Figure 4.60(a) shows a one-bit DAC which is implemented with MOS field-effect switches and a pair of capacitors. Quanta of charge are driven into or out of a virtual earth amplifier configured as an integrator by the switched capacitor action. Figure 4.60(b) shows the associated waveforms. Each data bit period is divided into two equal portions; that for which the clock is high, and that for which it is low. During the first half of the bit period, pulse P+ is generated if the data bit is a 1, or pulse P– is generated if the data bit is a 0. The reference input is a clean voltage corresponding to the gain required.

C1 is *discharged* during the second half of every cycle by the switches driven from the complemented clock. If the next bit is a 1, during the next high period of the clock the capacitor will be connected between the reference and the virtual earth. Current will flow into the virtual earth until the capacitor is charged. If the next bit is not a 1, the current through C1 will flow to ground.

C2 is *charged* to reference voltage during the second half of every cycle by the switches driven from the complemented clock. On the next high period of the clock, the reference end of C2 will be grounded, and so the op-amp end wil assume a negative reference voltage. If the next bit is a 0, this negative reference will be switched into the virtual earth, if not the capacitor will be discharged.

Thus on every cycle of the clock, a quantum of charge is either pumped into the integrator by C1 or pumped out by C2. The analog output therefore precisely reflects the ratio of ones to zeros.

4.22   One-bit noise-shaping ADCs

In order to overcome the DAC accuracy constraint of the sigma DPCM convertor, the sigma–delta convertor can be used as it has only one-bit internal resolution. A one-bit DAC cannot be non-linear by definition as it defines only two points on a transfer function. It can, however, suffer from other deficiencies such as DC offset and gain error although these are less offensive in audio. The one-bit ADC is a comparator.

As the sigma–delta convertor is only a one-bit device, clearly it must use a high oversampling factor and high-order noise shaping in order to have sufficiently good SNR for audio.[37] In practice the oversampling factor is limited not so much by the convertor technology as by the difficulty of computation in the decimator. A sigma–delta convertor has the advantage that the filter input 'words' are one bit long and this simplifies the filter design as multiplications can be replaced by selection of constants.

Conventional analysis of loops falls down heavily in the one-bit case. In particular the gain of a comparator is difficult to quantify, and the loop is highly non-linear so that considering the quantizing error as additive white noise in order to use a linear loop model gives rather optimistic results. In the absence of an accurate mathematical model, progress has been made empirically, with listening tests and by using simulation.

Single-bit sigma–delta convertors are prone to long idling patterns because the low resolution in the voltage domain requires more bits in the time domain to be integrated to cancel the error. Clearly the longer the period of an idling pattern, the more likely it is to enter the audio band as an objectional whistle or 'birdie'. They also exhibit threshold effects or deadbands where the output fails to react to an input change at certain levels. The problem is reduced by the order of the filter and the wordlength of the embedded DAC. Second- and third-order feedback loops are still prone to audible idling patterns and threshold effect.[38] The traditional approach to linearizing sigma–delta convertors is to use dither. Unlike conventional quantizers, the dither used was of a frequency outside the audio band and of considerable level. Square-wave dither has been used and it is advantageous to choose a frequency which is a multiple of the final output sampling rate as then the harmonics will coincide with the troughs in the stopband ripple of the decimator. Unfortunately the level of dither needed to linearize the convertor is high enough to cause premature clipping of high-level signals, reducing the dynamic range. This problem is overcome by using in-band white noise dither at low level.[39]

An advantage of the one-bit approach is that in the one-bit DAC, precision components are replaced by precise timing in switched capacitor networks. The same approach can be used to implement the loop filter in an ADC. Figure 4.61 shows a third-order sigma–delta modulator incorporating a DAC based on the principle of Figure 4.60. The loop filter is also implemented with switched capacitors.

4.23   Operating levels in digital audio

Analog tape recorders use operating levels which are some way below saturation. The range between the operating level and saturation is called the headroom. In this range, distortion becomes progressively worse and sustained recording in the headroom is avoided. However, transients may be recorded in the headroom as the ear cannot respond to distortion products unless they are sustained. The PPM level meter has an attack time constant which simulates the temporal distortion sensitivity of the ear. If a transient is too brief to deflect a PPM into the headroom, distortion will not be heard either.

Operating levels are used in two ways. On making a recording from a microphone, the gain is increased until distortion is just avoided, thereby obtaining a recording having the best SNR. In post-production the gain will be set to whatever level is required to obtain the desired subjective effect in the context of the program material. This is particularly important to broadcasters who require the relative loudness of different material to be controlled so that the listener does not need to make continuous adjustments to the volume control.

In order to maintain level accuracy, analog recordings are traditionally preceded by line-up tones at standard operating level. These are used to adjust the gain in various stages of dubbing and transfer along land lines so that no level changes occur to the program material.

Unlike analog recorders, digital recorders do not have headroom, as there is no progressive onset of distortion until convertor clipping, the equivalent of saturation, occurs at 0 dBFs. Accordingly many digital recorders have level meters which read in dBFs. The scales are marked

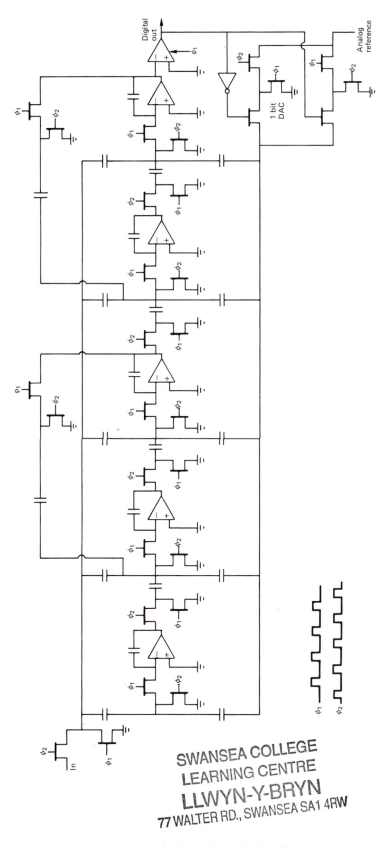

Figure 4.61 A third-order sigma–delta modulator using a switched capacitor loop filter.

with 0 at the clipping level and all operating levels are below that. This causes no difficulty provided the user is aware of the consequences.

However, in the situation where a digital copy of an analog tape is to be made, it is very easy to set the input gain of the digital recorder so that line-up tone from the analog tape reads 0 dB. This lines up digital clipping with the analog operating level. When the tape is dubbed, all signals in the headroom suffer convertor clipping.

In order to prevent such problems, manufacturers and broadcasters have introduced artificial headroom on digital level meters, simply by calibrating the scale and changing the analog input sensitivity so that 0 dB analog is some way below clipping. Unfortunately there has been little agreement on how much artificial headroom should be provided, and machines which have it are seldom labelled with the amount. There is an argument which suggests that the amount of headroom should be a function of the sample wordlength, but this causes difficulties when transferring from one wordlength to another. The EBU[40] concluded that a single relationship between analog and digital level was desirable. In sixteen-bit working, 12 dB of headroom is a useful figure, but now that eighteen- and twenty-bit convertors are available, the EBU recommends 18 dB.

References

1. Shannon, C.E., A mathematical theory of communication. *Bell Syst. Tech. J.*, **27**, 379 (1948)
2. Jerri, A.J., The Shannon sampling theorem – its various extensions and applications: a tutorial review. *Proc. IEEE*, **65**, 1565–1596 (1977)
3. Betts, J.A., *Signal Processing Modulation and Noise*, Sevenoaks: Hodder and Stoughton (1970)
4. Meyer, J., Time correction of anti-aliasing filters used in digital audio systems. *J. Audio Eng. Soc.*, **32**, 132–137 (1984)
5. Lipshitz, S.P., Pockock, M. and Vanderkooy, J., On the audibility of midrange phase distortion in audio systems. *J. Audio Eng. Soc.*, **30**, 580–595 (1982)
6. Preis, D. and Bloom, P.J., Perception of phase distortion in anti-alias filters. *J.Audio Eng. Soc.*, **32**, 842–848 (1984)
7. Lagadec, R. and Stockham, T.G., Jr, Dispersive models for A-to-D and D-to-A conversion systems. Presented at the 75th Audio Engineering Society Convention (Paris, 1984), Preprint 2097(H-8)
8. Blesser, B., Advanced A/D conversion and filtering: data conversion. In *Digital Audio*, edited by B.A. Blesser, B. Locanthi and T.G. Stockham Jr, pp. 37–53, New York: Audio Engineering Society (1983)
9. Lagadec, R., Weiss, D. and Greutmann, R., High-quality analog filters for digital audio. Presented at the 67th Audio Engineering Society Convention (New York, 1980), Preprint 1707(B–4)
10. Anon., AES recommended practice for professional digital audio applications employing pulse code modulation: preferred sampling frequencies. AES5–1984 (ANSI S4.28–1984), *J. Audio Eng. Soc.*, **32**, 781–785 (1984)
11. Pease, R., Understand capacitor soakage to optimise analog systems. *Electronics and Wireless World*, 832–835 (1992)

12. Harris, S., The effects of sampling clock jitter on Nyquist sampling analog to digital convertors and on oversampling delta-sigma ADCs. *J. Audio Eng. Soc.*, **38**, 537–542 (1990)

13. Nunn, J., Jitter specification and assessment in digital audio equipment. Presented at the 93rd Audio Engineering Society Convention. (San Francisco, 1992), Preprint No. 3361 (C–2)

14. Widrow, B., Statistical analysis of amplitude quantized sampled-data systems. *Trans. AIEE*, Part II, **79**, 555–568 (1961)

15. Lipshitz, S.P., Wannamaker, R.A. and Vanderkooy, J., Quantization and dither: a theoretical survey. *J. Audio Eng. Soc.*, **40**, 355–375 (1992)

16. Maher, R.C., On the nature of granulation noise in uniform quantization systems. *J. Audio Eng. Soc.*, **40**, 12–20 (1992)

17. Roberts, L.G., Picture coding using pseudo random noise. *IRE Trans. Inform. Theory*, **IT-8**, 145–154 (1962)

18. Schuchman, L., Dither signals and their effect on quantizing noise. *Trans. Commun. Technol.*, **COM-12**, 162–165 (1964)

19. Sherwood, D. T., Some theorems on quantization and an example using dither. In *Conf. Rec., 19th Asilomar Conf. on circuits, systems and computers*, (Pacific Grove, CA, 1985)

20. Gerzon, M. and Craven, P.G., Optimal noise shaping and dither of digital signals. Presented at the 87th Audio Engineers Society Convention (New York, 1989), Preprint No. 2822 (J-1)

21. Vanderkooy, J. and Lipshitz, S.P., Resolution below the least significant bit in digital systems with dither. *J. Audio Eng. Soc.*, **32**, 106–113 (1984)

22. Blesser, B., Advanced A-D conversion and filtering: data conversion. In *Digital Audio*, edited by B.A. Blesser, B. Locanthi, and T.G. Stockham Jr., pp. 37–53. New York: Audio Engineering Society. (1983)

23. Vanderkooy, J. and Lipshitz, S.P., Digital dither. Presented at the 81st Audio Engineering Society Convention (Los Angeles, 1986), Preprint 2412 (C-8)

24. Vanderkooy, J. and Lipshitz, S.P., Digital dither. In *Audio in Digital Times*, New York: Audio Engineering Society (1989)

25. v.d. Plassche, R.J., Dynamic element matching puts trimless convertors on chip. *Electronics*, 16 June 1983

26. v.d. Plassche, R.J. and Goedhart, D., A monolithic 14 bit D/A convertor. *IEEE J. Solid-State Circuits*, **SC-14**, 552–556 (1979)

27. Adams, R.W., Companded predictive delta modulation: a low-cost technique for digital recording. *J. Audio Eng. Soc.*, **32**, 659–672 (1984)

28. Hauser, M.W., Principles of oversampling A/D conversion. *J. Audio Eng. Soc.*, **39**, 3–26 (1991)

29. Cutler, C.C., Transmission systems employing quantization. US Pat. No. 2,927,962 (1960)

30. v.d. Plassche, R.J. and Dijkmans, E.C., A monolithic 16 bit D/A conversion system for digital audio. In *Digital Audio*, edited by B.A. Blesser, B. Locanthi and T.G. Stockham Jr, pp. 54–60. New York: Audio Engineering Society (1983)

31. Fielder, L.D., Human Auditory capabilities and their consequences in digital audio convertor design. In *Audio in Digital Times*, New York: Audio Engineering Society (1989)

32. Wannamaker, R.A., Psychoacoustically optimal noise shaping. *J. Audio Eng. Soc.*, **40**, 611–620 (1992)

33. Lipshitz, S.P., Wannamaker, R.A. and Vanderkooy, J., Minimally audible noise shaping. *J. Audio Eng. Soc.*, **39**, 836–852 (1991)

34. Adams, R.W., Design and implementation of an audio 18-bit A/D convertor using oversampling techniques. Presented at the 77th Audio Engineering Society Convention (Hamburg, 1985), preprint 2182

35. Adams, R.W., An IC chip set for 20 bit A/D conversion. In *Audio in Digital Times*, New York: Audio Engineering Society (1989)

36. Richards, M., Improvements in oversampling analogue to digital convertors. Presented at the 84th Audio Engineering Society Convention (Paris, 1988), Preprint 2588 (D-8)

37. Inose, H. and Yasuda, Y., A unity bit coding method by negative feedback. *Proc. IEEE*, **51**, 1524–1535 (1963)

38. Naus, P.J. *et al.*, Low signal level distortion in sigma-delta modulators. Presented at the 84th Audio Engineering Society Convention (Paris, 1988), Preprint 2584

39. Stikvoort, E., High order one bit coder for audio applications. Presented at the 84th Audio Engineering Society Convention (Paris, 1988), Preprint 2583(D-3)

40. Moller, L., Signal levels across the EBU/AES digital audio interface. In *Proc. 1st NAB Radio Montreux Symp.* (Montreux, 1992) 16–28

5

Compression

5.1 Introduction

Compression, bit rate reduction and data reduction are all terms which mean basically the same thing in this context. In essence the same (or nearly the same) audio information is carried using a smaller quantity or rate of data. It should be pointed out that in audio, *compression* traditionally means a process in which the dynamic range of the sound is reduced, typically by broadcasters wishing their station to sound louder. However, when bit rate reduction is employed, the dynamics of the decoded signal are unchanged. Provided the context is clear, the two meanings can co-exist without a great deal of confusion.

There are several reasons why compression techniques are popular:

(a) Compression extends the playing time of a given storage device.

(b) Compression allows miniaturization. With fewer data to store, the same playing time is obtained with smaller hardware. This is useful in portable and consumer devices.

(c) Tolerances can be relaxed. With fewer data to record, storage density can be reduced, making equipment which is more resistant to adverse environments and which requires less maintenance.

(d) In transmission systems, compression allows a reduction in bandwidth which will generally result in a reduction in cost. This may make possible some process which would be uneconomic without it.

(e) If a given bandwidth is available to an uncompressed signal, compression allows faster than real-time transmission within that bandwidth.

(f) If a given bandwidth is available, compression allows a better-quality signal within that bandwidth.

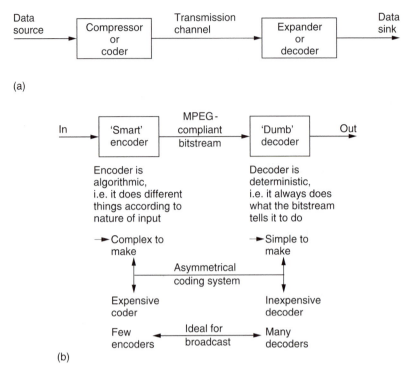

Figure 5.1 In (a) a compression system consists of compressor or coder, a transmission channel and a matching expander or decoder. The combination of coder and decoder is known as a codec. (b) MPEG is asymmetrical since the encoder is much more complex than the decoder.

Compression is summarized in Figure 5.1. It will be seen in (a) that the PCM audio data rate is reduced at source by the *compressor*. The compressed data are then passed through a communication channel and returned to the original audio rate by the *expander*. The ratio between the source data rate and the channel data rate is called the *compression factor*. The term *coding gain* is also used. Sometimes a compressor and expander in series are referred to as a *compander*. The compressor may equally well be referred to as a *coder* and the expander a *decoder* in which case the tandem pair may be called a *codec*.

Where the encoder is more complex than the decoder the system is said to be asymmetrical. Figure 5.1(b) shows that MPEG[1,2] audio coders work in this way, as do many others. The encoder needs to be algorithmic or adaptive whereas the decoder is 'dumb' and carries out fixed actions. This is advantageous in applications such as broadcasting where the number of expensive complex encoders is small but the number of simple inexpensive decoders is large. In point-to-point applications the advantage of asymmetrical coding is not so great. In MPEG audio coding the encoder is typically two or three times as complex as the decoder.

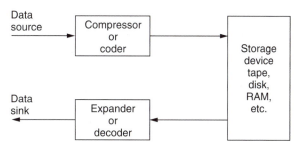

Figure 5.2 Compression can be used around a recording medium. The storage capacity may be increased or the access time reduced according to the application.

Figure 5.2 shows the use of a codec with a recorder. The playing time of the medium is extended in proportion to the compression factor. In the case of tapes, the access time is improved because the length of tape needed for a given recording is reduced and so it can be rewound more quickly. In some cases, compression may be used to improve the recorder quality. A lossless coder with a very light compression factor can be used to give a sixteen-bit DAT recorder eighteen- or twenty-bit performance.

In communications, the cost of data links is often roughly proportional to the data rate and so there is simple economic pressure to use a high compression factor. The use of heavy compression to allow audio to be sent over the Internet is an example of this.

In workstations designed for audio editing, the source material is stored on hard disks for rapid access. Whilst top-grade systems may function without compression, many systems use compression to offset the high cost of disk storage.

When a workstation is used for *off-line* editing, a high compression factor can be used and artifacts will be audible. This is of no consequence as these are only heard by the editor who uses the system to make an EDL (edit decision list) which is no more than a list of actions and the timecodes at which they occur. The original uncompressed material is then *conformed* to the EDL to obtain a high-quality edited work. When *on-line* editing is being performed, the output of the workstation is the finished product and clearly a lower compression factor will have to be used.

The cost of digital storage continues to fall and the pressure to use compression for recording purposes falls with it. Perhaps it is in broadcasting and the Internet where the use of compression will have its greatest impact. There is only one electromagnetic spectrum and pressure from other services such as cellular telephones makes efficient use of bandwidth mandatory. Analog broadcasting is an old technology and makes very inefficient use of bandwidth. Its replacement by a compressed digital transmission will be inevitable for the practical reason that the bandwidth is needed elsewhere.

Fortunately in broadcasting there is a mass market for decoders and these can be implemented as low-cost integrated circuits. Fewer encoders are needed and so it is less important if these are expensive. Whilst the cost of digital storage goes down year on year, the cost of electromagnetic spectrum goes up. Consequently in the future the pressure to use compression in recording will ease whereas the pressure to use it in radio communications will increase.

5.2 Lossless and perceptive coding

Although there are many different audio coding tools, all of them fall into one or other of these categories. In *lossless* coding, the data from the expander are identical bit-for-bit with the original source data. The so-called 'stacker' programs which increase the apparent capacity of disk drives in personal computers use lossless codecs. Clearly with computer programs the corruption of a single bit can be catastrophic. Lossless coding is generally restricted to compression factors of around 2:1.

It is important to appreciate that a lossless coder cannot guarantee a particular compression factor and the communications link or recorder used with it must be able to handle the variable output data rate. Audio material which results in poor compression factors on a given codec is described as *difficult*. It should be pointed out that the difficulty is often a function of the codec. In other words audio which one codec finds difficult may not be found difficult by another. Lossless codecs can be included in bit-error-rate testing schemes. It is also possible to cascade or *concatenate* lossless codecs without any special precautions.

In *lossy* coding, data from the decoder are not identical bit-for-bit with the source data and as a result comparing the input with the output is bound to reveal differences. Clearly lossy codecs are not suitable for computer data, but are used in many audio coders, MPEG included, as they allow greater compression factors than lossless codecs. The most successful lossy codecs are those in which the errors are arranged so that the listener finds them subjectively difficult to detect. Thus lossy codecs must be based on an understanding of psychoacoustic perception and are often called *perceptive* codes.

Perceptive coding relies on the principle of auditory masking, which was considered in Chapter 2. Masking causes the ear/brain combination to be less sensitive to sound at one frequency in the presence of another at a nearby frequency. If a first tone is present in the input, then it will mask signals of lower level at nearby frequencies. The quantizing of the first tone and of further tones at those frequencies can be made coarser. Fewer bits are needed and a coding gain results. The increased quantizing distortion is allowable if it is masked by the presence of the first tone.

In perceptive coding, the greater the compression factor required, the more accurately must the human senses be modelled. Perceptive coders can be forced to operate at a fixed compression factor. This is convenient for practical transmission applications where a fixed data rate is easier to handle than a variable rate. However, the result of a fixed compression factor is that the subjective quality can vary with the 'difficulty' of the input material. Perceptive codecs should not be concatenated indiscriminately especially if they use different algorithms. As the reconstructed signal from a perceptive codec is not bit-for-bit accurate, clearly such a codec cannot be included in any bit error rate testing system as the coding differences would be indistinguishable from real errors.

5.3   Compression principles

In a PCM audio system the bit rate is the product of the sampling rate and the number of bits in each sample and this is generally constant. Nevertheless the *information* rate of a real signal varies. In all real signals, part of the signal is obvious from what has gone before or what may come later and a suitable decoder can predict that part so that only the true information actually has to be sent. If the characteristics of a predicting decoder are known, the transmitter can omit parts of the message in the knowledge that the decoder has the ability to recreate it. Thus all encoders must contain a model of the decoder.

In a predictive codec there are two identical predictors, one in the coder and one in the decoder. Their job is to examine a run of previous data values and to extrapolate forward to estimate or predict what the next value will be. This is subtracted from the *actual* next value at the encoder to produce a prediction error or *residual* which is transmitted. The decoder then adds the prediction error to its own prediction to obtain the output code value again. Predictive coding can be applied to any type of information. In audio coders the information may be PCM samples, transform coefficients or even side-chain data such as scale factors.

Predictive coding has the advantage that provided the residual is transmitted intact, there is no loss of information.

One definition of information is that it is the unpredictable or surprising element of data. Newspapers are a good example of information because they only mention items which are surprising. Newspapers never carry items about individuals who have *not* been involved in an accident as this is the normal case. Consequently the phrase 'no news is good news' is remarkably true because if an information channel exists but nothing has been sent then it is most likely that nothing remarkable has happened.

The difference between the information rate and the overall bit rate is known as the redundancy. Compression systems are designed to eliminate as much of that redundancy as practicable or perhaps affordable. One way in which this can be done is to exploit statistical predictability in signals. The information content or *entropy* of a sample is a function of how different it is from the predicted value. Most signals have some degree of predictability. A sine wave is highly predictable, because all cycles look the same. According to Shannon's theory, any signal which is totally predictable carries no information. In the case of the sine wave this is clear because it represents a single frequency and so has no bandwidth.

At the opposite extreme a signal such as noise is completely unpredictable and as a result all codecs find noise *difficult*. There are two consequences of this characteristic. First, a codec which is designed using the statistics of real material should not be tested with random noise because it is not a representative test. Second, a codec which performs well with clean source material may perform badly with source material containing superimposed noise such as analog tape hiss. Practical compression units may require some form of pre-processing before the compression stage proper and appropriate noise reduction should be incorporated into the pre-processing if noisy signals are anticipated. It will also be necessary to restrict the degree of compression applied to noisy signals.

All real signals fall part-way between the extremes of total predictability and total unpredictability or noisiness. If the bandwidth (set by the sampling rate) and the dynamic range (set by the wordlength) of the transmission system are used to delineate an area, this sets a limit on the information capacity of the system. Figure 5.3(a) shows that most real signals only occupy part of that area. The signal may not contain all frequencies, or it may not have full dynamics at certain frequencies.

Entropy can be thought of as a measure of the actual area occupied by the signal. This is the area that *must* be transmitted if there are to be no subjective differences or *artifacts* in the received signal. The remaining area is called the *redundancy* because it adds nothing to the information conveyed. Thus an ideal coder could be imagined which miraculously sorts out the entropy from the redundancy and only sends the former. An ideal decoder would then recreate the original impression of the information quite perfectly.

As the ideal is approached, the coder complexity and the latency (delay) both rise. Figure 5.3(b) shows how complexity increases with compression factor. Figure 5.3(c) shows how increasing the codec latency can improve the compression factor. Obviously we would have to provide a channel which could accept whatever entropy the coder extracts in order to have transparent quality. As a result moderate coding

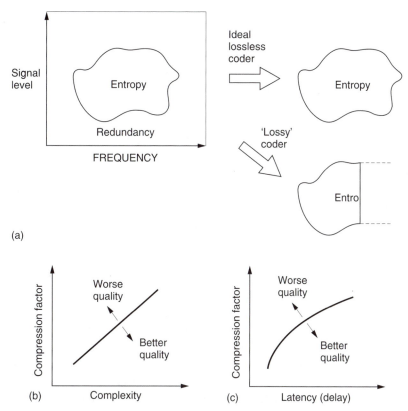

Figure 5.3 (a) A perfect coder removes only the redundancy from the input signal and results in subjectively lossless coding. If the remaining entropy is beyond the capacity of the channel some of it must be lost and the codec will then be lossy. An imperfect coder will also be lossy as it fails to keep all entropy. (b) As the compression factor rises, the complexity must also rise to maintain quality. (c) High compression factors also tend to increase latency or delay through the system.

gains which only remove redundancy need not in principle cause artifacts and can result in systems which are described as *subjectively lossless*. This assumes that such systems are well engineered, which may not be the case in actual hardware.

If the channel capacity is not sufficient for that, then the coder will have to discard some of the entropy and with it useful information. Larger coding gains which remove some of the entropy must result in artifacts. It will also be seen from Figure 5.3 that an imperfect coder will fail to separate the redundancy and may discard entropy instead, resulting in artifacts at a suboptimal compression factor.

A single variable rate transmission channel is inconvenient and unpopular with channel providers because it is difficult to police. The requirement can be overcome by combining several compressed channels into one constant rate transmission in a way which flexibly allocates data rate between the channels. Provided the material is unrelated, the probability of all channels reaching peak entropy at once is very small and so those channels which are at one instant passing easy material will free up transmission capacity for those channels which are handling difficult material. This is the principle of statistical multiplexing.

Where the same type of source material is used consistently, e.g. English text, then it is possible to perform a statistical analysis on the frequency with which particular letters are used. Variable-length coding is used in which frequently used letters are allocated short codes and letters which occur infrequently are allocated long codes. This results in a lossless code. The well-known Morse code used for telegraphy is an example of this approach. The letter e is the most frequent in English and is sent with a single dot.

An infrequent letter such as z is allocated a long complex pattern. It should be clear that codes of this kind which rely on a prior knowledge of the statistics of the signal are only effective with signals actually having those statistics. If Morse code is used with another language, the transmission becomes significantly less efficient because the statistics are quite different; the letter z, for example, is quite common in Czech.

The Huffman code[3] is one which is designed for use with a data source having known statistics and shares the same principles with the Morse code. The probability of the different code values to be transmitted is studied, and the most frequent codes are arranged to be transmitted with short wordlength symbols. As the probability of a code value falls, it will

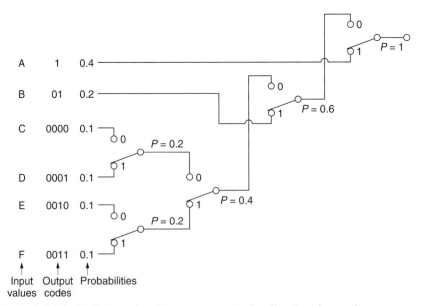

Figure 5.4 The Huffman code achieves compression by allocating short codes to frequent values. To aid deserializing the short codes are not prefixes of longer codes.

be allocated longer wordlength. The Huffman code is used in conjunction with a number of compression techniques and is shown in Figure 5.4.

The input or *source* codes are assembled in order of descending probability. The two lowest probabilities are distinguished by a single code bit and their probabilities are combined. The process of combining probabilities is continued until unity is reached and at each stage a bit is used to distinguish the path. The bit will be a zero for the most probable path and one for the least. The compressed output is obtained by reading the bits which describe which path to take going from right to left.

In the case of computer data, there is no control over the data statistics. Data to be recorded could be instructions, images, tables, text files and so on; each having their own code value distributions. In this case a coder relying on fixed source statistics will be completely inadequate. Instead a system is used which can learn the statistics as it goes along. The Lempel–Ziv–Welch (LZW) lossless codes are in this category. These codes build up a conversion table between frequent long source data strings and short transmitted data codes at both coder and decoder and initially their compression factor is below unity as the contents of the conversion tables are transmitted along with the data. However, once the tables are established, the coding gain more than compensates for the initial loss. In some applications, a continuous analysis of the frequency of code selection is made and if a data string in the table is no longer being used with sufficient frequency it can be deselected and a more common string substituted.

Lossless codes are less common in audio coding where perceptive codes are more popular. The perceptive codes often obtain a coding gain by shortening the wordlength of the data representing the signal waveform. This must increase the level of quantizing distortion and for good perceived quality the encoder must ensure that the resultant distortion is placed at frequencies where human senses are least able to perceive it. As a result although the received signal is measurably different from the source data, it can *appear* the same to the human listener under certain conditions. As these codes rely on the characteristics of human hearing, they can only fully be tested subjectively.

The compression factor of such codes can be set at will by choosing the wordlength of the compressed data. Whilst mild compression may be indetectible, with greater compression factors, artefacts become noticeable. Figure 5.3 shows that this is inevitable from entropy considerations.

5.4   Codec level calibration

The functioning of the ear is noticeably level dependent and perceptive coders take this into account. However, all signal processing takes place

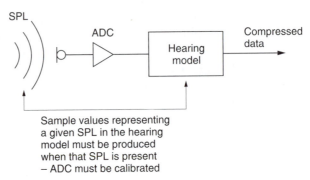

Sample values representing
a given SPL in the hearing
model must be produced
when that SPL is present
– ADC must be calibrated

Figure 5.5 Audio coders must be level calibrated so that the psychoacoustic decisions in the coder are based on correct sound pressure levels.

in the electrical or digital domain with respect to electrical or numerical levels whereas the hearing mechanism operates with respect to true sound pressure level. Figure 5.5 shows that in an ideal system the overall gain of the microphones and ADCs is such that the PCM codes have a relationship with sound pressure which is the same as that assumed by the model in the codec. Equally the overall gain of the DAC and loudspeaker system should be such that the sound pressure levels which the codec assumes are those actually heard. Clearly the gain control of the microphone and the volume control of the reproduction system must be calibrated if the hearing model is to function properly. If, for example, the microphone gain was too low and this was compensated by advancing the loudspeaker gain, the overall gain would be the same but the codec would be fooled into thinking that the sound pressure level was less than it really was and the masking model would not then be appropriate.

The above should come as no surprise as analog audio codecs such as the various Dolby systems have required and implemented line-up procedures and suitable tones. However obvious the need to calibrate coders may be, the degree to which this is recognized in the industry is almost negligible to date and this can only result in suboptimal performance.

5.5 Quality measurement

As has been seen, one way in which coding gain is obtained is to requantize sample values to reduce the wordlength. Since the resultant requantizing error is a distortion mechanism it results in energy moving from one frequency to another. The masking model is essential to estimate how audible the effect will be. The greater the degree of compression required, the more precise the model must be. If the masking model is inaccurate, then equipment based upon it may produce audible artifacts under some circumstances. Artifacts may also result if the model is not properly implemented. As a result, development of audio compression units requires careful listening tests with a wide range of source material[4,5] and precision loudspeakers. The presence of artifacts at a given compression factor indicates only that performance is below expectations; it does not distinguish between the implementation and the model. If the implementation is verified, then a more detailed model must be sought. Naturally comparative listening tests are only valid if all the codecs have been level calibrated and if the loudspeakers cause less loss of information than any of the codecs, a requirement which is frequently overlooked.

Properly conducted listening tests are expensive and time consuming, and alternative methods have been developed which can be used objectively to evaluate the performance of different techniques. The noise to masking ratio (NMR) is one such measurement.[6] Figure 5.6 shows how

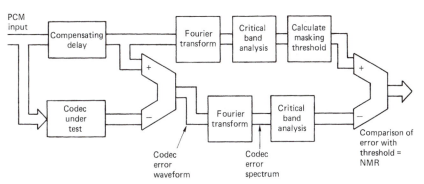

Figure 5.6 The noise-to-masking ratio is derived as shown here.

NMR is measured. Input audio signals are fed simultaneously to a data-reduction coder and decoder in tandem and to a compensating delay whose length must be adjusted to match the codec delay. At the output of the delay, the coding error is obtained by subtracting the codec output from the original. The original signal is spectrum-analysed into critical bands in order to derive the masking threshold of the input audio, and this is compared with the critical band spectrum of the error. The NMR in each critical band is the ratio between the masking threshold and the quantizing error due to the codec. An average NMR for all bands can be computed. A positive NMR in any band indicates that artifacts are potentially audible. Plotting the average NMR against time is a powerful technique, as with an ideal codec the NMR should be stable with different types of program material. If this is not the case the codec could perform quite differently as a function of the source material. NMR excursions can be correlated with the waveform of the audio input to analyse how the extra noise was caused and to redesign the codec to eliminate it.

Practical systems should have a finite NMR in order to give a degree of protection against difficult signals which have not been anticipated and against the use of post-codec equalization or several tandem codecs which could change the masking threshold. There is a strong argument that devices used for audio production should have a greater NMR than consumer or program delivery devices.

5.6   The limits

There are, of course, limits to all technologies. Eventually artifacts will be heard as the amount of compression is increased which no amount of detailed modelling will remove. The ear is only able to perceive a certain proportion of the information in a given sound. This could be called the perceptual entropy,[7] and all additional sound is redundant or irrelevant. Compression works by removing the redundancy, and clearly an ideal system would remove all of it, leaving only the entropy. Once this has been done, the masking capacity of the ear has been reached and the NMR has reached zero over the whole band. Assuming an ideal masking model, further reduction of the data rate must cause the level of distortion products to rise above the masking level equally at all frequencies rendering it audible.

The result is that the perceived quality of a codec suddenly falls at a critical bit rate. Figure 5.7 shows this effect which is variously known as a crash knee, graceless degradation or the cliff-edge effect. It is a simple consequence of human perception that a coder which keeps to the left of the crash knee 99 per cent of the time will still be marked down because the sudden failure for one per cent of the time causes irritation out of proportion to its duration.

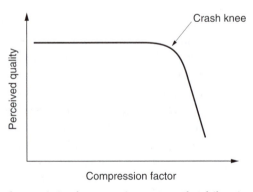

Figure 5.7 It is a characteristic of compression systems that failure is sudden.

In practice the audio bandwidth will have to be reduced in order to keep the distortion level acceptable. For example, in MPEG-1, pre-filtering allows data from higher sub-bands to be neglected. MPEG-2 has also introduced some low sampling rate options for this purpose. Thus there is a limit to the degree of compression which can be achieved even with an ideal coder. Systems which go beyond that limit are not appropriate for high-quality music, but are relevant in news gathering and communications where intelligibility of speech is the criterion.

Interestingly, the data rate out of a coder is virtually independent of the input sampling rate unless the sampling rate is very low. This is because the entropy of the sound is in the waveform, not in the number of samples carrying it.

It follows from the above that to obtain the highest audio quality for a given bit rate, every redundancy in the input signal must be explored. The more lossless coding tools which can be used, the less will be the extent to which the lossy tools operate. For example, MPEG Layers I and II audio coding don't employ prediction or buffering whereas Layer III uses buffering. MPEG-2 AAC[8] uses both prediction and buffering and can thus obtain better quality at a given bit rate or the same quality at a lower bit rate.

The compression factor of a coder is only part of the story. All codecs cause delay, and in general the greater the compression, the longer the delay. In some applications, such as telephony, a short delay is required.[9] In many applications, the compressed channel will have a constant bit rate, and so a constant compression factor is required. In real program material, the entropy varies and so the NMR will fluctuate. If greater delay can be accepted, as in a recording application, memory buffering can be used to allow the coder to operate at constant NMR and instantaneously variable data rate. The memory absorbs the instantaneous data rate differences of the coder and allows a constant rate in the channel. A higher effective compression factor will then be obtained. Near-constant quality can also be achieved using statistical multiplexing.

5.7   Some guidelines

Although compression techniques themselves are complex, there are some simple rules which can be used to avoid disappointment. Used wisely, audio compression has a number of advantages. Used in an inappropriate manner, disappointment is almost inevitable and the technology could get a bad name. The next few points are worth remembering.

- Compression technology may be exciting, but if it is not necessary it should not be used.
- If compression is to be used, the degree of compression should be as small as possible; i.e. use the highest practical bit rate.
- Cascaded compression systems cause loss of quality and the lower the bit rates, the worse this gets. Quality loss increases if any post-production steps are performed between compressions.
- Compression systems cause delay.
- Compression systems work best with clean source material. Noisy signals give poor results.
- Compressed data are generally more prone to transmission errors than non-compressed data. The choice of a compression scheme must consider the error characteristics of the channel.
- Audio codecs need to be level calibrated so that when sound pressure level-dependent decisions are made in the coder those levels actually exist at the microphone.
- Low bit rate coders should only be used for the final delivery of post-produced signals to the end user.
- Compression quality can only be assessed subjectively on precision loudspeakers. Codecs often sound fine on cheap speakers when in fact they are not.
- Compression works best in mono and less well in stereo and surround-sound systems where the imaging, ambience and reverb are frequently not well reproduced.
- Don't be browbeaten by the technology. You don't have to understand it to assess the results. Your ears are as good as anyone's so don't be afraid to criticize artifacts.

5.8 Audio compression tools

There are many different techniques available for audio compression, each having advantages and disadvantages. Real compressors will combine several techniques or tools in various ways to achieve different combinations of cost and complexity. Here it is intended to examine the tools separately before seing how they are used in actual compression systems.

The simplest coding tool is companding which is a digital parallel of the noise reducers used in analog tape recording. Figure 5.8(a) shows that in companding the input signal level is monitored. Whenever the input level falls below maximum, it is amplified at the coder. The gain which was applied at the coder is added to the data stream so that the decoder can apply an equal attenuation. The advantage of companding is that the signal is kept as far away from the noise floor as possible. In analog noise

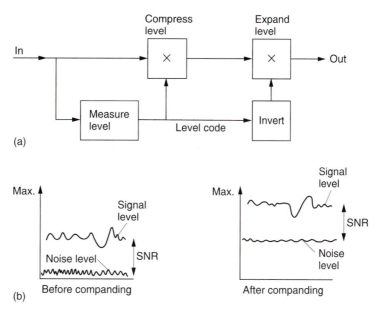

Figure 5.8 Digital companding. In (a) the encoder amplifies the input to maximum level and the decoder attenuates by the same amount. (b) In a companded system, the signal is kept as far as possible above the noise caused by shortening the sample wordlength.

reduction this is used to maximize the SNR of a tape recorder, whereas in digital compression it is used to keep the signal level as far as possible above the distortion introduced by various coding steps.

One common way of obtaining coding gain is to shorten the wordlength of samples so that fewer bits need to be transmitted. Figure 5.8(b) shows that when this is done, the distortion will rise by 6 dB for every bit removed. This is because removing a bit halves the number of quantizing intervals which then must be twice as large, doubling the error amplitude.

Clearly if this step follows the compander of (a), the audibility of the distortion will be minimized. As an alternative to shortening the wordlength, the uniform quantized PCM signal can be converted to a non-uniform format. In non-uniform coding, shown at (c), the size of the quantizing step rises with the magnitude of the sample so that the distortion level is greater when higher levels exist.

Companding is a relative of floating point coding shown in Figure 5.9 where the sample value is expressed as a mantissa and a binary exponent which determines how the mantissa needs to be shifted to have its correct absolute value on a PCM scale. The exponent is the equivalent of the gain setting or scale factor of a compander.

Clearly in floating point the signal-to-noise ratio is defined by the number of bits in the mantissa, and as shown in Figure 5.10, this will vary

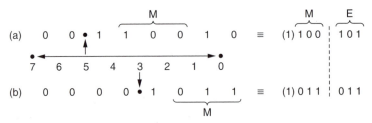

Figure 5.9 In this example of floating point notation, the radix point can have eight positions determined by the exponent E. The point is placed to the left of the first '1', and the next 4 bits to the right form the mantissa M. As the MSB of the mantissa is always 1, it need not always be stored.

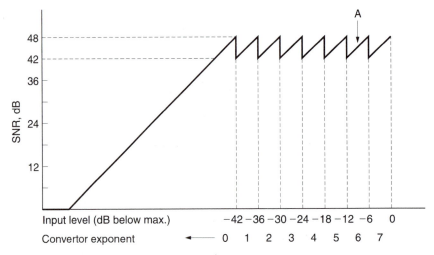

Figure 5.10 In this example of an eight-bit mantissa, three-bit exponent system, the maximum SNR is 6 dB × 8 = 48 dB with maximum input of 0 dB. As input level falls by 6 dB, the convertor noise remains the same, so SNR falls to 42 dB. Further reduction in signal level causes the convertor to shift range (point A in the diagram) by increasing the input analog gain by 6 dB. The SNR is restored, and the exponent changes from 7 to 6 in order to cause the same gain change at the receiver. The noise modulation would be audible in this simple system. A longer mantissa word is needed in practice.

as a sawtooth function of signal level, as the best value, obtained when the mantissa is near overflow, is replaced by the worst value when the mantissa overflows and the exponent is incremented. Floating-point notation is used within DSP chips as it eases the computational problems involved in handling long wordlengths. For example, when multiplying floating point numbers, only the mantissae need to be multiplied. The exponents are simply added.

A floating point system requires one exponent to be carried with each mantissa and this is wasteful because in real audio material the level does not change so rapidly and there is redundancy in the exponents. A better alternative is floating point block coding, also known as near-instanta-

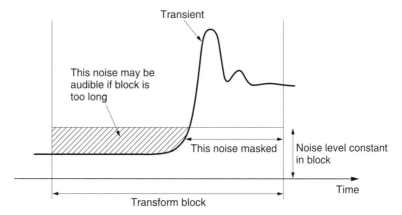

Figure 5.11 If a transient occurs towards the end of a transform block, the quantizing noise will still be present at the beginning of the block and may result in a pre-echo where the noise is audible before the transient.

neous companding, where the magnitude of the largest sample in a block is used to determine the value of an exponent which is valid for the whole block. Sending one exponent per block requires a lower data rate than in true floating point.[10]

In block coding the requantizing in the coder raises the quantizing error, but it does so over the entire duration of the block. Figure 5.11 shows that if a transient occurs towards the end of a block, the decoder will reproduce the waveform correctly, but the quantizing noise will start at the beginning of the block and may result in a burst of distortion products (also called pre-noise or pre-echo) which is audible before the transient. Temporal masking may be used to make this inaudible. With a 1 ms block, the artifacts are too brief to be heard.

Another solution is to use a variable time window according to the transient content of the audio waveform. When musical transients occur, short blocks are necessary and the coding gain will be low.[11] At other times the blocks become longer allowing a greater coding gain.

Whilst the above systems used alone do allow coding gain, the compression factor has to be limited because little benefit is obtained from masking. This is because the techniques above produce distortion which may be found anywhere over the entire audio band. If the audio input spectrum is narrow, this noise will not be masked.

Sub-band coding[12] splits the audio spectrum into many different frequency bands. Once this has been done, each band can be individually processed. In real audio signals many bands will contain lower-level signals than the loudest one. Individual companding of each band will be more effective than broadband companding. Sub-band coding also allows the level of distortion products to be raised selectively so that distortion is created only at frequencies where spectral masking will be effective.

It should be noted that the result of reducing the wordlength of samples in a sub-band coder is often referred to as noise. Strictly, noise is an unwanted signal which is decorrelated from the wanted signal. This is not generally what happens in audio compression. Although the original audio conversion would have been correctly dithered, the linearizing random element in the low-order bits will be some way below the end of the shortened word. If the word is simply rounded to the nearest integer the linearizing effect of the original dither will be lost and the result will be quantizing distortion. As the distortion takes place in a bandlimited system the harmonics generated will alias back within the band. Where the requantizing process takes place in a sub-band, the distortion products will be confined to that sub-band as shown in Figure 5.12. Such distortion is anharmonic.

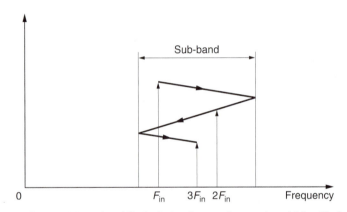

Figure 5.12 Requantizing a band-limited signal causes harmonics which will always alias back within the band.

Following any perceptive coding steps, the resulting data may be further subjected to lossless binary compression tools such as prediction, Huffman coding or a combination of both.

Audio is usually considered to be a time-domain waveform as this is what emerges from a microphone. As has been seen in Chapter 3, spectral analysis allows any periodic waveform to be represented by a set of harmonically related components of suitable amplitude and phase. In theory it is perfectly possible to decompose a periodic input waveform into its constituent frequencies and phases, and to record or transmit the transform. The transform can then be inverted and the original waveform will be precisely recreated.

Although one can think of exceptions, the transform of a typical audio waveform changes relatively slowly much of the time. The slow speech of an organ pipe or a violin string or the slow decay of most musical sounds allow the rate at which the transform is sampled to be reduced, and a coding gain results. At some frequencies the level will be below maximum and a shorter wordlength can be used to describe the coefficient. Further coding gain will be achieved if the coefficients describing frequencies which will experience masking are quantized more coarsely.

In practice there are some difficulties, real sounds are not periodic, but contain transients which transformation cannot accurately locate in time. The solution to this difficulty is to cut the waveform into short segments and then to transform each individually. The delay is reduced, as is the computational task, but there is a possibility of artifacts arising because of the truncation of the waveform into rectangular time windows. A solution is to use window functions, and to overlap the segments as shown in Figure 5.13. Thus every input sample appears in just two transforms, but with variable weighting depending upon its position along the time axis.

Figure 5.13 Transform coding can only be practically performed on short blocks. These are overlapped using window functions in order to handle continuous waveforms.

The DFT (discrete frequency transform) does not produce a continuous spectrum, but instead produces coefficients at discrete frequencies. The frequency resolution (i.e. the number of different frequency coefficients) is equal to the number of samples in the window. If overlapped windows are used, twice as many coefficients are produced as are theoretically necessary. In addition, the DFT requires intensive computation, owing to the requirement to use complex arithmetic to render the phase of the components as well as the amplitude. An alternative is to use discrete cosine transforms (DCT) or the modified discrete cosine transform (MDCT) which has the ability to eliminate the overhead of coefficients due to overlapping the windows and return to the critically sampled domain.[13] Critical sampling is a term which means that the number of coefficients does not exceed the number which would be obtained with non-overlapping windows.

5.9   Sub-band coding

Sub-band coding takes advantage of the fact that real sounds do not have uniform spectral energy. The wordlength of PCM audio is based on the dynamic range required and this is generally constant with frequency although any pre-emphasis will affect the situation. When a signal with an uneven spectrum is conveyed by PCM, the whole dynamic range is occupied only by the loudest spectral component, and all the other components are coded with excessive headroom. In its simplest form, sub-band coding works by splitting the audio signal into a number of frequency bands and companding each band according to its own level. Bands in which there is little energy result in small amplitudes which can be transmitted with short wordlength. Thus each band results in variable-length samples, but the sum of all the sample wordlengths is less than that of PCM and so a coding gain can be obtained. Sub-band coding is not restricted to the digital domain; the analog Dolby noise-reduction systems use it extensively.

The number of sub-bands to be used depends upon what other compression tools are to be combined with the sub-band coding. If it is intended to optimize compression based on auditory masking, the sub-bands should preferably be narrower than the critical bands of the ear, and therefore a large number will be required. This requirement is frequently not met: ISO/MPEG Layers I and II use only 32 sub-bands. Figure 5.14 shows the critical condition where the masking tone is at the top edge of the sub-band. It will be seen that the narrower the sub-band, the higher the requantizing 'noise' that can be masked. The use of an excessive number of sub-bands will, however, raise complexity and the coding delay, as well as risking pre-ringing on transients which may exceed the temporal masking.

On the other hand, if used in conjunction with predictive sample coding, relatively few bands are required. The apt-X100 system, for example, uses only four sub-bands as simulations showed that a greater number gave diminishing returns.[14]

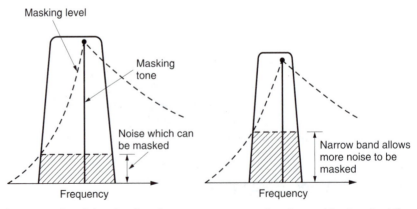

Figure 5.14 In sub-band coding the worst case occurs when the masking tone is at the top edge of the sub-band. The narrower the band, the higher the noise level which can be masked.

The bandsplitting process is complex and requires a lot of computation. One bandsplitting method which is useful is quadrature mirror filtering.[15] The QMF is is a kind of twin FIR filter which converts a PCM sample stream into to two sample streams of half the input sampling rate, so that the output data rate equals the input data rate. The frequencies in the lower half of the audio spectrum are carried in one sample stream, and the frequencies in the upper half of the spectrum are carried in the other. Whilst the lower-frequency output is a PCM band-limited representation of the input waveform, the upper frequency output isn't. A moment's thought will reveal that it could not be so because the sampling rate is not high enough. In fact the upper half of the input spectrum has been heterodyned down to the same frequency band as the lower half by the clever use of aliasing. The waveform is unrecognizable, but when heterodyned back to its correct place in the spectrum in an inverse step, the correct waveform will result once more.

Sampling theory states that the sampling rate needed must be at least twice the bandwidth in the signal to be sampled. If the signal is band limited, the sampling rate need only be more than twice the signal *bandwidth* not the signal *frequency*. Downsampled signals of this kind can be reconstructed by a reconstruction or *synthesis* filter having a bandpass response rather than a low pass response. As only signals within the passband can be output, it is clear from Figure 5.15 that the waveform which will result is the original as the intermediate aliased waveform lies outside the passband.

Figure 5.16 shows the operation of a simple QMF. At (a) the input spectrum of the PCM audio is shown, having an audio baseband extending up to half the sampling rate and the usual lower sideband extending down from there up to the sampling frequency. The input is passed through a FIR low-pass filter which cuts off at one quarter of the sampling rate to give the spectrum shown at (b). The input also passes in parallel through a second FIR filter which is physically identical, but the coefficients are different. The impulse response of the FIR LPF is multiplied by a cosinusoidal waveform which amplitude modulates it.

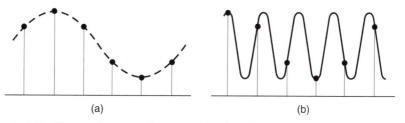

(a)   (b)

Figure 5.15 The sample stream shown would ordinarily represent the waveform shown in (a), but if it is known that the original signal could exist only between two frequencies then the waveform in (b) must be the correct one. A suitable bandpass reconstruction filter, or synthesis filter, will produce the waveform in (b).

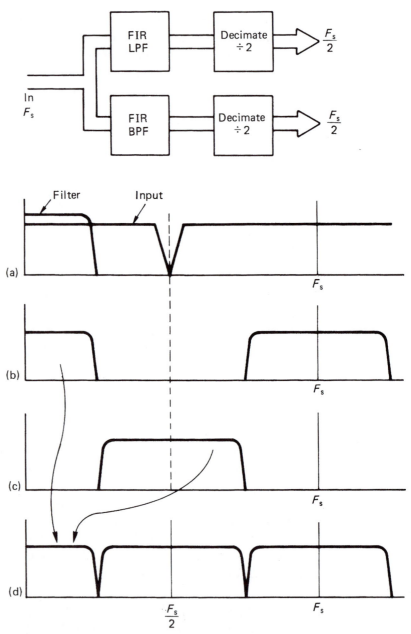

Figure 5.16 The quadrature mirror filter. At (a) the input spectrum has an audio baseband extending up to half the sampling rate. The input is passed through an FIR low-pass filter which cuts off at one-quarter of the sampling rate to give the spectrum shown at (b). The input also passes in parallel through a second FIR filter whose impulse response has been multiplied by a cosinusoidal waveform in order to amplitude-modulate it. The resultant impulse gives the filter a mirror image frequency response shown at (c). The spectra of both (b) and (c) show that both are oversampled by a factor of two because they are half empty. As a result both can be decimated by a factor of two, resulting at (d) in two identical Nyquist-sampled frequency bands of half the original width.

The resultant impulse gives the filter a frequency response shown at (c). This is a mirror image of the LPF response. If certain criteria are met, the overall frequency response of the two filters is flat. The spectra of both (b) and (c) show that both are oversampled by a factor of 2 because they are half-empty. As a result both can be decimated by a factor of two, which is the equivalent of dropping every other sample. In the case of the lower half of the spectrum, nothing remarkable happens. In the case of the upper half of the spectrum, it has been resampled at half the original frequency as shown at (d). The result is that the upper half of the audio spectrum aliases or heterodynes to the lower half.

An inverse QMF will recombine the bands into the original broadband signal. It is a feature of a QMF/inverse QMF pair that any energy near the band edge which appears in both bands due to inadequate selectivity in the filtering reappears at the correct frequency in the inverse filtering process provided that there is uniform quantizing in all the sub-bands. In practical coders, this criterion is not met, but any residual artifacts are sufficiently small to be masked.

The audio band can be split into as many bands as required by cascading QMFs in a tree. However, each stage can only divide the input spectrum in half. In some coders certain sub-bands will have passed through one splitting stage more than others and will be half their bandwidth.[16] A delay is required in the wider sub-band data for time alignment.

A simple quadrature mirror is computationally intensive because sample values are calculated which are later decimated or discarded, and an alternative is to use polyphase pseudo-QMF filters[17] or wave filters[18] in which the filtering and decimation process is combined. Only wanted sample values are computed. A polyphase QMF operates in a manner not unlike the polyphase operation of a FIR filter used for interpolation in sampling rate conversion (see Chapter 4). In a polyphase filter a set of samples is shifted into position in the transversal register and then these are multiplied by different sets of coefficients and accumulated in each of several phases to give the value of a number of different samples between input samples. In a polyphase QMF, the same approach is used.

Figure 5.17 shows an example of a 32-band polyphase QMF having a 512 sample window. With 32 sub-bands, each band will be decimated to $\frac{1}{32}$ of the input sampling rate. Thus only one sample in 32 will be retained after the combined filter/decimate operation. The polyphase QMF only computes the value of the sample which is to be retained in each sub-band. The filter works in 32 different phases with the same samples in the transversal register. In the first phase, the coefficients will describe the impulse response of a low-pass filter, the so-called prototype filter, and the result of 512 multiplications will be accumulated to give a single

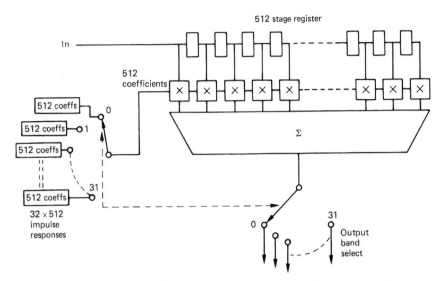

Figure 5.17 In polyphase QMF the same input samples are subject to computation using coefficient sets in many different time-multiplexed phases. The decimation is combined with the filtering so only wanted values are computed.

sample in the first band. In the second phase the coefficients will be obtained by multiplying the impulse response of the prototype filter by a cosinusoid at the centre frequency of the second band. Once more 512 multiply accumulates will be required to obtain a single sample in the second band. This is repeated for each of the 32 bands, and in each case a different centre frequency is obtained by multiplying the prototype impulse by a different modulating frequency. Following 32 such computations, 32 output samples, one in each band, will have been computed. The transversal register then shifts 32 samples and the process repeats.

The principle of the polyphase QMF is not so different from the techniques used to compute a frequency transform and effectively blurs the distinction between sub-band coding and transform coding.

The QMF technique is restricted to bands of equal width. It might be throught that this is a drawback because the critical bands of the ear are non-uniform. In fact this is only a problem when very high compression factors are required. In all cases it is the masking model of hearing which must have correct critical bands. This model can then be used to determine how much masking and therefore coding gain is possible within the actual sub-bands used. Uniform-width sub-bands will not be able to obtain as much masking as bands which are matched to critical bands, but for many applications the additional coding gain is not worth the added filter complexity.

5.10   Transform coding

Many transform coders use the discrete cosine transform described in section 3.31. The DCT works on blocks of samples which are windowed. For simplicity the following example uses a very small block of only eight samples whereas a real encoder might use several hundred.

Figure 5.18 shows the table of basis functions or *wave table* for an eight-point DCT. Adding these two-dimensional waveforms together in different proportions will give any combination of the original eight PCM audio samples. The coefficients of the DCT simply control the proportion of each wave which is added in the inverse transform. The top-left wave

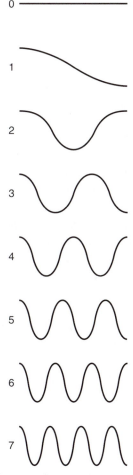

Figure 5.18 A table of basis functions for an eight-point DCT. If these waveforms are added together in various proportions, any original waveform can be reconstructed. In practice these waveforms are stored as samples, but after reconstruction to the analog domain they would appear as shown here.

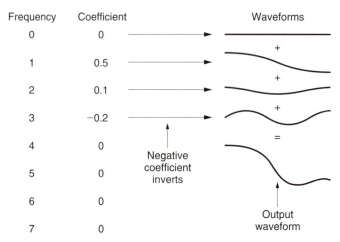

Figure 5.19 An example of an inverse DCT. The coefficients determine the amplitudes of the waves from the table in Figure 5.18 which are to be added together. Note that coefficient 3 is negative so that the wave is inverted.

has no modulation at all because it conveys the DC component of the block. Increasing the DC coefficient adds a constant amount to every sample in the block.

Moving to the right the coefficients represent increasing frequencies. All these coefficients are bipolar, where the polarity indicates whether the original waveform at that frequency was inverted.

Figure 5.19 shows an example of an inverse transform. The DC coefficient produces a constant level throughout the sample block. The remaining waves in the table are AC coefficients. A zero coefficient would result in no modulation, leaving the DC level unchanged. The wave next to the DC component represents the lowest frequency in the transform which is half a cycle per block. A positive coefficient would increase the signal voltage at the left side of the block whilst reducing it on the right, whereas a negative coefficient would do the opposite. The magnitude of the coefficient determines the amplitude of the wave which is added. Figure 5.19 also shows that the next wave has a frequency of one cycle per block, i.e. the waveform is made more positive at both sides and more negative in the middle.

Consequently an inverse DCT is no more than a process of mixing various waveforms from the wave table where the relative amplitudes and polarity of these patterns are controlled by the coefficients. The original transform is simply a mechanism which finds the coefficient amplitudes from the original PCM sample block.

The DCT itself achieves no compression at all. The number of coefficients which are output always equals the number of audio samples in the block. However, in typical program material, not all coefficients will have significant values; there will often be a few dominant coefficients. The coefficients representing the higher frequencies will often be zero or of small value, due to the typical energy distribution of audio.

Coding gain (the technical term for reduction in the number of bits needed) is achieved by transmitting the low-valued coefficients with shorter wordlengths. The zero-valued coefficients need not be transmitted at all. Thus it is not the DCT which compresses the audio, it is the subsequent processing. The DCT simply expresses the audio samples in a form which makes the subsequent processing easier.

Higher compression factors require the coefficient wordlength to be further reduced using requantizing. Coefficients are divided by some factor which increases the size of the quantizing step. The smaller number of steps which results permits coding with fewer bits, but of course with an increased quantizing error. The coefficients will be multiplied by a reciprocal factor in the decoder to return to the correct magnitude.

Further redundancy in transform coefficients can also be identified. This can be done in various ways. Within a transform block, the coefficients may be transmitted using differential coding so that the first coefficient is sent in an absolute form whereas the remainder are transmitted as differences with respect to the previous one. Some coders attempt to predict the value of a given coefficient using the value of the same coefficient in typically the two previous blocks. The prediction is subtracted from the actual value to produce a prediction error or residual which is transmitted to the decoder. Another possibility is to use prediction within the transform block. The predictor scans the coefficients from, say, the low-frequency end upwards and tries to predict the value of the next coefficient in the scan from the values of the earlier coefficients. Again a residual is transmitted.

Inter-block prediction works well for stationary material, whereas intra-block prediction works well for transient material. An intelligent coder may select a prediction technique using the input entropy in the same way that it selects the window size.

Inverse transforming a requantized coefficient means that the frequency it represents is reproduced in the output with the wrong amplitude. The difference between the original and the reconstructed amplitude is considered to be a noise added to the wanted data. The audibility of such noise depends on the degree of masking prevailing.

5.11   Compression formats

There are numerous formats intended for audio compression and these can be divided into international standards and proprietary designs.

The ISO (International Standards Organization) and the IEC (International Electrotechnical Commission) recognized that compression would have an important part to play and in 1988 established the ISO/IEC/MPEG (Moving Picture Experts Group) to compare and assess various coding schemes in order to arrive at an international standard for compressing video. The terms of reference were extended the same year to include audio and the MPEG/Audio group was formed.

MPEG audio coding is used for DAB (digital audio broadcasting) and for the audio content of digital television broadcasts to the DVB standard.

In the USA, it has been proposed to use an alternative compression technique for the audio content of ATSC (advanced television systems committee) digital television broadcasts. This is the AC-3[19] system developed by Dolby Laboratories. The MPEG transport stream structure has also been standardized to allow it to carry AC-3 coded audio. The digital video disk (DVD) can also carry AC-3 or MPEG audio coding.

Other popular proprietary codes include apt-X which is a mild compression factor/short delay codec and ATRAC which is the codec used in MiniDisc.

5.12   MPEG Audio compression

The subject of audio compression was well advanced when the MPEG/Audio group was formed. As a result it was not necessary for the group to produce *ab initio* codecs because existing work was considered suitable. As part of the Eureka 147 project, a system known as MUSICAM[20] (Masking pattern adapted Universal Sub-band Integrated Coding And Multiplexing) was developed jointly by CCETT in France, IRT in Germany and Philips in the Netherlands. MUSICAM was designed to be suitable for DAB (digital audio broadcasting).

As a parallel development, the ASPEC[21] (Adaptive Spectral Perceptual Entropy Coding) system was developed from a number of earlier systems as a joint proposal by AT&T Bell Labs, Thomson, the Fraunhofer Society and CNET. ASPEC was designed for use at high compression factors to allow audio transmission on ISDN.

These two systems were both fully implemented by July 1990 when comprehensive subjective testing took place at the Swedish Broadcasting Corporation.[4,22,23] As a result of these tests, the MPEG/Audio group combined the attributes of both ASPEC and MUSICAM into a standard[1,24] having three levels of complexity and performance.

These three different levels, which are known as *layers*, are needed because of the number of possible applications. Audio coders can be operated at various compression factors with different quality expectations. Stereophonic classical music requires different quality criteria from monophonic speech. The complexity of the coder will be reduced with a smaller compression factor. For moderate compression, a simple codec will be more cost effective. On the other hand, as the compression factor is increased, it will be necessary to employ a more complex coder to maintain quality.

MPEG Layer 1 is a simplified version of MUSICAM which is appropriate for the mild compression applications at low cost. Layer II is identical to MUSICAM and is used for DAB and for the audio content of DVB digital television broadcasts. Layer III is a combination of the best features of ASPEC and MUSICAM and is mainly applicable to telecommunications where high compression factors are required.

The approach of the ISO to standardization in MPEG Audio is novel because the encoder is not completely specified. Figure 5.20(a) shows that instead the way in which a decoder shall interpret the bitstream is defined. A decoder which can successfully interpret the bitstream is said to be *compliant*. Figure 5.20(b) shows that the advantage of standardizing the decoder is that over time encoding algorithms, particularly masking models, can improve yet compliant decoders will continue to function with them.

Manufacturers can supply encoders using algorithms which are proprietary and their details do not need to be published. A useful result is that there can be competition between different encoder designs which means that better designs will evolve. The user will have greater choice because different levels of cost and complexity can exist in a range of coders yet a compliant decoder will operate with them all.

MPEG is, however, much more than a compression scheme as it also standardizes the protocol and syntax under which it is possible to combine or multiplex audio data with video data to produce a digital equivalent of a television program. Many such programs can be combined in a single multiplex and MPEG defines the way in which such multiplexes can be created and transported. The definitions include the metadata which decoders require to demultiplex correctly and which users will need to locate programs of interest.

At each layer, MPEG Audio coding allows input sampling rates of 32, 44.1 and 48 kHz and supports output bit rates of 32, 48, 56, 64, 96, 112, 128, 192, 256 and 384 kbits/s. The transmission can be mono, dual-channel (e.g. bilingual), or stereo. Another possibility is the use of joint stereo mode in which the audio becomes mono above a certain frequency. This allows a lower bit rate with the obvious penalty of reduced stereo fidelity.

The layers of MPEG Audio coding (I, II and III) should not be confused with the MPEG-1 and MPEG-2 television coding standards. MPEG-1 and MPEG-2 flexibly define a range of systems for video and audio coding, whereas the layers define types of audio coding.

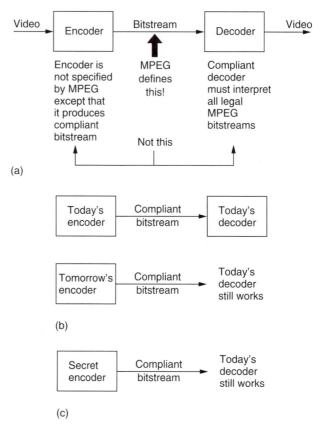

Figure 5.20 (a) MPEG defines the protocol of the bitstream between encoder and decoder. The decoder is defined by implication, the encoder is left very much to the designer. (b) This approach allows future encoders of better performance to remain compatible with existing decoders. (c) This approach also allows an encoder to produce a standard bitstream while its technical operation remains a commercial secret.

The earlier MPEG-1 standard compresses audio and video into about 1.5 Mbits/s. The audio coding of MPEG-1 may be used on its own to encode one or two channels at bit rates up to 448 kbits/s. MPEG-2 allows the number of channels to increase to five: Left, Right, Centre, Left surround, Right surround and Subwoofer. In order to retain reverse compatibility with MPEG-1, the MPEG-2 coding converts the five-channel input to a compatible two-channel signal, L_o, R_o, by matrixing[25] as shown in Figure 5.21. The data from these two channels are encoded in a standard MPEG-1 audio frame, and this is followed in MPEG-2 by an ancillary data frame which an MPEG-1 decoder will ignore. The ancillary frame contains data for another three audio channels. Figure 5.22 shows that there are eight modes in which these three channels can be obtained. The encoder will select the mode which gives the least data rate for the prevailing distribution of energy in the input channels. An MPEG-2

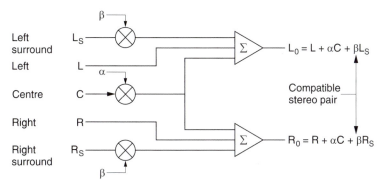

Figure 5.21 To allow compatibility with two-channel systems, a stereo signal pair is derived from the five surround signals in this manner.

L	R	C
L	R_S	C
L_S	R	C
L_S	R_S	C
L	R	L_S
L	R	R_S
L_S	R	R_S
L	R_S	L_S

Figure 5.22 In addition to sending the stereo compatible pair, one of the above combinations of signals can be sent. In all cases a suitable inverse matrix can recover the original five channels.

decoder will extract those three channels in addition to the MPEG-1 frame and then recover all five original channels by an inverse matrix which is steered by mode select bits in the bitstream.

The requirement for MPEG-2 Audio to be backward compatible with MPEG-1 audio coding was essential for some markets, but did compromise the performance because certain useful coding tools could not be used. Consequently the MPEG Audio group evolved a multi-channel standard which was not backward compatible because it incorporated additional coding tools in order to achieve higher performance. This came to be known as MPEG-2 AAC (advanced audio coding).

5.13   MPEG Layer I

Figure 5.23 shows a block diagram of a Layer I coder which is a simplified version of that used in the MUSICAM system. A polyphase filter divides

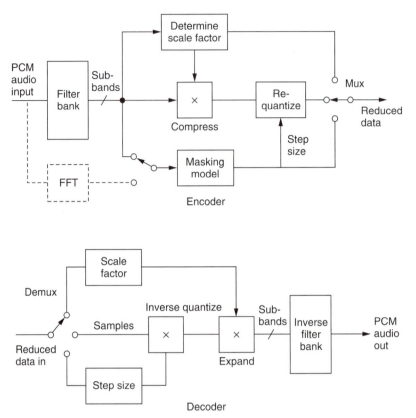

Figure 5.23 A simple sub-band coder. The bit allocation may come from analysis of the sub-band energy, or, for greater reduction, from a spectral analysis in a side chain.

the audio spectrum into 32 equal sub-bands. The output of the filter bank is critically sampled. In other words the output data rate is no higher than the input rate because each band has been heterodyned to a frequency range from zero upwards.

Sub-band compression takes advantage of the fact that real sounds do not have uniform spectral energy. The wordlength of PCM audio is based on the dynamic range required and this is generally constant with frequency although any pre-emphasis will affect the situation. When a signal with an uneven spectrum is conveyed by PCM, the whole dynamic range is occupied only by the loudest spectral component, and all the other components are coded with excessive headroom. In its simplest form, sub-band coding works by splitting the audio signal into a number of frequency bands and companding each band according to its own level. Bands in which there is little energy result in small amplitudes which can be transmitted with short wordlength. Thus each band results in variable-length samples, but the sum of all the sample wordlengths is less than that of the PCM input and so a degree of coding gain can be obtained.

A Layer I-compliant encoder, i.e. one whose output can be understood by a standard decoder, can be made which does no more than this. Provided the syntax of the bitstream is correct, the decoder is not concerned with how the coding decisions were made. However, higher compression factors require the distortion level to be increased and this should only be done if it is known that the distortion products will be masked. Ideally the sub-bands should be narrower than the critical bands of the ear. Figure 5.14 showed the critical condition where the masking tone is at the top edge of the sub-band. The use of an excessive number of sub-bands will, however, raise complexity and the coding delay. The use of 32 equal sub-bands in MPEG Layers I and II is a compromise. Efficient polyphase bandsplitting filters can only operate with equal-width sub-bands and the result, in an octave-based hearing model, is that sub-bands are too wide at low frequencies and too narrow at high frequencies.

To offset the lack of accuracy in the sub-band filter a parallel fast Fourier transform is used to drive the masking model. The standard suggests masking models, but compliant bitstreams can result from other models. In Layer I a 512-point FFT is used. The output of the FFT is used to determine the masking threshold which is the sum of all masking sources. Masking sources include at least the threshold of hearing which may locally be raised by the frequency content of the input audio. The degree to which the threshold is raised depends on whether the input audio is sinusoidal or atonal (broadband, or noise-like).

In the case of a sine wave, the magnitude and phase of the FFT at each frequency will be similar from one window to the next, whereas if the sound is atonal the magnitude and phase information will be chaotic.

The masking threshold is effectively a graph of just noticeable noise as a function of frequency. Figure 5.24(a) shows an example. The masking threshold is calculated by convolving the FFT spectrum with the cochlea spreading function (see section 2.11) with corrections for tonality. The level of the masking threshold cannot fall below the absolute masking threshold which is the threshold of hearing.

The masking threshold is then superimposed on the actual frequencies of each sub-band so that the allowable level of distortion in each can be established. This is shown in Figure 5.24(b).

Constant-size input blocks are used, containing 384 samples. At 48 kHz, 384 samples corresponds to a period of 8 ms. After the sub-band filter each band contains 12 samples per block. The block size is too long to avoid the pre-masking phenomenon of Figure 5.11. Consequently the masking model must ensure that heavy requantizing is not used in a block which contains a large transient following a period of quiet. This

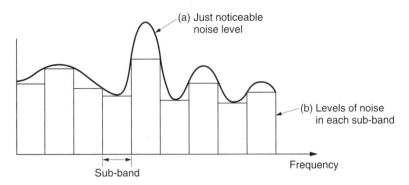

Figure 5.24 A continuous curve (a) of the just-noticeable noise level is calculated by the masking model. The levels of noise in each sub-band (b) must be set so as not to exceed the level of the curve.

can be done by comparing parameters of the current block with those of the previous block as a significant difference will indicate transient activity.

The samples in each sub-band block or *bin* are companded according to the peak value in the bin. A six-bit scale factor is used for each sub-band which applies to all 12 samples. The gain step is 2 dB and so with a six-bit code over 120 dB of dynamic range is available.

A fixed-output bit rate is employed, and as there is no buffering the size of the coded output block will be fixed. The wordlengths in each bin will have to be such that the sum of the bits from all the sub-bands equals the size of the coded block. Thus some sub-bands can have long wordlength coding if others have short wordlength coding. The process of determining the requantization step size, and hence the wordlength in each sub-band, is known as bit allocation. In Layer I all sub-bands are treated in the same way and fourteen different requantization classes are used. Each one has an odd number of quantizing intervals so that all codes are referenced to a precise zero level.

Where masking takes place, the signal is quantized more coarsely until the distortion level is raised to just below the masking level. The coarse quantization requires shorter wordlengths and allows a coding gain. The bit allocation may be iterative as adjustments are made to obtain an equal NMR across all sub-bands. If the allowable data rate is adequate, a positive NMR will result and the decoded quality will be optimal. However, at lower bit rates and in the absence of buffering a temporary increase in bit rate is not possible. The coding distortion cannot be masked and the best the encoder can do is to make the (negative) NMR equal across the spectrum so that artifacts are not emphasized unduly in any one sub-band. It is possible that in some sub-bands there will be no data at all, either because such frequencies were absent in the program material or because the encoder has discarded them to meet a low bit rate.

The samples of differing wordlength in each bin are then assembled into the output coded block. Unlike a PCM block, which contains samples of fixed wordlength, a coded block contains many different wordlengths which may vary from one sub-band to the next. In order to deserialize the block into samples of various wordlengths and demultiplex the samples into the appropriate frequency bins, the decoder has to be told what bit allocations were used when it was packed, and some synchronizing means is needed to allow the beginning of the block to be identified.

The compression factor is determined by the bit-allocation system. It is trivial to change the output block size parameter to obtain a different compression factor. If a larger block is specified, the bit allocator simply iterates until the new block size is filled. Similarly the decoder need only deserialize the larger block correctly into coded samples and then the expansion process is identical except for the fact that expanded words contain less noise. Thus codecs with varying degrees of compression are available which can perform different bandwidth/performance tasks with the same hardware.

Figure 5.25(a) shows the format of the Layer I elementary stream. The frame begins with a sync pattern to reset the phase of deserialization, and a header which describes the sampling rate and any use of pre-emphasis. Following this is a block of 32 four-bit allocation codes. These specify the wordlength used in each sub-band and allow the decoder to deserialize the sub-band sample block. This is followed by a block of 32 six-bit scale factor indices, which specify the gain given to each band during companding. The last block contains 32 sets of 12 samples. These samples vary in wordlength from one block to the next, and can be from 0 to 15 bits long. The deserializer has to use the 32 allocation information codes

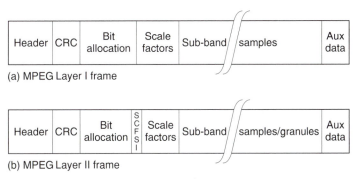

(a) MPEG Layer I frame

(b) MPEG Layer II frame

Figure 5.25 (a) The MPEG Layer I data frame has a simple structure. (b) in the Layer II frame, the compression of the scale factors requires the additional SCFSI code described in the text.

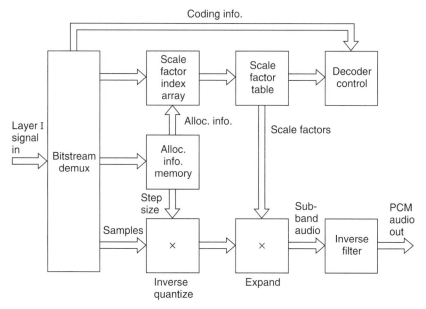

Figure 5.26 The Layer I decoder. See text for details.

to work out how to deserialize the sample block into individual samples of variable length.

The Layer I MPEG decoder is shown in Figure 5.26. The elementary stream is deserialised using the sync pattern and the variable-length samples are assembled using the allocation codes. The variable-length samples are returned to fifteen-bit wordlength by adding zeros. The scale factor indices are then used to determine multiplication factors used to return the waveform in each sub-band to its original amplitude. The 32 sub-band signals are then merged into one spectrum by the synthesis filter. This is a set of bandpass filters which heterodynes every sub-band to the correct place in the audio spectrum and then adds them to produce the audio output.

5.14   MPEG Layer II

MPEG Layer II audio coding is identical to MUSICAM. The same 32-band filterbank and the same block companding scheme as Layer I is used. In order to give the masking model better spectral resolution, the side-chain FFT has 1024 points. The FFT drives the masking model which may be the same as is suggested for Layer I. The block length is increased to 1152 samples. This is three times the block length of Layer I, corresponding to 24 ms at 48 kHz.

Figure 5.25(b) shows the Layer II elementary stream structure. Following the sync pattern the bit-allocation data are sent. The requantizing process of Layer II is more complex than in Layer I. The sub-bands are categorized into three frequency ranges, low, medium and high, and the requantizing in each range is different. Low-frequency samples can be quantized into 15 different wordlengths, mid-frequencies into seven different wordlengths and high frequencies into only three different wordlengths. Accordingly the bit-allocation data use words of four, three and two bits depending on the sub-band concerned. This reduces the amount of allocation data to be sent. In each case one extra combination exists in the allocation code. This is used to indicate that no data are being sent for that sub-band.

The 1152-sample block of Layer II is divided into three blocks of 384 samples so that the same companding structure as Layer I can be used. The 2 dB step size in the scale factors is retained. However, not all the scale factors are transmitted, because they contain a degree of redundancy. In real program material, the difference between scale factors in successive blocks in the same band exceeds 2 dB less than 10 per cent of the time. Layer II coders analyse the set of three successive scale factors in each sub-band. On stationary program, these will be the same and only one scale factor out of three is sent. As the transient content increases in a given sub-band, two or three scale factors will be sent. A two-bit code known as SCFSI (scale factor select information) must be sent to allow the decoder to determine which of the three possible scale factors have been sent for each sub-band. This technique effectively halves the scale factor bit rate.

As for Layer I, the requantizing process always uses an odd number of steps to allow a true centre zero step. In long wordlength codes this is not a problem, but when three, five or nine quantizing intervals are used, binary is inefficient because some combinations are not used. For example, five intervals needs a three-bit code having eight combinations leaving three unused.

The solution is that when three,-five-or nine-level coding is used in a sub-band, sets of three samples are encoded into a *granule*. Figure 5.27 shows how granules work. Continuing the example of five quantizing intervals, each sample could have five different values, therefore all combinations of three samples could have 125 different values. As 128 values can be sent with a seven-bit code, it will be seen that this is more efficient than coding the samples separately as three five-level codes would need nine bits. The three requantized samples are used to address a look-up table which outputs the granule code. The decoder can establish that granule coding has been used by examining the bit-allocation data.

The requantized samples/granules in each sub-band, bit allocation data, scale factors and scale factor select codes are multiplexed into the output bitstream.

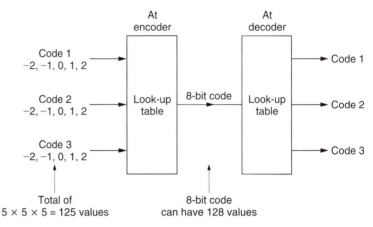

Figure 5.27 Codes having ranges smaller than a power of two are inefficient. Here three codes with a range of five values which would ordinarily need 3 × 3 bits can be carried in a single eight-bit word.

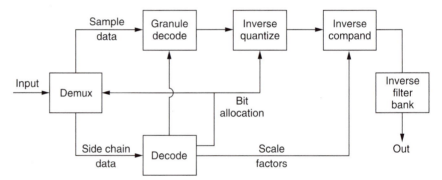

Figure 5.28 A Layer II decoder is slightly more complex than the Layer I decoder because of the need to decode granules and scale factors.

The Layer II decoder is shown in Figure 5.28. This is not much more complex than the Layer I decoder. The demultiplexing will separate the sample data from the side information. The bit-allocation data will specify the wordlength or granule size used so that the sample block can be deserialized and the granules decoded. The scale factor select information will be used to decode the compressed scale factors to produce one scale factor per block of 384 samples. Inverse quantizing and inverse sub-band filtering takes place as for Layer I.

5.15 MPEG Layer III

Layer III is the most complex layer, and is only really necessary when the most severe data rate constraints must be met. It is also known as MP3 in its application of music delivery over the Internet. It is a transform code based on the ASPEC system with certain modifications to give a degree of commonality with Layer II. The original ASPEC coder used a direct MDCT on the input samples. In Layer III this was modified to use a hybrid transform incorporating the existing polyphase 32-band QMF of Layers I and II and retaining the block size of 1152 samples. In Layer III, the 32 sub-bands from the QMF are further processed by a critically sampled MDCT.

The windows overlap by two to one. Two window sizes are used to reduce pre-echo on transients. The long window works with 36 sub-band samples corresponding to 24 ms at 48 kHz and resolves 18 different frequencies, making 576 frequencies altogether. Coding products are spread over this period which is acceptable in stationary material but not in the vicinity of transients. In this case the window length is reduced to 8 ms. Twelve sub-band samples are resolved into six different frequencies making a total of 192 frequencies. This is the Heisenberg inequality: by increasing the time resolution by a factor of three, the frequency resolution has fallen by the same factor.

Figure 5.29 shows the available window types. In addition to the long and short symmetrical windows there is a pair of transition windows, know as start and stop windows which allow a smooth transition between the two window sizes. In order to use critical sampling, MDCTs must resolve into a set of frequencies which is a multiple of four. Switching between 576 and 192 frequencies allows this criterion to be met. Note that an 8 ms window is still too long to

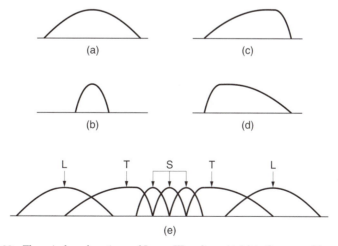

Figure 5.29 The window functions of Layer III coding. At (a) is the normal long window, whereas (b) shows the short window used to handle transients. Switching between window sizes requires transition windows (c) and (d). An example of switching using transition windows is shown in (e).

eliminate pre-echo. Pre-echo is eliminated using buffering. The use of a short window minimizes the size of the buffer needed.

Layer III provides a suggested (but not compulsory) pycho-acoustic model which is more complex than that suggested for Layers I and II, primarily because of the need for window switching. Pre-echo is associated with the entropy in the audio rising above the average value and this can be used to switch the window size. The perceptive model is used to take advantage of the high-frequency resolution available from the DCT which allows the noise floor to be shaped much more accurately than with the 32 sub-bands of Layers I and II. Although the MDCT has high-frequency resolution, it does not carry the phase of the waveform in an identifiable form and so is not useful for discriminating between tonal and atonal inputs. As a result a side FFT which gives conventional amplitude and phase data is still required to drive the masking model.

Non-uniform quantizing is used, in which the quantizing step size becomes larger as the magnitude of the coefficient increases. The quantized coefficients are then subject to Huffman coding. This is a technique where the most common code values are allocated the shortest wordlength. Layer III also has a certain amount of buffer memory so that pre-echo can be avoided during entropy peaks despite a constant output bit rate.

Figure 5.30 shows a Layer III encoder. The output from the sub-band filter is 32 continuous band-limited sample streams. These are subject to 32 parallel MDCTs. The window size can be switched individually in each sub-band as required by the characteristics of the input audio. The parallel FFT drives the masking model which decides on window sizes as well as producing the masking threshold for the coefficient quantizer. The distortion control loop iterates until the available output data capacity is reached with the most uniform NMR.

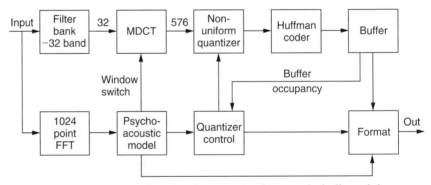

Figure 5.30 The Layer III coder. Note the connection between the buffer and the quantizer which allows different frames to contain different amounts of data.

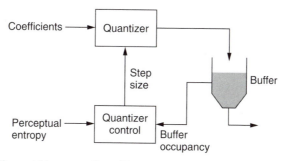

Figure 5.31 The variable rate coding of Layer III. An approaching transient via the perceptual entropy signal causes the coder to quantize more heavily in order to empty the buffer. When the transient arrives, the quantizing can be made more accurate and the increased data can be accepted by the buffer.

The available output capacity can vary owing to the presence of the buffer. Figure 5.31 shows that the buffer occupancy is fed back to the quantizer. During stationary program material, the buffer contents are deliberately run down by slight coarsening of the quantizing. The buffer empties because the output rate is fixed but the input rate has been reduced. When a transient arrives, the large coefficients which result can be handled by filling the buffer, avoiding raising the output bit rate whilst also avoiding the pre-echo which would result if the coefficients were heavily quantized.

In order to maintain synchronism between encoder and decoder in the presence of buffering, headers and side information are sent synchronously at frame rate. However, the position of boundaries between the main data blocks which carry the coefficients can vary with respect to the position of the headers in order to allow a variable frame size. Figure 5.32

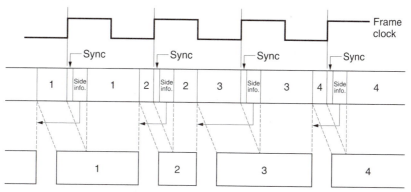

Figure 5.32 In Layer III, the logical frame rate is constant and is transmitted by equally spaced sync patterns. The data blocks do not need to coincide with sync. A pointer after each sync pattern specifies where the data block starts. In this example block 2 is smaller whereas 1 and 3 have enlarged.

shows that the frame begins with an unique sync pattern which is followed by the side information. The side information contains a parameter called *main data begin* which specifies where the main data for the present frame began in the transmission. This parameter allows the decoder to find the coefficient block in the decoder buffer. As the frame headers are at fixed locations, the main data blocks may be interrupted by the headers.

5.16   MPEG-2 AAC

The MPEG standards system subsequently developed an enhanced system known as advanced audio coding (AAC).[8,26] This was intended to be a standard which delivered the highest possible performance using newly developed tools that could not be used in any standard which was backward compatible. AAC will also form the core of the audio coding of MPEG-4.

AAC supports up to 48 audio channels with default support of monophonic, stereo and 5.1 channel (3/2) audio. The AAC concept is based on a number of coding tools known as *modules* which can be combined in different ways to produce bitstreams at three different profiles.

The main profile requires the most complex encoder which makes use of all the coding tools. The low-complexity (LC) profile omits certain tools and restricts the power of others to reduce processing and memory requirements. The remaining tools in LC profile coding are identical to those in main profile such that a main profile decoder can decode LC profile bitstreams.

The scaleable sampling rate (SSR) profile splits the input audio into four equal frequency bands each of which results in a self-contained bitstream. A simple decoder can decode only one, two or three of these bitstreams to produce a reduced bandwidth output. Not all the AAC tools are available to SSR profile.

The increased complexity of AAC allows the introduction of lossless coding tools. These allow a lower bit rate for the same quality or improved quality at a given bit rate where the reliance on lossy coding is reduced. There is greater attention given to the interplay between time-domain and frequency-domain precision in the human hearing system.

Figure 5.33 shows a block diagram of an AAC main profile encoder. The audio signal path is straight through the centre. The formatter assembles any side-chain data along with the coded audio data to produce a compliant bitstream. The input signal passes to the filter bank and the perceptual model in parallel.

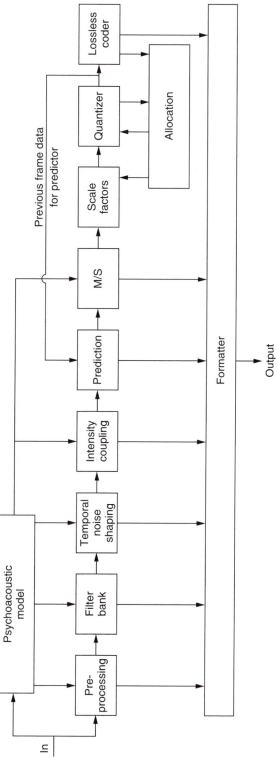

Figure 5.33 The AAC encoder. Signal flow is from left to right whereas side-chain data flow is vertical.

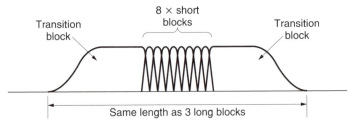

Figure 5.34 In AAC short blocks must be used in multiples of 8 so that the long block phase is undisturbed. This keeps block synchronism in multichannel systems.

The filter bank consists of a 50 per cent overlapped critically sampled MDCT which can be switched between block lengths of 2048 and 256 samples. At 48 kHz the filter allows resolutions of 23 Hz and 21 ms or 187 Hz and 2.6 ms. As AAC is a multichannel coding system, block length switching cannot be done indiscriminately as this would result in loss of block phase between channels. Consequently if short blocks are selected, the coder will remain in short block mode for integer multiples of eight blocks. This is illustrated in Figure 5.34 which also shows the use of transition windows between the block sizes as was done in Layer III.

The shape of the window function interferes with the frequency selectivity of the MDCT. In AAC it is possible to select either a sine window or a Kaiser–Bessel-derived (KBD) window as a function of the input audio spectrum. As was seen in Chapter 3, filter windows allow different compromises between bandwidth and rate of roll-off. The KBD window rolls off later but is steeper and thus gives better rejection of frequencies more than about 200 Hz apart whereas the sine window rolls off earlier but less steeply and so gives better rejection of frequencies less than 70 Hz.

Following the filter bank is the intra-block predictive coding module. When enabled this module finds redundancy between the coefficients within one transform block. In Chapter 3 the concept of transform duality was introduced, in which a certain characteristic in the frequency domain would be accompanied by a dual characteristic in the time domain and vice versa. Figure 5.35 shows that in the time domain, predictive coding works well on stationary signals but fails on transients. The dual of this characteristic is that in the frequency domain, predictive coding works well on transients but fails on stationary signals.

Equally, a predictive coder working in the time domain produces an error spectrum which is related to the input spectrum. The dual of this characteristic is that a predictive coder working in the frequency domain produces a prediction error which is related to the input time-domain signal.

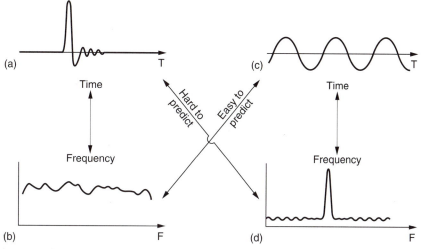

Figure 5.35 Transform duality suggests that predictability will also have a dual characteristic. A time predictor will not anticipate the transient in (a), whereas the broad spectrum of signal (a), shown in (b), will be easy for a predictor advancing down the frequency axis. In contrast, the stationary signal (c) is easy for a time predictor, whereas in the spectrum of (c) shown at (d) the spectral spike will not be predicted.

This explains the use of the term *temporal noise shaping* (TNS) used in the AAC documents.[27] When used during transients, the TNS module produces a distortion which is time-aligned with the input such that pre-echo is avoided. The use of TNS also allows the coder to use longer blocks more of the time. This module is responsible for a significant amount of the increased performance of AAC.

Figure 5.36 shows that the coefficients in the transform block are serialized by a commutator. This can run from the lowest frequency to the highest or in reverse. The prediction method is a conventional forward predictor structure in which the result of filtering a number of earlier coefficients (20 in main profile) is used to predict the current one. The

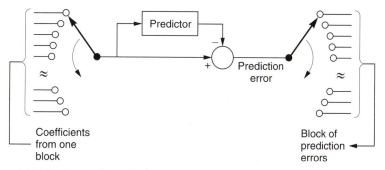

Figure 5.36 Predicting along the frequency axis is performed by running along the coefficients in a block and attempting to predict the value of the current coefficient from the values of some earlier ones. The prediction error is transmitted.

prediction is subtracted from the actual value to produce a prediction error or residual which is transmitted. At the decoder, an identical predictor produces the same prediction from earlier coefficient values and the error in this is cancelled by adding the residual.

Following the intra-block prediction, an optional module known as the intensity/coupling stage is found. This is used for very low bit rates where spatial information in stereo and surround formats is discarded to keep down the level of distortion. Effectively over at least part of the spectrum a mono signal is transmitted along with amplitude codes which allow the signal to be panned in the spatial domain at the decoder.

The next stage is the inter-block prediction module. Whereas the intra-block predictor is most useful on transients, the inter-block predictor module explores the redundancy between successive blocks on stationary signals.[28] This prediction only operates on coefficients below 16 kHz. For each DCT coefficient in a given block, the predictor uses the quantized coefficients from the same locations in two previous blocks to estimate the present value. As before, the prediction is subtracted to produce a residual which is transmitted. Note that the use of quantized coefficients to drive the predictor is necessary because this is what the decoder will have to do.

The predictor is adaptive and calculates its own coefficients from the signal history. The decoder uses the same algorithm so that the two predictors always track. The predictors run all the time whether prediction is enabled or not in order to keep the prediction coefficients adapted to the signal.

Audio coefficients are associated into sets known as *scale factor bands* for later companding. Within each scale factor band inter-block prediction can be turned on or off depending on whether a coding gain results.

Protracted use of prediction makes the decoder prone to bit errors and drift and removes decoding entry points from the bitstream. Consequently the prediction process is reset cyclically. The predictors are assembled into groups of 30 and after a certain number of a frames a different group is reset until all have been reset. Predictor reset codes are transmitted in the side data. Reset will also occur if short frames are selected.

In stereo and 3/2 surround formats there is less redundancy because the signals also carry spatial information. The effecting of masking may be up to 20 dB less when distortion products are at a different location in the stereo image from the masking sounds. As a result stereo signals require much higher bit rate than two mono channels, particularly on transient material which is rich in spatial clues.

In some cases a better result can be obtained by converting the signal to a mid-side (M/S) or sum/difference format before quantizing. In surround-sound the M/S coding can be applied to the front L/R pair and

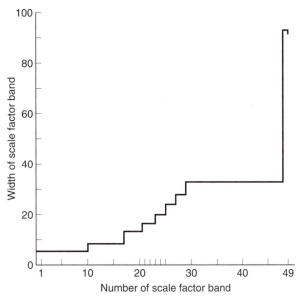

Figure 5.37 In AAC the fine-resolution coefficients are grouped together to form scale factor bands. The size of these varies to loosely mimic the width of critical bands.

the rear L/R pair of signals. The M/S format can be selected on a block-by-block basis for each scale factor band.

Next comes the lossy stage of the coder where distortion is selectively introduced as a function of frequency as determined by the masking threshold. This is done by a combination of amplification and requantizing. As mentioned, coefficients (or residuals) are grouped into scale factor bands. As Figure 5.37 shows, the number of coefficients varies in order to divide the coefficients into approximate critical bands. Within each scale factor band, all coefficients will be multiplied by the same scale factor prior to requantizing. Coefficients which have been multiplied by a large scale factor will suffer less distortion by the requantizer whereas those which have been multiplied by a small scale factor will have more distortion. Using scale factors, the psychoacoustic model can shape the distortion as a function of frequency so that it remains masked. The scale factors allow gain control in 1.5 dB steps over a dynamic range equivalent to 24-bit PCM and are transmitted as part of the side data so that the decoder can re-create the correct magnitudes.

The scale factors are differentially coded with respect to the first one in the block and the differences are then Huffman coded.

The requantizer uses non-uniform steps which give better coding gain and has a range of ±8191. The global step size (which applies to all scale factor bands) can be adjusted in 1.5 dB steps. Following requantizing the coefficients are Huffman coded.

There are many ways in which the coder can be controlled and any which results in a compliant bitstream is acceptable although the highest performance may not be reached. The requantizing and scale factor stages will need to be controlled in order to make best use of the available bit rate and the buffering. This is non-trivial because of the use of Huffman coding after the requantizer makes it impossible to predict the exact amount of data which will result from a given step size. This means that the process must iterate.

Whatever bit rate is selected, a good encoder will produce consistent quality by selecting window sizes, intra- or inter-frame prediction and using the buffer to handle entropy peaks. This suggests a connection between buffer occupancy and the control system. The psychoacoustic model will analyse the incoming audio entropy and during periods of average entropy it will empty the buffer by slightly raising the quantizer step size so that the bit rate entering the buffer falls. By running the buffer down, the coder can temporarily support a higher bit rate to handle transients or difficult material.

Simply stated, the scale factor process is controlled so that the distortion spectrum has the same shape as the masking threshold and the quantizing step size is controlled to make the level of the distortion spectrum as low as possible within the allowed bit rate. If the bit rate allowed is high enough, the distortion products will be masked.

5.17   apt-X

The apt-X100 codec[14] uses predictive coding in four sub-bands to achieve compression to 0.25 of the original bit rate. The sub-bands are derived with quadrature mirror filters, but in each sub-band a continous predictive coding takes place which is matched by a continuous decoding at the receiver. Blocks are not used for coding, but only for packing the difference values for transmission. The output block consists of 2048 bits and commences with a synchronizing pattern which enables the decoder to correctly assemble difference values and attribute them to the appropriate sub-band. The decoder must see three sync patterns at the correct spacing before locking is considered to have occurred. The synchronizing system is designed so that four compressed data streams can be compressed into one sixteen-bit channel and correctly demultiplexed at the decoders.

With a continuous DPCM coder there is no reliance on temporal masking, but adaptive coders which vary the requantizing step size will need to have a rapid step size attack in order to avoid clipping on transients. Following the transient, the signal will often decay more quickly than the step size, resulting in excessively coarse requantization.

During this period, temporal masking prevents audibility of the noise. As the process is waveform based rather than spectrum based, neither an accurate model of auditory masking nor a large number of sub-bands are necessary. As a result, apt-X100 can operate over a wide range of sampling rates without adjustment whereas in the majority of coders changing the sampling rate means that the sub-bands have different frequencies and will require different masking parameters. A further salient advantage of the predictive approach is that the delay through the codec is less than 4 ms, which is advantageous for live (rather than recorded) applications.

5.18   Dolby AC-3

Dolby AC-3[19] is in fact a family of transform coders based on time-domain aliasing cancellation (TDAC) which allow various compromises between coding delay and bit rate to be used. In the modified discrete cosine transform (MDCT), windows with 50 per cent overlap are used. Thus twice as many coefficients as necessary are produced. These are sub-sampled by a factor of two to give a critically sampled transform, which results in potential aliasing in the frequency domain. However, by making a slight change to the transform, the alias products in the second half of a given window are equal in size but of opposite polarity to the alias products in the first half of the next window, and so will be cancelled on reconstruction. This is the principle of TDAC.

Figure 5.38 shows the generic block diagram of the AC-3 coder. Input audio is divided into 50 per cent overlapped blocks of 512 samples. These are subject to a TDAC transform which uses alternate modified sine and cosine transforms. The transforms produce 512 coefficients per block, but these are redundant and after the redundancy has been removed there are 256 coefficients per block.

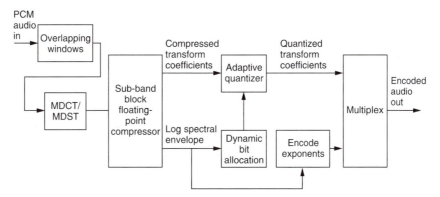

Figure 5.38   Block diagram of the Dolby AC-3 coder. See text for details.

The input waveform is constantly analysed for the presence of transients and if these are present the block length will be halved to prevent pre-noise. This halves the frequency resolution but doubles the temporal resolution.

The coefficients have high-frequency resolution and are selectively combined in sub-bands which approximate the critical bands. Coefficients in each sub-band are normalized and expressed in floating point block notation with common exponents. The exponents in fact represent the logarithmic spectral envelope of the signal and can be used to drive the perceptive model which operates the bit allocation. The mantissae of the transform coefficients are then requantized according to the bit allocation.

The output bitstream consists of the requantized coefficients and the log spectral envelope in the shape of the exponents. There is a great deal of redundancy in the exponents. In any block, only the first exponent, corresponding to the lowest frequency, is transmitted absolutely. Remaining coefficients are transmitted differentially. Where the input has a smooth spectrum the exponents in several bands will be the same and the differences will then be zero. In this case exponents can be grouped using flags.

Further use is made of temporal redundancy. An AC-3 sync frame contains six blocks. The first block of the frame contains absolute exponent data, but where stationary audio is encountered, successive blocks in the frame can use the same exponents.

The receiver uses the log spectral envelope to deserialize the mantissae of the coefficients into the correct wordlengths. The highly redundant exponents are decoded starting with the lowest-frequency coefficient in the first block of the frame and adding differences to create the remainder. The exponents are then used to return the coefficients to fixed point notation. Inverse transforms are then computed, followed by a weighted overlapping of the windows to obtain PCM data.

5.19   ATRAC

The ATRAC (Adaptive TRansform Acoustic Coder) coder was developed by Sony and is used in MiniDisc. ATRAC uses a combination of sub-band coding and modified discrete cosine transform (MDCT) coding. Figure 5.39 shows a block diagram of an ATRAC coder. The input is sixteen-bit PCM audio. This passes through a quadrature mirror filter which splits the audio band into two halves. The lower half of the spectrum is split in half once more, and the upper half passes through a compensating delay. Each frequency band is formed into blocks, and each block is then subject to a modified discrete cosine transform. The frequencies of the DCT are

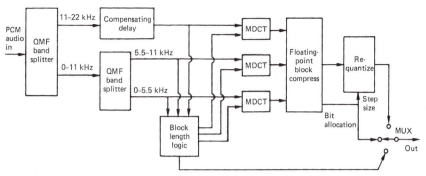

Figure 5.39 The ATRAC coder uses variable-length blocks and MDCT in three sub-bands.

grouped into a total of 52 frequency bins which are of varying bandwidth according to the width of the critical bands in the hearing mechanism. The coefficients in each frequency bin are then companded and requantized. The requantizing is performed once more on a bit-allocation basis using a masking model.

In order to prevent pre-echo, ATRAC selects blocks as short as 1.45 ms in the case of large transients, but the block length can increase in steps up to a maximum of 11.6 ms when the waveform has stationary characteristics. The block size is selected independently in each of the three bands.

The coded data include side-chain parameters which specify the block size and the wordlength of the coefficients in each frequency bin.

Decoding is straightforward. The bitstream is deserialized into coefficients of various wordlengths and block durations according to the side-chain data. The coefficients are then used to control inverse DCTs which recreate time-domain waveforms in the three sub-bands. These are recombined in the output filter to produce the conventional PCM output. In MiniDisc, the ATRAC coder compresses 44.1 kHz sixteen-bit PCM to 0.2 of the original data rate.

References

1. ISO/IEC JTC1/SC29/WG11 MPEG, International standard ISO 11172–3, Coding of moving pictures and associated audio for digital storage media up to 1.5 Mbits/s, Part 3: Audio (1992)
2. MPEG Video Standard: ISO/IEC 13818–2: Information technology – generic coding of moving pictures and associated audio information: Video (1996) (aka ITU-T Rec. H-262) (1996)
3. Huffman, D.A. A method for the construction of minimum redundancy codes. *Proc. IRE*, **40**, 1098–1101 (1952)
4. Grewin, C. and Ryden, T., Subjective assessments on low bit-rate audio codecs. *Proc. 10th. Int. Audio Eng. Soc. Conf.*, 91–102, New York: Audio Engineeringse Society (1991)

5. Gilchrist, N.H.C., Digital sound: the selection of critical programme material and preparation of the recordings for CCIR tests on low bit rate codecs. *BBC Research Dept Report*, RD 1993/1

6. Colomes, C. and Faucon, G., A perceptual objective measurement system (POM) for the quality assessment of perceptual codecs. Presented at the 96th Audio Engineering Society Convention (Amsterdam, 1994), Preprint No. 3801 (P4.2)

7. Johnston, J., Estimation of perceptual entropy using noise masking criteria. *ICASSP*, 2524–2527 (1988)

8. ISO/iec 13818-7, Information Technology – Generic coding of moving pictures and associated audio, Part 7: Advanced audio coding (1997)

9. Gilchrist, N.H.C., Delay in broadcasting operations. Presented at the 90th Audio Engineering Society Convention (1991), Preprint 3033

10. Caine, C.R., English, A.R. and O'Clarey, J.W.H., NICAM-3: near-instantaneous companded digital transmission for high-quality sound programmes. *J. IERE*, **50**, 519–530 (1980)

11. Davidson, G.A. and Bosi, M., AC-2: High quality audio coding for broadcast and storage, in *Proc. 46th Ann. Broadcast Eng. Conf.*, Las Vegas, 98–105 (1992)

12. Crochiere, R.E., Sub-band coding. *Bell System Tech. J.*, **60**, 1633–1653 (1981)

13. Princen, J.P., Johnson, A. and Bradley, A.B., Sub-band/transform coding using filter bank designs based on time domain aliasing cancellation. *Proc. ICASSP*, 2161–2164 (1987)

14. Smyth, S.M.F. and McCanny, J.V., 4-bit Hi-Fi: High quality music coding for ISDN and broadcasting applications. *Proc. ASSP*, 2532–2535 (1988)

15. Jayant, N.S. and Noll, P., *Digital Coding of Waveforms: Principles and applications to speech and video*, Englewood Cliffs: Prentice Hall (1984)

16. Theile, G., Stoll, G. and Link, M., Low bit rate coding of high quality audio signals: an introduction to the MASCAM system. *EBU Tech. Review*, No. 230, 158–181 (1988)

17. Chu, P.L., Quadrature mirror filter design for an arbitrary number of equal bandwidth channels. *IEEE Trans. ASSP*, **ASSP-33**, 203–218 (1985)

18. Fettweis, A., Wave digital filters: Theory and practice. *Proc. IEEE*, **74**, 270–327 (1986)

19. Davis, M.F., The AC-3 multichannel coder. Presented at the 95th Audio Engineering Society Convention, Preprint 2774.

20. Wiese, D., MUSICAM: flexible bitrate reduction standard for high quality audio. Presented at the Digital Audio Broadcasting Conference (London, March 1992)

21. Brandenburg, K., ASPEC coding. *Proc. 10th. Audio Eng. Soc. Int. Conf.*, 81–90, New York: Audio Engineering Society (1991)

22. ISO/IEC JTC1/SC2/WG11 N0030: MPEG/AUDIO test report, Stockholm (1990)

23. ISO/IEC JTC1/SC2/WG11 MPEG 91/010, The SR report on: The MPEG/AUDIO subjective listening test, Stockholm (1991)

24. Brandenburg, K. and Stoll, G., ISO-MPEG-1 Audio: A generic standard for coding of high quality audio. *JAES*, **42**, 780–792 (1994)

25. Bonicel, P. *et al.*, A real time ISO/MPEG2 Multichannel decoder. Presented at the 96th Audio Engineering Society Convention (1994), Preprint No. 3798 (P3.7)4.30

26. Bosi. M. *et al.*, ISO/IEC MPEG-2 Advanced Audio Coding *JAES*, **45**, 789–814 (1997)

27. Herre, J. and Johnston, J.D., Enhancing the performance of perceptual audio coders by using temporal noise shaping (TNS). Presented at the 101st Audio Engineering Society Convention, Preprint 4384 (1996)

28. Fuchs, H., Improving MPEG audio coding by backward adaptive linear stereo prediction. Presented at the 99th Audio Engineering Society Convention (1995), Preprint 4086

6

Digital recording and transmission principles

Recording and transmission are quite different tasks, but they have a great deal in common and have always been regarded as being different applications of the same art. Digital transmission consists of converting data into a waveform suitable for the path along which it is to be sent. Digital recording is basically the process of recording a digital transmission waveform on a suitable medium. Although the physics of the recording or transmission processes are unaffected by the meaning attributed to signals, digital techniques are rather different from those used with analog signals, although often the same phenomenon shows up in a different guise. In this chapter the fundamentals of digital recording and transmission are introduced along with descriptions of the coding techniques used in practical applications. The parallel subject of error correction is dealt with in the next chapter.

6.1 Introduction to the channel

Data can be recorded on many different media and conveyed using many forms of transmission. The generic term for the path down which the information is sent is the *channel*. In a transmission application, the channel may be no more than a length of cable. In a recording application the channel will include the record head, the medium and the replay head. In analog systems, the characteristics of the channel affect the signal directly. It is a fundamental strength of digital audio that by pulse code modulating an audio waveform the quality can be made independent of the channel. The dynamic range required by the programme material no longer directly decides the dynamic range of the channel.

In digital circuitry there is a great deal of noise immunity because the signal has only two states, which are widely separated compared with the amplitude of noise. In both digital recording and transmission this is not always the case. In magnetic recording, noise immunity is a function of track width and reduction of the working SNR of a digital track allows the same information to be carried in a smaller area of the medium, improving economy of operation. In broadcasting, the noise immunity is a function of the transmitter power and reduction of working SNR allows lower power to be used with consequent economy. These reductions also increase the random error rate, but, as was seen in Chapter 1, an error-correction system may already be necessary in a practical system and it is simply made to work harder.

In real channels, the signal may *originate* with discrete states which change at discrete times, but the channel will treat it as an analog waveform and so it will not be *received* in the same form. Various loss mechanisms will reduce the amplitude of the signal. These attenuations will not be the same at all frequencies. Noise will be picked up in the channel as a result of stray electric fields or magnetic induction. As a result the voltage received at the end of the channel will have an infinitely varying state along with a degree of uncertainty due to the noise. Different frequencies can propagate at different speeds in the channel; this is the phenomenon of group delay. An alternative way of considering group delay is that there will be frequency-dependent phase shifts in the signal and these will result in uncertainty in the timing of pulses.

In digital circuitry, the signals are generally accompanied by a separate clock signal which reclocks the data to remove jitter as was shown in Chapter 1. In contrast, it is generally not feasible to provide a separate clock in recording and transmission applications. In the transmission case, a separate clock line would not only raise cost, but is impractical because at high frequency it is virtually impossible to ensure that the clock cable propagates signals at the same speed as the data cable except over short distances. In the recording case, provision of a separate clock track is impractical at high density because mechanical tolerances cause phase errors between the tracks. The result is the same; timing differences between parallel channels which are known as skew.

The solution is to use a self-clocking waveform and the generation of this is a further essential function of the coding process. Clearly if data bits are simply clocked serially from a shift register in so-called direct recording or transmission this characteristic will not be obtained. If all the data bits are the same, for example all zeros, there is no clock when they are serialized.

It is not the channel which is digital; instead the term describes the way in which the received signals are *interpreted*. When the receiver makes discrete decisions from the input waveform it attempts to reject the uncertainties in voltage and time. The technique of channel coding is one where transmitted waveforms are restricted to those which still allow the receiver to make discrete decisions despite the degradations caused by the analog nature of the channel.

6.2   Types of transmission channel

Transmission can be by electrical conductors, radio or optical fibre. Although these appear to be completely different, they are in fact just different examples of electromagnetic energy travelling from one place to another. If the energy is made to vary is some way, information can be carried.

Even today electromagnetism is not fully understood, but sufficiently good models, based on experimental results, exist so that practical equipment can be made. It is not actually necessary to fully understand a process in order to harness it; it is only necessary to be able to reliably predict what will happen in given circumstances.

Electromagnetic energy propagates in a manner which is a function of frequency, and our partial understanding requires it to be considered as electrons, waves or photons so that we can predict its behaviour in given circumstances.

At DC and at the low frequencies used for power distribution, electromagnetic energy is called electricity and it is remarkably aimless stuff which needs to be transported completely inside conductors. It has to have a complete circuit to flow in, and the resistance to current flow is determined by the cross-sectional area of the conductor. The insulation around the conductor and the spacing between the conductors has no effect on the ability of the conductor to pass current. At DC an inductor appears to be a short circuit, and a capacitor appears to be an open circuit.

As frequency rises, resistance is exchanged for impedance. Inductors display increasing impedance with frequency, capacitors show falling impedance. Electromagnetic energy becomes increasingly desperate to leave the conductor. The first symptom is that the current flows only in the outside layer of the conductor effectively causing the resistance to rise. This is the skin effect and gives rise to such techniques as Litz wire which has as much surface area as possible per unit cross-section, and to silver-plated conductors in which the surface has lower resistivity than the interior.

As the energy is starting to leave the conductors, the characteristics of the space between them become important. This determines the imped-ance. A change of impedance causes reflections in the energy flow and some of it heads back towards the source. Constant impedance cables with fixed conductor spacing are necessary, and these must be suitably terminated to prevent reflections. The most important characteristic of the insulation is its thickness as this determines the spacing between the conductors.

As frequency rises still further, the energy travels less in the conductors and more in the insulation between them. Their composition becomes important and they begin to be called dielectrics. A poor dielectric like PVC absorbs high-frequency energy and attenuates the signal. So-called low-loss dielectrics such as PTFE are used, and one way of achieving low loss is to incorporate as much air in the dielectric as possible by making it in the form of a foam or extruding it with voids.

Further rise in frequency causes the energy to start behaving more like waves and less like electron movement. As the wavelength falls it becomes increasingly directional. The transmission line becomes a waveguide and microwaves are sufficiently directional that they can keep going without any conductor at all. Microwaves are simply low-frequency radiant heat, which is itself low-frequency light. All three are reflected well by electrical conductors, and can be refracted at the boundary between media having different propagation speeds. A waveguide is the microwave equivalent of an optical fibre.

This frequency-dependent behaviour is the most important factor in deciding how best to harness electromagnetic energy flow for information transmission. It is obvious that the higher the frequency, the greater the possible information rate, but in general, losses increase with frequency, and flat frequency response is elusive. The best that can be managed is that over a narrow band of frequencies, the response can be made reasonably constant with the help of equalization. Unfortunately raw data when serialized have an unconstrained spectrum. Runs of identical bits can produce frequencies much lower than the bit rate would suggest. One of the essential steps in a transmission system is to modify the spectrum of the data into something more suitable.

At moderate bit rates, say a few megabits per second, and with moderate cable lengths, say a few metres, the dominant effect will be the capacitance of the cable due to the geometry of the space between the conductors and the dielectric between. The capacitance behaves under these conditions as if it were a single capacitor connected across the signal. Figure 6.1 shows the equivalent circuit.

The effect of the series source resistance and the parallel capacitance is that signal edges or transitions are turned into exponential curves as the capacitance is effectively being charged and discharged through the source impedance. This effect can be observed on the AES/EBU interface with short cables. Although the position where the edges cross the centreline is displaced, the signal eventually reaches the same amplitude as it would at DC.

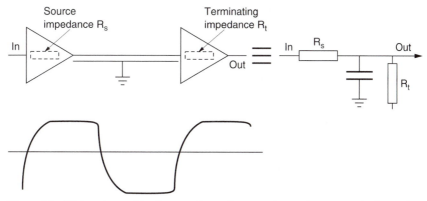

Figure 6.1 With a short cable, the capacitance between the conductors can be lumped as if it were a discrete component. The effect of the parallel capacitor is to slope off the edges of the signal.

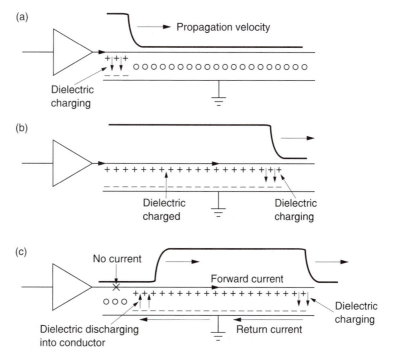

Figure 6.2 A transmission line conveys energy packets which appear to alternate with respect to the dielectric. In (a) the driver launches a pulse which charges the dielectric at the beginning of the line. As it propagates the dielectric is charged further along as in (b). When the driver ends the pulse, the charged dielectric discharges into the line. A current loop is formed where the current in the return loop flows in the opposite direction to the current in the 'hot' wire.

As cable length increases, the capacitance can no longer be lumped as if it were a single unit; it has to be regarded as being distributed along the cable. With rising frequency, the cable inductance also becomes significant, and it too is distributed.

The cable is now a transmission line and pulses travel down it as current loops which roll along as shown in Figure 6.2. If the pulse is positive, as it is launched along the line, it will charge the dielectric locally as at (a). As the pulse moves along, it will continue to charge the local dielectric as at (b). When the driver finishes the pulse, the trailing edge of the pulse follows the leading edge along the line. The voltage of the dielectric charged by the leading edge of the pulse is now higher than the voltage on the line, and so the dielectric discharges into the line as at (c). The current flows forward as it is in fact the same current which is flowing into the dielectric at the leading edge. There is thus a loop of current rolling down the line flowing forward in the 'hot' wire and backwards in the return. The analogy with the tracks of a Caterpillar tractor is quite good. Individual plates in the track find themselves being lowered to the ground at the front and raised again at the back.

The constant to-ing and fro-ing of charge in the dielectric results in dielectric loss of signal energy. Dielectric loss increases with frequency and so a long transmission line acts as a filter. Thus the term 'low-loss' cable refers primarily to the kind of dielectric used.

Transmission lines which transport energy in this way have a characteristic impedance caused by the interplay of the inductance along the conductors with the parallel capacitance. One consequence of that transmission mode is that correct termination or matching is required between the line and both the driver and the receiver. When a line is correctly matched, the rolling energy rolls straight out of the line into the load and the maximum energy is available. If the impedance presented by the load is incorrect, there will be reflections from the mismatch. An open circuit will reflect all the energy back in the same polarity as the original, whereas a short circuit will reflect all the energy back in the opposite polarity. Thus impedances above or below the correct value will have a tendency towards reflections whose magnitude depends upon the degree of mismatch and whose polarity depends upon whether the load is too high or too low. In practice it is the need to avoid reflections which is the most important reason to terminate correctly.

Reflections at impedance mismatches have practical applications; electricity companies inject high-frequency pulses into faulty cables and the time taken until the reflection from the break or short returns can be used to locate the source of damage. The same technique can be used to find wiring breaks in large studio complexes.

A perfectly square pulse contains an indefinite series of harmonics, but the higher ones suffer progressively more loss. A square pulse at the

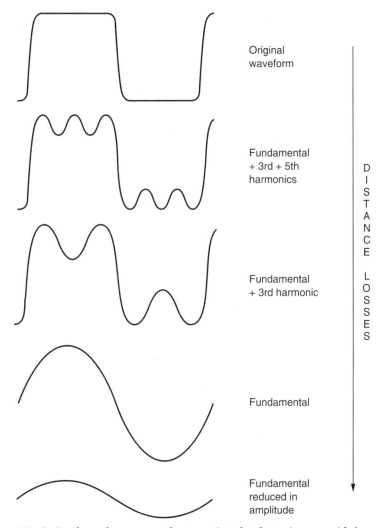

Figure 6.3 A signal may be square at the transmitter, but losses increase with frequency and as the signal propagates, more of the harmonics are lost until only the fundamental remains. The amplitude of the fundamental then falls with further distance.

driver becomes less and less square with distance as Figure 6.3 shows. The harmonics are progressively lost until in the extreme case all that is left is the fundamental. A transmitted square wave is received as a sine wave. Fortunately data can still be recovered from the fundamental signal component.

Once all the harmonics have been lost, further losses cause the amplitude of the fundamental to fall. The effect worsens with distance and it is necessary to ensure that data recovery is still possible from a signal of unpredictable level.

6.3   Types of recording medium

There is considerably more freedom of choice for digital media than was the case for analog signals. Once converted to the digital domain, audio is no more than data and can take advantage of the research expended in computer data recording.

Digital media do not need to have linear transfer functions, nor do they need to be noise-free or continuous. All they need to do is to allow the player to be able to distinguish the presence or absence of replay events, such as the generation of pulses, with reasonable (rather than perfect) reliability. In a magnetic medium, the event will be a flux change from one direction of magnetization to another. In an optical medium, the event must cause the pickup to perceive a change in the intensity of the light falling on the sensor. In CD, the apparent contrast is obtained by interference. In some disks it will be through selective absorption of light by dyes. In magneto-optical disks the recording itself is magnetic, but it is made and read using light.

6.4   Magnetism

Magnetism is vital to digital audio recording. Hard disks and tapes store magnetic patterns and media are driven by motors which themselves rely on magnetism.

A magnetic field can be created by passing a current through a solenoid, which is no more than a coil of wire. When the current ceases, the magnetism disappears. However, many materials, some quite common, display a permanent magnetic field with no apparent power source. Magnetism of this kind results from the spin of electrons within atoms. Atomic theory describes atoms as having nuclei around which electrons orbit, spinning as they go. Different orbits can hold a different number of electrons. The distribution of electrons determines whether the element is diamagnetic (non-magnetic) or paramagnetic (magnetic characteristics are possible). Diamagnetic materials have an even number of electrons in each orbit, and according to the Pauli exclusion principle half of them spin in each direction. The opposed spins cancel any resultant magnetic moment. Fortunately there are certain elements, the transition elements, which have an odd number of electrons in certain orbits. The magnetic moment due to electronic spin is not cancelled out in these paramagnetic materials.

Figure 6.4 shows that paramagnetism materials can be classified as antiferromagnetic, ferrimagnetic and ferromagnetic. In some materials alternate atoms are anti-parallel and so the magnetic moments are cancelled. In ferrimagnetic materials there is a certain amount of

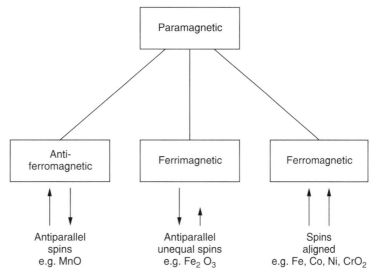

Figure 6.4 The classification of paramagnetic materials. The ferromagnetic materials exhibit the strongest magnetic behaviour.

antiparallel cancellation, but a net magnetic moment remains. In ferromagnetic materials such as iron, cobalt or nickel, all the electron spins can be aligned and as a result the most powerful magnetic behaviour is obtained.

It is not immediately clear how a material in which electron spins are parallel could ever exist in an unmagnetized state or how it could be partially magnetized by a relatively small external field. The theory of magnetic domains has been developed to explain what is observed in practice. Figure 6.5(a) shows a ferromagnetic bar which is demagnetized. It has no net magnetic moment because it is divided into domains or volumes which have equal and opposite moments. Ferromagnetic material divides into domains in order to reduce its magnetostatic energy. Figure 6.5(b) shows a domain wall which is around 0.1 micrometre thick. Within the wall the axis of spin gradually rotates from one state to another. An external field of quite small value is capable of disturbing the equilibrium of the domain wall by favouring one axis of spin over the other. The result is that the domain wall moves and one domain becomes larger at the expense of another. In this way the net magnetic moment of the bar is no longer zero as shown in (c).

For small distances, the domain wall motion is linear and reversible if the change in the applied field is reversed. However, larger movements are irreversible because heat is dissipated as the wall jumps to reduce its energy. Following such a domain wall jump, the material remains magnetized after the external field is removed and an opposing external field must be applied which must do further work to bring the domain

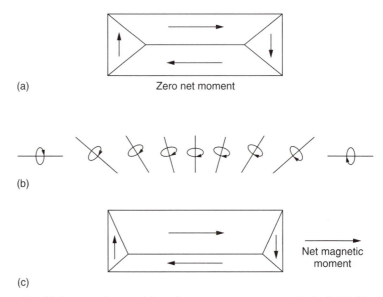

(a)   Zero net moment

(b)

(c)   Net magnetic moment

Figure 6.5 (a) A magnetic material can have a zero net moment if it is divided into domains as shown here. Domain walls (b) are areas in which the magnetic spin gradually changes from one domain to another. The stresses which result store energy. When some domains dominate, a net magnetic moment can exist as in (c).

wall back again. This is a process of hysteresis where work must be done to move each way. Were it not for this, non-linear mechanism magnetic recording would be impossible. If magnetic materials were linear, tapes would return to the demagnetized state immediately after leaving the field of the head and this book would be a good deal thinner.

Figure 6.6 shows a hysteresis loop which is obtained by plotting the magnetization M when the external field H is swept to and fro. On the macroscopic scale, the loop appears to be a smooth curve, whereas on a small scale it is in fact composed of a large number of small jumps. These were first discovered by Barkhausen. Starting from the unmagnetized state at the origin, as an external field is applied, the response is initially linear and the slope is given by the susceptibility. As the applied field is increased a point is reached where the magnetization ceases to increase. This is the saturation magnetization M_s. If the applied field is removed, the magnetization falls, not to zero, but the the remanent magnetization M_d. This remanence is the magnetic memory mechanism which makes recording and permanent magnets possible. The ratio of M_r to M_d is called the squareness ratio. In recording media squareness is beneficial as it increases the remanent magnetization.

If an increasing external field is applied in the opposite direction, the curve continues to the point where the magnetization is zero. The field required to achieve this is called the intrinsic coercive force $_mH_c$. A small

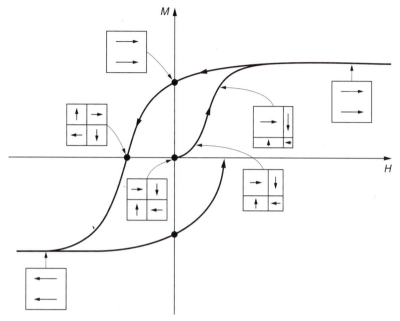

Figure 6.6 A hysteresis loop which comes about because of the non-linear behaviour of magnetic materials. If this characteristic were absent, magnetic recording would not exist.

increase in the reverse field reaches the point where, if the field were to be removed, the remanent magnetization would become zero. The field required to do this is the remanent coercive force, $_rH_c$.

As the external field H is swept to and fro, the magnetization describes a major hysteresis loop. Domain wall transit causes heat to be dissipated on every cycle around the loop and the dissipation is proportional to the loop area. For a recording medium, a large loop is beneficial because the replay signal is a function of the remanence and high coercivity resists erasure. The same is true for a permanent magnet. Heating is not an issue.

For a device such as a recording head, a small loop is beneficial. Figure 6.7(a) shows the large loop of a hard magnetic material used for recording media and for permanent magnets. Figure 6.7(b) shows the small loop of a soft magnetic material which is used for recording heads and transformers.

According to the Nyquist noise theorem, anything which dissipates energy when electrical power is supplied must generate a noise voltage when in thermal equilibrium. Thus magnetic recording heads have a noise mechanism which is due to their hysteretic behaviour. The smaller the loop, the less the hysteretic noise. In conventional heads, there are a large number of domains and many small domain wall jumps. In thin film heads there are fewer domains and the jumps must be larger. The

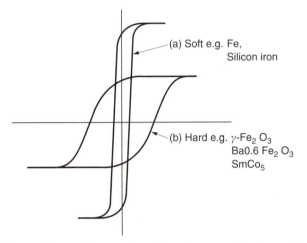

Figure 6.7 The recording medium requires a large loop area (a) whereas the head requires a small loop area (b) to cut losses.

noise this causes is known as Barkhausen noise, but as the same mechanism is responsible it is not possible to say at what point hysteresis noise should be called Barkhausen noise.

6.5 Magnetic recording

Magnetic recording relies on the hysteresis of certain magnetic materials. After an applied magnetic field is removed, the material remains magnetized in the same direction. By definition the process is non-linear, and analog magnetic recorders have to use bias to linearize it. Digital recorders are not concerned with the non-linearity, and HF bias is unnecessary.

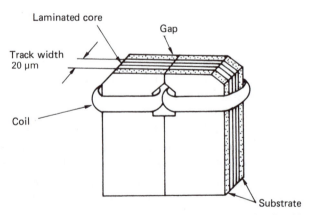

Figure 6.8 A digital record head is similar in principle to an analog head but uses much narrower tracks.

Figure 6.8 shows the construction of a typical digital record head, which is not disimilar to an analog record head. A magnetic circuit carries a coil through which the record current passes and generates flux. A non-magnetic gap forces the flux to leave the magnetic circuit of the head and penetrate the medium. The current through the head must be set to suit the coercivity of the tape, and is arranged to almost saturate the track. The amplitude of the current is constant, and recording is performed by reversing the direction of the current with respect to time. As the track passes the head, this is converted to the reversal of the magnetic field left on the tape with respect to distance. The magnetic recording is therefore bipolar. Figure 6.9 shows that the recording is actually made just after the trailing pole of the record head where the flux strength from the gap is falling. As in analog recorders, the width of the gap is generally made quite large to ensure that the full thickness of the magnetic coating is recorded, although this cannot be done if the same head is intended to replay.

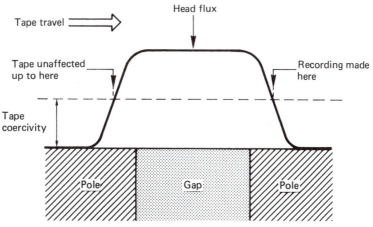

Figure 6.9   The recording is actually made near the trailing pole of the head where the head flux falls below the coercivity of the tape.

Figure 6.10 shows what happens when a conventional inductive head, i.e. one having a normal winding, is used to replay the bipolar track made by reversing the record current. The head output is proportional to the rate of change of flux and so only occurs at flux reversals. In other words, the replay head differentiates the flux on the track. The polarity of the resultant pulses alternates as the flux changes and changes back. A circuit is necessary which locates the peaks of the pulses and outputs a signal corresponding to the original record current waveform. There are two ways in which this can be done.

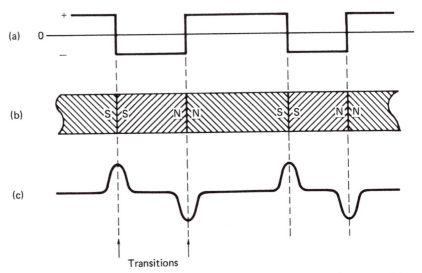

Figure 6.10 Basic digital recording. At (a) the write current in the head is reversed from time to time, leaving a binary magnetization pattern shown at (b). When replayed, the waveform at (c) results because an output is only produced when flux in the head changes. Changes are referred to as transitions.

The amplitude of the replay signal is of no consequence and often an AGC system is used to keep the replay signal constant in amplitude. What matters is the time at which the write current, and hence the flux stored on the medium, reverses. This can be determined by locating the peaks of the replay impulses, which can conveniently be done by differentiating the signal and looking for zero crossings. Figure 6.11

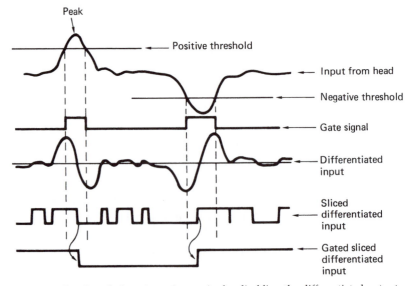

Figure 6.11 Gated peak detection rejects noise by disabling the differentiated output between transitions.

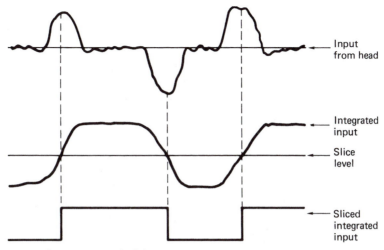

Figure 6.12 Integration method for re-creating write-current waveform.

shows that this results in noise between the peaks. This problem is overcome by the gated peak detector, where only zero crossings from a pulse which exceeds the threshold will be counted. The AGC system allows the thresholds to be fixed. As an alternative, the record waveform can also be restored by integration, which opposes the differentiation of the head as in Figure 6.12.[1]

The head shown in Figure 6.8 has a frequency response shown in Figure 6.13. At DC there is no change of flux and no output. As a result, inductive heads are at a disadvantage at very low speeds. The output rises with frequency until the rise is halted by the onset of thickness loss. As the frequency rises, the recorded wavelength falls and flux from the

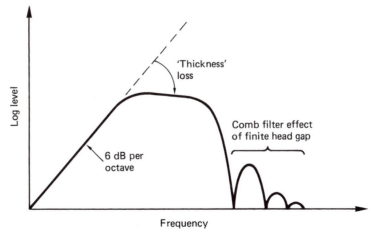

Figure 6.13 The major mechanism defining magnetic channel bandwidth.

shorter magnetic patterns cannot be picked up so far away. At some point, the wavelength becomes so short that flux from the back of the tape coating cannot reach the head and a decreasing thickness of tape contributes to the replay signal.[2] In digital recorders using short wavelengths to obtain high density, there is no point in using thick coatings. As wavelength further reduces, the familiar gap loss occurs, where the head gap is too big to resolve detail on the track. The construction of the head results in the same action as that of a two-point transversal filter, as the two poles of the head see the tape with a small delay interposed due to the finite gap. As expected, the head response is like a comb filter with the well-known nulls where flux cancellation takes place across the gap. Clearly the smaller the gap, the shorter the wavelength of the first null. This contradicts the requirement of the record head to have a large gap. In quality analog audio recorders, it is the norm to have different record and replay heads for this reason, and the same will be true in digital machines which have separate record and playback heads. Clearly where the same pair of heads are used for record and play, the head gap size will be determined by the playback requirement.

As can be seen, the frequency response is far from ideal, and steps must be taken to ensure that recorded data waveforms do not contain frequencies which suffer excessive losses.

A more recent development is the magneto-resistive (M-R) head. This is a head which measures the flux on the tape rather than using it to generate a signal directly. Flux measurement works down to DC and so offers advantages at low tape speeds. Unfortunately flux-measuring heads are not polarity conscious but sense the modulus of the flux and if used directly they respond to positive and negative flux equally, as shown in Figure 6.14. This is overcome by using a small extra winding in the head carrying a constant current. This creates a steady bias field which adds to the flux from the tape. The flux seen by the head is now unipolar and changes between two levels and a more useful output waveform results.

Recorders which have low head-to-medium speed, such as DCC (digital compact cassette) use M-R heads, whereas recorders with high speeds, such as DASH (digital audio stationary head), DAT (digital audio tape) and magnetic disk drives use inductive heads.

Heads designed for use with tape work in actual contact with the magnetic coating. The tape is tensioned to pull it against the head. There will be a wear mechanism and need for periodic cleaning.

In the hard disk, the rotational speed is high in order to reduce access time, and the drive must be capable of staying on line for extended periods. In this case the heads do not contact the disk surface, but are supported on a boundary layer of air. The presence of the air film causes

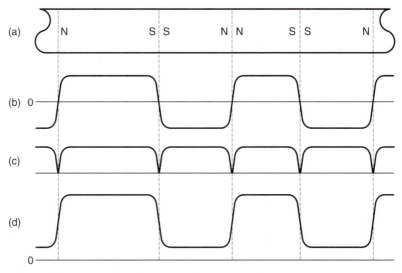

Figure 6.14 The sensing element in a magneto-resistive head is not sensitive to the polarity of the flux, only the magnitude. At (a) the track magnetization is shown and this causes a bidirectional flux variation in the head as at (b), resulting in the magnitude output at (c). However, if the flux in the head due to the track is biased by an additional field, it can be made unipolar as at (d) and the correct output waveform is obtained.

spacing loss, which restricts the wavelengths at which the head can replay. This is the penalty of rapid access.

Digital audio recorders must operate at high density in order to offer a reasonable playing time. This implies that shortest possible wavelengths will be used. Figure 6.15 shows that when two flux changes, or transitions, are recorded close together, they affect each other on replay.

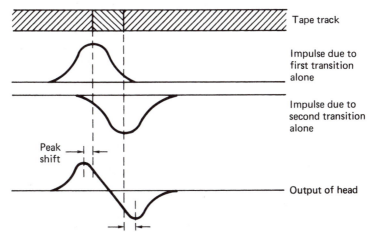

Figure 6.15 Readout pulses from two closely recorded transitions are summed in the head and the effect is that the peaks of the waveform are moved outwards. This is known as peak-shift distortion and equalization is necessary to reduce the effect.

The amplitude of the composite signal is reduced, and the position of the peaks is pushed outwards. This is known as inter-symbol interference, or peak-shift distortion and it occurs in all magnetic media.

The effect is primarily due to high-frequency loss and it can be reduced by equalization on replay, as is done in most tapes, or by pre-compensation on record as is done in hard disks.

6.6   Azimuth recording and rotary heads

Figure 6.16(a) shows that in azimuth recording, the transitions are laid down at an angle to the track by using a head which is tilted. Machines using azimuth recording must always have an even number of heads, so that adjacent tracks can be recorded with opposite azimuth angle. The two track types are usually referred to as A and B. Figure 6.16(b) shows the effect of playing a track with the wrong type of head. The playback process suffers from an enormous azimuth error. The effect of azimuth error can be understood by imagining the tape track to be made from many identical parallel strips.

In the presence of azimuth error, the strips at one edge of the track are played back with a phase shift relative to strips at the other side. At some wavelengths, the phase shift will be 180°, and there will be no output; at other wavelengths, especially long wavelengths, some output will reappear. The effect is rather like that of a comb filter, and serves to attenuate crosstalk due to adjacent tracks so that no guard bands are required. Since no tape is wasted between the tracks, more efficient use is

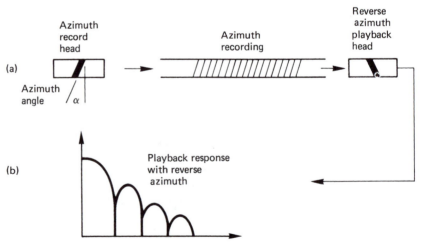

Figure 6.16 In azimuth recording (a), the head gap is tilted. If the track is played with the same head, playback is normal, but the response of the reverse azimuth head is attenuated (b).

made of the tape. The term guard-band-less recording is often used instead of, or in addition to, the term azimuth recording. The failure of the azimuth effect at long wavelengths is a characteristic of azimuth recording, and it is necessary to ensure that the spectrum of the signal to be recorded has a small low-frequency content. The signal will need to pass through a rotary transformer to reach the heads, and cannot therefore contain a DC component.

In recorders such as DAT there is no separate erase process, and erasure is achieved by overwriting with a new waveform. Overwriting is only successful when there are no long wavelengths in the earlier recording, since these penetrate deeper into the tape, and the short wavelengths in a new recording will not be able to erase them. In this case the ratio between the shortest and longest wavelengths recorded on tape should be limited. Restricting the spectrum of the code to allow erasure by overwrite also eases the design of the rotary transformer.

6.7 Optical disks

Optical recorders have the advantage that light can be focused at a distance whereas magnetism cannot. This means that there need be no physical contact between the pickup and the medium and no wear mechanism. In the same way that the recorded wavelength of a magnetic recording is limited by the gap in the replay head, the density of optical recording is limited by the size of light spot which can be focused on the medium. This is controlled by the wavelength of the light used and by the aperture of the lens. When the light spot is as small as these limits allow, it is said to be diffraction limited. The recorded details on the disk are minute, and could easily be obscured by dust particles. In practice the information layer needs to be protected by a thick transparent coating. Light enters the coating well out of focus over a large area so that it can pass around dust particles, and comes to a focus within the thickness of the coating. Although the number of bits per unit area is high in optical recorders the number of bits per unit volume is not as high as that of tape because of the thickness of the coating.

Figure 6.17 shows the principle of readout of the Compact Disc which is a read-only disk manufactured by pressing. The track consists of raised bumps separated by flat areas. The entire surface of the disk is metallized, and the bumps are one quarter of a wavelength in height. The player spot is arranged so that half of its light falls on top of a bump, and half on the surrounding surface. Light returning from the flat surface has travelled half a wavelength further than light returning from the top of the bump, and so there is a phase reversal between the two components of the reflection. This causes destructive interference, and light cannot return to

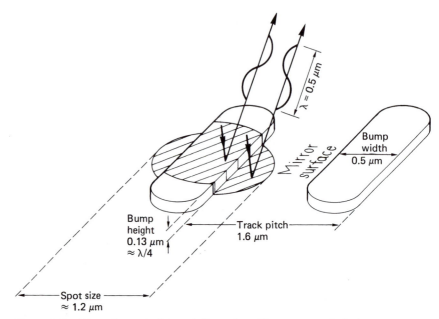

Figure 6.17 CD readout principle and dimensions. The presence of a bump causes destructive interference in the reflected light.

the pickup. It must reflect at angles which are outside the aperture of the lens and be lost. Conversely, when light falls on the flat surface between bumps, the majority of it is reflected back to the pickup. The pickup thus sees a disk *apparently* having alternately good or poor reflectivity. The sensor in the pickup responds to the incident intensity and so the replay signal is unipolar and varies between two levels in a manner similar to the output of a M-R head.

Some disks can be recorded once, but not subsequently erased or rerecorded. These are known as WORM (Write Once Read Many) disks. One type of WORM disk uses a thin metal layer which has holes punched in it on recording by heat from a laser. Others rely on the heat raising blisters in a thin metallic layer by decomposing the plastic material beneath. Yet another alternative is a layer of photo-chemical dye which darkens when struck by the high-powered recording beam. Whatever the recording principle, light from the pickup is reflected more or less, or absorbed more or less, so that the pickup senses a change in reflectivity. Certain WORM disks can be read by conventional CD players and are thus called recordable CDs, or CD-R, whereas others will only work in a particular type of drive.

All optical disks need mechanisms to keep the pickup following the track and sharply focused on it. These will be discussed in Chapter 12 and need not be treated here.

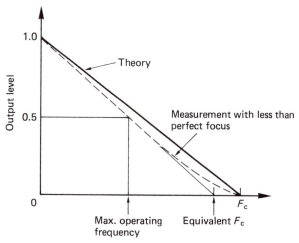

Figure 6.18 Frequency response of laser pickup. Maximum operating frequency is about half of cut-off frequency *E*.

The frequency response of an optical disk is shown in Figure 6.18. The response is best at DC and falls steadily to the optical cut-off frequency. Although the optics work down to DC, this cannot be used for the data recording. DC and low frequencies in the data would interfere with the focus and tracking servos and, as will be seen, difficulties arise when attempting to demodulate a unipolar signal. In practice the signal from the pickup is split by a filter. Low frequencies go to the servos, and higher frequencies go to the data circuitry. As a result the optical disk channel has the same inability to handle DC as does a magnetic recorder, and the same techniques are needed to overcome it.

6.8 Magneto-optical disks

When a magnetic material is heated above its Curie temperature, it becomes demagnetized, and on cooling will assume the magnetization of an applied field which would be too weak to influence it normally. This is the principle of magneto-optical recording used in the Sony MiniDisc. The heat is supplied by a finely focused laser and the field is supplied by a coil which is much larger.

Figure 6.19 shows that the medium is initially magnetized in one direction only. In order to record, the coil is energized with a current in the opposite direction. This is too weak to influence the medium in its normal state, but when it is heated by the recording laser beam the heated area will take on the magnetism from the coil when it cools. Thus a magnetic recording with very small dimensions can be made even though the magnetic circuit involved is quite large in comparison.

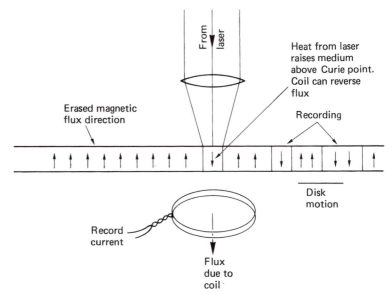

Figure 6.19 The thermomagneto-optical disk uses the heat from a laser to allow magnetic field to record on the disk.

Readout is obtained using the Kerr effect or the Faraday effect, which are phenomena whereby the plane of polarization of light can be rotated by a magnetic field. The angle of rotation is very small and needs a sensitive pickup. The pickup contains a polarizing filter before the sensor. Changes in polarization change the ability of the light to get through the polarizing filter and results in an intensity change which once more produces a unipolar output.

The magneto-optic recording can be erased by reversing the current in the coil and operating the laser continuously as it passes along the track. A new recording can then be made on the erased track.

A disadvantage of magneto-optical recording is that all materials having a Curie point low enough to be useful are highly corrodible by air and need to be kept under an effectively sealed protective layer.

The magneto-optical channel has the same frequency response as that shown in Figure 6.18.

6.9   Equalization

The characteristics of most channels are that signal loss occurs which increases with frequency. This has the effect of slowing down rise times and thereby sloping off edges. If a signal with sloping edges is sliced, the time at which the waveform crosses the slicing level will be changed, and this causes jitter. Figure 6.20 shows that slicing a sloping waveform in the presence of baseline wander causes more jitter.

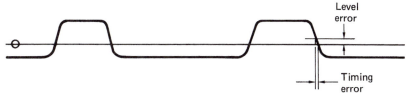

Figure 6.20 A DC offset can cause timing errors.

On a long cable, high-frequency rolloff can cause sufficient jitter to move a transition into an adjacent bit period. This is called inter-symbol interference and the effect becomes worse in signals which have greater asymmetry, i.e. short pulses alternating with long ones. The effect can be reduced by the application of equalization, which is typically a high-frequency boost, and by choosing a channel code which has restricted asymmetry.

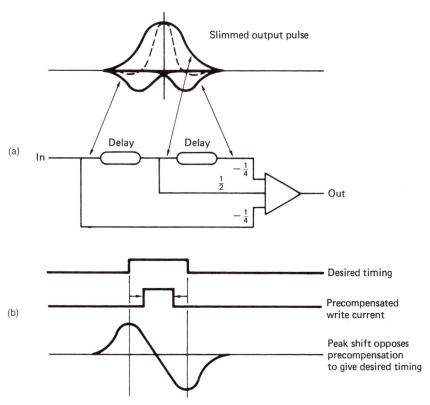

Figure 6.21 Peak-shift distortion is due to the finite width of replay pulses. The effect can be reduced by the pulse slimmer shown in (a) which is basically a transversal filter. The use of a linear operational amplifier emphasizes the analog nature of channels. Instead of replay pulse slimming, transitions can be written with a displacement equal and opposite to the anticipated peak shift as shown in (b).

Compensation for peak shift distortion in recording requires equalization of the channel,[3] and this can be done by a network after the replay head, termed an equalizer or pulse sharpener,[4] as in Figure 6.21(a). This technique uses transversal filtering to oppose the inherent transversal effect of the head. As an alternative, pre-compensation in the record stage can be used as shown in (b). Transitions are written in such a way that the anticipated peak shift will move the readout peaks to the desired timing.

6.10   Data separation

The important step of information recovery at the receiver or replay circuit is known as data separation. The data separator is rather like an analog-to-digital convertor because the two processes of sampling and quantizing are both present. In the time domain, the sampling clock is derived from the clock content of the channel waveform. In the voltage domain, the process of *slicing* converts the analog waveform from the channel back into a binary representation. The slicer is thus a form of quantizer which has only one-bit resolution. The slicing process makes a discrete decision about the voltage of the incoming signal in order to reject noise. The sampler makes discrete decisions along the time axis in order to reject jitter. These two processes will be described in detail.

6.11   Slicing

The slicer is implemented with a comparator which has analog inputs but a binary output. In a cable receiver, the input waveform can be sliced directly. In an inductive magnetic replay system, the replay waveform is differentiated and must first pass through a peak detector (Figure 6.11) or an integrator (Figure 6.12). The signal voltage is compared with the midway voltage, known as the threshold, baseline or slicing level by the comparator. If the signal voltage is above the threshold, the comparator outputs a high level, if below, a low level results.

Figure 6.22 shows some waveforms associated with a slicer. At (a) the transmitted waveform has an uneven duty cycle. The DC component, or average level, of the signal is received with high amplitude, but the pulse amplitude falls as the pulse gets shorter. Eventually the waveform cannot be sliced. At (b) the opposite duty cycle is shown. The signal level drifts to the opposite polarity and once more slicing is impossible. The phenomenon is called baseline wander and will be observed with any signal whose average voltage is not the same as the slicing level. At (c) it will be seen that if the transmitted waveform has a relatively constant

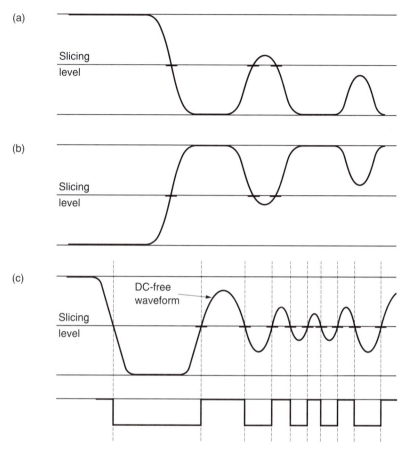

Figure 6.22 Slicing a signal which has suffered losses works well if the duty cycle is even. If the duty cycle is uneven, as in (a), timing errors will become worse until slicing fails. With the opposite duty cycle, the slicing fails in the opposite direction as in (b). If, however, the signal is DC free, correct slicing can continue even in the presence of serious losses, as (c) shows.

average voltage, slicing remains possible up to high frequencies even in the presence of serious amplitude loss, because the received waveform remains symmetrical about the baseline.

It is clearly not possible to simply serialize data in a shift register for so-called direct transmission, because successful slicing can only be obtained if the number of ones is equal to the number of zeros; there is little chance of this happening consistently with real data. Instead, a modulation code or channel code is necessary. This converts the data into a waveform which is DC-free or nearly so for the purpose of transmission.

The slicing threshold level is naturally zero in a bipolar system such as magnetic inductive replay or a cable. When the amplitude falls it does so symmetrically and slicing continues. The same is not true of M-R heads and optical pickups, which both respond to intensity and therefore

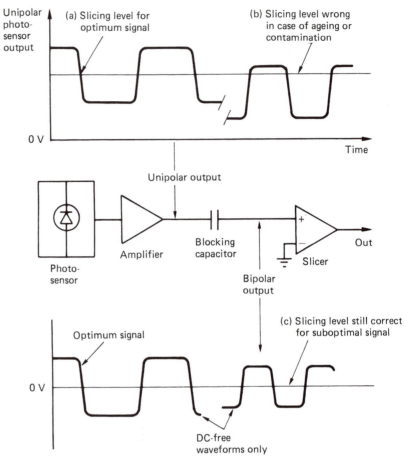

Figure 6.23 (a) Slicing a unipolar signal requires a non-zero threshold. (b) If the signal amplitude changes, the threshold will then be incorrect. (c) If a DC-free code is used, a unipolar waveform can be converted to a bipolar waveform using a series capacitor. A zero threshold can be used and slicing continues with amplitude variations.

produce a unipolar output. If the replay signal is sliced directly, the threshold cannot be zero, but must be some level approximately half the amplitude of the signal as shown in Figure 6.23(a). Unfortunately when the signal level falls it falls towards zero and not towards the slicing level. The threshold will no longer be appropriate for the signal as can be seen at (b). This can be overcome by using a DC-free coded waveform. If a series capacitor is connected to the unipolar signal from an optical pickup, the waveform is rendered bipolar because the capacitor blocks any DC component in the signal. The DC-free channel waveform passes through unaltered. If an amplitude loss is suffered, Figure 6.23(c) shows that the resultant bipolar signal now reduces in amplitude about the slicing level and slicing can continue.

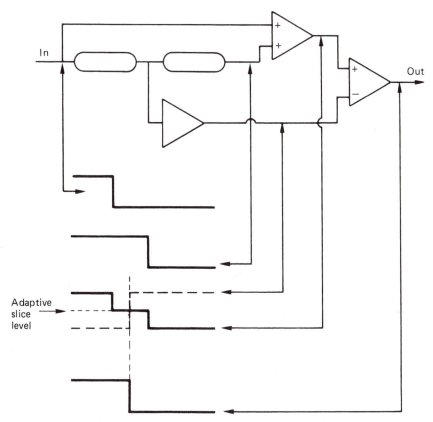

Figure 6.24 An adaptive slicer uses delay lines to produce a threshold from the waveform itself. Correct slicing will then be possible in the presence of baseline wander. Such a slicer can be used with codes which are not DC-free.

Whilst cables and optical recording channels need to be DC-free, some channel waveforms used in magnetic recording have a reduced DC component, but are not completely DC-free. As a result the received waveform will suffer from baseline wander. If this is moderate, an adaptive slicer which can move its threshold can be used. As Figure 6.24 shows, the adaptive slicer consists of a pair of delays. If the input and output signals are linearly added together with equal weighting, when a transition passes, the resultant waveform has a plateau which is at the half-amplitude level of the signal and can be used as a threshold voltage for the slicer. The coding of the DASH format is not DC-free and a slicer of this kind is employed.

6.12 Jitter rejection

The binary waveform at the output of the slicer will be a replica of the transmitted waveform, except for the addition of jitter or time uncertainty

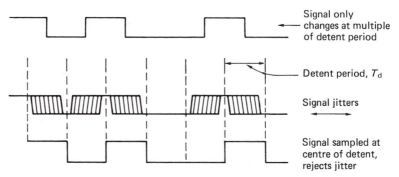

Figure 6.25 A certain amount of jitter can be rejected by changing the signal at multiples of the basic detent period T_d.

in the position of the edges due to noise, baseline wander, intersymbol interference and imperfect equalization.

Binary circuits reject noise by using discrete voltage levels which are spaced further apart than the uncertainty due to noise. In a similar manner, digital coding combats time uncertainty by making the time axis discrete using events, known as transitions, spaced apart at integer multiples of some basic time period, called a detent, which is larger than the typical time uncertainty. Figure 6.25 shows how this jitter-rejection mechanism works. All that matters is to identify the detent in which the transition occurred. Exactly where it occurred within the detent is of no consequence.

As ideal transitions occur at multiples of a basic period, an oscilloscope, which is repeatedly triggered on a channel-coded signal carrying random data, will show an eye pattern if connected to the output of the equalizer. Study of the eye pattern reveals how well the coding used suits the channel. In the case of transmission, with a short cable, the losses will be small, and the eye opening will be virtually square except for some edge sloping due to cable capacitance. As cable length increases, the harmonics are lost and the remaining fundamental gives the eyes a diamond shape. The same eye pattern will be obtained with a recording channel where it is uneconomic to provide bandwidth much beyond the fundamental.

Noise closes the eyes in a vertical direction, and jitter closes the eyes in a horizontal direction, as in Figure 6.26. If the eyes remain sensibly open, data separation will be possible. Clearly more jitter can be tolerated if there is less noise, and vice versa. If the equalizer is adjustable, the optimum setting will be where the greatest eye opening is obtained.

In the centre of the eyes, the receiver must make binary decisions at the channel bit rate about the state of the signal, high or low, using the slicer output. As stated, the receiver is sampling the output of the slicer, and it needs to have a sampling clock in order to do that. In order to give the

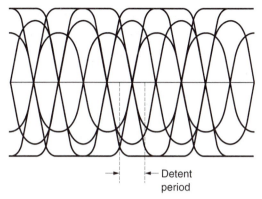

Figure 6.26 A transmitted waveform which is generated according to the principle of Figure 6.25 will appear like this on an oscilloscope as successive parts of the waveform are superimposed on the tube. When the waveform is rounded off by losses, diamond-shaped eyes are left in the centre, spaced apart by the detent period.

best rejection of noise and jitter, the clock edges which operate the sampler must be in the centre of the eyes.

As has been stated, a separate clock is not practicable in recording or transmission. A fixed-frequency clock at the receiver is of no use as even if it was sufficiently stable, it would not know what phase to run at.

The only way in which the sampling clock can be obtained is to use a phase-locked loop to regenerate it from the clock content of the self-clocking channel coded waveform. In phase-locked loops, the voltage-controlled oscillator is driven by a phase error measured between the output and some reference, such that the output eventually has the

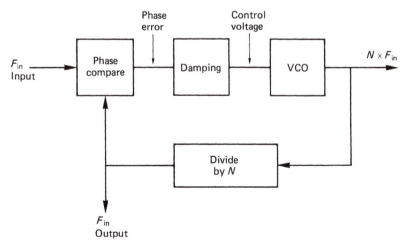

Figure 6.27 A typical phase-locked loop where the VCO is forced to run at a multiple of the input frequency. If the input ceases, the output will continue for a time at the same frequency until it drifts.

same frequency as the reference. If a divider is placed between the VCO and the phase comparator, as in Figure 6.27, the VCO frequency can be made to be a multiple of the reference. This also has the effect of making the loop more heavily damped. If a channel-coded waveform is used as a reference to a PLL, the loop will be able to make a phase comparison whenever a transition arrives and will run at the channel bit rate. When there are several detents between transitions, the loop will *flywheel* at the last known frequency and phase until it can rephase at a subsequent transition. Thus a continuous clock is recreated from the clock content of the channel waveform. In a recorder, if the speed of the medium should change, the PLL will change frequency to follow. Once the loop is locked, clock edges will be phased with the average phase of the jittering edges of the input waveform. If, for example, rising edges of the clock are phased to input transitions, then falling edges will be in the centre of the eyes. If these edges are used to clock the sampling process, the maximum jitter and noise can be rejected. The output of the slicer when sampled by the PLL edge at the centre of an eye is the value of a channel bit. Figure 6.28 shows the complete clocking system of a channel code from encoder to data separator. Clearly data cannot be separated if the PLL is not locked, but it cannot be locked until it has seen transitions for a reasonable period. In recorders, which have discontinuous recorded blocks to allow editing, the solution is to precede each data block with a pattern of transitions whose sole purpose is to provide a timing reference for synchronizing the phase-locked loop. This pattern is known as a preamble. In interfaces, the transmission can be continuous and there is no difficulty

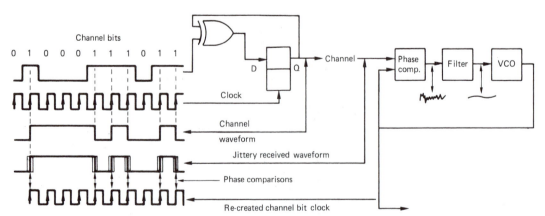

Figure 6.28 The clocking system when channel coding is used. The encoder clock runs at the channel bit rate, and any transitions in the channel must coincide with encoder clock edges. The reason for doing this is that, at the data separator, the PLL can lock to the edges of the channel signal, which represent an intermittent clock, and turn it into a continuous clock. The jitter in the edges of the channel signal causes noise in the phase error of the PLL, but the damping acts as a filter and the PLL runs at the average phase of the channel bits, rejecting the jitter.

remaining in lock indefinitely. There will simply be a short delay on first applying the signal before the receiver locks to it.

One potential problem area which is frequently overlooked is to ensure that the VCO in the receiving PLL is correctly centred. If it is not, it will be running with a static phase error and will not sample the received waveform at the centre of the eyes. The sampled bits will be more prone to noise and jitter errors. VCO centring can simply be checked by displaying the control voltage. This should not change significantly when the input is momentarily interrupted.

6.13   Channel coding

In summary, it is not practicable simply to serialize raw data in a shift register for the purpose of recording or for transmission except over relatively short distances. Practical systems require the use of a modulation scheme, known as a channel code, which expresses the data as waveforms which are self-clocking in order to reject jitter, separate the received bits and to avoid skew on separate clock lines. The coded waveforms should further be DC-free or nearly so to enable slicing in the presence of losses and have a narrower spectrum than the raw data to make equalization possible.

Jitter causes uncertainty about the time at which a particular event occurred. The frequency response of the channel then places an overall limit on the spacing of events in the channel. Particular emphasis must be placed on the interplay of bandwidth, jitter and noise, which will be shown here to be the key to the design of a successful channel code.

Figure 6.29 shows that a channel coder is necessary prior to the record stage, and that a decoder, known as a data separator, is necessary after the replay stage. The output of the channel coder is generally a logic level signal which contains a 'high' state when a transition is to be generated. The waveform generator produces the transitions in a signal whose level and impedance is suitable for driving the medium or channel. The signal may be bipolar or unipolar as appropriate.

Some codes eliminate DC entirely, which is advantageous for optical media and for rotary head recording. Some codes can reduce the channel bandwidth needed by lowering the upper spectral limit. This permits higher linear density, usually at the expense of jitter rejection. Other codes narrow the spectrum by raising the lower limit. A code with a narrow spectrum has a number of advantages. The reduction in asymmetry will reduce peak shift and data separators can lock more readily because the range of frequencies in the code is smaller. In theory the narrower the spectrum, the less noise will be suffered, but this is only achieved if

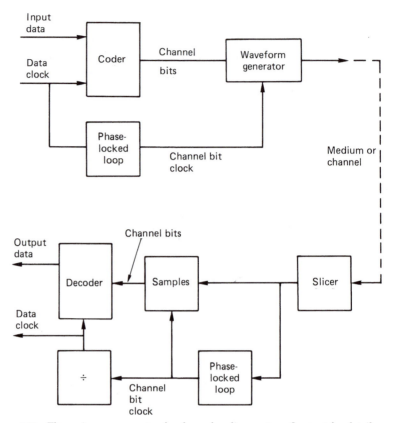

Figure 6.29   The major components of a channel coding system. See text for details.

filtering is employed. Filters can easily cause phase errors which will nullify any gain.

A convenient definition of a channel code (for there are certainly others) is: 'A method of modulating real data such that they can be reliably received despite the shortcomings of a real channel, while making maximum economic use of the channel capacity.'

The basic time periods of a channel-coded waveform are called positions or detents, in which the transmitted voltage will be reversed or stay the same. The symbol used for the units of channel time is T_d.

There are many ways of creating such a waveform, but the most convenient is to convert the raw data bits to a larger number of *channel bits* which are output from a shift register to the waveform generator at the detent rate. The coded waveform will then be high or low according to the state of a channel bit which describes the detent.

Channel coding is the art of converting real data into channel bits. It is important to appreciate that the convention most commonly used in coding is one in which a channel-bit one represents a voltage change, whereas a zero represents no change. This convention is used because it is possible to assemble sequential groups of channel bits together without worrying about whether the polarity of the end of the last group matches the beginning of the next. The polarity is unimportant in most codes and all that matters is the length of time between transitions. It should be stressed that channel bits are not recorded. They exist only in a circuit technique used to control the waveform generator. In many media, for example CD, the channel bit rate is beyond the frequency response of the channel and so it *cannot* be recorded.

One of the fundamental parameters of a channel code is the density ratio (DR). One definition of density ratio is that it is the worst-case ratio of the number of data bits recorded to the number of transitions in the channel. It can also be thought of as the ratio between the Nyquist rate of the data (one-half the bit rate) and the frequency response required in the channel. The storage density of data recorders has steadily increased due to improvements in medium and transducer technology, but modern storage densities are also a function of improvements in channel coding. Figure 6.30(a) shows how density ratio has improved as more sophisticated codes have been developed.

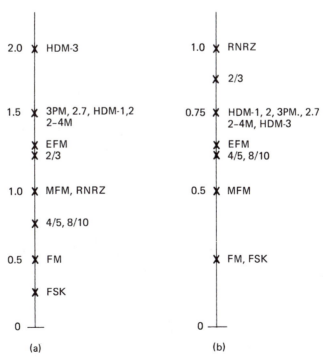

Figure 6.30 (a) Comparison of codes by density ratio; (b) comparison of codes by figure of merit. Note how 4/5, 2/3, 8/10 and RNRZ move up because of good jitter performance; HDM-3 moves down because of jitter sensitivity.

As jitter is such an important issue in digital recording and transmission, a parameter has been introduced to quantify the ability of a channel code to reject time instability. This parameter, the jitter margin, also known as the window margin or phase margin (T_w), is defined as the permitted range of time over which a transition can still be received correctly, divided by the data bit-cell period (T).

Since equalization is often difficult in practice, a code which has a large jitter margin will sometimes be used because it resists the effects of inter-symbol interference well. Such a code may achieve a better performance in practice than a code with a higher density ratio but poor jitter performance.

A more realistic comparison of code performance will be obtained by taking into account both density ratio and jitter margin. This is the purpose of the figure of merit (FoM), which is defined as DR $\times T_w$. Figure 6.30(b) shows a number of codes compared by FoM.

6.14   Recording-oriented codes

Many channel codes are sufficiently versatile that they have been used in recording, electrical or optical cable transmission and radio transmission. Others are more specialized and are intended for only one of these categories. Channel coding has roots in computers, in telemetry and in Telex services, but has for some time been considered a single subject. These starting points will be considered here.

In magnetic recording, the first digital recordings were developed for early computers and used very simple techniques. Figure 6.31(a) shows that in Return to Zero (RZ) recording, the record current has a zero state between bits and flows in one direction to record a one and in the opposite direction to record a zero. Thus every bit contains two flux changes which replay as a pair of pulses, one positive and one negative. The signal is self-clocking because pulses always occur. The order in which they occur determines the state of the bit. RZ recording cannot erase by overwrite because there are times when no record current flows. Additionally the signal amplitude is only one half of what is possible. These problems were overcome in the Non-Return to Zero code shown in Figure 6.31(b). As the name suggests, the record current does not cease between bits, but flows at all times in one direction or the other dependent on the state of the bit to be recorded. This results in a replay pulse only when the data bits change from state to another. As a result if one pulse was missed, the subsequent bits would be inverted. This was avoided by adapting the coding such that the record current would change state or invert whenever a data one occurred, leading to the term Non-Return to Zero Invert or NRZI shown in Figure 6.31(c). In NRZI a

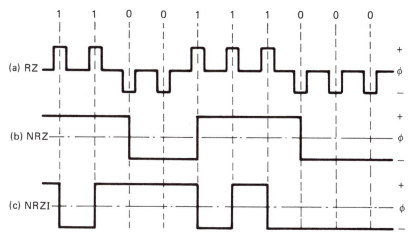

Figure 6.31 Early magnetic recording codes. RZ shown at (a) had poor signal-to-noise ratio and poor overwrite capability. NRZ at (b) overcame these problems but suffered error propagation. NRZI at (c) was the final result where a transition represented a one. NRZI is not self-clocking.

replay pulse occurs whenever there is a data one. Clearly neither NRZ or NRZI are self-clocking, but require a separate clock track. Skew between tracks can only be avoided by working at low density and so the system cannot be used for digital audio. However, virtually all the codes used for magnetic recording are based on the principle of reversing the record current to produce a transition.

6.15   Transmission-oriented codes

In cable transmission, also known as line signalling, and in telemetry, the starting point was often the speech bandwidth available in existing telephone lines and radio links. There was no DC response, just a range of frequencies available. Figure 6.32(a) shows that a pair of frequencies can be used, one for each state of a data bit. The result is frequency shift keying (FSK) which is the same as would be obtained from an analog frequency modulator fed with a two-level signal. This is exactly what happens when two-level pseudo-video from a PCM adaptor is fed to a VCR and is the technique used in units such as the PCM F-1 and the PCM-1630. PCM adaptors have also been used to carry digital audio over a video landline or microwave link. Clearly FSK is DC-free and self-clocking.

Instead of modulating the frequency of the signal, the phase can be modulated or shifted instead, leading to the generic term of phase shift keying or PSK. This method is highly suited to broadcast as it is easily applied to a radio frequency carrier. The simplest technique is selectively

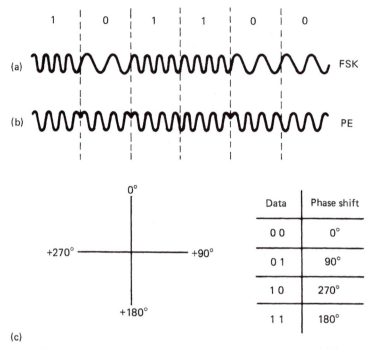

Figure 6.32 Various communications oriented codes are shown here: at (a) frequency shift keying (FSK), at (b) phase encoding and at (c) differential quadrature phase shift keying (DQPSK).

to invert the carrier phase according to the data bit as in Figure 6.32(b). There can be many cycles of carrier in each bit period. This technique is known as phase encoding (PE) and is used in GPS (Global Positioning System) broadcasts. The receiver in a PE system is a well-damped phase-locked loop which runs at the average phase of the transmission. Phase changes will then result in phase errors in the loop and so the phase error is the demodulated signal.

6.16 General-purpose codes

Despite the different origins of codes, there are many similarities between them. If the two frequencies in an FSK system are one octave apart, the limiting case in which the highest data rate is obtained is when there is one half-cycle of the lower frequency or a whole cycle of the high frequency in one bit period. This gives rise to the frequency modulation (FM). In the same way, the limiting case of phase encoding is where there is only one cycle of carrier per bit. In recording, this what is meant by phase encoding. These approaches can be contrasted in Figure 6.33.

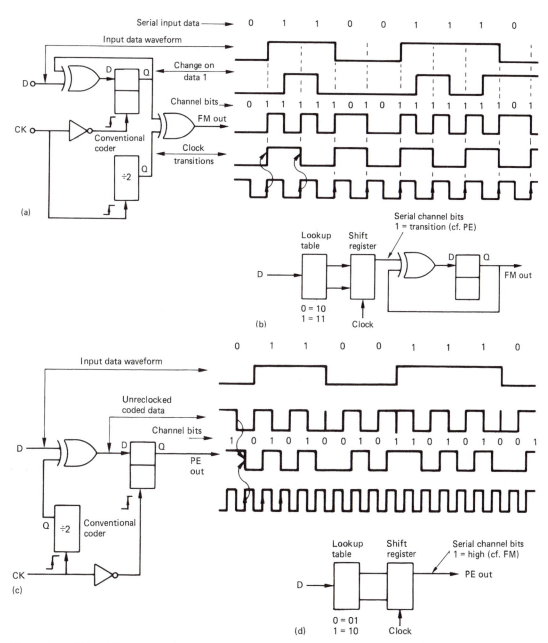

Figure 6.33 FM and PE contrasted. In (a) are the FM waveform and the channel bits which may be used to describe transitions in it. The FM coder is shown in (b). The PE waveform is shown in (c). As PE is polarity conscious, the channel bits must describe the signal level rather than the transitions. The coder is shown in (d).

The FM code, also known as Manchester code or bi-phase mark code, shown in Figure 6.33(a) was the first practical self-clocking binary code and it is suitable for both transmission and recording. It is DC-free and very easy to encode and decode. It is the code specified for the AES/EBU digital audio interconnect standard which will be described in Chapter 8. In the field of recording it remains in use today only where density is not of prime importance, for example in SMPTE/EBU timecode for professional audio and video recorders and in floppy disks.

In FM there is always a transition at the bit-cell boundary which acts as a clock. For a data one, there is an additional transition at the bit-cell centre. Figure 6.33(a) shows that each data bit can be represented by two channel bits. For a data zero, they will be 10, and for a data one they will be 11. Since the first bit is always one, it conveys no information, and is responsible for the density ratio of only one-half. Since there can be two transitions for each data bit, the jitter margin can only be half a bit, and the resulting FoM is only 0.25. The high clock content of FM does, however, mean that data recovery is possible over a wide range of speeds; hence the use for timecode. The lowest frequency in FM is due to a stream of zeros and is equal to half the bit rate. The highest frequency is due to a stream of ones, and is equal to the bit rate. Thus the fundamentals of FM are within a band of one octave. Effective equalization is generally possible over such a band. FM is not polarity conscious and can be inverted without changing the data.

Figure 6.33(b) shows how an FM coder works. Data words are loaded into the input shift register which is clocked at the data bit rate. Each data bit is converted to two channel bits in the code book or look-up table. These channel bits are loaded into the output register. The output register is clocked twice as fast as the input register because there are twice as many channel bits as data bits. The ratio of the two clocks is called the code rate, in this case it is a rate one-half code. Ones in the serial channel bit output represent transitions whereas zeros represent no change. The channel bits are fed to the waveform generator which is a one-bit delay, clocked at the channel bit rate, and an exclusive-OR gate. This changes state when a channel bit one is input. The result is a coded FM waveform where there is always a transition at the beginning of the data bit period, and a second optional transition whose presence indicates a one.

In PE there is always a transition in the centre of the bit but Figure 6.33(c) shows that the transition between bits is dependent on the data values. Although its origins were in line coding, phase encoding can be used for optical and magnetic recording as it is DC-free and self-clocking. It has the same DR and T_w as FM, and the waveform can also be described using channel bits, but with a different notation. As PE is polarity sensitive, the channel bits determine the level of the encoded signal rather than causing a transition. Figure 6.33(d) shows that the allowable channel bit patterns are now 10 and 01.

In modified frequency modulation (MFM) also known as Miller code,[5] the highly redundant clock content of FM was reduced by the use of a phase-locked loop in the receiver which could flywheel over missing clock transitions. This technique is implicit in all the more advanced codes. Figure 6.34(a) shows that the bit-cell centre transition on a data one

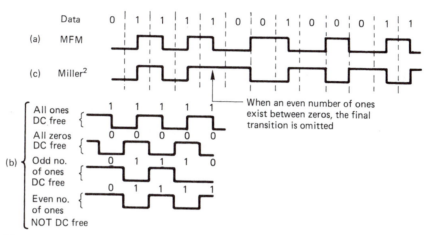

Figure 6.34 MFM or Miller code is generated as shown here. The minimum transition spacing is twice that of FM or PE. MFM is not always DC free as shown at (b). This can be overcome by the modification of (c) which results in the Miller² code.

was retained, but the bit-cell boundary transition is now only required between successive zeros. There are still two channel bits for every data bit, but adjacent channel bits will never be one, doubling the minimum time between transitions, and giving a DR of 1. Clearly the coding of the current bit is now influenced by the preceding bit. The maximum number of prior bits which affect the current bit is known as the constraint length L_c, measured in data-bit periods. For MFM $L_c = T$. Another way of considering the constraint length is that it assesses the number of data bits which may be corrupted if the receiver misplaces one transition. If L_c is long, all errors will be burst errors.

MFM doubled the density ratio compared to FM and PE without changing the jitter performance; thus the FoM also doubles, becoming 0.5. It was adopted for many rigid disks at the time of its development, and remains in use on double-density floppy disks. It is not, however, DC-free. Figure 6.34(b) shows how MFM can have DC content under certain conditions.

6.17   Miller^2 code

The Miller^2 code is derived from MFM, and Figure 6.34(c) shows that the DC content is eliminated by a slight increase in complexity.[6,7] Wherever an even number of ones occurs between zeros, the transition at the last one is omitted. This creates two additional, longer run lengths and increases the T_{max} of the code. The decoder can detect these longer run lengths in order to re-insert the suppressed ones. The FoM of Miller^2 is 0.5 as for MFM. Miller^2 was used in early 3M stationary head digital audio recorders, in high rate instrumentation recorders and in the D-2 DVTR format.

6.18   Group codes

Further improvements in coding rely on converting patterns of real data to patterns of channel bits with more desirable characteristics using a conversion table known as a codebook. If a data symbol of m bits is considered, it can have 2^m different combinations. As it is intended to discard undesirable patterns to improve the code, it follows that the number of channel bits n must be greater than m. The number of patterns which can be discarded is:

$$2^n - 2^m$$

One name for the principle is group code recording (GCR), and an important parameter is the code rate, defined as:

$$R = \frac{m}{n}$$

It will be evident that the jitter margin T_w is numerically equal to the code rate, and so a code rate near to unity is desirable. The choice of patterns which are used in the codebook will be those which give the desired balance between clock content, bandwidth and DC content.

Figure 6.35 shows that the upper spectral limit can be made to be some fraction of the channel bit rate according to the minimum distance between ones in the channel bits. This is known as T_{min}, also referred to as the minimum transition parameter M and in both cases is measured in data bits T. It can be obtained by multiplying the number of channel detent periods between transitions by the code rate. Unfortunately, codes are measured by the number of consecutive zeros in the channel bits, given the symbol d, which is always one less than the number of detent periods. In fact $T_{\min}$ is numerically equal to the density ratio.

$$T_{\min} = M = DR = \frac{(d + 1) \times m}{n}$$

It will be evident that choosing a low code rate could increase the density ratio, but it will impair the jitter margin. The figure of merit is:

$$FoM = DR \times T_w = \frac{(d + 1) \times m^2}{n^2}$$

since $T_w = m/n$

Figure 6.35 also shows that the lower spectral limit is influenced by the maximum distance between transitions $T_{\max}$. This is also obtained by multiplying the maximum number of detent periods between transitions by the code rate. Again, codes are measured by the maximum number of zeros between channel ones, k, and so:

$$T_{\max} = \frac{(k + 1) \times m}{n}$$

and the maximum/minimum ratio P is:

$$P = \frac{(k + 1)}{(d + 1)}$$

The length of time between channel transitions is known as the *run length*. Another name for this class is the run-length-limited (RLL) codes.[8] Since m data bits are considered as one symbol, the constraint length L_c will be increased in RLL codes to at least m. It is, however, possible for a code to have run-length limits without it being a group code.

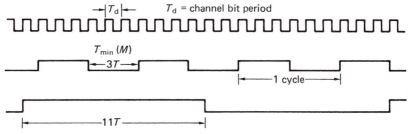

Figure 6.35 A channel code can control its spectrum by placing limits on $T_{\min}$ (M) and $T_{\max}$ which define upper and lower frequencies. The ratio of $T_{\max}/T_{\min}$ determines the asymmetry of waveform and predicts DC content and peak shift. Example shown is EFM.

In practice, the junction of two adjacent channel symbols may violate run-length limits, and it may be necessary to create a further codebook of symbol size $2n$ which converts violating code pairs to acceptable patterns. This is known as merging and follows the golden rule that the substitute $2n$ symbol must finish with a pattern which eliminates the possibility of a subsequent violation. These patterns must also differ from all other symbols.

Substitution may also be used to different degrees in the same nominal code in order to allow a choice of maximum run length, e.g. 3PM.[9] The maximum number of symbols involved in a substitution is denoted by r.[10,11] There are many RLL codes and the parameters d,k,m,n, and r are a way of comparing them.

Sometimes the code rate forms the name of the code, as in 2/3, 8/10 and EFM; at other times the code may be named after the d,k parameters, as in 2,7 code. Various examples of group codes will be given to illustrate the principles involved.

6.19   4/5 code of MADI

In the MADI (multi-channel audio interface) standard[12], a four-fifths rate code is used where groups of four data bits are represented by groups of five channel bits.

Four bits have 16 combinations whereas five bits have 32 combinations. Clearly only 16 out of these 32 are necessary to convey all the possible data. Figure 6.36 shows that the 16 channel bit patterns chosen are those

4 bit data	5 bit encoded data
0000	11110
0001	01001
0010	10100
0011	10101
0100	01010
0101	01011
0110	01110
0111	01111
1000	10010
1001	10011
1010	10110
1011	10111
1100	11010
1101	11011
1110	11100
1111	11101
SYNC	{ 11000
	{ 10001

Figure 6.36   The codebook of the 4/5 code of MADI. Note that a one represents a transition in the channel.

which have the least DC component combined with a high clock content. Adjacent ones are permitted in the channel bits, so there can be no violation of T_{min} at the boundary of two symbols. T_{max} is determined by the worst case run of zeros at a symbol boundary and as $k = 3$, T_{max} is $16/5 = 3.2T$. The code is thus described as 0,3,4,5,1 and $L_c = 4T$.

The jitter resistance of a group code is equal to the code rate. For example, in 4/5 transitions cannot be closer than 0.8 of a data bit apart and so this represents the peak to peak jitter which can be rejected. The density ratio is also 0.8 so the FoM is 0.64; an improvement over FM.

A further advantage of group coding is that it is possible to have codes which have no data meaning. In MADI further channel bit patterns are used for packing and synchronizing. Packing is where dummy data are sent when the real data rate is low in order to keep the channel frequencies constant. This is necessary so that fixed equalization can be used. The packing pattern does not decode to data and so it can be easily discarded at the receiver. Further details of MADI can be found in Chapter 8.

6.20   2/3 code

Figure 6.37(a) shows the code book of an optimized code which illustrates one merging technique. This is a 1,7,2,3,2 code known as 2/3. It is designed to have a good jitter window in order to resist peak shift distortion in disk drives, but it also has a good density ratio.[13] In 2/3 code, pairs of data bits create symbols of three channel bits. For bandwidth reduction, codes having adjacent ones are eliminated so that $d = 1$. This halves the upper spectral limit and the DR is improved accordingly:

$$\text{DR} = \frac{(d + 1) \times m}{n} = \frac{2 \times 2}{3} = 1.33$$

In Figure 6.37(b) it will be seen that some group combinations cause violations. To avoid this, pairs of three-channel bit symbols are replaced with a new six-channel bit symbol. L_c is thus 4T, the same as for the 4/5 code. The jitter window is given by:

$$T_w = \frac{m}{n} = \frac{2}{3}T$$

and the FoM is:

$$\frac{2}{3} \times \frac{4}{3} = \frac{8}{9}$$

Data	Code
0 0	1 0 1
0 1	1 0 0
1 0	0 0 1
1 1	0 1 0

(a)

Data	Illegal code	Substitution
0 0 0 0	1 0 1 1 0 1	1 0 1 0 0 0
0 0 0 1	1 0 1 1 0 0	1 0 0 0 0 0
1 0 0 0	0 0 1 1 0 1	0 0 1 0 0 0
1 0 0 1	0 0 1 1 0 0	0 1 0 0 0 0

(b)

Figure 6.37 2/3 code. In (a) two data bits (m) are expressed as three channel bits (n) without adjacent transitions ($d = 1$). Violations are dealt with by substitution

$$DR = \frac{(d + 1)m}{n} = \frac{2 \times 2}{3} = 1.33$$

Adjacent data pairs can break the encoding rule; in these cases substitutions are made, as shown in (b).

This is an extremely good figure for an RLL code , and is some 10 per cent better than the FoM of 3PM[14] and 2,7 and as a result 2/3 has been highly successful in Winchester disk drives.

6.21   EFM code in CD

This section is concerned solely with the channel coding of CD. A more comprehensive discussion of how the coding is designed to suit the specific characteristics of an optical disk is given in Chapter 12. Figure 6.38 shows the 8,14 code (EFM) used in the Compact Disc. Here eight-bit symbols are represented by 14-bit channel symbols.[15] There are 256 combinations of eight data bits, whereas 14 bits have 16K combinations. Of these only 267 satisfy the criteria that the maximum runlength shall not exceed 11 channel bits ($k = 10$) nor be less than three

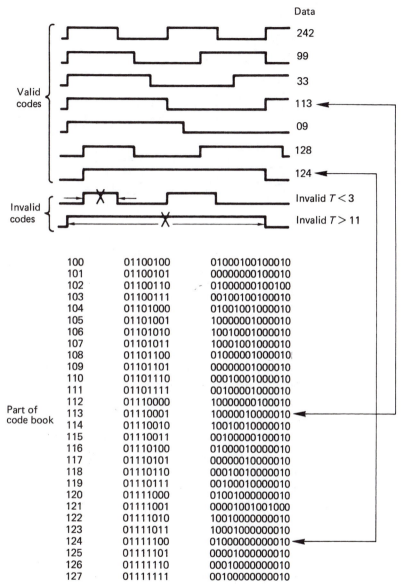

Figure 6.38 EFM code: $d = 2$, $k = 10$. Eight data bits produce 14 channel bits plus three packing bits. Code rate is 8/17. DR = $(3 \times 8)/17 = 1.41$.

channel bits ($d = 2$). A section of the codebook is shown in the figure. In fact 258 of the the 267 possible codes are used because two unique patterns are used to synchronize the subcode blocks (see Chapter 12). It is not possible to prevent violations betwen adjacent symbols by substitution, and extra merging bits having no data meaning are placed between the symbols. Two merging bits would be adequate to prevent violations, but in practice three are used because a further task of the

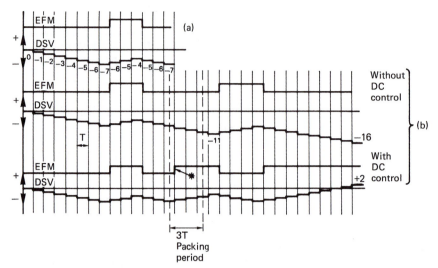

Figure 6.39 (a) Digital sum value example calculated from EFM waveform. (b) Two successive 14T symbols without DC control (upper) give DSV of –16. Additional transition (*) results in DSV of +2, anticipating negative content of next symbol.

merging bits is to control the DC content of the waveform. The merging bits are selected by computing the digital sum value (DSV) of the waveform. The DSV is computed as shown in Figure 6.39(a). One is added to a count for every channel bit period where the waveform is in a high state, and one is subtracted for every channel bit period spent in a low state. Figure 6.39(b) shows that if two successive channel symbols have the same sense of DC offset, these can be made to cancel one another by placing an extra transition in the merging period. This has the effect of inverting the second pattern and reversing its DC content. The DC-free code can be high-pass filtered on replay and the lower-frequency signals are then used by the focus and tracking servos without noise due to the DC content of the audio data. Encoding EFM is complex, but was acceptable when CD was launched because only a few encoders are necessary in comparison with the number of players. Decoding is simpler as no DC content decisions are needed and a look-up table can be used. The code book was computer optimized to permit the implementation of a programmable logic array (PLA) decoder with the minimum complexity.

Owing to the inclusion of merging bits, the code rate becomes 8/17, and the density ratio becomes:

$$\frac{3 \times 8}{17} = 1.41$$

and the FoM is:

$$\frac{3 \times 8^2}{17^2} = 0.66$$

The code is thus a 2,10,8,17, r system where r has meaning only in the context of DC control.[16] The constraints d and k can still be met with $r = 1$ because of the merging bits. The figure of merit is less useful for optical media because the straight-line frequency response does not produce peak shift and the rigid, non-contact medium has good speed stability. The density ratio and the freedom from DC are the most important factors.

6.22   The 8/10 group code of DAT

The essential feature of the channel code of DAT is that it must be able to work well in an azimuth recording system. There are many channel codes available, but few of them are suitable for azimuth recording because of the large amount of crosstalk. The crosstalk cancellation of azimuth recording fails at low frequencies, so a suitable channel code must not only be free of DC, but it must suppress low frequencies as well. A further issue is that erasure is by overwriting, and as the heads are optimized for short-wavelength working, best erasure will be when the ratio between the longest and shortest wavelengths in the recording is small.

In Figure 6.40, some examples from the 8/10 group code of DAT are shown.[17] Clearly a channel waveform which spends as much time high as low has no net DC content, and so all ten-bit patterns which meet this criterion of zero disparity can be found. As was seen in section 6.21, the

Eight-bit dataword	Ten-bit codeword	DSV	Alternative codeword	DSV
00010000	1101010010	0		
00010001	0100010010	2	1100010010	−2
00010010	0101010010	0		
00010011	0101110010	0		
00010100	1101110001	2	0101110001	−2
00010101	1101110011	2	0101110011	−2
00010110	1101110110	2	0101110110	−2
00010111	1101110010	0		

Figure 6.40 Some of the 8/10 codebook for non-zero DSV symbols (two entries) and zero DSV symbols (one entry).

term used to measure DC content is called the digital sum value (DSV). For every bit the channel spends high, the DSV will increase by one; for every bit the channel spends low, the DSV will decrease by one. As adjacent channel ones are permitted, the window margin and DR will be 0.8, comparing favourably with the figure of 0.5 for MFM, giving an FoM of 0.64. Unfortunately there are not enough DC-free combinations in ten channel bits to provide the 256 patterns necessary to record eight data bits. A further constraint is that it is desirable to restrict the maximum run length to improve overwrite capability and reduce peak shift. In the 8/10 code of DAT, no more than three channel zeros are permitted between channel ones, which makes the longest wavelength only four times the shortest. There are only 153 ten-bit patterns which are within this maximum run length and which have a DSV of zero.

The remaining 103 data combinations are recorded using channel patterns that have non-zero DSV. Two channel patterns are allocated to each of the 103 data patterns. One of these has a DSV of +2, the other has a DSV of −2. For simplicity, the only difference between them is that the first channel bit is inverted. The choice of which channel-bit pattern to use is based on the DSV due to the previous code.

For example, if several bytes have been recorded with some of the 153 DC-free patterns, the DSV of the code will be zero. The first data byte is then found which has no zero disparity pattern. If the +2 DSV pattern is used, the code at the end of the pattern will also become +2 DSV. When the next pattern of this kind is found, the code having the DSV of −2 will automatically be selected to return the channel DSV to zero. In this way the code is kept DC-free, but the maximum distance between transitions can be shortened. A code of this kind is known as a low-disparity code.

In order to reduce the complexity of encoding logic, it is usual in group codes to computer-optimize the relationship between data patterns and code patterns. This has been done for 8/10 so that the conversion can be performed in a programmed logic array. The Boolean expressions for calculating the channel bits from data can be seen in Figure 6.41(a). Only DC-free or DSV = +2 patterns are produced by the logic, since the DSV = − 2 pattern can be obtained by reversing the first bit. The assessment of DSV is performed in an interesting manner. If in a pair of channel bits the second bit is one, the pair must be DC-free because each detent has a different value. If the five even channel bits in a ten-bit pattern are checked for parity and the result is one, the pattern could have a DSV of 0, ± 4 or ± If the result is zero, the DSV could be ± 2, ±6 or ±10. However, the codes used are known to be either zero or +2 DSV, so the state of the parity bit discriminates between them. Figure 6.41(b) shows the encoding circuit. The lower set of XOR gates calculate parity on the latest pattern to be recorded, and store the DSV bit in the latch. The next data byte to be

$a = A + CZ + Y\,(\overline{C} \oplus \overline{F}\,(G + H))$

$b = A\,(B + D\overline{E}) + \overline{A}\,(\overline{B} + \overline{C})$

$c = \overline{A}C + A\,(\overline{D} + E) + BDE$

$d = A\,(C + BD\overline{E}) + CDE + \overline{C}Z + (\overline{A}\overline{B} \oplus \overline{F}\overline{G}HY)$

$\overline{e} = (AB + \overline{D})\,\overline{E} + \overline{A}BCDE + Y\overline{F}\,(\overline{G} + \overline{H})$

$f = \overline{A}\,\overline{E}\,[C + (B \oplus D)] + [(\overline{D} + C\overline{E}) \oplus F\,(\overline{G} + \overline{H})]$

$\overline{g} = \overline{F}\,\overline{G} + Y + (B + C)\,Z$

$h = FG\overline{H} + \overline{F}\,\overline{Y}$

$i = H + FG + \overline{F}\overline{Y}$ $\qquad$ where $Y = \overline{A}\,(\overline{B} + C)\,D\overline{E}$

$j = F\overline{G} + \overline{F}\,\overline{Y}$ $\qquad\qquad\qquad Z = \overline{A}\,\overline{D}\overline{E}\,F\,(\overline{G} + \overline{H})$

(a)

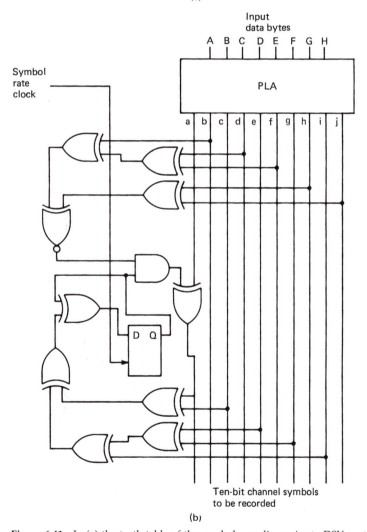

(b)

Figure 6.41 In (a) the truth table of the symbol encoding prior to DSV control. In (b) this circuit controls code disparity by remembering non-zero DSV in the latch and selecting a subsequent symbol with opposite DSV.

recorded is fed to the PLA, which outputs a ten-bit pattern. If this is a zero disparity code, it passes to the output unchanged. If it is a DSV = +2 code, this will be detected by the upper XOR gates. If the latch is set, this means that a previous pattern had been +2 DSV, and so the first bit of the channel pattern is inverted by the XOR gate in that line, and the latch will be cleared because the DSV of the code has been returned to zero.

Decoding is simpler, because there is a direct relationship between ten-bit codes and eight-bit data.

6.23 Tracking signals

Many recorders use track following systems to help keep the head(s) aligned with the narrow tracks used in digital media. These can operate by sensing low-frequency tones which are recorded along with the data. Whilst this can be done by linearly adding the tones to the coder output, this requires a linear record amplifier. An alternative is to use the DC content group codes. A code is devised where for each data pattern, several code patterns exist having a range of DC components. By choosing groups with a suitable sequence of DC offsets, a low frequency can be added to the coded signal. This can be filtered from the data waveform on replay.

6.24 Convolutional RLL codes

It has been mentioned that a code can be run-length limited without being a group code. An example of this is the HDM-1 code used in DASH format (digital audio stationary head – see Chapter 9) recorders. The coding is best described as convolutional, and is rather complex, as Figure 6.42 shows.[18] The DR of 1.5 is achieved by treating the input sequence of 0,1 as a single symbol which has a transition recorded at the centre of the one. The code then depends upon whether the data continue with ones or revert to zeros. The shorter run lengths are used to describe sequential ones; the longer run lengths describe sequential zeros, up to a maximum run length of 4.5 T, with a constraining length of 5.5 T. In HDM-2, a derivative, the maximum run length is reduced to 4 T with the penalty that L_c becomes 7.5 T.

The 2/4M code used by the Mitsubishi ProDigi quarter-inch format recorders[19] is also convolutional, and has an identical density ratio and window margin to HDM-1. T_{max} is eight bits. Neither HDM-1 nor 2/4M are DC-free, but this is less important in stationary head recorders and an adaptive slicer as shown in section 6.11 can be used. The encoding of 2/4M is just as complex as that of HDM-1 and is shown in Figure 6.43.

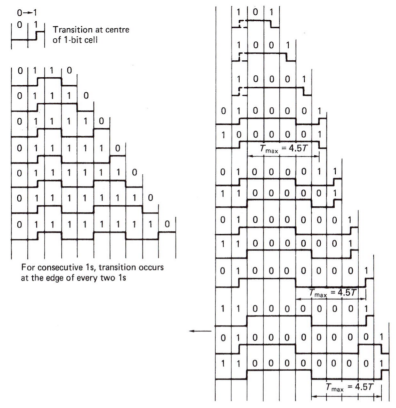

Figure 6.42 HDM-1 code of the DASH format is encoded according to the above rules. Transitions will never be closer than 1.5 bits, nor further apart than 4.5 bits.

Two data bits form a group, and result in four channel bits where there are always two channel zeros between ones, to obtain a DR of 1.5. There are numerous exceptions required to the coding to prevent violation of the run-length limits and this requires a running sample of ten data bits to be examined. Thus the code is convolutional although it has many of the features of a substituting group code.

6.25   Graceful degradation

In all the channel codes described here all data bits are asumed to be equally important and if the characteristics of the channel degrade, there is an equal probability of corruption of any bit. In digital audio samples the bits are not equally important. Loss of a high-order bit causes greater degradation than loss of a low-order bit. For applications where the bandwidth of the channel is unpredictable, or where it may improve as technology matures, a different form of channel

X X X X E4 E3 E2 E1 D D L1 L2 L3 L4 X X X X

Running sample of
ten data bits

DD = current set of data bits
E(N) = earlier data bits
L(N) = later data bits

(a)

Data bits DD	Channel bits C_1 C_2 C_3 C_4	Exceptions and substitutions
00	0 1 0 0	E4 E3 = 10
	0 0 0 0	E4 E3 $\neq$ 10 and E2 E2 = 10 and L1 L2 $\neq$ 01
	0 0 0 1	E4 E3 $\neq$ 10 and E2 E1 = 10 and L1 L2 = 01
01	0 0 1 0	
10	Y 0 0 1	E2 E1 $\neq$ 10 and L1 L2 = 00
	0 1 0 0	E2 E1 = 10 and L1 L2 = 10 and L3 L4 = 00
	0 0 0 1	E2 E1 = 10 and L1 L2 = 00
	0 0 0 0	E2 E1 = 10 and L1 L2 = 10 and L3 L4 = 00
11	Y 0 0 0	

Y = XNOR of C_3 C_4 of previous DD

(b)

Figure 6.43 Coding rules for 2/4M code. In (a) a running sample is made of two data bits DD and earlier and later bits. In (b) the two data bits become the four channel bits shown except when the substitutions specified are made.

coding has been proposed[20] where the probability of corruption of bits is not equal. The channel spectrum is divided in such a way that the least significant bits ocupy the highest frequencies and the most significant bits occupy the lower frequencies. When the bandwidth of the channel is reduced, the eye pattern is degraded such that certain eyes are indeterminate, but others remain open, guaranteeing reception and clocking of high-order bits. In PCM audio the result would be sensibly the same waveform but an increased noise level. Any error-correction techniques would need to consider the unequal probability of error possibly by assembling codewords from bits of the same significance.

6.26 Randomizing

NRZ has a DR of 1 and a jitter window of 1 and so has a FoM of 1 which is better than the group codes. It does, however, suffer from an unconstrained spectrum and poor clock content. This can be overcome using randomizing. At the encoder, a pseudo-random sequence (see

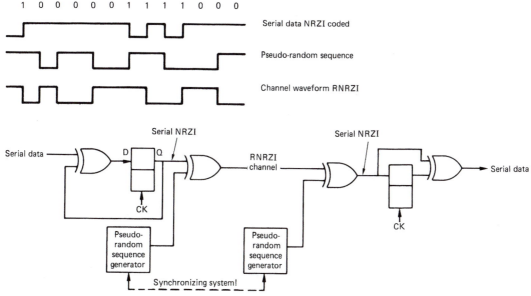

Figure 6.44 When randomizing is used, the same pseudo-random sequence must be provided at both ends of the channel with bit synchronism.

Chapter 3) is added modulo 2 to the serial data and the resulting ones generate transitions in the channel. This process drastically reduces T_{max} and reduces DC content. Figure 6.44 shows that at the receiver the transitions are converted back to a serial bitstream to which the same pseudo-random sequence is again added modulo 2. As a result the random signal cancels itself out to leave only the serial data, provided that the two pseudo-random sequences are synchronized to bit accuracy.

Randomizing with NRZI (RNRZI) is used in the D-1 DVTR. Randomizing can also be used in addition to any other channel coding or modulation scheme. It is employed in NICAM 728 and in DAB as will be seen in the next section.

6.27 Communications codes

Since the original FSK and PSK codes were developed, advances in circuit techniques have allowed more complex signalling techniques to be used. The common goal of all of these is to minimize the channel bandwidth needed for a given bit rate whilst offering immunity from multipath reception and interference. This is the equivalent of the DR in recording, but is measured in bits/s/Hz.

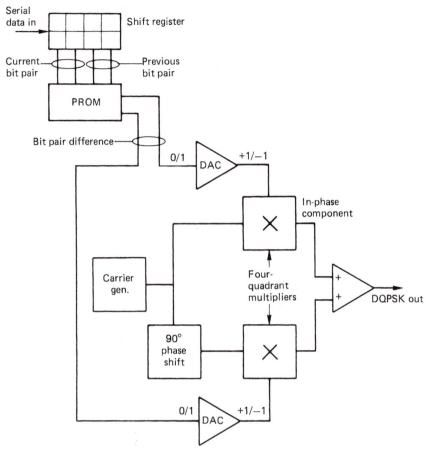

Figure 6.45 A DQPSK coder conveys two bits for each modulation period. See text for details.

In PSK it is possible to use more than two distinct phases. When four phases in quadrature are used, the result is quadrature phase shift keying or QPSK. Each period of the transmitted waveform can have one of four phases and therefore conveys the value of two data bits. In order to resist reflections in broadcasting, QPSK can be modified so that a knowledge of absolute phase is not needed at the receiver. Instead of encoding the signal phase, the data determine the magnitude of a phase shift. This is known as differential quadrature phase shift keying or DQPSK and is the modulation scheme used for NICAM 728 digital TV sound. A DQPSK coder is shown in Figure 6.45 and as before two bits are conveyed for each transmitted period. It will be seen that one bit pattern results in no phase change. If this pattern is sustained the entire transmitter power will be concentrated in the carrier. This can cause patterning on the associated TV pictures. The randomizing technique of section 6.26 is used to overcome the problem. The effect is to spread the signal energy uniformly throughout the allowable channel bandwidth so that it has less energy at a given frequency. This reduces patterning on the analog video signal in addition to making the signal more resistant to multipath reception which tends to remove notches from the spectrum.

A pseudo-random sequence generator as described in Chapter 3 is used to generate the randomizing sequence used in NICAM. A nine-bit device has a sequence length of 511, and is preset to a standard value of all ones at the beginning of each frame. The serialized data are XORed with the LSB of the Galois field, which randomizes the output which then goes to the modulator. The spectrum of the transmission is now determined by the spectrum of the psendo-random sequence. This was shown in Chapter 3 to have a spikey $\sin x/x$ envelope. The frequencies beyond the first nulls are filtered out at the transmitter, leaving the characteristic 'dead hedgehog' shape seen on a spectrum analyser.

On reception, the de-randomizer must contain the identical ring counter which must also be set to the starting condition to bit accuracy. Its output is then added to the data stream from the demodulator. The randomizing will effectively then have been added twice to the data in modulo 2, and as a result is cancelled out leaving the original serial data.

Where an existing wide-band channel having a DC response and a good SNR is being used for digital signalling, an increase in data rate can be had using multi-level signalling or m-ary coding instead of binary. This is the basis of the sound-in-syncs technique used by broadcasters to convey PCM audio along baseband video routes by inserting data bursts in the analog video sync pulses. Figure 6.46 shows the four-level waveform of the UK DSIS (Dual Channel Sound in Syncs) system which

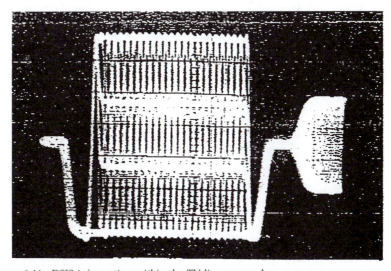

Figure 6.46   DSIS information within the TV line sync pulse.

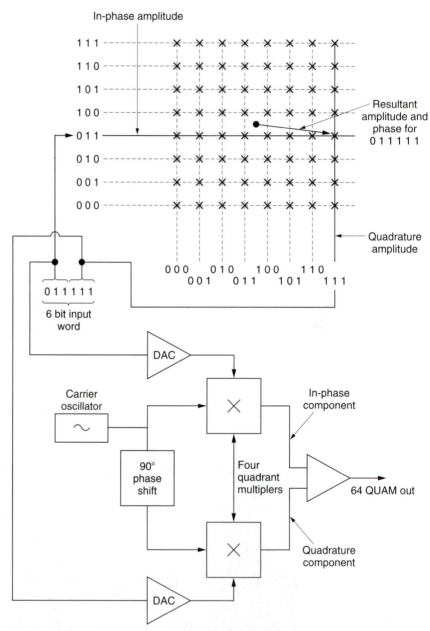

Figure 6.47 In 64-QUAM, two carriers are generated with a quadrature relationship. These are independently amplitude-modulated to eight discrete levels in four quadrant multipliers. Adding the signals produces a QUAM signal having 64 unique combinations of amplitude and phase. Decoding requires the waveform to be sampled in quadrature like a colour TV subcarrier.

is used to carry stereo audio to NICAM-equipped transmitters. Clearly the data separator must have a two-bit ADC which can resolve the four signal levels. The gain and offset of the signal must be precisely set so that the quantizing levels register precisely with the centres of the eyes.

Where the maximum data rate is needed for economic reasons as in Digital Audio Broadcasting (DAB) or digital television broadcasts, multi-level signalling can be combined with PSK to obtain multi-level Quadrature Amplitude Modulation (QUAM). Figure 6.47 shows the example of 64-QUAM. Incoming six-bit data words are split into two three-bit words and each is used to amplitude modulate a pair of sinusoidal carriers which are generated in quadrature. The modulators are four-quadrant devices such that 2^3 amplitudes are available, four which are in phase with the carrier and four which are antiphase. The two AM carriers are linearly added and the result is a signal which has 2^6 or 64 combinations of amplitude and phase. There is a great deal of similarity between QUAM and the colour subcarrier used in analog television in which the two colour difference signals are encoded into one amplitude and phase modulated waveform. On reception, the waveform is sampled twice per cycle in phase with the two original carriers and the result is a pair of eight-level signals.

The data are randomized by addition to a pseudo-random sequence before being fed to the modulator. The resultant spectrum has once again the $\sin x/x$ shape with nulls at multiples of the randomizer clock rate. As a result, a large number of carriers can be spaced at multiples of the randomizer clock frequency such that each carrier centre frequency coincides with the nulls of all the adjacent carriers. The result is referred to as COFDM or coded orthogonal frequency division multiplexing.[21]

6.28 Convolutional randomizing

The randomizing in NICAM is block based, since this matches the one millisecond block structure of the transmission. Where there is no obvious block structure, convolutional, or endless randomizing can be used. This is the approach used in the Scrambled Serial digital video interconnect which allows composite or component video of up to ten-bit wordlength to be sent serially along with digital audio channels.

In convolutional randomizing, the signal sent down the channel is the serial data waveform which has been convolved with the impulse response of a digital filter. On reception the signal is deconvolved to restore the original data. Figure 6.48(a) shows that the filter is an infinite impulse response (IIR) filter which has recursive paths from the output back to the input. As it is a one-bit filter its output cannot decay, and once excited, it runs indefinitely. The filter is followed by a transition generator which consists of a one-bit delay and an exclusive-OR gate. An input 1 results in an output transition on the next clock edge. An input 0 results in no transition.

A result of the infinite impulse response of the filter is that frequent transitions are generated in the channel which result in sufficient clock content for the phase-locked loop in the receiver.

Transitions are converted back to 1s by a differentiator in the receiver. This consists of a one-bit delay with an exclusive-OR gate comparing the input and the output. When a transition passes through the delay, the input and the output will be different and the gate outputs a 1 which enters the deconvolution circuit.

Figure 6.48(b) shows that in the deconvolution circuit a data bit is simply the exclusive-OR of a number of channel bits at a fixed spacing. The deconvolution is implemented with a shift register having the exclusive-OR gates connected in a reverse pattern to that in the encoder. The same effect as block randomizing is obtained, in that long runs are broken up and the DC content is reduced, but it has the advantage over

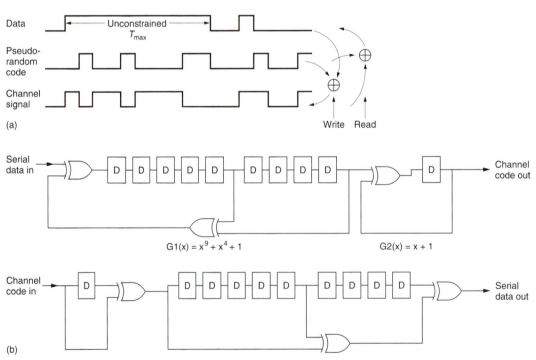

Figure 6.48 (a) Modulo-2 addition with a pseudo-random code removes unconstrained runs in real data. Identical process must be provided on replay. (b) Convolutional randomizing encoder, at top, transmits exclusive OR of three bits at a fixed spacing in the data. One-bit delay, far right, produces channel transitions from data ones. Decoder, below, has opposing one-bit delay to return from transitions to data levels, followed by an opposing shift register which exactly reverses the coding process.

block randomizing that no synchronizing is required to remove the randomizing, although it will still be necessary for deserialization. Clearly the system will take a few clock periods to produce valid data after commencement of transmission, but this is no problem on a permanent wired connection where the transmission is continuous.

6.29 Synchronizing

Once the PLL in the data separator has locked to the clock content of the transmission, a serial channel bitstream and a channel bit clock will emerge from the sampler. In a group code, it is essential to know where a group of channel bits begins in order to assemble groups for decoding to data bit groups. In a randomizing system it is equally vital to know at what point in the serial data stream the words or samples commence. In serial transmission and in recording, channel bit groups or randomized data words are sent one after the other, one bit at a time, with no spaces in between, so that although the designer knows that a data block contains, say, 128 bytes, the receiver simply finds 1024 bits in a row. If the exact position of the first bit is not known, then it is not possible to put all the bits in the right places in the right bytes; a process known as deserializing. The effect of sync slippage is devastating, because a one-bit disparity between the bit count and the bitstream will corrupt every symbol in the block.[22]

The synchronization of the data separator and the synchronization to the block format are two distinct problems, which are often solved by the same sync pattern. Deserializing requires a shift register which is fed with serial data and read out once per word. The sync detector is simply a set of logic gates which are arranged to recognize a specific pattern in the register. The sync pattern is either identical for every block or has a restricted number of versions and it will be recognized by the replay circuitry and used to reset the bit count through the block. Then by counting channel bits and dividing by the group size, groups can be deserialized and decoded to data groups. In a randomized system, the pseudo-random sequence generator is also reset. Then counting derandomized bits from the sync pattern and dividing by the wordlength enables the replay circuitry to deserialize the data words.

In digital audio the two's complement coding scheme is universal and traditionally no codes have been reserved for synchronizing; they are all available for sample values. It would in any case be impossible to reserve all ones or all zeros as these are in the centre of the range in two's complement. Even if a specific code were excluded from the recorded data so it could be used for synchronizing, this cannot ensure that the same pattern cannot be falsely created at the junction between two

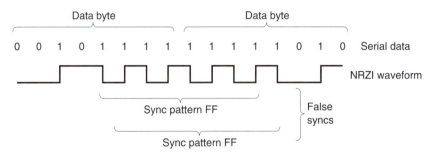

Figure 6.49 Concatenation of two words can result in the accidental generation of a word which is reserved for synchronizing.

allowable data words. Figure 6.49 shows how false synchronizing can occur due to concatenation. It is thus not practical to use a bit pattern which is a data code value in a simple synchronizing recognizer. The problem is overcome in NICAM 728 by using the fact that sync patterns occur exactly once per millisecond or 728 bits. The sync pattern of NICAM 728 is just a bit pattern and no steps are taken to prevent it from appearing in the randomized data. If the pattern is seen by the recognizer, the recognizer is disabled for the rest of the frame and only enabled when the next sync pattern is expected. If the same pattern recurs every millisecond, a genuine sync condition exists. If it does not, there was a false sync and the recognizer will be enabled again. As a result it will take a few milliseconds before sync is achieved, but once achieved it should not be lost unless the transmission is interrupted. This is fine for the application and no-one objects to the short mute of the NICAM sound during a channel switch. The principle cannot, however, be used for recording because channel interruptions are more frequent due to head switches and dropouts and loss of several blocks of data due to a single dropout is unacceptable.

In run-length-limited codes this is not a problem. The sync pattern is no longer a data bit pattern but is a specific waveform. If the sync waveform contains run lengths which violate the normal coding limits, there is no way that these run lengths can occur in encoded data, nor any possibility that they will be interpreted as data. They can, however, be readily detected by the replay circuitry. The sync patterns of the AES/EBU interface are shown in Figure 6.50. It will be seen from Figure 6.33 that the maximum run length in FM coded data is one bit. The sync pattern begins with a run length of one and a half bits which is unique. There are three types of sync pattern in the AES/EBU interface, as will be seen in Chapter 8. These are distinguished by the position of a second pulse after the run-length violation. Note that the sync patterns are also DC-free like the FM code.

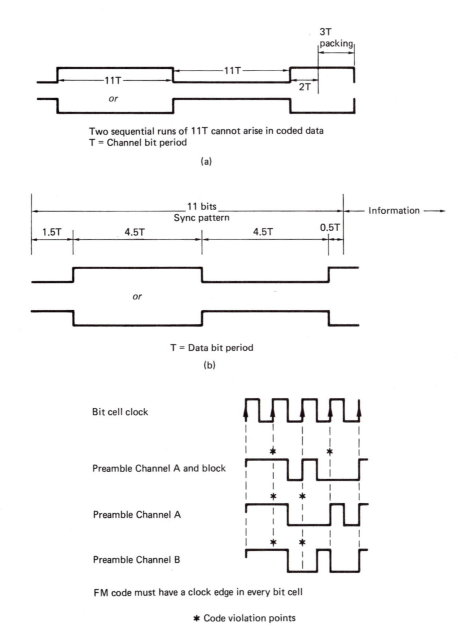

Figure 6.50 Sync patterns in various applications. In (a) the sync pattern of CD violates EFM coding rules, and is uniquely identifiable. In (b) the sync pattern of DASH stays within the run length of HDM-1. (c) The sync patterns of AES/EBU interconnect.

In a group code there are many more combinations of channel bits than there are combinations of data bits. Thus after all data bit patterns have been allocated group patterns, there are still many unused group patterns which cannot occur in the data. With care, group patterns can be found which cannot occur due to the concatenation of any pair of groups representing data. These are then unique and can be used for synchronizing.

In MADI, this approach is used as will be seen in Chapter 8. A similar approach is used in CD. Here the sync pattern does not violate a run length limit, but consists of two sequential maximum run lengths of 11 channel bit periods each as in Figure 6.50(a). This pattern cannot occur in the data because the data symbols are only 14 channel bits long and the packing bit generator can be programmed to exclude accidental sync pattern generation due to concatenation.

References

1. Deeley, E.M., Integrating and differentiating channels in digital tape recording. *Radio Electron. Eng.,* **56** 169–173 (1986)
2. Mee, C.D., *The Physics of Magnetic Recording*, Amsterdam and New York: Elsevier–North Holland Publishing (1978)
3. Jacoby, G.V., Signal equalization in digital magnetic recording. *IEEE Trans. Magn.,* **MAG-11**, 302–305 (1975)
4. Schneider, R.C., An improved pulse-slimming method for magnetic recording. *IEEE Trans. Magn.,* **MAG-11**, 1240–1241 (1975)
5. Miller, A., US Patent. No. 3,108,261 (1960)
6. Mallinson, J.C. and Miller, J.W., Optimum codes for digital magnetic recording. *Radio and Electron. Eng.,* **47**, 172–176 (1977)
7. Miller, J.W., DC-free encoding for data transmission system. US Patent 4,027,335 (1977)
8. Tang, D.T., Run-length-limited codes. IEEE International Symposium on Information Theory (1969)
9. Cohn, M. and Jacoby, G., Run-length reduction of 3PM code via lookahead technique. *IEEE Trans. Magn.,* **18**, 1253–1255 (1982)
10. Horiguchi, T. and Morita, K., On optimisation of modulation codes in digital recording. *IEEE Trans. Magn.,* **12**, 740–742 (1976)
11. Franaszek, P.A., Sequence state methods for run-length linited coding. *IBM J. Res. Dev.,* **14**, 376–383 (1970)
12. AES Recommended practice for Digital Audio Engineering – Serial Multichannel Audio Digital Interface (MADI). *J. Audio Eng. Soc.,* **39**, No.5, 371–377 (1991)
13. Jacoby, G.V. and Kost, R., Binary two-thirds-rate code with full word lookahead. *IEEE Trans. Magn.,* **20**, 709–714 (1984)
14. Jacoby, G.V., A new lookahead code for increased data density. *IEEE Trans. Magn.,* **13**, 1202–1204 (1977)
15. Ogawa, H. and Schouhamer Immink, K.A., EFM – the modulation method for the Compact Disc digital audio system. In *Digital Audio*, edited by B. Blesser, B. Locanthi and T.G. Stockham Jr, pp. 117–124, New York: Audio Engineering Society (1982)
16. Schouhamer Immink, K.A. and Gross, U., Optimisation of low frequency properties of eight-to-fourteen modulation. *Radio Electron. Eng.,* **53**, 63–66 (1983)
17. Fukuda, S., Kojima, Y., Shimpuku, Y. and Odaka, K., 8/10 modulation codes for digital magnetic recording. *IEEE Trans. Magn.,* **MAG–22**, 1194–1196 (1986)

18. Doi, T.T., Channel codings for digital audio recordings. *J. Audio Eng. Soc.*, **31**, 224–238 (1983)
19. Anon., PD format for stationary head type 2-channel digital audio recorder. Mitsubishi (January 1986)
20. Schouhamer Immink, K.A., Graceful degradation of digital audio transmission systems. Presented at the 82nd Audio Engineering Society Convention (London, 1987), Preprint 2434(C-3)
21. Alard. M. and Lasalle, R. Principles of modulation and channel coding for digital broadcasting for mobile receivers. *EBU Review*, **224**, 168–190 (1987)
22. Griffiths, F.A., A digital audio recording system. Presented at the 65th Audio Engineering Society Convention (London, 1980), Preprint 1580(C1)

7

Error correction

The subject of error correction is almost always described in mathematical terms by specialists for the benefit of other specialists. Such mathematical approaches are quite inappropriate for a proper understanding of the concepts of error correction and only become necessary to analyse the quantitative behaviour of a system. The description below will use the minimum possible amount of mathematics, and it will then be seen that error correction is, in fact, quite straightforward.

7.1   Sensitivity of message to error

Before attempting to specify any piece of equipment, it is necessary to quantify the problems to be overcome and how effectively they need to be overcome. For a digital recording system the causes of errors must be studied to quantify the problem, and the sensitivity of the destination to errors must be assessed. In audio the sensitivity to errors must be subjective. In PCM, the effect of a single bit in error depends upon the significance of the bit. If the least significant bit of a sample is wrong, the chances are that the effect will be lost in the noise. Advantage is taken of this in NICAM 728 which does not detect low-order bit errors. Conversely, if a high-order bit is in error, a massive transient will be added to the sound waveform. The effect of uncorrected errors in PCM audio is rather like that of vehicle ignition interference on a radio.

The effect of errors in delta-modulated data is smaller as every bit has the same significance and the information content of each bit is lower as was explained in Chapter 4. In some applications, a delta-modulated system can be used without error correction when this would be impossible with PCM.

Whilst the exact BER (bit error rate) which can be tolerated will depend on the application, digital audio is less tolerant of errors than digital video and more tolerant than computer data.

As might be expected, when compresssion is used, as in DCC, DAB and MiniDisc, much of the redundancy is removed from the data and as a result sensitivity to bit errors inevitably increases. In all these cases, if the maximum error rate which the destination can tolerate is likely to be exceeded by the unaided channel, some form of error handling will be necessary.

There are a number of terms which have idiomatic meanings in error correction. The raw BER is the error rate of the medium, whereas the residual or uncorrected BER is the rate at which the error-correction system fails to detect or miscorrects errors. In practical digital audio systems, the residual BER is negligibly small. If the error correction is turned off, the two figures become the same.

7.2   Error mechanisms

There are many different types of recording and transmission channel and consequently there will be many different error mechanisms. In magnetic recording, data can be corrupted by mechanical problems such as media dropout and poor tracking or head contact, or Gaussian thermal noise in replay circuits and heads. In optical recording, contamination of the medium interrupts the light beam. Warped disks and birefringent pressings cause defocussing. Inside equipment, data are conveyed on short wires and the noise environment is under the designer's control. With suitable design techniques, errors can be made effectively negligible. In communication systems, there is considerably less control of the electromagnetic environment. In cables, crosstalk and electromagnetic interference occur and can corrupt data, although optical fibres are resistant to interference of this kind. In data networks, errors can be caused if two devices on the same cable inadvertently start transmitting at the same instant.

In long-distance cable transmission the effects of lightning and exchange switching noise must be considered. In DAB, multipath reception causes notches in the received spectrum where signal cancellation takes place. In MOS memories the datum is stored in a tiny charge well which acts as a capacitor (see Chapter 3) and natural radioactive decay produces alpha particles which have enough energy to discharge a well, resulting in a single bit error. This only happens once every few decades in a single chip, but when large numbers of chips are assembled in computer memories the probability of error rises to once every few minutes. In Chapter 6 it was seen that when group codes are used, a single defect in a group changes the group symbol and may cause errors up to the size of the group. Single-bit errors are therefore less common in group-coded channels.

Irrespective of the cause, all these mechanisms cause one of two effects. There are large isolated corruptions, called error bursts, where numerous bits are corrupted all together in an area which is otherwise error-free, and there are random errors affecting single bits or symbols. Whatever the mechanism, the result will be that the received data will not be exactly the same as those sent. It is a tremendous advantage of digital audio that the discrete data bits will each be either right or wrong. A bit cannot be off-colour as it can only be interpreted as 0 or 1. Thus the subtle degradations of analog systems are absent from digital recording and transmission channels and will only be found in convertors. Equally if a binary digit is known to be wrong, it is only necessary to invert its state and then it must be right and indistinguishable from its original value! Thus error correction itself is trivial; the hard part is reliably working out *which* bits need correcting.

In Chapter 3 the Gaussian nature of noise probability was discussed. Some conclusions can be drawn from the Gaussian distribution of noise.[1] First, it is not possible to make error-free digital recordings, because however high the signal-to-noise ratio of the recording, there is still a small but finite chance that the noise can exceed the signal. Measuring the signal-to-noise ratio of a channel establishes the noise power, which determines the width of the noise-distribution curve relative to the signal amplitude. When in a binary system the noise amplitude exceeds the signal amplitude, a bit error will occur. Knowledge of the shape of the Gaussian curve allows the conversion of signal-to-noise ratio into bit error rate (BER). It can be predicted how many bits will fail due to noise in a given recording, but it is not possible to say *which* bits will be affected. Increasing the SNR of the channel will not eliminate errors, it just reduces their probability. The logical solution is to incorporate an error-correction system.

7.3 Basic error correction

Error correction works by adding some bits to the data which are calculated from the data. This creates an entity called a codeword which spans a greater length of time than one bit alone. In recording, the requirement is to spread the codeword over an adequate area of the medium. The statistics of noise means that whilst one bit may be lost in a codeword, the loss of the rest of the codeword because of noise is highly improbable. As will be described later in this chapter, codewords are designed to be able to correct totally a finite number of corrupted bits.

The greater the timespan or area over which the coding is performed, the greater will be the reliability achieved, although this does mean that greater encoding and decoding delays will have to be accepted.

Shannon[2] proved that a message can be transmitted to any desired degree of accuracy provided that it is spread over a sufficient timespan or area of the medium. Engineers have to compromise, because excessive coding delay is not acceptable. For example, most short digital audio cable interfaces do not employ error correction because the build-up of coding delays in large systems is unacceptable.

If error correction is necessary as a practical matter, it is then only a small step to put it to maximum use. All error correction depends on adding bits to the original message, and this, of course, increases the number of bits to be recorded, although it does not increase the information recorded. It might be imagined that error correction is going to reduce storage or transmission capacity, because space has to be found for all the extra bits. Nothing could be further from the truth. Once an error-correction system is used, the signal-to-noise ratio of the channel can be reduced, because the raised BER of the channel will be overcome by the error-correction system. Reduction of the SNR by 3 dB in a magnetic tape track can be achieved by halving the track width, provided that the system is not dominated by head or preamplifier noise. This doubles the recording density, making the storage of the additional bits needed for error correction a trivial matter. By a similar argument, digital radio transmitters can use less power. In short, error correction is not a nuisance to be tolerated; it is a vital tool needed to maximize the efficiency of recorders. Digital audio would not be economically viable without it.

7.4 Error handling

Figure 7.1 shows the broad subdivisions of error handling. The first stage might be called error avoidance and includes such measures as creating bad block files on hard disks or using verified media. The data pass through the channel, which causes whatever corruptions it feels like. On receipt of the data the occurrence of errors is first detected, and this

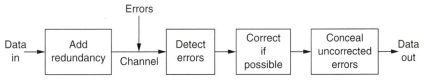

Figure 7.1 The basic stages of an error-correction system. Of these the most critical is the detection stage, since this controls the subsequent actions.

process must be extremely reliable, as it does not matter how effective the correction or how good the concealment algorithm if it is not known that they are necessary! The detection of an error then results in a course of action being decided.

A retry is not possible if the data are required in real time for replay purposes. However, in the case of an audio file transfer in a disk-based network, real-time operation is not required. A transmission error due to a network collision or interference will result in a retransmission. If the disk drive detects a read error a retry is easy as the disk is turning at several thousand rpm and will quickly re-present the data. An error due to a dust particle may not occur on the next revolution. Many magnetic tape systems have *read after write*. During recording, offtape data are immediately checked for errors. If an error is detected, the tape will abort the recording, reverse to the beginning of the current block and erase it. The data from that block are then recorded further down the tape.

7.5   Concealment by interpolation

There are some practical differences between data recording for audio and the general computer data-recording application. Although audio recorders seldom have time for retries, they have the advantage that there is a certain amount of redundancy in the information conveyed. In audio systems, if an error cannot be corrected, then it can be concealed. If a sample is lost, it is possible to obtain an approximation to it by interpolating between the samples before and after the missing one. Clearly concealment of any kind cannot be used with computer data.

In NICAM 728 errors are relatively infrequent and correction is not used. There is simply an error-detecting system which causes samples in error to be concealed. This is described in greater detail in Chapter 8. Momentary interpolations are not serious, but sustained use of interpolation can result in aliasing if high frequencies are present in the recording.

In systems which use compression, bit errors are serious because they cause loss of synchronization in variable-length coding, leading to an audible error much larger than the actual data loss. This is known as error-propagation and to avoid it, compressed systems must use reliable error-correction systems. Concealment is also more difficult in compression systems. In advanced concealment systems, a spectral analysis of the sound is made, and if correct sample values are not available, samples having the same spectral characteristics are substituted. This concealment method can conceal greater damage than simple interpolation because the spectral shape changes quite slowly compared to the voltage domain signal.

If there is too much corruption for concealment, the only course in audio is to mute the output as large numbers of uncorrected errors reaching the analog domain cause noise which can be of a high level.

If use is to be made of concealment on replay, the data must generally be reordered or shuffled prior to recording. To take a simple example, odd-numbered samples are recorded in a different area of the medium from even-numbered samples. On playback, if a gross error occurs on the tape, depending on its position, the result will be either corrupted odd samples or corrupted even samples, but it is most unlikely that both will be lost. Interpolation is then possible if the power of the correction system is exceeded.

It should be stressed that corrected data are indistinguishable from the original and thus there can be no audible artifacts. In contrast, concealment is only an approximation to the original information and could be audible. In practical equipment, concealment occurs infrequently unless there is a defect requiring attention.

7.6   Parity

The error-detection and error-correction processes are closely related and will be dealt with together here. The actual correction of an error is simplified tremendously by the adoption of binary. As there are only two symbols, 0 and 1, it is enough to know that a symbol is wrong, and the correct value is obvious. Figure 7.2 shows a minimal circuit required for correction once the bit in error has been identified. The XOR (exclusive-OR) gate shows up extensively in error correction and the figure also shows the truth table. One way of remembering the characteristics of this useful device is that there will be an output when the inputs are different. Inspection of the truth table will show that there is an even number of ones in each row (zero is an even number) and so the device could also be called an even parity gate. The XOR gate is also a adder in modulo 2 (see Chapter 3).

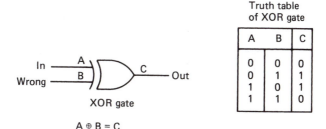

$$A \oplus B = C$$

Figure 7.2   Once the position of the error is identified, the correction process in binary is easy.

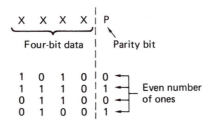

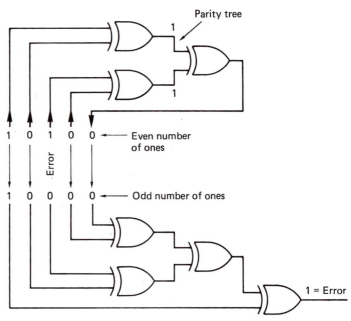

Figure 7.3 Parity checking adds up the number of ones in a word using, in this example, parity trees. One error bit and odd numbers of errors are detected. Even numbers of errors cannot be detected.

Parity is a fundamental concept in error detection. In Figure 7.3, the example is given of a four-bit data word which is to be protected. If an extra bit is added to the word which is calculated in such a way that the total number of ones in the five-bit word is even, this property can be tested on receipt. The generation of the parity bit in Figure 7.3 can be performed by a number of the ubiquitous XOR gates configured into what is known as a parity tree. In the figure, if a bit is corrupted, the received message will be seen no longer to have an even number of ones. If two bits are corrupted, the failure will be undetected. This example can be used to introduce much of the terminology of error correction. The extra bit added to the message carries no information of its own, since it is calculated from the other bits. It is therefore called a *redundant* bit. The addition of the redundant bit gives the message a special property, i.e. the number of ones is even. A message having some special property *irrespective of the actual data content* is called a *codeword*. All error correction relies on adding redundancy to real data to form codewords for transmission. If any corruption occurs, the intention is that the received message will not have the special property; in other words if the received message is not a codeword there has definitely been an error. The receiver can check for the special property without any prior knowledge of the data content. Thus the same check can be made on all received data. If the received message is a codeword, there probably has not been an error. The word 'probably' must be used because the figure shows that two bits in error will cause the received message to be a codeword, which cannot be discerned from an error-free message. If it is known that generally the only failure mechanism in the channel in question is loss of a single bit, it is *assumed* that receipt of a codeword means that there has been no error. If there is a probability of two error bits, that becomes very nearly the probability of failing to detect an error, since all odd numbers of errors will be detected, and a four-bit error is much less likely.

It is paramount in all error-correction systems that the protection used should be appropriate for the probability of errors to be encountered. An inadequate error-correction system is actually worse than not having any correction. Error correction works by trading probabilities. Error-free performance with a certain error rate is achieved at the expense of performance at higher error rates. Figure 7.4 shows the effect of an error-correction system on the residual BER for a given raw BER. It will be seen that there is a characteristic knee in the graph. If the expected raw BER has been misjudged, the consequences can be disastrous. Another result

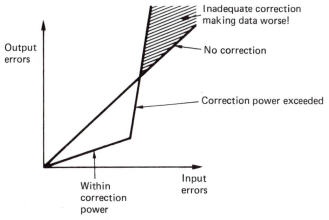

Figure 7.4 An error-correction system can only reduce errors at normal error rates at the expense of increasing errors at higher rates. It is most important to keep a system working to the left of the knee in the graph.

demonstrated by the example is that we can only guarantee to detect the same number of bits in error as there are redundant bits.

7.7   Block and convolutional codes

Figure 7.5(a) shows that in a crossword, or product, code the data are formed into a two-dimensional array, in which each location can be a single bit or a multi-bit symbol. Parity is then generated on both rows and columns. If a single bit or symbol fails, one row parity check and one column parity check will fail, and the failure can be located at the intersection of the two failing checks. Although two symbols in error confuse this simple scheme, using more complex coding in a two-dimensional structure is very powerful, and further examples will be given throughout this chapter.

The example of Figure 7.5(a) assembles the data to be coded into a block of finite size and then each codeword is calculated by taking different set of symbols. This should be contrasted with the operation of the circuit of Figure 7.5(b). Here the data are not in a block, but form an endless stream. A shift register allows four symbols to be available simultaneously to the encoder. The action of the encoder depends upon the delays. When symbol 3 emerges from the first delay, it will be added (modulo 2) to symbol 6. When this sum emerges from the second delay, it will be added to symbol 9 and so on. The codeword produced is shown in Figure 7.5(c) where it will be seen to be bent such that it has a vertical section and a diagonal section. Four symbols later the next codeword will be created one column further over in the data.

This is a convolutional code because the coder always takes parity on the same pattern of symbols which is convolved with the data stream on an endless basis. Figure 7.5(c) also shows that if an error occurs, it will cause a parity error in two codewords. The error will be on the diagonal part of one codeword and on the vertical part of the other so that it can uniquely be located at the intersection and corrected by parity.

Comparison with the block code of Figure 7.5(a) *will* show that the convolutional code needs less redundancy for the same single-symbol location and correction performance as only a single redundant symbol is required for every four data symbols. Convolutional codes are computed on an endless basis which makes them inconvenient in recording applications where editing is anticipated. Here the block code is more appropriate as it allows edit gaps to be created between codes. In the case of uncorrectable errors, the convolutional principle causes the syndromes to be affected for some time afterwards and results in miscorrections of symbols which were not actually in error. This is a further example of error propagation and is a characteristic of

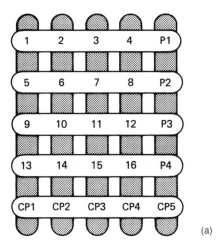

(a)

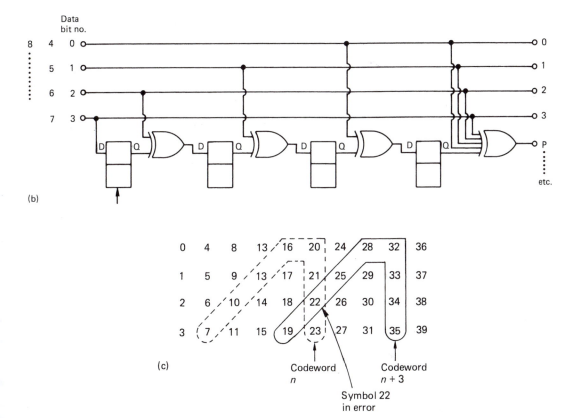

Figure 7.5 A block code is shown in (a). Each location in the block can be a bit or a word. Horizontal parity checks are made by adding P1, P2, etc., and cross-parity or vertical checks are made by adding CP1, CP2, etc. Any symbol in error will be at the intersection of the two failing codewords. In (b) a convolutional coder is shown. Symbols entering are subject to different delays which result in the codewords in (c) being calculated. These have a vertical part and a diagonal part. A symbol in error will be at the intersection of the diagonal part of one code and the vertical part of another.

convolutional codes. Recording media tend to produce somewhat variant error statistics because media defects and mechanical problems cause errors which do not fit the classical additive noise channel. Convolutional codes can easily be taken beyond their correcting power if used with real recording media.

In transmission and broadcasting, the error statistics are more stable and the editing requirement is absent. As a result, convolutional codes are used in DAB and DVB whereas block codes are used in recording. Convolutional codes are not restricted to the simple parity example given here, but can be used in conjuction with more sophisticated redundancy techniques such as the Reed–Solomon codes.

7.8   Hamming code

In a one-dimensional code, the position of the failing bit can be determined by using more parity checks. In Figure 7.6, the four data bits have been used to compute three redundancy bits, making a seven-bit codeword. The four data bits are examined in turn, and each bit which is a one will cause the corresponding row of a generator matrix to be added to an exclusive-OR sum. For example, if the data were 1001, the top and bottom rows of the matrix would be XORed. The matrix used is known as an identity matrix, because the data bits in the codeword are identical to the data bits to be conveyed. This is useful because the original data can be stored unmodified, and the check bits are simply attached to the end to make a so-called systematic codeword. Almost all digital recording equipment uses systematic codes. The way in which the redundancy bits are calculated is simply that they do not all use every data bit. If a data bit has not been included in a parity check, it can fail without affecting the outcome of that check. The position of the error is deduced from the pattern of successful and unsuccessful checks in the check matrix. This pattern is known as a syndrome.

In the figure the example of a failing bit is given. Bit three fails, and because this bit is included in only two of the checks, there are two ones in the failure pattern, 011. As some care was taken in designing the matrix pattern for the generation of the check bits, the syndrome, 011, is the address of the failing bit. This is the fundamental feature of the Hamming codes due to Richard Hamming.[3] The performance of this seven-bit codeword can be assessed. In seven bits there can be 128 combinations, but in four data bits there are only sixteen combinations. Thus out of 128 possible received messages, only sixteen will be codewords, so if the message is completely trashed by a gross corruption, it will still be possible to detect that this has happened 112 times out of 127, as in these cases the syndrome will be non-zero (the 128th case is the correct data).

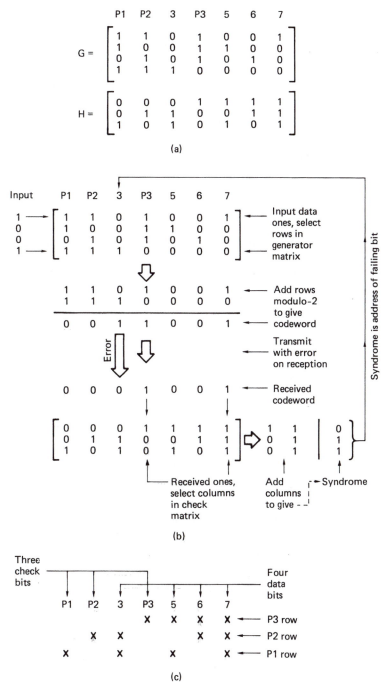

Figure 7.6 (a) The generator and check matrices of a Hamming code. The data and check bits are arranged as shown because this causes the syndrome to be the binary address of the failing bit. (b) An example of Hamming-code generation and error correction. (c) Another way of looking at Hamming code is to say that the rows of crosses in this chart are calculated to have even parity. If bit 3 fails, parity check P3 is not affected, but parity checks P1 and P2 both include bit 3 and will fail.

There is thus only a probability of detecting that all of the message is corrupt. In an idle moment it is possible to work out, in a similar way, the number of false codewords which can result from different numbers of bits being assumed to have failed. For fewer than three bits, the failure will always be detected, because there are three check bits. Returning to the example, if two bits fail, there will be a non-zero syndrome, but if this is used to point to a bit in error, a miscorrection will result. From these results can be deduced another important feature of error codes. The power of detection is always greater than the power of correction, which is also fortunate, since if the correcting power is exceeded by an error it will at least be a known problem, and steps can be taken to prevent any undesirable consequences.

The efficiency of the example given is not very high because three check bits are needed for every four data bits. Since the failing bit is located with a binary-split mechanism, it is possible to double the code length by adding a single extra check bit. Thus with four-bit syndromes there are fifteen non-zero codes and so the codeword will be fifteen bits long. Four bits are redundant and eleven are data. Using five bits of redundancy, the code can be 31 bits long and contain 26 data bits. Thus provided that the number of errors to be detected stays the same, it is more efficient to use long codewords. Error-correcting memories use typically four or eight data bytes plus redundancy. A drawback of long codes is that if it is desired to change a single memory byte it is necessary to read the entire codeword, modify the desired data byte and re-encode, the so-called read–modify–write process.

The Hamming code shown is limited to single-bit correction, but by addition of another bit of redundancy can be made to correct one-bit and detect two-bit errors. This is ideal for error-correcting MOS memories where the SECDED (single-error correcting double-error detecting) characteristic matches the type of failures experienced.

The correction of one bit is of little use in the presence of burst errors, but a Hamming code can be made to correct burst errors by using interleaving. Figure 7.7 shows that if several codewords are calculated beforehand and woven together as shown before they are sent down the channel, then a burst of errors which corrupts several bits will become a number of single-bit errors in separate codewords upon de-interleaving.

Interleaving is used extensively in digital recording and transmission, and will be discussed in greater detail later in this chapter.

7.9 Hamming distance

It is useful at this point to introduce the concept of Hamming distance. It is not a physical distance but is a specific measure of the difference between two binary numbers. Hamming distance is defined in the

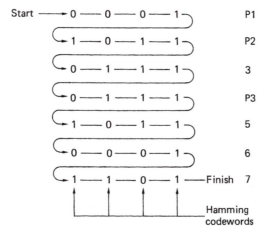

Figure 7.7 The vertical columns of this diagram are all codewords generated by the matrix of Figure 7.6, which can correct a single-bit error. If these words are recorded in the order shown, a burst error of up to four bits will result in one single-bit error in each codeword, which is correctable. Interleave requires memory, and causes delay. De-interleave requires the same.

general case as the number of bit positions in which a pair of words differ. The Hamming distance of a code is defined as the minimum number of bits that must be changed in any codeword in order to turn it into another codeword. This is an important yardstick because if errors convert one codeword into another, it will have the special characteristic of the code and so the corruption will not even be detected.

Figure 7.8 shows Hamming distance diagrammatically. A three-bit codeword is used with two data bits and one parity bit. With three bits, a received code could have eight combinations, but only four of these will be codewords. The valid codewords are shown in the centre of each of the disks, and these will be seen to be identical to the rows of the truth table in Figure 7.2. At the perimeter of the disks are shown the received words which would result from a single-bit error, i.e. they have a Hamming distance of one from codewords. It will be seen that the same received word (on the vertical bars) can be obtained from a different single-bit corruption of any three codewords. It is thus not possible to tell which codeword was corrupted, so although all single-bit errors can be detected, correction is not possible. This diagram should be compared with that of Figure 7.9, which is a Venn diagram where there is a set in which the MSB is 1 (upper circle), a set in which the middle bit is 1 (lower left circle) and a set in which the LSB is 1 (lower right circle). Note that in crossing any boundary only one bit changes, and so each boundary represents a Hamming distance change of one. The four codewords of Figure 7.8 are repeated here, and it will be seen that single-bit errors in any codeword produce a non-codeword, and so single-bit errors are always detectable.

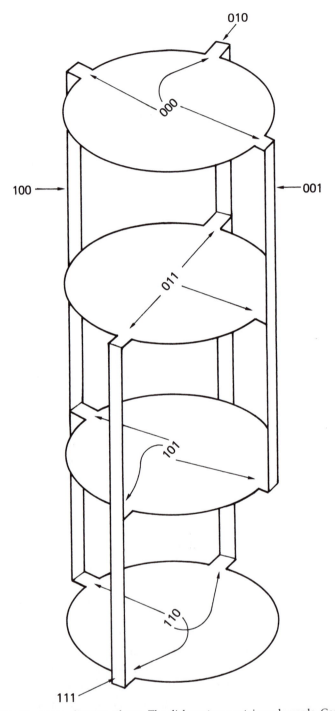

Figure 7.8 Hamming distance of two. The disk centres contain codewords. Corrupting each bit in turn produces the distance 1 values on the vertical members. In order to change one codeword to another, two bits must be changed, so the code has a Hamming distance of two.

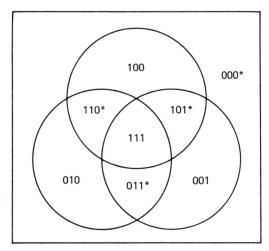

Figure 7.9   Venn diagram shows a one-bit change crossing any boundary which is a Hamming distance of one. Compare with Figure 7.8. Codewords marked*.

Correction is possible if the number of non-codewords is increased by increasing the number of redundant bits. This means that it is possible to spread out the actual codewords in Hamming distance terms.

Figure 7.10(a) shows a distance 2 code, where there is only one redundancy bit, and so half of the possible words will be codewords. There will be non-codewords at distance 1 which can be produced by altering a single bit in either of two codewords. In this case it is not possible to tell what the original codeword was in the case of a single-bit error.

Figure 7.10(b) shows a distance 3 code, where there will now be at least two non-codewords between codewords. If a single-bit error occurs in a codeword, the resulting non-codeword will be at distance 1 from the original codeword. This same non-codeword could also have been produced by changing *two* bits in a different codeword. If it is known that the failure mechanism is a single bit, it can be *assumed* that the original codeword was the one which is closest in Hamming distance to the received bit pattern, and so correction is possible. If, however, our assumption about the error mechanism proved to be wrong, and in fact a two-bit error had occurred, this assumption would take us to the wrong codeword, turning the event into a three-bit error. This is an illustration of the knee in the graph of Figure 7.4, where if the power of the code is exceeded it makes things worse.

Figure 7.10(c) shows a distance 4 code. There are now three non-codewords between codewords, and clearly single-bit errors can still be corrected by choosing the nearest codeword. Double-bit errors will be detected, because they result in non-codewords equidistant in Hamming

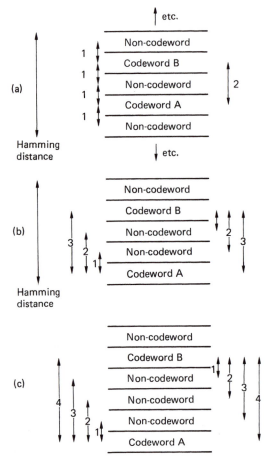

Figure 7.10 (a) Distance 2 code; non-codewords are at distance 1 from two possible codewords so it cannot be deduced what the correct one is. (b) Distance 3 code; non-codewords which have *single-bit errors* can be attributed to the nearest codeword. Breaks down in presence of double-bit errors. (c) Distance 4 code; non-codewords which have single-bit errors can be attributed to the nearest codeword, AND double-bit errors form *different* non-codewords, and can thus be detected but not corrected.

terms from codewords, but it is not possible to determine what the original codeword was.

7.10 Cyclic codes

The parallel implementation of a Hamming code can be made very fast using parity trees, which is ideal for memory applications where access time is increased by the correction process. However, in digital audio recording applications, the data are stored serially on a track, and it is desirable to use relatively large data blocks to reduce the amount of the medium devoted to preambles, addressing and synchronizing. Where large data blocks are to be handled, the use of a look-up table or tree has to be abandoned because it would become impossibly large. The principle of codewords having a special characteristic will still be employed, but they will be generated and checked algorithmically by equations. The syndrome will then be converted to the bit(s) in error not by looking them up, but by solving an equation.

Where data can be accessed serially, simpler circuitry can be used because the same gate will be used for many XOR operations. Unfortunately the reduction in component count is only paralleled by an increase in the difficulty of explaining what takes place.

The circuit of Figure 7.11 is a kind of shift register, but with a particular feedback arrangement which leads it to be known as a twisted-ring counter. If seven message bits A–G are applied serially to this circuit, and each one of them is clocked, the outcome can be followed in the diagram. As bit A is presented and the system is clocked, bit A will enter the left-hand latch. When bits B and C are presented, A moves across to the right. Both XOR gates will have A on the upper input from the right-hand latch, the left one has D on the lower input and the right one has B on the lower input. When clocked, the left latch will thus be loaded with the XOR of A and D, and the right one with the XOR of A and B. The remainder of the sequence can be followed, bearing in mind that when the same term

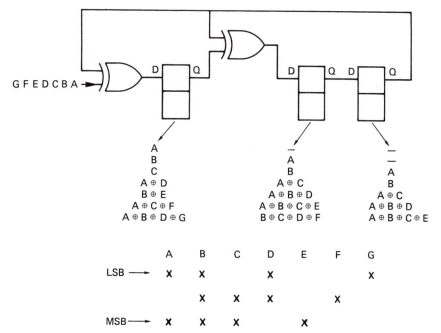

Figure 7.11 When seven successive bits A–G are clocked into this circuit, the contents of the three latches are shown for each clock. The final result is a parity-check matrix.

appears on both inputs of an XOR gate, it goes out, as the exclusive-OR of something with itself is nothing. At the end of the process, the latches contain three different expressions. Essentially, the circuit makes three parity checks through the message, leaving the result of each in the three stages of the register. In the figure, these expressions have been used to draw up a check matrix. The significance of these steps can now be explained.

The bits A B C and D are four data bits, and the bits E F and G are redundancy. When the redundancy is calculated, bit E is chosen so that there are an even number of ones in bits A B C and E; bit F is chosen such that the same applies to bits B C D and F, and similarly for bit G. Thus the four data bits and the three check bits form a seven-bit codeword. If there is no error in the codeword, when it is fed into the circuit shown, the result of each of the three parity checks will be zero and every stage of the shift register will be cleared. As the register has eight possible states, and one of them is the error-free condition, then there are seven remaining states, hence the seven-bit codeword. If a bit in the codeword is corrupted, there will be a non-zero result. For example, if bit D fails, the check on bits A B D and G will fail, and a one will appear in the left-hand latch. The check on bits B C D F will also fail, and the centre latch will set. The check on bits A B C E will not fail, because D is not involved in it, making the right-hand bit zero. There will be a syndrome of 110 in the register, and this will be seen from the check matrix to correspond to an error in bit D. Whichever bit fails, there will be a different three-bit syndrome which uniquely identifies the failed bit. As there are only three latches, there can be eight different syndromes. One of these is zero, which is the error-free condition, and so there are seven remaining error syndromes. The length of the codeword cannot exceed seven bits, or there would not be enough syndromes to correct all the bits. This can also be made to tie in with the generation of the check matrix. If fourteen bits, A to N, were fed into the circuit shown, the result would be that the check matrix repeated twice, and if a syndrome of 101 were to result, it could not be determined whether bit D or bit K failed. Because the check repeats every seven bits, the code is said to be a cyclic redundancy check (CRC) code.

In Figure 7.6 an example of a Hamming code was given. Comparison of the check matrix of Figure 7.11 with that of Figure 7.6 will show that the only difference is the order of the matrix columns. The two different processes have thus achieved exactly the same results, and the performance of both must be identical. This is not true in general, but a very small cyclic code has been used for simplicity and to allow parallels to be seen. In practice CRC code blocks will be much longer than the blocks used in Hamming codes.

It has been seen that the circuit shown makes a matrix check on a received word to determine if there has been an error, but the same circuit

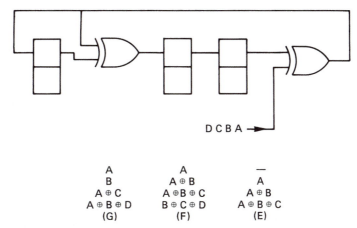

DCBA →

A	A	—
B	A ⊕ B	A
A ⊕ C	A ⊕ B ⊕ C	A ⊕ B
A ⊕ B ⊕ D	B ⊕ C ⊕ D	A ⊕ B ⊕ C
(G)	(F)	(E)

Figure 7.12 By moving the insertion point three places to the right, the calculation of the check bits is completed in only four clock periods and they can follow the data immediately. This is equivalent to premultiplying the data by x^3.

can also be used to generate the check bits. To visualize how this is done, examine what happens if only the data bits A B C and D are known, and the check bits E F and G are set to zero. If this message, ABCD000, is fed into the circuit, the left-hand latch will afterwards contain the XOR of A B C and zero, which is, of course, what E should be. The centre latch will contain the XOR of B C D and zero, which is what F should be and so on. This process is not quite ideal, however, because it is necessary to wait for three clock periods after entering the data before the check bits are available. Where the data are simultaneously being recorded and fed into the encoder, the delay would prevent the check bits being easily added to the end of the data stream. This problem can be overcome by slightly modifying the encoder circuit as shown in Figure 7.12. By moving the position of the input to the right, the operation of the circuit is advanced so that the check bits are ready after only four clocks. The process can be followed in the diagram for the four data bits A B C and D. On the first clock, bit A enters the left two latches, whereas on the second clock, bit B will appear on the upper input of the left XOR gate, with bit A on the lower input, causing the centre latch to load the XOR of A and B and so on.

The way in which the cyclic codes work has been described in engineering terms, but it can be described mathematically if analysis is contemplated.

Just as the position of a decimal digit in a number determines the power of ten (whether that digit means one, ten or a hundred), the position of a binary digit determines the power of two (whether it means one, two or four). It is possible to rewrite a binary number so that it is expressed as a list of powers of two. For example, the binary number 1101 means $8 + 4 + 1$, and can be written:

$$2^3 + 2^2 + 2^0$$

In fact, much of the theory of error correction applies to symbols in number bases other than 2, so that the number can also be written more generally as

$$x^3 + x^2 + 1 \ (2^0 = 1)$$

which also looks much more impressive. This expression, containing as it does various powers, is of course a polynomial, and the circuit of Figure 7.11 which has been seen to construct a parity-check matrix on a codeword can also be described as calculating the remainder due to dividing the input by a polynomial using modulo-2 arithmetic. In modulo-2 there are no borrows or carries, and addition and subtraction are replaced by the XOR function, which makes hardware implementation very easy. In Figure 7.13 it will be seen that the circuit of Figure 7.11 actually divides the codeword by a polynomial which is

$$x^3 + x + 1 \text{ or } 1011$$

This can be deduced from the fact that the right-hand bit is fed into two lower-order stages of the register at once. Once all the bits of the message have been clocked in, the circuit contains the remainder. In mathematical terms, the special property of a codeword is that it is a polynomial which yields a remainder of zero when divided by the generating polynomial. The receiver will make this division, and the result should be zero in the error-free case. Thus the codeword itself disappears from the division. If an error has occurred it is considered that this is due to an error polynomial which has been added to the codeword polynomial. If a codeword divided by the check polynomial is zero, a non-zero syndrome must represent the error polynomial divided by the check polynomial. Thus if the syndrome is multiplied by the check polynomial, the latter will be cancelled out and the result will be the error polynomial. If this is added modulo-2 to the received word, it will cancel out the error and leave the corrected data.

Some examples of modulo-2 division are given in Figure 7.13 which can be compared with the parallel computation of parity checks according to the matrix of Figure 7.11.

The process of generating the codeword from the original data can also be described mathematically. If a codeword has to give zero remainder when divided, it follows that the data can be converted to a

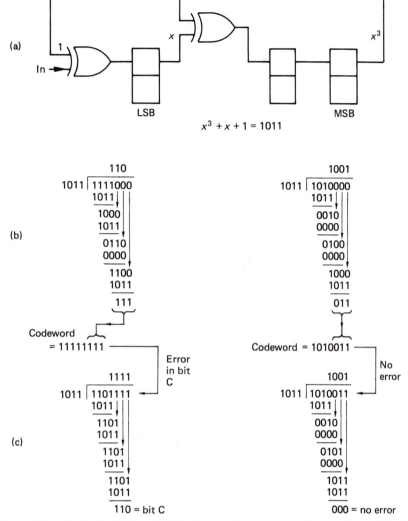

Figure 7.13 (a) Circuit of Figure 7.11 divides by $x^3 + x + 1$ to find remainder. At (b) this is used to calculate check bits. At (c) right, zero syndrome, no error.

codeword by adding the remainder when the data are divided. Generally speaking, the remainder would have to be subtracted, but in modulo-2 there is no distinction. This process is also illustrated in Figure 7.13. The four data bits have three zeros placed on the right-hand end, to make the wordlength equal to that of a codeword, and this word is then divided by the polynomial to calculate the remainder. The remainder is added to the zero-extended data to form a codeword. The modified circuit of Figure 7.12 can be described as premultiplying the data by x^3 before dividing.

CRC codes are of primary importance for detecting errors, and several have been standardized for use in digital communications. The most common of these are:

$$x^{16} + x^{15} + x^2 + 1 \text{ (CRC-16)}$$

$$x^{16} + x^{12} + x^5 + 1 \text{ (CRC-CCITT)}$$

The implementation of the cyclic codes is much easier if all the necessary logic is present in one integrated circuit. The Fairchild 9401 was found in early digital audio equipment because it implemented a variety of polynomials including the two above. A feature of the chip is that the feedback register can be configured to work backwards if required. The desired polynomial is selected by a three-bit control code as shown in Figure 7.14. The code is implemented by switching in a particular feedback configuration stored in ROM. During recording or transmission, the serial data are clocked in whilst the control input CWE (check word enable) is held true. At the end of the serial data, this input is made false and this has the effect of disabling the feedback so that the device becomes a conventional shift register and the CRCC is clocked out of the Q output and appended to the data. On playback, the entire message is clocked into the device with CWE once more true. At the end, if the register contains all zeros, the message was a codeword. If not, there has been an error.

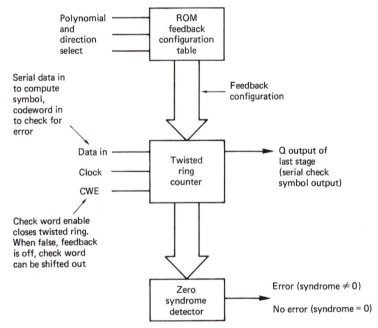

Figure 7.14 Simplified block of CRC chip which can implement several polynomials, and both generate and check redundancy.

7.11 Punctured codes

The sixteen-bit cyclic codes have codewords of length $2^{16} - 1$ or 65 535 bits long. This may be too long for the application. Another problem with very long codes is that with a given raw BER, the longer the code, the more errors will occur in it. There may be enough errors to exceed the power of the code. The solution in both cases is to shorten or *puncture* the code. Figure 7.15 shows that in a punctured code, only the end of the

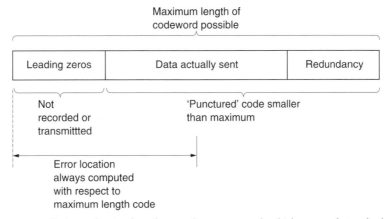

Figure 7.15 Codewords are often shortened, or punctured, which means that only the end of the codeword is actually transmitted. The only precaution to be taken when puncturing codes is that the computed position of an error will be from the beginning of the codeword, not from the beginning of the message.

codeword is used, and the data and redundancy are preceded by a string of zeros. It is not necessary to record these zeros, and, of course, errors cannot occur in them. Implementing a punctured code is easy. If a CRC generator starts with the register cleared and is fed with serial zeros, it will not change its state. Thus it is not necessary to provide the zeros, and encoding can begin with the first data bit. In the same way, the leading zeros need not be provided during playback. The only precaution needed is that if a syndrome calculates the location of an error, this will be from the beginning of the codeword not from the beginning of the data. Where codes are used for detection only, this is of no consequence.

7.12 Applications of cyclic codes

The AES/EBU digital audio interface described in Chapter 8 uses an eight-bit cyclic code to protect the channel-status data. The polynomial used and a typical circuit for generating it can be seen in Figure 7.16. The full codeword length is 255 bits but it is punctured to 192 bits, or 24 bytes

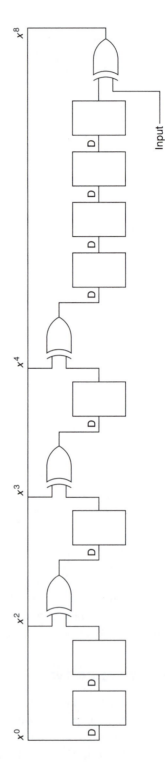

Figure 7.16 The CRCC in the AES/EBU interface is generated by premultiplying the data by x^8 and dividing by $x^8 + x^4 + x^3 + x^2 + 1$. The process can be performed on a serial input by the circuit shown. Premultiplication is achieved by connecting the input at the most significant end of the system. If the output of the right-hand XOR gate is 1 then a 1 is fed back to all of the powers shown, and the polynomial process required is performed. At the end of 23 data bytes, the CRCC will be in the eight latches. At the end of an error-free 24 byte message, the latches will be all zero.

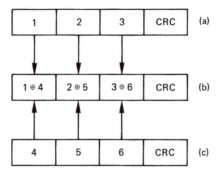

Figure 7.17 The simple crossword code of the PCM-1610/1630 format. Horizontal codewords are cyclic polynomials; vertical codewords are simple parity. Cyclic code detects errors and acts as erasure pointer for parity correction. For example, if word 2 fails, CRC (a) fails, and 1, 2 and 3 are all erased. The correct values are computed from (b) and (c) such that:

$$1 = (1 \oplus 4) \oplus 4$$
$$2 = (2 \oplus 5) \oplus 5$$
$$3 = (3 \oplus 6) \oplus 6$$

which is the length of the AES/EBU channel status block. The CRCC is placed in the last byte.

The Sony PCM-1610/1630 CD mastering recorders used a sixteen-bit cyclic code for error detection. Figure 7.17 shows that in this system, two sets of three sixteen-bit audio samples have a CRCC added to form punctured codewords 64 bits long. The PCM-1610 used the 9401 chip of Figure 7.14 to perform the calculation. Three parity words are formed by taking the XOR of the two sets of samples and a CRCC is added to this also. The three codewords are then recorded. If an error should occur, one of the cyclic codes will have a non-zero remainder, and *all* the samples in that codeword are deemed to be in error. The samples can be restored by taking the XOR of the remaining two codewords. If the error is in the parity words, no action is necessary. Further details of these recorders can be found in section 9.2. There is 100 per cent redundancy in this unit, but it is designed to work with an existing video cassette recorder whose bandwidth is predetermined and so in this application there is no penalty. The CRCC simply detects errors and acts as a pointer to a further correction means. This technique is often referred to as correction by erasure. The failing data is set to zero, or erased, since in some correction schemes the erroneous data will interfere with the calculation of the correct values.

7.13   Burst correction

Figure 7.18 lists all the possible codewords in the code of Figure 7.11. Examination will show that it is necessary to change at least three bits in one codeword before it can be made into another. Thus the code has a

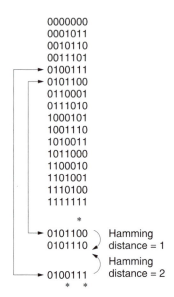

Figure 7.18 All possible codewords of $x^3 + x + 1$ are shown, and the fact that a double error in one codeword can produce the same pattern as a single error in another. Thus double errors cannot be corrected.

Hamming distance of three and cannot detect three-bit errors. The single-bit error correction limit can also be deduced from the figure. In the example given, the codeword 0101100 suffers a single-bit error marked * which converts it to a non-codeword at a Hamming distance of 1. No other codeword can be turned into this word by a single-bit error; therefore the codeword which is the shortest Hamming distance away must be the correct one. The code can thus reliably correct single-bit errors. However, the codeword 0100111 can be made into the same failure word by a two-bit error, also marked *, and in this case the original codeword cannot be found by selecting the one which is nearest in Hamming distance. A two-bit error cannot be corrected and the system will miscorrect if it is attempted.

The concept of Hamming distance can be extended to explain how more than one bit can be corrected. In Figure 7.19 the example of two bits in error is given. If a codeword four bits long suffers a single-bit error, it could produce one of four different words. If it suffers a two-bit error, it could produce one of $3 + 2 + 1$ different words as shown in the figure (the error bits are underlined). The total number of possible words of Hamming distance 1 or 2 from a four-bit codeword is thus:

$$4 + 3 + 2 + 1 = 10$$

If the two-bit error is to be correctable, no other codeword can be allowed to become one of this number of error patterns because of a two-bit error

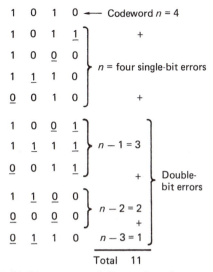

Figure 7.19 Where double-bit errors occur, the number of patterns necessary is $(n - 1) + (n - 2) + (n - 3) + \ldots$ Total necessary is $1 + n + (n - 1) + (n - 2) + (n - 3) + \ldots$ etc. Example here is of four bits, and all possible patterns up to a Hamming distance of two are shown (errors underlined).

of its own. Thus every codeword requires space for itself plus all possible error patterns of Hamming distance 2 or 1, which is eleven patterns in this example. Clearly there are only sixteen patterns available in a four-bit code, and thus no data can be conveyed if two-bit protection is necessary.

The number of different patterns possible in a word of n bits is

$$1 + n + (n-1) + (n-2) + (n-3) + \ldots$$

and this pattern range has to be shared between the ranges of each codeword without overlap. For example, an eight-bit codeword could result in $1 + 8 + 7 + 6 + 5 + 4 + 3 + 2 + 1 = 37$ patterns. As there are only 256 patterns in eight bits, it follows that only 256/37 pieces of information can be conveyed. The nearest integer below is six, and the nearest power of two below is four, which corresponds to two data bits and six check bits in the eight-bit word. The amount of redundancy necessary to correct *any* two bits in error is large, and as the number of bits to be corrected grows, the redundancy necessary becomes enormous and impractical. A further problem is that the more redundancy is added, the greater the probability of an error in a codeword. Fortunately, in practice errors occur in bursts, as has already been described, and it is a happy consequence that the number of patterns that result from the corruption of a codeword by *adjacent* two-bit errors is much smaller.

It can be deduced that the number of redundant bits necessary to correct a burst error is twice the number of bits in the burst for a perfect code. This is done by working out the number of received messages which could result from corruption of the codeword by bursts of from one bit up to the largest burst size allowed, and then making sure that there are enough redundant bits to allow that number of combinations in the received message.

Some codes, such as the Fire code due to Philip Fire,[4] are designed to correct single bursts, whereas later codes such as the B-adjacent code due to Bossen[5] could correct two bursts. The Reed–Solomon codes (Irving Reed and Gustave Solomon[6]) have the advantage that an arbitrary number of bursts can be corrected by choosing the appropriate amount of redundancy at the design stage.

7.14 Introduction to the Reed–Solomon codes

The Reed–Solomon codes are inherently burst correcting because they work on multi-bit symbols rather than individual bits. The R–S codes are also extremely flexible in use. One code may be used both to detect and correct errors and the number of bursts which are correctable can be chosen at the design stage by the amount of redundancy. A further advantage of the R–S codes is that they can be used in conjunction with a separate error-detection mechanism in which case they perform only the correction by erasure. R–S codes operate at the theoretical limit of correcting efficiency. In other words, no more efficient code can be found.

In the simple CRC system described in section 7.10, the effect of the error is detected by ensuring that the codeword can be divided by a polynomial. The CRC codeword was created by adding a redundant symbol to the data. In the Reed–Solomon codes, several errors can be isolated by ensuring that the codeword will divide by a number of polynomials. Clearly if the codeword must divide by, say, two polynomials, it must have two redundant symbols. This is the minimum case of an R–S code. On receiving an R–S coded message there will be two syndromes following the division. In the error-free case, these will both be zero. If both are not zero, there is an error.

It has been stated that the effect of an error is to add an error polynomial to the message polynomial. The number of terms in the error polynomial is the same as the number of errors in the codeword. The codeword divides to zero and the syndromes are a function of the error only. There are two syndromes and two equations. By solving these simultaneous equations it is possible to obtain two unknowns. One of these is the position of the error, known as the *locator* and the other is the error bit pattern, known as the *corrector*. As the locator is the same size as the code symbol, the length of the codeword is determined by the size of the symbol. A symbol size of eight bits is commonly used because it fits in conveniently with both sixteen-bit audio samples and byte-oriented computers. An eight-bit syndrome results in a locator of the same wordlength. Eight bits have 2^8 combinations, but one of these is the error-free condition, and so the locator can specify one of only 255 symbols. As each symbol contains eight bits, the codeword will be $255 \times 8 = 2040$ bits long.

As further examples, five-bit symbols could be used to form a codeword 31 symbols long, and three-bit symbols would form a codeword seven symbols long. This latter size is small enough to permit some worked examples, and will be used further here. Figure 7.20 shows that in the seven-symbol codeword, five symbols of three bits each, A–E, are the data, and P and Q are the two redundant symbols. This simple example will

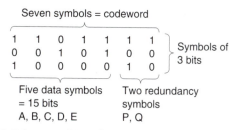

Figure 7.20 A Reed–Solomon codeword. As the symbols are of three bits, there can only be eight possible syndrome values. One of these is all zeros, the error-free case, and so it is only possible to point to seven errors; hence the codeword length of seven symbols. Two of these are redundant, leaving five data symbols.

locate and correct a single symbol in error. It does not matter, however, how many bits in the symbol are in error.

The two check symbols are solutions to the following equations:

$$A \oplus B \oplus C \oplus D \oplus E \oplus P \oplus Q = 0 \ [\oplus = \text{XOR symbol}]$$

$$a^7 A \oplus a^6 B \oplus a^5 C \oplus a^4 D \oplus a^3 E \oplus a^2 P \oplus aQ = 0$$

where a is a constant. The original data A–E followed by the redundancy P and Q pass through the channel.

The receiver makes two checks on the message to see if it is a codeword. This is done by calculating syndromes using the following expressions, where the (′) implies the received symbol which is not necessarily correct:

$$S^0 = A' \oplus B' \oplus C' \oplus D' \oplus E' \oplus P' \oplus Q'$$

(This is in fact a simple parity check)

$$S^1 = a^7 A' \oplus a^6 B' \oplus a^5 C' \oplus a^4 D' \oplus a^3 E' \oplus a^2 P' \oplus aQ'$$

If two syndromes of all zeros are not obtained, there has been an error. The information carried in the syndromes will be used to correct the error. For the purpose of illustration, let it be considered that D' has been corrupted before moving to the general case. D' can be considered to be the result of adding an error of value E to the original value D such that $D' = D \oplus E$.

As $A \oplus B \oplus C \oplus D \oplus E \oplus P \oplus Q = 0$

then $A \oplus B \oplus C \oplus (D \oplus E) \oplus E \oplus P \oplus Q = E = S_0$

As $D' = D \oplus E$

then $D = D' \oplus E = D' \oplus S_0$

Thus the value of the corrector is known immediately because it is the same as the parity syndrome S_0. The corrected data symbol is obtained simply by adding S_0 to the incorrect symbol.

At this stage, however, the corrupted symbol has not yet been identified, but this is equally straightforward:

As $a^7 A \oplus a^6 B \oplus a^5 C \oplus a^4 D \oplus a^3 E \oplus a^2 P \oplus aQ = 0$

Then:

$$a^7 A \oplus a^6 B \oplus a^5 C \oplus a^4 (D \oplus E) \oplus a^3 E \oplus a^2 P \oplus aQ = a^4 E = S_1$$

Thus the syndrome S_1 is the error bit pattern E, but it has been raised to a power of a which is a function of the position of the error symbol in the block. If the position of the error is in symbol k, then k is the locator value and:

$$S_0 \times a^k = S_1$$

Hence:

$$a^k = \frac{S_1}{S_0}$$

The value of k can be found by multiplying S_0 by various powers of a until the product is the same as S_1. Then the power of a necessary is equal to k. The use of the descending powers of a in the codeword calculation is now clear because the error is then multiplied by a different power of a dependent upon its position, known as the locator, because it gives the position of the error. The process of finding the error position by experiment is known as a Chien search.[7]

7.15   R–S Calculations

Whilst the expressions above show that the values of P and Q are such that the two syndrome expressions sum to zero, it is not yet clear how P and Q are calculated from the data. Expressions for P and Q can be found by solving the two R–S equations simultaneously. This has been done in Appendix 7.1. The following expressions must be used to calculate P and Q from the data in order to satisfy the codeword equations. These are:

$$P = a^6 \, A \oplus aB \oplus a^2 \, C \oplus a^5 \, D \oplus a^3 \, E$$

$$Q = a^2 \, A \oplus a^3 \, B \oplus a^6 \, C \oplus a^4 \, D \oplus aE$$

In both the calculation of the redundancy shown here and the calculation of the corrector and the locator it is necessary to perform numerous multiplications and raising to powers. This appears to present a formidable calculation problem at both the encoder and the decoder. This would be the case if the calculations involved were conventionally executed. However, they can be simplified by using logarithms. Instead of multiplying two numbers, their logarithms are added. In order to find the cube of a number, its logarithm is added three times. Division is performed by subtracting the logarithms. Thus all the manipulations necessary can be achieved with addition or subtraction, which is straightforward in logic circuits.

The success of this approach depends upon simple implementation of log tables. As was seen in Chapter 3, raising a constant, *a*, known as the *primitive element*, to successively higher powers in modulo 2 gives rise to a Galois field. Each element of the field represents a different power *n* of *a*. It is a fundamental of the R–S codes that all the symbols used for data, redundancy and syndromes are considered to be elements of a Galois

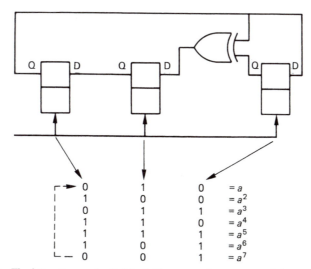

Figure 7.21 The bit patterns of a Galois field expressed as powers of the primitive element *a*. This diagram can be used as a form of log table in order to multiply binary numbers. Instead of an actual multiplication, the appropriate powers of *a* are simply added.

field. The number of bits in the symbol determines the size of the Galois field, and hence the number of symbols in the codeword.

Figure 7.21 repeats a Galois field deduced in Chapter 3. The binary values of the elements are shown alongside the power of a they represent. In the R–S codes, symbols are no longer considered simply as binary numbers, but also as equivalent powers of a. In Reed–Solomon coding and decoding, each symbol will be multiplied by some power of a. Thus if the symbol is also known as a power of a it is only necessary to add the two powers. For example, if it is necessary to multiply the data symbol 100 by a^3, the calculation proceeds as follows, referring to Figure 7.21:

$$100 = a^2 \text{ so } 100 \times a^3 = a^{(2+3)} = a^5 = 111$$

Note that the results of a Galois multiplication are quite different from binary multiplication. Because all products must be elements of the field, sums of powers which exceed seven wrap around by having seven subtracted. For example:

$$a^5 \times a^6 = a^{11} = a^4 = 110$$

Figure 7.22 shows some examples of circuits which will perform this kind of multiplication. Note that they require a minimum amount of logic.

Figure 7.23 shows an example of the Reed–Solomon encoding process. The Galois field shown in Figure 7.21 has been used, having the primitive element $a = 010$. At the beginning of the calculation of P, the symbol A is multiplied by a^6. This is done by converting A to a power of a. According

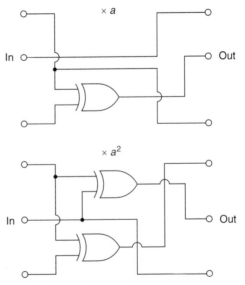

Figure 7.22 Some examples of GF multiplier circuits.

A	101	$a^6A = 111$	$a^2A = 010$
B	100	$a\ B = 011$	$a^3B = 111$
C	010	$a^2C = 011$	$a^6C = 001$
D	100	$a^5D = 001$	$a^4D = 101$
E	111	$a^3E = 010$	$a\ E = 101$
P	100 ◄	————100	————100
Q	100 ◄		

Input data { A B C D E

Check symbols { P Q

Codeword {

A	101	$a^7A = 101$
B	100	$a^6B = 010$
C	010	$a^5C = 101$
D	100	$a^4D = 101$
E	111	$a^3E = 010$
P	100	$a^2P = 110$
Q	100	$a\ Q = 011$
$S_0 = 000$		$S_1 = 000$ ◄——— Both syndromes zero

Figure 7.23 Five data symbols A–E are used as terms in the generator polynomials derived in Appendix 7.1 to calculate two redundant symbols P and Q. An example is shown at the top. Below is the result of using the codeword symbols A–Q as terms in the checking polynomials. As there is no error, both syndromes are zero.

to Figure 7.21, $101 = a^6$ and so the product will be $a^{(6+6)} = a^{12} = a^5 = 111$. In the same way, B is multiplied by a, and so on, and the products are added modulo-2. A similar process is used to calculate Q.

Figure 7.24 shows a circuit which can calculate P or Q. The symbols A–E are presented in succession, and the circuit is clocked for each one. On the first clock, a^6 A is stored in the left-hand latch. If B is now

A, B, C, D, E

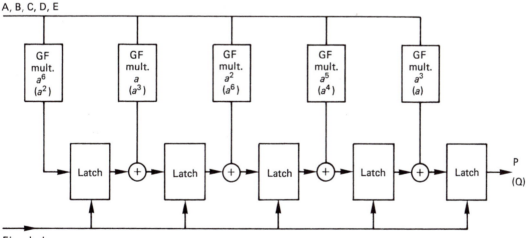

Five clock pulses

Figure 7.24 If the five data symbols of Figure 7.23 are supplied to this circuit in sequence, after five clocks, one of the check symbols will appear at the output. Terms without brackets will calculate P, bracketed terms calculate Q.

provided at the input, the second GF multiplier produces aB and this is added to the output of the first latch and when clocked will be stored in the second latch which now contains a^6A + aB. The process continues in this fashion until the complete expression for P is available in the right-hand latch. The intermediate contents of the right-hand latch are ignored.

The entire codeword now exists, and can be recorded or transmitted. Figure 7.23 also demonstrates that the codeword satisfies the checking equations. The modulo 2 sum of the seven symbols, S_0, is 000 because each column has an even number of ones. The calculation of S_1 requires multiplication by descending powers of a. The modulo-2 sum of the products is again zero. These calculations confirm that the redundancy calculation was properly carried out.

Figure 7.25 gives three examples of error correction based on this codeword. The erroneous symbol is marked with a dash. As there has been an error, the syndromes S_0 and S_1 will not be zero.

Figure 7.26 shows circuits suitable for parallel calculation of the two syndromes at the receiver. The S_0 circuit is a simple parity checker which accumulates the modulo-2 sum of all symbols fed to it. The S_1 circuit is more subtle, because it contains a Galois field (GF) multiplier in a feedback loop, such that early symbols fed in are raised to higher powers

7	A	101	a^7A = 101	$\dfrac{S_1}{S_0} = \dfrac{a^4}{1} = a^4$
6	B	100	a^6B = 010	
5	C	010	a^5C = 101	
4	D'	101	a^4D' = 011	$k = 4$
3	E	111	a^3E = 010	
2	P	100	a^2P = 110	D' + S_0 = 101 + 001
1	Q	100	a Q = 011	D = 100
	S_0 =	001	S_1 = 110	

7	A	101	a^7A = 101	$\dfrac{S_1}{S_0} = \dfrac{1}{a^2} = \dfrac{1}{a^2} \times \dfrac{a^5}{a^5} = a^5$
6	B	100	a^6B = 010	
5	C'	110	a^5C = 100	
4	D	100	a^4D = 101	$k = 5$
3	E	111	a^3E = 010	
2	P	100	a^2P = 110	C' + S_0 = 110 + 100
1	Q	100	a Q = 011	C = 010
	S_0 =	100	S_1 = 001	

7	A'	111	a^7A = 111	$\dfrac{S_1}{S_0} = \dfrac{a}{a} = 001 = a^7$
6	B	100	a^6B = 010	
5	C	010	a^5C = 101	
4	D	100	a^4D = 101	$k = 7$
3	E	111	a^3E = 010	
2	P	100	a^2P = 110	A' + S_0 = 111 + 010
1	Q	100	a Q = 011	A = 101
	S_0 =	010	S_1 = 010	

Figure 7.25 Three examples of error location and correction. The number of bits in error in a symbol is irrelevant; if all three were wrong, S_0 would be 111, but correction is still possible.

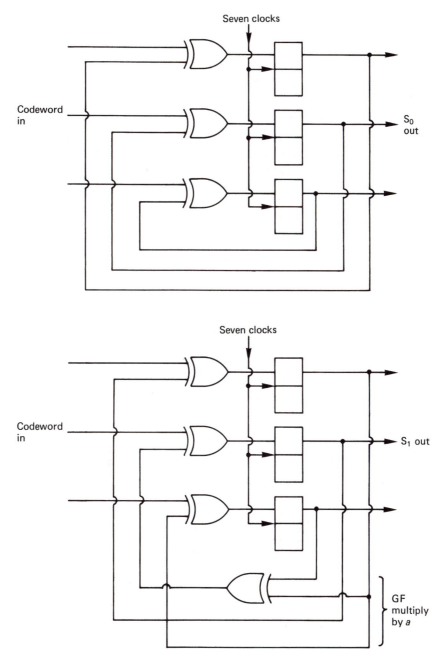

Figure 7.26 Circuits for parallel calculation of syndromes S_0, S_1. S_0 is a simple parity check. S_1 has a GF multiplication by a in the feedback, so that A is multiplied by a^7, B is multiplied by a^6, etc., and all are summed to give S_1.

than later symbols because they have been recirculated through the GF multiplier more often. It is possible to compare the operation of these circuits with the example of Figure 7.25 and with subsequent examples to confirm that the same results are obtained.

7.16   Correction by erasure

In the examples of Figure 7.25, two redundant symbols P and Q have been used to locate and correct one error symbol. If the positions of errors are known by some separate mechanism (see product codes, section 7.18) the locator need not be calculated. The simultaneous equations may instead be solved for two correctors. In this case the number of symbols which can be corrected is equal to the number of redundant symbols. In Figure 7.27(a) two errors have taken place, and it is known that they are in symbols C and D. Since S_0 is a simple parity check, it will reflect the modulo-2 sum of the two errors. Hence $S_0 = E_C \oplus E_D$.

The two errors will have been multiplied by different powers in S_1, such that:

$$S_1 = a^5 E_C \oplus a^4 E_D$$

These two equations can be solved, as shown in the figure, to find E_C and E_D, and the correct value of the symbols will be obtained by adding these correctors to the erroneous values. It is, however, easier to set the values of the symbols in error to zero. In this way the nature of the error is rendered irrelevant and it does not enter the calculation. This setting of symbols to zero gives rise to the term erasure. In this case,

$$S_0 = C \oplus D$$

$$S_1 = a^5 C \oplus a^4 D$$

Erasing the symbols in error makes the errors equal to the correct symbol values and these are found more simply as shown in Figure 7.27(b).

Practical systems will be designed to correct more symbols in error than in the simple examples given here. If it is proposed to correct by erasure an arbitrary number of symbols in error given by t, the codeword must be divisible by t different polynomials. Alternatively if the errors must be located and corrected, $2t$ polynomials will be needed. These will be of the form $(x + a^n)$ where n takes all values up to t or $2t$. a is the primitive element discussed in Chapter 3.

Where four symbols are to be corrected by erasure, or two symbols are to be located and corrected, four redundant symbols are necessary, and the codeword polynomial must then be divisible by

$$(x + a^0)(x + a^1)(x + a^2)(x + a^3)$$

$$
\begin{array}{llll}
A & 1\,01 & a^7 A = & 101 \\
B & 1\,00 & a^6 B = & 010 \\
(C \oplus E_C) & 0\,01 & a^5\,(C \oplus E_C) & 111 \\
(D \oplus E_D) & 0\,10 & a^4\,(D \oplus E_D) & 111 \\
E & 1\,11 & a^3 E = & 010 \\
P & 1\,00 & a^2 P = & 110 \\
Q & 1\,00 & a\,Q = & \underline{011} \\
S_1 \;\; = & \overline{101} & S_1 \;\; = & \overline{000}
\end{array}
$$

$$
S_0 \;\; = E_C \oplus E_D \qquad S_1 \;\; = a^5 E_C \oplus a^4 E_D
$$

$$
S_1 \;\; = a^5 E_C \oplus a^4\,(S_0 \oplus E_C)
$$

$$
= a^5 E_C \oplus a^4 S_0 \oplus a^4 E_C
$$

$$
\therefore E_C = \frac{S_1 \oplus a^4 S_0}{a^5 \oplus a^4} = \frac{000 \oplus 011}{001} = 011
$$

$$
C \;\; = (C \oplus E_C) \oplus E_C = 001 \oplus 011 = \underline{010}
$$

$$
S_1 \;\; = a^5\,(S_0 \ominus E_D) \oplus a^4 E_D
$$

$$
= a^5 S_0 \oplus a^5 E_D \oplus a^4 E_D
$$

$$
\therefore E_D = \frac{S_1 \oplus a^5 S_0}{a^5 \oplus a^4} = \frac{000 \oplus 110}{001} = 110
$$

$$
D \;\; = (D \oplus E_D) + E_D = 010 \oplus 110 = \underline{100} \qquad \textbf{(a)}
$$

$$
\begin{array}{lll}
A & 101 & a^7 A = 101 \\
B & 100 & a^6 B = 010 \qquad S_0 = C \oplus D \\
C & \underline{000} & a^5 C = \overline{000} \\
D & \underline{000} & a^4 D = \overline{000} \qquad S_1 = a^5 C \oplus a^4 D \\
E & 111 & a^3 E = 010 \\
P & 100 & a^2 P = 110 \\
Q & 100 & a\,Q = 011 \\
S_0 & = 100 & S_1 \;\; = 000
\end{array}
$$

$$
S_1 = a^5 S_0 \oplus a^5 D \oplus a^4 D = a^5 S_0 \oplus D
$$

$$
\therefore D = S_1 \oplus a^5 S_0 = 000 \oplus 100 = \underline{100}
$$

$$
S_1 = a^5 C \oplus a^4 C \oplus a^4 S_0 = C \oplus a^4 S_0
$$

$$
\therefore C = S_1 \oplus a^4 S_0 = 000 \oplus 010 = \underline{010}
$$

(b)

Figure 7.27 If the location of errors is known, then the syndromes are a known function of the two errors as shown in (a). It is, however, much simpler to set the incorrect symbols to zero, i.e. to *erase* them as in (b). Then the syndromes are a function of the wanted symbols and correction is easier.

Upon receipt of the message, four syndromes must be calculated, and the four correctors or the two error patterns and their positions are determined by solving four simultaneous equations. This generally requires an iterative procedure, and a number of algorithms have been developed for the purpose.[8-10] Modern digital audio formats such as CD and DAT use eight-bit R–S codes and erasure extensively. The primitive polynomial commonly used with GF(256) is

$$x^8 + x^4 + x^3 + x^2 + 1$$

The codeword will be 255 bytes long but will often be shortened by puncturing. The larger Galois fields require less redundancy, but the computational problem increases. LSI chips have been developed specifically for R–S decoding in many high-volume formats.[11,12] As an alternative to dedicated circuitry, it is also possible to perform Reed–Solomon calculations in software using general-purpose processors.[13] This may be more economical in small-volume products.

7.17 Interleaving

The concept of bit interleaving was introduced in connection with a single-bit correcting code to allow it to correct small bursts. With burst-correcting codes such as Reed–Solomon, bit interleave is unnecessary. In most channels, particularly high-density recording channels used for digital audio, the burst size may be many bytes rather than bits, and to rely on a code alone to correct such errors would require a lot of redundancy. The solution in this case is to employ symbol interleaving, as shown in Figure 7.28. Several codewords are encoded from input data, but these are not recorded in the order they were input, but are physically reordered in the channel, so that a real burst error is split into smaller bursts in several codewords. The size of the burst seen by each codeword is now determined primarily by the parameters of the interleave, and Figure 7.29 shows that the probability of occurrence of bursts with respect to the burst length in a given codeword is modified. The number of bits in the interleave word can be made equal to the burst-correcting ability of the code in the knowledge that it will be exceeded only very infrequently.

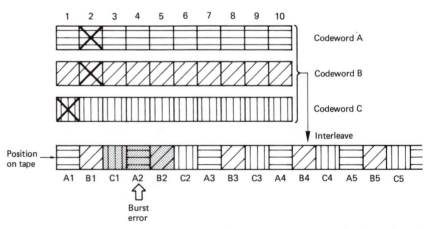

Figure 7.28 The interleave controls the size of burst errors in individual codewords.

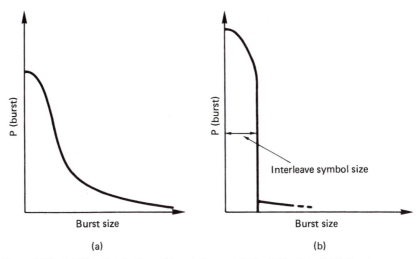

Figure 7.29 (a) The distribution of burst sizes might look like this. (b) Following interleave, the burst size within a codeword is controlled to that of the interleave symbol size, except for gross errors which have low probability.

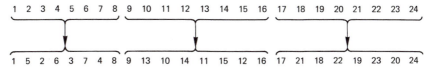

Figure 7.30 In block interleaving, data are scrambled within blocks which are themselves in the correct order.

There are a number of different ways in which interleaving can be performed. Figure 7.30 shows that in block interleaving, words are reordered within blocks which are themselves in the correct order. This approach is attractive for rotary-head recorders, such as DAT, because the scanning process naturally divides the tape up into blocks. The block interleave is achieved by writing samples into a memory in sequential address locations from a counter, and reading the memory with non-sequential addresses from a sequencer. The effect is to convert a one-dimensional sequence of samples into a two-dimensional structure having rows and columns.

Rotary-head recorders naturally interleave spatially on the tape. Figure 7.31 shows that a single large tape defect becomes a series of small defects owing to the geometry of helical scanning.

The alternative to block interleaving is convolutional interleaving where the interleave process is endless. In Figure 7.32 symbols are assembled into short blocks and then delayed by an amount proportional to the position in the block. It will be seen from the figure that the delays have the effect of shearing the symbols so that columns on the left side of

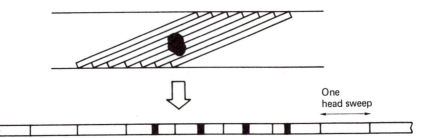

Figure 7.31 Helical-scan recorders produce a form of mechanical interleaving, because one large defect on the medium becomes distributed over several head sweeps.

the diagram become diagonals on the right. When the columns on the right are read, the convolutional interleave will be obtained. Convolutional interleave works well with stationary head recorders where there is no natural track break and with CD where the track is a continuous spiral. Convolutional interleave has the advantage of requiring less memory to implement than a block code. This is because a block code requires the entire block to be written into the memory before it can be read, whereas a convolutional code requires only enough memory to cause the required delays. Now that RAM is relatively inexpensive, convolutional interleave is less popular.

It is possible to make a convolutional code of finite size by making a loop. Figure 7.33(a) shows that symbols are written in columns on the outside of a cylinder. The cylinder is then sheared or twisted, and the columns are read. The result is a block-completed convolutional interleave shown at (b). This technique is used in the digital audio blocks of the Video-8 format.

7.18 Product codes

In the presence of burst errors alone, the system of interleaving works very well, but it is known that in most practical channels there are also uncorrelated errors of a few bits due to noise. Figure 7.34 shows an interleaving system where a dropout-induced burst error has occurred which is at the maximum correctable size. All three codewords involved are working at their limit of one symbol. A random error due to noise in the vicinity of a burst error will cause the correction power of the code to be exceeded. Thus a random error of a single bit causes a further entire symbol to fail. This is a weakness of an interleave solely designed to handle dropout-induced bursts. Practical high-density equipment must address the problem of noise-induced or random errors and burst errors occurring at the same time. This is done by forming codewords both before and after the interleave process. In block interleaving, this results

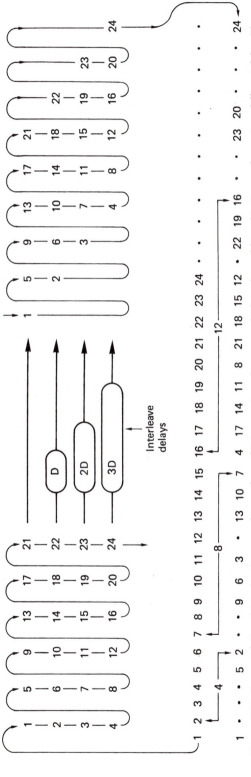

Figure 7.32 In convolutional interleaving, samples are formed into a rectangular array, which is sheared by subjecting each row to a different delay. The sheared array is read in vertical columns to provide the interleaved output. In this example, samples will be found at 4, 8 and 12 places away from their original order.

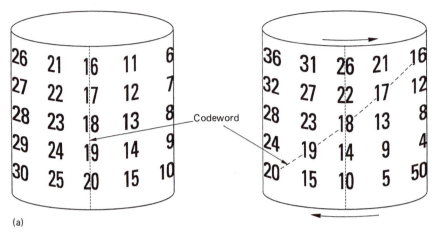

(a)

Figure 7.33 (a) A block-completed convolutional interleave can be considered to be the result of shearing a cylinder.

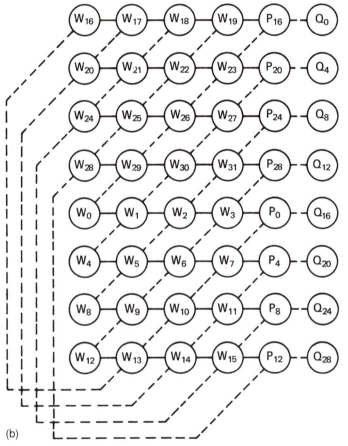

(b)

Figure 7.33 (b) A block completed convolutional interleave results in horizontal and diagonal codewords as shown here.

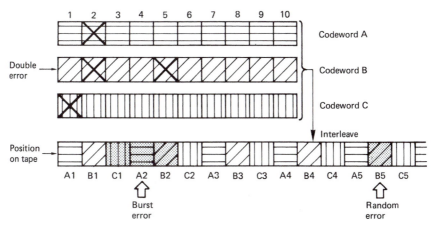

Figure 7.34 The interleave system falls down when a random error occurs adjacent to a burst.

in a *product code*, whereas in the case of convolutional interleave the result is called *cross-interleaving*.[14]

Figure 7.35 shows that in a product code the redundancy calculated first and checked last is called the outer code, and the redundancy calculated second and checked first is called the inner code. The inner code is formed along tracks on the medium. Random errors due to noise are corrected by the inner code and do not impair the burst-correcting power of the outer code. Burst errors are declared uncorrectable by the inner code which flags the bad samples on the way into the de-interleave memory. The outer code reads the error flags in order to correct the flagged symbols by erasure. The error flags are also known as erasure flags. As it does not have to compute the error locations, the outer code needs half as much redundancy for the same correction power. Thus the inner code redundancy does not raise the code overhead. The combination of codewords with interleaving in several dimensions yields an error-protection strategy which is truly synergistic, in that the end result is more powerful than the sum of the parts. Needless to say, the technique is used extensively in modern formats such DAT and DCC. The error-correction strategy of DAT is treated in the next section as a representative example of a modern product code.

An alternative to the product block code is the convolutional cross-interleave, shown in Figure 7.32. In this system, the data are formed into an endless array and the codewords are produced on columns and diagonals. The Compact Disc and DASH formats use such a system. The original advantage of the cross-interleave is that it needed less memory than a product code. This advantage is no longer so significant now that memory prices have fallen so much. It has the disadvantage that editing is more complicated. The error-correction system of CD is discussed in detail in Chapter 12.

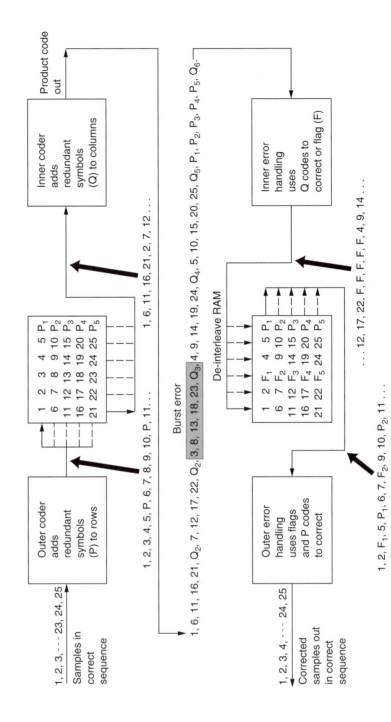

Figure 7.35 In addition to the redundancy P on rows, inner redundancy Q is also generated on columns. On replay, the Q code checker will pass on flags F if it finds an error too large to handle itself. The flags pass through the de-interleave process and are used by the outer error correction to identify which symbol in the row needs correcting with P redundancy. The concept of crossing two codes in this way is called a product code.

7.19   Introduction to error correction in DAT

The interleave and error-correction systems of DAT will now be discussed. Figure 7.36 is a conceptual block diagram of the system which shows that DAT uses a product code formed by producing Reed–Solomon codewords at right angles across an array. The array is formed in a memory, and the layout used in the case of 48 kHz sampling can be seen in Figure 7.37.

There are two recorded tracks for each drum revolution and incoming samples for that period of time are routed to a pair of memory areas of 4 bytes capacity, one for each track. These memories are structured as 128 columns of 32 bytes each. The error correction works with eight-bit symbols, and so each sample is divided into high byte and low byte and occupies two locations in memory. Figure 7.37 shows only one of the two memories. Incoming samples are written across the memory in rows, with the exception of an area in the centre, 24 bytes wide. Each row of data in the RAM is used as the input to the Reed–Solomon encoder for the outer code. The encoder starts at the left-hand column, and then takes a byte from every fourth column, finishing at column 124 with a total of 26 bytes. Six bytes of redundancy are calculated to make a 32 byte outer codeword. The redundant bytes are placed at the top of columns 52, 56, 60, etc. The encoder then makes a second pass through the memory, starting in the second column and taking a byte from every fourth column finishing at column 125. A further six bytes of redundancy are calculated and put into the top of columns 53, 57, 61, and so on. This process is performed four times for each row in the memory, except for the last eight rows where only two passes are necessary because odd-numbered columns have sample bytes only down to row 23. The total number of outer codewords produced is 112.

In order to encode the inner codewords to be recorded, the memory is read in columns. Figure 7.38 shows that, starting at top left, bytes from the sixteen even-numbered rows of the first column, and from the first twelve even-numbered rows of the second column, are assembled and fed to the inner encoder. This produces four bytes of redundancy which are written into the memory in the areas marked P1. Four bytes P1, when added to the 28 bytes of data, makes an inner codeword 32 bytes long. The second inner code is encoded by making a second pass through the first two columns of the memory to read the samples on odd-numbered rows. Four bytes of redundancy are placed in memory in locations marked P2. Each column of memory is then read completely and becomes one sync block on tape. Two sync blocks contain two interleaved inner codes such that the inner redundancy for both is at the end of the second sync block. The effect is that adjacent symbols

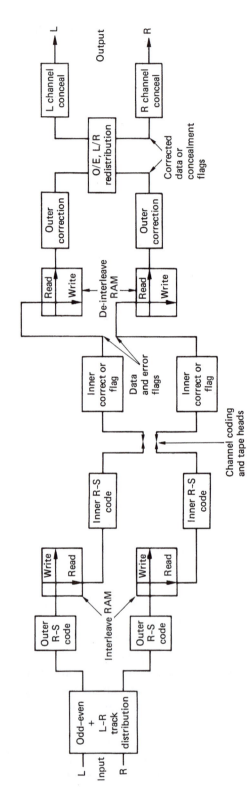

Figure 7.36 The error-protection strategy of DAT. To allow concealment on replay, an odd/even, left/right track distribution is used. Outer codes are generated on RAM rows, inner codes on columns. On replay, inner codes correct random errors. Flags pass through de-interleave RAM to outer codes which use them as erasure pointers. Uncorrected errors can be concealed after redistribution to real-time sequence.

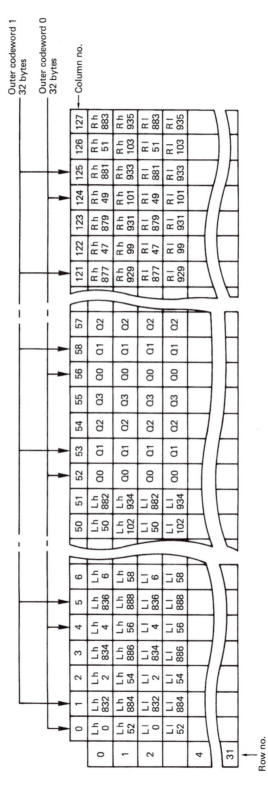

Figure 7.37 Left even/right odd interleave memory. Incoming samples are split into high byte (h) and low byte (l), and written across the memory rows using first the even columns for L 0–830 and R 1–831, and then the odd columns for L 832–1438 and R 833–1439. For 44.1 kHz working, the number of samples is reduced from 1440 to 1323, and fewer locations are filled.

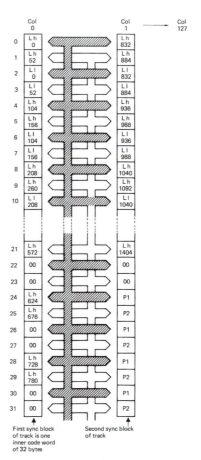

Figure 7.38 The columns of memory are read out to form inner codewords. First, even bytes from the first two columns make one codeword and then odd bytes from the first two columns. As there are 128 columns, there will be 128 sync blocks in one audio segment.

in a sync block are not in the same codeword. The process then repeats down the next two columns in the memory and so on until 128 blocks have been written to the tape.

Upon replay, the sync blocks will suffer from a combination of random errors and burst errors. The effect of interleaving is that the burst errors will be converted to many single-symbol errors in different outer codewords.

As there are four bytes of redundancy in each inner codeword, a theoretical maximum of two bytes can be corrected. The probability of miscorrection in the inner code is minute for a single-byte error, because all four syndromes will agree on the nature of the error, but the probability of miscorrection on a double-byte error is much higher. The inner code logic is exposed to random noise during dropout and mistracking conditions, and the probability of noise producing what appears to be only a two-symbol error is too great. If more than one byte is in error in an inner code it is more reliable to declare all bytes bad by attaching flags to them as they enter the de-interleave memory. The interleave of the inner codes over two sync blocks is necessary because of the use of a group code. In the 8/10 code described in Chapter 6, a single mispositioned transition will change one ten-bit group into another, potentially corrupting up to eight data bits. A small disturbance at the boundary between two groups could corrupt up to sixteen bits. By interleaving the inner codes at symbol level, the worst case of a disturbance at the boundary of two groups is to produce a single-symbol error in two different inner codes. Without the inner code interleave, the entire contents of an inner code could be caused to be flagged bad by a single small defect. The inner code interleave halves the error propagation of the group code, which increases the chances of random errors being corrected by the inner codes instead of impairing the burst-error correction of the outer codes.

After de-interleave, any uncorrectable inner codewords will show up as single-byte errors in many different outer codewords accompanied by error flags. To guard against miscorrections in the inner code, the outer code will calculate syndromes even if no error flags are received from the inner code. If two bytes or less in error are detected, the outer code will correct them even though they were due to inner code miscorrections. This can be done with high reliability because the outer code has three-byte detecting and correcting power which is never used to the full. If more than two bytes are in error in the outer codeword, the correction process uses the error flags from the inner code to correct up to six bytes in error.

The reasons behind the complex interleaving process now become clearer. Because of the four-way interleave of the outer code, four entire sync blocks can be destroyed, but only one byte will be corrupted in a given outer codeword. As an outer codeword can correct up to six bytes in error by erasure, it follows that a burst error of up to 24 sync blocks could be corrected. This corresponds to a length of track of just over 2.5 mm, and is more than enough to cover the tenting effect due to a particle of debris lifting the tape away from the head. In practice the interleave process is a little more complicated than this description would suggest, owing to the requirement to produce recognizable sound in shuttle. This process will be detailed in Chapter 9.

7.20   Editing interleaved recordings

The interleave, de-interleave, time-compression and timebase-correction processes cause delay and this is evident in the time taken before audio emerges after starting a digital machine. Confidence replay takes place later than the distance between record and replay heads would indicate. In DASH format recorders, confidence replay is about one tenth of a second behind the input. Processes such as editing and synchronous recording require new techniques to overcome the effect of the delays.

In analog recording, there is a direct relationship between the distance down the track and the time through the recording and it is possible to mark and cut the tape at a particular time. A further consequence of interleaving in digital recorders is that the reordering of samples means that this relationship is lost.

Editing must be undertaken with care. In a block-based interleave, edits can be made at block boundaries so that coded blocks are not damaged, but these blocks are usually too large for accurate audio editing. In a convolutional interleave, there are no blocks and an edit or splice will damage diagonal codewords over a constraint length near the edit as shown in Figure 7.39.

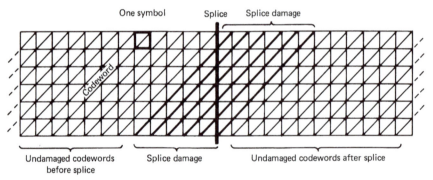

Figure 7.39 Although interleave is a powerful weapon against burst errors, it causes greater data loss when tape is spliced because many codewords are replayed in two unrelated halves.

The only way in which audio can be edited satisfactorily in the presence of interleave is to use a read–modify–write approach, where an entire frame is read into memory and de-interleaved to the real-time sample sequence. Any desired part of the frame can be replaced with new material before it is re-interleaved and re-recorded. In recorders which can only record or play at one time, an edit of this kind would take a long time because of all the tape repositioning needed. With extra heads read–modify–write editing can be performed dynamically. The sequence is shown in Figure 7.40 for a rotary-head machine but is equally applicable to stationary head transports. The replay head plays back the existing recording, and this is de-interleaved to the normal sample sequence, a process which introduces a delay. The sample

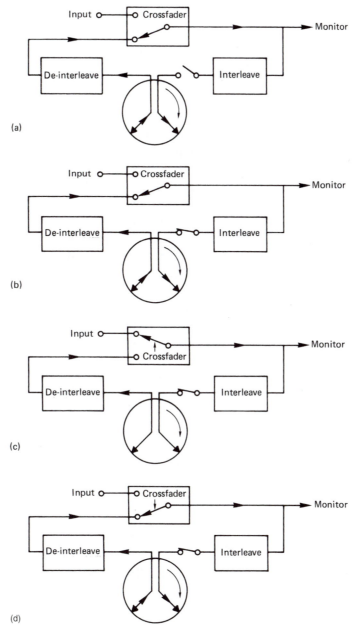

Figure 7.40 In the most sophisticated version of audio editing, there are advanced replay heads on the scanner, which allow editing to be performed on de-interleaved data. An insert sequence is shown. In (a) the replay-head signal is decoded and fed to the encoder which, after some time, will produce an output representing what is already on the tape. In (b), at a sector boundary, the write circuits are turned on, and the machine begins to re-record. In (c) the crossfade is made to the insert material. In (d) the insert ends with a crossfade back to the signal from the advanced replay heads. After this, the write heads will once again be recording what is already on the tape, and the write circuits can be disabled at a sector boundary. An assemble edit consists of the first three of these steps only.

stream now passes through a crossfader which at this stage will be set to accept only the offtape signal. The output of the crossfader is then fed to the record interleave stage which introduces further delay. This signal passes to the record heads which must be positioned so that the original recording on the tape reaches them at the same time that the re-encoded signal arrives, despite the encode and decode delays. In a rotary-head recorder this can be done by positioning the record heads at a different height to the replay heads so that they reach the same tracks on different revolutions. With this arrangement it is possible to enable the record heads at the beginning of a frame, and they will then re-record what is already on the tape. Next, the crossfader can be operated to fade across to new material, at any desired crossfade speed. Following the interleave stage, the new recording will update only the new samples in the frame and re-record those which do not need changing. After a short time, the recording will only be a function of the new input. If the edit is an insert, it is possible to end the process by crossfading back to the replay signal and allowing the replay data to be re-recorded. Once this re-recording has taken place for a short time, the record process can be terminated at the end of a frame. There is no limit to the crossfade periods which can be employed in this operating technique, in fact the crossfade can be manually operated so that it can be halted at a suitable point to allow, for example, a commentary to be superimposed upon a recording.

One important point to appreciate about read–modify–write editing is that the physical frames at which the insert begins and ends are independent of the in- and out-points of the audio edit, because the former are in areas where re-recording of the existing data takes place. Electronic editing and tape-cut editing of digital recordings is discussed in Chapter 11.

Appendix 7.1 Calculation of Reed–Solomon generator polynomials

For a Reed–Solomon codeword over $GF(2^3)$, there will be seven three-bit symbols. For location and correction of one symbol, there must be two redundant symbols P and Q, leaving A–E for data.

The following expressions must be true, where a is the primitive element of $x^3 \oplus x \oplus 1$ and $\oplus$ is XOR throughout:

$$A \oplus B \oplus C \oplus D \oplus E \oplus P \oplus Q = 0 \qquad (1)$$

$$a^7A \oplus a^6B \oplus a^5C \oplus a^4D \oplus a^3E \oplus a^2P \oplus aQ = 0 \qquad (2)$$

Dividing equation (2) by a:

$$a^6\text{A} \oplus a^5\text{B} \oplus a^4\text{C} \oplus a^3\text{D} \oplus a^2\text{E} \oplus a\text{P} \oplus \text{Q} = 0$$

$$= \text{A} \oplus \text{B} \oplus \text{C} \oplus \text{D} \oplus \text{E} \oplus \text{P} \oplus \text{Q}$$

Cancelling Q, and collecting terms:

$$(a^6 \oplus 1)\text{A} \oplus (a^5 \oplus 1)\text{B} \oplus (a^4 \oplus 1)\text{C} \oplus (a^3 \oplus 1)\text{D} \oplus (a^2 \oplus 1)\text{E}$$

$$= (a + 1)\text{P}$$

Using Figure 7.21 to calculate $(a^n + 1)$, e.g. $a^6 + 1 = 101 + 001 = 100 = a^2$:

$$a^2\text{A} \oplus a^4\text{B} \oplus a^5\text{C} \oplus a\text{D} \oplus a^6\text{E} = a^3\text{P}$$

$$a^6\text{A} \oplus a\text{B} \oplus a^2\text{C} \oplus a^5\text{D} \oplus a^3\text{E} = \text{P}$$

Multiply equation (1) by a^2 and equating to equation (2):

$$a^2\text{A} \oplus a^2\text{B} \oplus a^2\text{C} \oplus a^2\text{D} \oplus a^2\text{E} \oplus a^2\text{P} \oplus a^2\text{Q} = 0$$

$$= a^7\text{A} \oplus a^6\text{B} \oplus a^5\text{C} \oplus a^4\text{D} \oplus a^3\text{E} \oplus a^2\text{P} \oplus a\text{Q}$$

Cancelling terms $a^2\text{P}$ and collecting terms (remember $a^2 \oplus a^2 = 0$):

$$(a^7 \oplus a^2)\text{A} \oplus (a^6 \oplus a^2)\text{B} \oplus (a^5 \oplus a^2)\text{C} \oplus (a^4 \oplus a^2)\text{D} \oplus$$

$$(a^3 \oplus a^2)\text{E} = (a^2 \oplus a)\text{Q}$$

Adding powers according to Figure 7.21, e.g.

$$a^7 \oplus a^2 = 001 \oplus 100 = 101 = a^6:$$

$$a^6\text{A} \oplus \text{B} \oplus a^3\,\text{C} \oplus a\text{D} \oplus a^5\text{E} = a^4\text{Q}$$

$$a^2\text{A} \oplus a^3\text{B} \oplus a^6\text{C} \oplus a^4\text{D} \oplus a\text{E} = \text{Q}$$

References

1. Michaels, S.R., Is it Gaussian? *Electronics World and Wireless World*, 72–73 (January 1993)
2. Shannon, C.E., A mathematical theory of communication. *Bell System Tech. J.*, **27**, 379 (1948)
3. Hamming, R.W., Error-detecting and error-correcting codes. *Bell System Tech. J.*, **26**, 147–160 (1950)

4. Fire, P., A class of multiple-error correcting codes for non-independent errors. *Sylvania Reconnaissance Systems Lab. Report*, RSL-E-2 (1959)

5. Bossen, D.C., B-adjacent error correction. *IBM J. Res. Dev.*, **14**, 402–408 (1970)

6. Reed, I.S. and Solomon, G., Polynomial codes over certain finite fields. *J. Soc. Indust. Appl. Math.*, **8**, 300–304 (1960)

7. Chien, R.T., Cunningham, B.D. and Oldham, I.B., Hybrid methods for finding roots of a polynomial – with application to BCH decoding. *IEEE Trans. Inf. Theory.*, **IT-15**, 329–334 (1969)

8. Berlekamp, E.R., *Algebraic Coding Theory*, New York: McGraw-Hill (1967). Reprint edition: Laguna Hills, CA: Aegean Park Press (1983)

9. Sugiyama, Y. *et al.*, An erasures and errors decoding algorithm for Goppa codes. *IEEE Trans. Inf. Theory*, **IT-22** (1976)

10. Peterson, W.W. and Weldon, E.J., *Error Correcting Codes*, 2nd edn., Cambridge, MA: MIT Press (1972)

11. Onishi, K., Sugiyama, K., Ishida, Y., Kusonoki, Y. and Yamaguchi, T., An LSI for Reed–Solomon encoder/decoder. Presented at the 80th Audio Engineering Society Convention (Montreux, 1986), Preprint 2316(A-4)

12. Anon. *Digital Audio Tape Deck Operation Manual*, Sony Corporation (1987)

13. van Kommer, R., Reed–Solomon coding and decoding by digital signal processors. Presented at the 84th Audio Engineering Society Convention (Paris, 1988), Preprint 2587(D-7)

14. Doi, T.T., Odaka, K., Fukuda, G. and Furukawa, S. Crossinterleave code for error correction of digital audio systems. *J. Audio Eng. Soc.*, **27**, 1028 (1979)

8

Transmission

8.1 Introduction

The distances involved in transmission vary from that of a short cable between adjacent units to communication anywhere on earth via data networks or radio communication. This chapter must consider a correspondingly wide range of possibilities. The importance of direct digital interconnection between audio devices was realized early, and numerous incompatible (and now obsolete) methods were developed by various manufacturers until standardization was reached in the shape of the AES/EBU digital audio interface for professional equipment and the SPDIF interface for consumer equipment. These standards were extended to produce the MADI standard for multi-channel interconnects. All of these work on uncompressed PCM audio.

As digital audio and computers continue to converge, computer networks are also being used for audio purposes. Audio may be transmitted on networks such as Ethernet, ISDN, ATM and Internet. Here compression may or may not be used, and non-real-time transmission may also be found according to economic pressures.

Digital audio is now being broadcast in its own right as DAB, alongside traditional analog television as NICAM digital audio and as MPEG or AC-3 coded signals in digital television broadcasts. Many of the systems described here rely upon coding principles described in Chapters 6 and 7.

Whatever the transmission medium, one universal requirement is a reliable synchronization system. In PCM systems, synchronization of the sampling rate between sources is necessary for mixing. In packet-based networks, synchronization allows the original sampling rate to be established at the receiver despite the intermittent transfer of a real packet systems. In digital television systems, synchronization between vision and sound is a further requirement.

8.2 Introduction to AES/EBU interface

The AES/EBU digital audio interface, originally published in 1985,[1] was proposed to embrace all the functions of existing formats in one standard. The goal was to ensure interconnection of professional digital audio equipment irrespective of origin. The EBU ratified the AES proposal with the proviso that the optional transformer coupling was made mandatory and led to the term AES/EBU interface, also called EBU/AES in some European countries. The contribution of the BBC to the development of the interface must be mentioned here. Alongside the professional format, Sony and Philips developed a similar format now known as SPDIF (Sony Philips Digital Interface) intended for consumer use. This offers different facilities to suit the application, yet retains sufficient compatibility with the professional interface so that, for many purposes, consumer and professional machines can be connected together.[2,3]

The AES concerns itself with professional audio and accordingly has had little to do with the consumer interface. Thus the recommendations to standards bodies such as the IEC (International Electrotechnical Commission) regarding the professional interface came primarily through the AES whereas the consumer interface input was primarily from industry, although based on AES professional proposals. The IEC and various national standards bodies naturally tended to combine the two into one standard such as IEC 958[4] which refers to the professional interface and the consumer interface. This process has been charted by Finger.[5]

Understandably with so many standards relating to the same subject differences in interpretation arise leading to confusion in what should or should not be implemented, and indeed what the interface should be called. This chapter will refer generically to the professional interface as the AES/EBU interface and the consumer interface as SPDIF.

Getting the best results out of the AES/EBU interface, or indeed any digital interface, requires some care. Section 13.9 treats this subject in some detail

8.3 The electrical interface

During the standardization process it was considered desirable to be able to use existing analog audio cabling for digital transmission. Existing professional analog signals use nominally 600 Ω impedance balanced line screened signalling, with one cable per audio channel, or in some cases one twisted pair per channel with a common screen. The 600 Ω standard came from telephony where long distances are involved in comparison with electrical audio wavelengths. The distances likely to be found within a studio complex are short compared to audio electrical wavelengths and as a result at audio frequency the impedance of cable is high and the 600 ohm figure is that of the source and termination. Such a cable has a different impedance at the frequencies used for digital audio, around 110 Ω.

If a single serial channel is to be used, the interconnect has to be self-clocking and self-synchronizing, i.e. the single signal must carry enough information to allow the boundaries between individual bits, words and blocks to be detected reliably. To fulfil these requirements, the AES/EBU and SPDIF interfaces use FM channel code (see Chapter 6) which is DC-free, strongly self-clocking and capable of working with a changing sampling rate. Synchronization of deserialization is achieved by violating the usual encoding rules.

The use of FM means that the channel frequency is the same as the bit rate when sending data ones. Tests showed that in typical analog audio cabling installations, sufficient bandwidth was available to convey two digital audio channels in one twisted pair. The standard driver and receiver chips for RS-422A[6] data communication (or the equivalent CCITT-V.11) are employed for professional use, but work by the BBC[7] suggested that equalization and transformer coupling were desirable for longer cable runs, particularly if several twisted pairs occupy a common shield. Successful transmission up to 350 m has been achieved with these techniques.[8] Figure 8.1 shows the standard configuration. The output impedance of the drivers will be about 110 ohms, and the impedance of the cable used should be similar at the frequencies of interest. The driver was specified in AES-3–1985 to produce between 3 and 10 V peak-to-peak into such an impedance but this was changed to between 2 and 7 volts in AES-3–1992 to better reflect the characteristics of actual RS-422 driver chips.

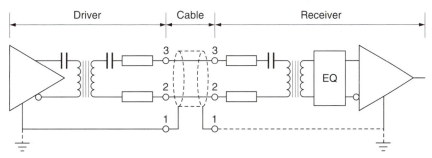

Figure 8.1 Recommended electrical circuit for use with the standard two-channel interface.

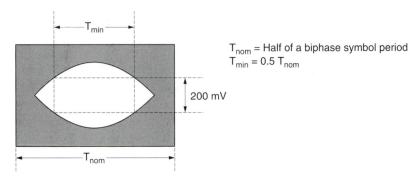

T_{nom} = Half of a biphase symbol period
T_{min} = 0.5 T_{nom}

200 mV

Figure 8.2 The minimum eye pattern acceptable for correct decoding of standard two-channel data.

The original receiver impedance was set at a high $250\,\Omega$, with the intention that up to four receivers could be driven from one source. This was found to be inadvisable because of reflections caused by impedance mismatches and AES-3–1992 is now a point-to-point interface with source, cable and load impedance all set at $110\,\Omega$. Whilst analog audio cabling was adequate for digital signalling, cable manufacturers have subsequently developed cables which are more appropriate for new digital installations, having lower loss factors allowing greater transmission distances.

In Figure 8.2, the specification of the receiver is shown in terms of the minimum eye pattern (see Chapter 6) which can be detected without error. It will be noted that the voltage of 200 mV specifies the height of the eye opening at a width of half a channel bit period. The actual signal amplitude will need to be larger than this, and even larger if the signal contains noise. Figure 8.3 shows the recommended equalization characteristic which can be applied to signals received over long lines.

As an adequate connector in the shape of the XLR is already in wide service, the connector made to IEC 268 Part 12 has been adopted for digital audio use. Effectively, existing analog audio cables having XLR connectors can be used without alteration for digital connections. The AES/EBU standard does, however, require that suitable labelling should be used so that it is clear that the connections on a particular unit are digital. Whilst the XLR connector was never designed to have constant impedance in the megaHertz range, it is capable of towing an outside broadcast vehicle without unlatching.

The need to drive long cables does not generally arise in the domestic environment, and so a low-impedance balanced signal was not considered necessary. The electrical interface of the consumer format uses a 0.5 V peak single-ended signal, which can be conveyed down conventional audio-grade coaxial cable connected with RCA 'phono' plugs. Figure 8.4 shows the resulting consumer interface as specified by IEC 958.

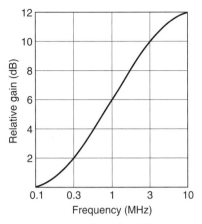

Figure 8.3 EQ characteristic recommended by the AES to improve reception in the case of long lines.

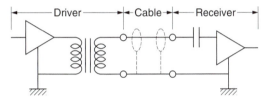

Figure 8.4 The consumer electrical interface.

There is the additional possibility[9] of a professional interface using coaxial cable and BNC connectors for distances of around 1000 m. This is simply the AES/EBU protocol but with a 75 Ω coaxial cable carrying a one-volt signal so that it can be handled by analog video distribution amplifiers. Impedance converting transformers are commercially available allowing balanced 110 ohm to unbalanced 75 Ω matching.

8.4   Frame structure

In Figure 8.5 the basic structure of the professional and consumer formats can be seen. One subframe consists of 32 bit-cells, of which four will be used by a synchronizing pattern. Subframes from the two audio channels, A and B, alternate on a time-division basis. Up to 24-bit sample wordlength can be used, which should cater for all conceivable future developments, but normally 20-bit maximum length samples will be available with four auxiliary data bits, which can be used for a voice-grade channel in a professional application. In a consumer DAT machine, subcode can be transmitted in bits 4–11, and the sixteen-bit audio in bits 12–27.

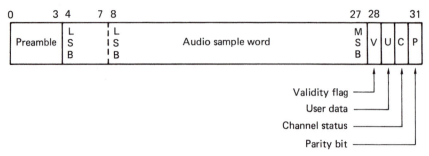

Figure 8.5   The basic subframe structure of the AES/EBU format. Sample can be 20 bits with four auxiliary bits, or 24 bits. LSB is transmitted first.

Preceding formats sent the most significant bit first. Since this was the order in which bits were available in successive approximation convertors it has become a *de-facto* standard for inter-chip transmission inside equipment. In contrast, this format sends the least significant bit first. One advantage of this approach is that simple serial arithmetic is then possible on the samples because the carries produced by the operation on a given bit can be delayed by one bit period and then included in the operation on the next higher-order bit. There is additional complication, however, if it is proposed to build adaptors from one of the manufacturers' formats to the new format because of the word reversal. This problem is a temporary issue, as new machines are designed from the outset to have the standard connections.

The format specifies that audio data must be in two's complement coding. Whilst pure binary could accept various alignments of different wordlengths with only a level change, this is not true of two's complement. If different wordlengths are used, the MSBs must always be in the same bit position otherwise the polarity will be misinterpreted. Thus the MSB has to be in bit 27 irrespective of wordlength. Shorter words are leading zero filled up to the 20-bit capacity. The channel status data included from AES-3–1992 signalling of the actual audio wordlength used so that receiving devices could adjust the digital dithering level needed to shorten a received word which is too long or pack samples onto a disk more efficiently.

Four status bits accompany each subframe. The validity flag will be reset if the associated sample is reliable. Whilst there have been many aspirations regarding what the V bit could be used for, in practice a single bit cannot specify much, and if combined with other V bits to make a word, the time resolution is lost. AES-3–1992 described the V bit as indicating that the information in the associated subframe is 'suitable for conversion to an analog signal'. Thus it might be reset if the interface was being used for non-audio data as is done, for example, in CD-I players.

The parity bit produces even parity over the subframe, such that the total number of ones in the subframe is even. This allows for simple detection of an odd number of bits in error, but its main purpose is that it makes successive sync patterns have the same polarity, which can be used to improve the probability of detection of sync. The user and channel-status bits are discussed later.

Two of the subframes described above make one frame, which repeats at the sampling rate in use. The first subframe will contain the sample from channel A, or from the left channel in stereo working. The second subframe will contain the sample from channel B, or the right channel in stereo. At 48 kHz, the bit rate will be 3.072 MHz, but as the sampling rate can vary, the clock rate will vary in proportion.

In order to separate the audio channels on receipt the synchronizing patterns for the two subframes are different as Figure 8.6 shows. These sync patterns begin with a run length of 1.5 bits which violates the FM channel coding rules and so cannot occur due to any data combination. The type of sync pattern is denoted by the position of the second transition which can be 0.5, 1.0 or 1.5 bits away from the first. The third transition is designed to make the sync patterns DC-free.

The channel status and user bits in each subframe form serial data streams with one bit of each per audio channel per frame. The channel status bits are given a block structure and synchronized every 192 frames, which at 48 kHz gives a block rate of 250 Hz, corresponding to a period of four milliseconds. In order to synchronize the channel-status blocks, the channel A sync pattern is replaced for one frame only

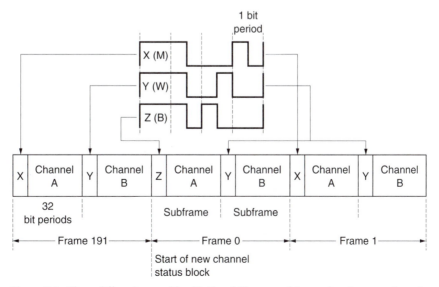

Figure 8.6 Three different preambles (X, Y and Z) are used to synchronize a receiver at the starts of subframes.

by a third sync pattern which is also shown in Figure 8.6. The AES standard refers to these as X,Y and Z whereas IEC 958 calls them M,W and B. As stated, there is a parity bit in each subframe, which means that the binary level at the end of a subframe will always be the same as at the beginning. Since the sync patterns have the same characteristic, the effect is that sync patterns always have the same polarity and the receiver can use that information to reject noise. The polarity of transmission is not specified, and indeed an accidental inversion in a twisted pair is of no consequence, since it is only the transition that is of importance, not the direction.

8.5 Talkback in auxiliary data

When 24-bit resolution is not required, which is most of the time, the four auxiliary bits can be used to provide talkback.

This was proposed by broadcasters[10] to allow voice coordination between studios as well as program exchange on the same cables. Twelve-bit samples of the talkback signal are taken at one third the main sampling rate. Each twelve-bit sample is then split into three nibbles (half a byte, for gastronomers) which can be sent in the auxiliary data slot of three successive samples in the same audio channel. As there are 192 nibbles per channel status block period, there will be exactly 64 talkback samples in that period. The reassembly of the nibbles can be synchronized by the channel status sync pattern as shown in Figure 8.7. Channel status byte 2 reflects the use of auxiliary data in this way.

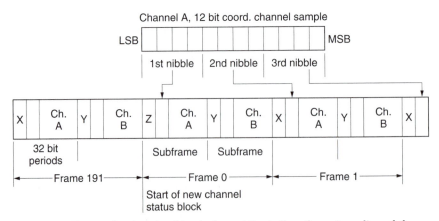

Figure 8.7 The coordination signal is of a lower bit rate than the main audio and thus may be inserted in the auxiliary nibble of the interface subframe, taking three subframes per coordination sample.

8.6   Professional channel status

In the both the professional and consumer formats, the sequence of channel-status bits over 192 subframes builds up a 24-byte channel-status block. However, the contents of the channel status data are completely different between the two applications. The professional channel status structure is shown in Figure 8.8. Byte 0 determines the use of emphasis and the sampling rate, with details in Figure 8.9. Byte 1 determines the channel usage mode, i.e. whether the data transmitted are a stereo pair, two unrelated mono signals or a single mono signal, and details the user bit handling. Figure 8.10 gives details. Byte 2 determines wordlength as in Figure 8.11. This was made more comprehensive in AES-3–1992. Byte 3 is applicable only to multichannel applications. Byte 4 indicates the suitability of the signal as a sampling rate reference and will be discussed in more detail later in this chapter.

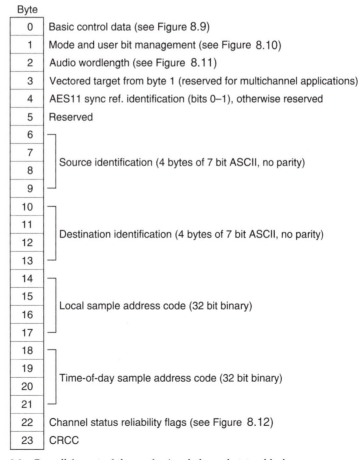

Byte

0	Basic control data (see Figure 8.9)
1	Mode and user bit management (see Figure 8.10)
2	Audio wordlength (see Figure 8.11)
3	Vectored target from byte 1 (reserved for multichannel applications)
4	AES11 sync ref. identification (bits 0–1), otherwise reserved
5	Reserved
6	
7	Source identification (4 bytes of 7 bit ASCII, no parity)
8	
9	
10	
11	Destination identification (4 bytes of 7 bit ASCII, no parity)
12	
13	
14	
15	Local sample address code (32 bit binary)
16	
17	
18	
19	Time-of-day sample address code (32 bit binary)
20	
21	
22	Channel status reliability flags (see Figure 8.12)
23	CRCC

Figure 8.8 Overall format of the professional channel status block.

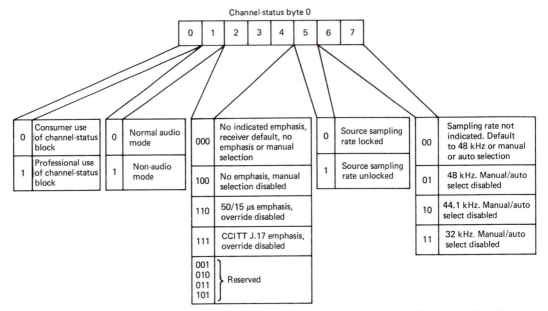

Figure 8.9 The first byte of the channel-status information in the AES/EBU standard deals primarily with emphasis and sampling-rate control.

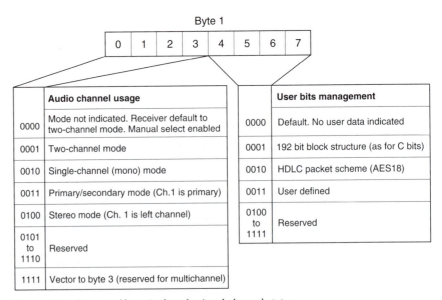

Figure 8.10 Format of byte 1 of professional channel status.

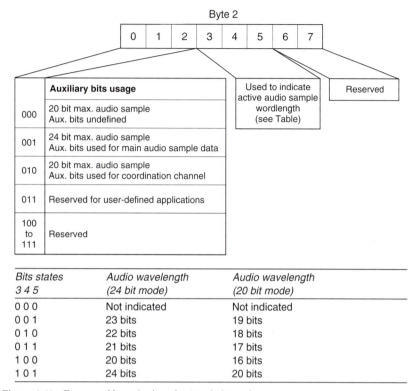

Figure 8.11 Format of byte 2 of professional channel status.

There are two slots of four bytes each which are used for alphanumeric source and destination codes. These can be used for routing. The bytes contain seven-bit ASCII characters (printable characters only) sent LSB first with the eighth bit set to zero acording to AES-3–1992. The destination code can be used to operate an automatic router, and the source code will allow the origin of the audio and other remarks to be displayed at the destination.

Bytes 14–17 convey a 32-bit sample address which increments every channel status frame. It effectively numbers the samples in a relative manner from an arbitrary starting point. Bytes 18–21 convey a similar number, but this is a time-of-day count, which starts from zero at midnight. As many digital audio devices do not have real-time clocks built in, this cannot be relied upon.

AES-3–92 specified that the time-of-day bytes should convey the real time at which a recording was made, making it rather like timecode. There are enough combinations in 32 bits to allow a sample count over 24 hours at 48 kHz. The sample count has the advantage that it is universal and independent of local supply frequency. In theory if the sampling rate is known, conventional hours, minutes, seconds, frames timecode can be

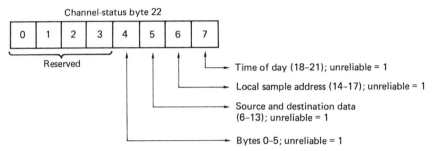

Figure 8.12 Byte 22 of channel status indicates if some of the information in the block is unreliable.

calculated from the sample count, but in practice it is a lengthy computation and users have proposed alternative formats in which the data from EBU or SMPTE timecode are transmitted directly in these bytes. Some of these proposals are in service as *de-facto* standards.

The penultimate byte contains four flags which indicate that certain sections of the channel-status information are unreliable (see Figure 8.12). This allows the transmission of an incomplete channel-status block where the entire structure is not needed or where the information is not available. For example, setting bit 5 to a logical one would mean that no origin or destination data would be interpreted by the receiver, and so it need not be sent.

The final byte in the message is a CRCC which converts the entire channel-status block into a codeword (see Chapter 7). The channel status message takes 4 ms at 48 kHz and in this time a router could have switched to another signal source. This would damage the transmission, but will also result in a CRCC failure so the corrupt block is not used. Error correction is not necessary, as the channel status data are either stationary, i.e. they stay the same, or change at a predictable rate, e.g. timecode. Stationary data will only change at the receiver if a good CRCC is obtained.

8.7 Consumer channel status

For consumer use, a different version of the channel-status specification is used. As new products come along, the consumer subcode expands its scope.

Figure 8.13 shows that the serial data bits are assembled into twelve words of sixteen bits each. In the general format, the first six bits of the first word form a control code, and the next two bits permit a mode select for future expansion. At the moment only mode zero is standardized, and the three remaining codes are reserved.

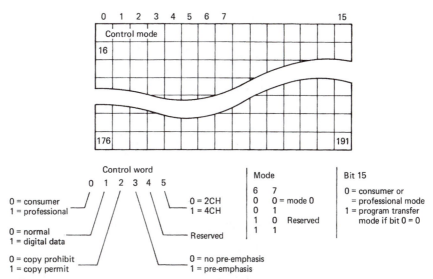

Figure 8.13 The general format of the consumer version of channel status. Bit 0 has the same meaning as in the professional format for compatibility. Bits 6–7 determine the consumer format mode, and presently only mode 0 is defined (see Figure 8.14).

Figure 8.14 shows the bit allocations for mode zero. In addition to the control bits, there are a category code, a simplified version of the AES/EBU source field, a field which specifies the audio channel number for multichannel working, a sampling-rate field, and a sampling-rate tolerance field.

Originally the consumer format was incompatible with the professional format, since bit zero of channel status would be set to a one by a four-channel consumer machine, and this would confuse a professional receiver because bit zero specifies professional format. The EBU proposed to the IEC that the four-channel bit be moved to bit 5 of the consumer format, so that bit zero would always then be zero. This proposal is incorporated into the bit definitions of Figures 8.13 and 8.14.

The category code specifies the type of equipment which is transmitting, and its characteristics. In fact each category of device can output one of two category codes, depending on whether bit 15 is or is not set. Bit 15 is the 'L-bit' and indicates whether the signal is from an original recording (0) or from a first-generation copy (1) as part of the SCMS (Serial Copying Management System) first implemented to resolve the stalemate over the sale of consumer DAT machines. In conjunction with the copyright flag, a receiving device can determine whether to allow or disallow recording. There were originally four categories; general purpose, two-channel CD player, two-channel PCM adaptor and two-channel digital tape recorder (DAT), but the list has now extended as Figure 8.14 shows.

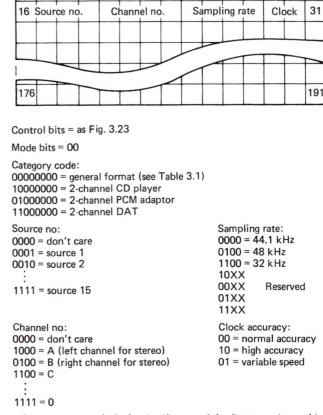

Figure 8.14 In consumer mode 0, the significance of the first two sixteen-bit channel-status words is shown here. The category codes are expanded in Tables 8.1 and 8.2.

Table 8.1 illustrates the format of the subframes in the general-purpose category. When used with CD players, Table 8.2 applies. In this application, the extensive subcode data of a CD recording (see Chapter 12) can be conveyed down the interface. In every CD sync block, there are twelve audio samples, and eight bits of subcode, P–W. The P flag is not transmitted, since it is solely positioning information for the player; thus only Q–W are sent. Since the interface can carry one user bit for every sample, there is surplus capacity in the user-bit channel for subcode. A CD subcode block is built up over 98 sync blocks, and has a repetition rate of 75 Hz. The start of the subcode data in the user bitstream will be seen in Figure 8.15 to be denoted by a minimum of sixteen zeros, followed by a start bit which is always set to one. Immediately after the start bit, the receiver expects to see seven subcode

Table 8.1 The general category code causes the subframe structure of the transmission to be interpreted as below (see Figure 8.5) and the stated channel-status bits are valid

Category code
00000000 = two-channel general format

Subframe structure

Two's complement, MSB in position 27, max 20 bits/sample
User bit channel = not used
V bit optional
Channel status left = Channel status right, unless channel number (Figure 8.14) is non-zero

Control bits in channel status
Emphasis = bit 3
Copy permit = bit 2

Sampling-rate bits in channel status
Bits 4–27 = according to rate in use

Clock-accuracy bits in channel status
Bits 28–29 = according to source accuracy

Table 8.2 In the CD category, the meaning below is placed on the transmission. The main difference from the general category is the use of user bits for subcode as specified in Figure 8.15

Category code
10000000 = two-channel CD player

Subframe structure

Two's complement MSB in position 27, 16 bits/sample
Use bit channel = CD subcode (see Figure 8.15)
V bit optional

Control bits in channel status
Derived from Q subcode control bits (see Chapter 12)

Sampling-rate bits in channel status
Bits 24–27 = 0000 = 44.1 kHz

Clock-accuracy bits in channel status
Bits 28–29 = according to source accuracy and use of variable speed

bits, Q–W. Following these, another start bit and another seven bits may follow immediately, or a space of up to eight zeros may be left before the next start bit. This sequence repeats 98 times, when another sync pattern will be expected. The ability to leave zeros between the subcode symbols simplifies the handling of the disparity between user

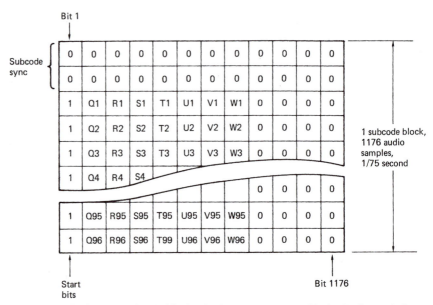

Figure 8.15 In CD, one subcode block is built up over 98 sync blocks. In this period there will be 1176 audio samples, and so there are 1176 user bits available to carry the subcode. There is insufficient subcode information to fill this capacity, and zero packing is used.

bit capacity and subcode bit rate. Figure 8.16 shows a representative example of a transmission from a CD player.

In a PCM adaptor, there is no subcode, and the only ancillary information available from the recording consists of copy-protect and emphasis bits. In other respects, the format is the same as the general-purpose format.

When a DAT player is used with the interface, the user bits carry several items of information.[11] Once per drum revolution, the user bit in one subframe is raised when that subframe contains the first sample of the interleave block (see Chapter 9). This can be used to synchronize several DAT machines together for editing purposes. Immediately following the sync bit, start ID will be transmitted when the player has found the code on the tape. This must be asserted for 300 ± 30 drum revolutions, or about 10 seconds. In the third bit position the skip ID is transmitted when the player detects a skip command on the tape. This indicates that the player will go into fast forward until it detects the next start ID. The skip ID must be transmitted for 33 ± 3 drum rotations. Finally DAT supports an end-of-skip command which terminates a skip when it is detected. This allows jump editing (see Chapter 11) to omit short sections of the recording. DAT can also transmit the track number (TNO) of the track being played down the user bitstream.

Notes	Subframe no.	Preamble SYNC	Aux		LSB	Audio samples		MSB	V	U	C	P
Channel status →	1	CS	0000	0000	XXXX	XXXX	XXXX	XXXX	0	0	C0L	P
block sync	2	B	0000	0000	XXXX	XXXX	XXXX	XXXX	0	0	C0R	P
	3	A	0000	0000	XXXX	XXXX	XXXX	XXXX	0	0	C1L	P
	4	B	0000	0000	XXXX	XXXX	XXXX	XXXX	0	0	C1R	P
A = left →	5	A	0000	0000	XXXX	XXXX	XXXX	XXXX	0	0	C2L	P
channel sample	6	B	0000	0000	XXXX	XXXX	XXXX	XXXX	0	0	C2R	P
	7	A	0000	0000	XXXX	XXXX	XXXX	XXXX	0	0	C3L	P
	8	B	0000	0000	XXXX	XXXX	XXXX	XXXX	0	0	C3R	P
	9	A	0000	0000	XXXX	XXXX	XXXX	XXXX	0	0	C4L	P
B = right →	10	B	0000	0000	XXXX	XXXX	XXXX	XXXX	0	0	C4R	P
channel sample	11	A	0000	0000	XXXX	XXXX	XXXX	XXXX	0	0	C5L	P
	12	B	0000	0000	XXXX	XXXX	XXXX	XXXX	0	0	C5R	P
	13	A	0000	0000	XXXX	XXXX	XXXX	XXXX	0	0	C6L	P
	14	B	0000	0000	XXXX	XXXX	XXXX	XXXX	0	0	C6R	P
	15	A	0000	0000	XXXX	XXXX	XXXX	XXXX	0	0	C7L	P
	16	B	0000	0000	XXXX	XXXX	XXXX	XXXX	0	0	C7R	P
16 zeros	17	A	0000	0000	XXXX	XXXX	XXXX	XXXX	0	0	C8L	P
in user bits	18	B	0000	0000	XXXX	XXXX	XXXX	XXXX	0	0	C8R	P
= subcode	19	A	0000	0000	XXXX	XXXX	XXXX	XXXX	0	0	C9L	P
sync word	20	B	0000	0000	XXXX	XXXX	XXXX	XXXX	0	0	C9R	P
	21	A	0000	0000	XXXX	XXXX	XXXX	XXXX	0	0	C10L	P
	22	B	0000	0000	XXXX	XXXX	XXXX	XXXX	0	0	C10R	P
	23	A	0000	0000	XXXX	XXXX	XXXX	XXXX	0	0	C11L	P
	24	B	0000	0000	XXXX	XXXX	XXXX	XXXX	0	0	C11R	P
1 in user →	25	A	0000	0000	XXXX	XXXX	XXXX	XXXX	0	1	C12L	P
bits = subcode	26	B	0000	0000	XXXX	XXXX	XXXX	XXXX	0	Q1	C12R	P
start bit	27	A	0000	0000	XXXX	XXXX	XXXX	XXXX	0	R1	C13L	P
	28	B	0000	0000	XXXX	XXXX	XXXX	XXXX	0	S1	C13R	P
U = Subcode	29	A	0000	0000	XXXX	XXXX	XXXX	XXXX	0	T1	C14L	P
	30	B	0000	0000	XXXX	XXXX	XXXX	XXXX	0	U1	C14R	P
	31	A	0000	0000	XXXX	XXXX	XXXX	XXXX	0	V1	C15L	P
	32	B	0000	0000	XXXX	XXXX	XXXX	XXXX	0	W1	C15R	P
	33	A	0000	0000	XXXX	XXXX	XXXX	XXXX	0	0	C16L	P
Subcode	34	B	0000	0000	XXXX	XXXX	XXXX	XXXX	0	0	C16R	P
space	35	A	0000	0000	XXXX	XXXX	XXXX	XXXX	0	0	C17L	P
	36	B	0000	0000	XXXX	XXXX	XXXX	XXXX	0	0	C17R	P
Start bit	37	A	0000	0000	XXXX	XXXX	XXXX	XXXX	0	1	C18L	P
	38	B	0000	0000	XXXX	XXXX	XXXX	XXXX	0	Q2	C18R	P
	39	A	0000	0000	XXXX	XXXX	XXXX	XXXX	0	R2	C19L	P
	40	B	0000	0000	XXXX	XXXX	XXXX	XXXX	0	S2	C19R	P
U = Subcode	41	A	0000	0000	XXXX	XXXX	XXXX	XXXX	0	T2	C20L	P
	42	B	0000	0000	XXXX	XXXX	XXXX	XXXX	0	U2	C20R	P
	43	A	0000	0000	XXXX	XXXX	XXXX	XXXX	0	V2	C21L	P
	44	B	0000	0000	XXXX	XXXX	XXXX	XXXX	0	W2	C21R	P
	45	A	0000	0000	XXXX	XXXX	XXXX	XXXX	0	0	C22L	P
	46	B	0000	0000	XXXX	XXXX	XXXX	XXXX	0	0	C22R	P
	47	A	0000	0000	XXXX	XXXX	XXXX	XXXX	0	0	C23L	P
	48	B	0000	0000	XXXX	XXXX	XXXX	XXXX	0	0	C23R	P

Figure 8.16 Compact Disc subcode transmitted in user bits of serial interface.

8.8   User bits

The user channel consists of one bit per audio channel per sample period. Unlike channel status, which only has a 192-bit frame structure, the user channel can have a flexible frame length. Figure 8.10 showed how byte 1 of the channel status frame describes the state of the user channel. Many professional devices do not use the user channel at all and would set the all-zeros code. If the user channel frame has the same length as the channel status frame then code 0001 can be set. One user channel format which is standardized is the data packet scheme of AES18–1992.[12,13] This was developed from proposals to employ the user channel for labelling in an asynchronous format.[14] A computer industry standard protocol known as HDLC (High-level Data Link Control)[15] is employed in order to take advantage of readily available integrated circuits.

The frame length of the user channel can be conveniently made equal to the frame period of an associated device. For example, it may be locked to Film, TV or DAT frames. The frame length may vary in NTSC as there are not an integer number of samples in a frame.

8.9   MADI – Multi-channel audio digital interface

Whilst the AES/EBU digital interface excels for the interconnection of stereo equipment, it is at a disadvantage when a large number of channels is required. MADI[16] was designed specifically to address the requirement for digital connection between multitrack recorders and mixing consoles by a working group set up jointly by Sony, Mitsubishi, Neve and SSL.

The standard provides for 56 simultaneous digital audio channels which are conveyed point-to-point on a single $75\,\Omega$ coaxial cable fitted with BNC connectors (as used for analog video) along with a separate synchronization signal. A distance of at least 50 m can be achieved.

Essentially MADI takes the subframe structure of the AES/EBU interface and multiplexes 56 of these into one sample period rather than the original two. Clearly this will result in a considerable bit rate, and the FM channel code of the AES/EBU standard would require excessive bandwidth. A more efficient code is used for MADI. In the AES/EBU interface the data rate is proportional to the sampling rate in use. Losses will be greater at the higher bit rate of MADI, and the use of a variable bit rate in the channel would make the task of achieving optimum equalization difficult. Instead the data bit rate is made a constant 100 megabits per second, irrespective of sampling rate. At lower sampling rates, the audio data are padded out to maintain the channel rate.

The MADI standard is effectively a superset of the AES/EBU interface in that the subframe data content is identical. This means that a number of separate AES/EBU signals can be fed into a MADI channel and recovered in their entirety on reception. The only caution required with such an application is that all channels must have the same synchronized sampling rate. The primary application of MADI is to multitrack recorders, and in these machines the sampling rates of all tracks are intrinsically synchronous. When the replay speed of such machines is varied, the sampling rate of all channels will change by the same amount, so they will remain synchronous.

At one extreme, MADI will accept a 32 kHz recorder playing $12\frac{1}{2}$ per cent slow, and at the other extreme a 48 kHz recorder playing $12\frac{1}{2}$ per

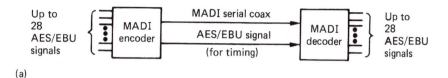

(a)

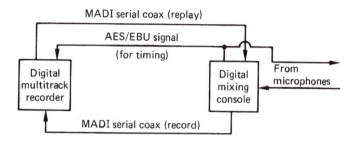

(b)

Figure 8.17 Some typical MADI applications. In (a) a large number of two-channel digital signals are multiplexed into the MADI cable to achieve economy of cabling. Note the separate timing signal. In (b) a pair of MADI links is necessary to connect a recorder to a mixing console. A third MADI link could be used to feed microphones into the desk from remote convertors.

cent fast. This is almost a factor of 2:1. Figure 8.17 shows some typical MADI configurations.

8.10   MADI data transmission

The data transmission of MADI is made using a group code, where groups of four data bits are represented by groups of five channel bits. Four bits have sixteen combinations, whereas five bits have 32 combinations. Clearly only 16 out of these 32 are necessary to convey all possible data. It is then possible to use some of the remaining patterns when it is required to pad out the data rate. The padding symbols will not correspond to a valid data symbol and so they can be recognized and thrown away on reception. A further use of this coding technique is that the 16 patterns of 5 bits which represent real data are chosen to be those which will have the best balance between high and low states, so that DC offsets at the receiver can be minimized. Chapter 6 discussed the coding rules of 4/5 in MADI. The 4/5 code adopted is the same one used for a computer transmission format known as FDDI, so that existing hardware can be used.

8.11   MADI frame structure

Figure 8.18(a) shows the frame structure of MADI. In one sample period, 56 time slots are available, and these each contain eight 4/5 symbols, corresponding to 32 data bits or 40 channel bits. Depending on the

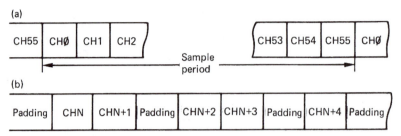

Figure 8.18 In (a) all 56 channels are sent in numerical order serially during the sample period. For simplicity no padding symbols are shown here. In (b) the use of padding symbols is illustrated. These are necessary to maintain the channel bit rate at 125 M bits/s irrespective of the sample rate in use. Padding can be inserted flexibly, but it must only be placed between the channels.

sampling rate in use, more or less padding symbols will need to be inserted in the frame to maintain a constant channel bit rate. Since the receiver does not interpret the padding symbols as data, it is effectively blind to them, and so there is considerable latitude in the allowable positions of the padding. Figure 8.18(b) shows some possibilities. The padding must not be inserted within a channel, only between channels, but the channels need not necessarily be separated by padding. At one extreme, all channels can be butted together, followed by a large padding area, or the channels can be evenly spaced throughout the frame. Although this sounds rather vague, it is intended to allow freedom in the design of associated hardware. Multitrack recorders generally have some form of internal multiplexed data bus, and these have various architectures and protocols. The timing flexibility allows an existing bus timing structure to be connected to a MADI link with the minimum of hardware. Since the channels can be inserted at a variety of places within the frame, the necessity of a separate synchronizing link between transmitter and receiver becomes clear.

8.12   MADI Audio channel format

Figure 8.19 shows the MADI channel format, which should be compared with the AES/EBU subframe shown in Figure 8.5. The last 28 bits are identical, and differences are only apparent in the synchronizing area. In order to remain transparent to an AES/EBU signal, which can contain two audio channels, MADI must tell the receiver whether a particular channel contains the A leg or B leg, and when the AES/EBU channel status block sync occurs. Bits 2 and 3 perform these functions. As the 56 channels of MADI follow one another in numerical order, it is necessary to identify channel zero so that the channels are not mixed up. This is the function of bit 0, which is set in channel zero and reset in all other

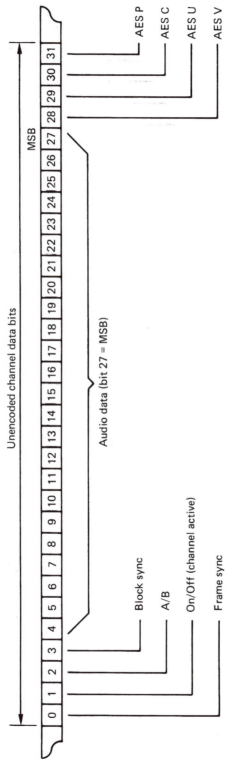

Figure 8.19 The MADI channel data are shown here. The last 28 bits are identical in every way to the AES/EBU interface, but the synchronizing in the first four bits differs. There is a frame sync bit to identify channel 0, and a channel active bit. The A/B leg of a possible AES/EBU input to MADI is conveyed, as is the channel status block sync.

channels. Finally bit 1 is set to indicate an active channel, for the case when less than 56 channels are being fed down the link. Active channels have bit 1 set, and must be consecutive starting at channel zero. Inactive channels have all bits set to zero, and must follow the active channels.

8.13   Fibre-optic interfacing

Whereas a parallel bus is ideal for a distributed multichannel system, for a point-to-point connection, the use of fibre optics is feasible, particularly as distance increases. An optical fibre is simply a glass filament which is encased in such a way that light is constrained to travel along it. Transmission is achieved by modulating the power of an LED or small laser coupled to the fibre. A phototransistor converts the received light back to an electrical signal.

Optical fibres have numerous advantages over electrical cabling. The bandwidth available is staggering. Optical fibres neither generate, nor are prone to, electromagnetic interference and, as they are insulators, ground loops cannot occur.[17] The disadvantage of optical fibres is that the terminations of the fibre where transmitters and receivers are attached suffer optical losses, and while these can be compensated in point-to-point links, the use of a bus structure is not really feasible. Fibre-optic links are already in service in digital audio mixing consoles.[18] The fibre implementation by Toshiba and known as the TOSLink is popular in consumer products, and the protocol is identical to the consumer electrical format.

8.14   Synchronizing

When digital audio signals are to be assembled from a variety of sources, either for mixing down or for transmission through a TDM (time-division multiplexing) system, the samples from each source must be synchronized to one another in both frequency and phase. The source of samples must be fed with a reference sampling rate from some central generator, and will return samples at that rate. The same will be true if digital audio is being used in conjunction with VTRs. As the scanner speed and hence the audio block rate is locked to video, it follows that the audio sampling rate must be locked to video. Such a technique has been used since the earliest days of television in order to allow vision mixing, but now that audio is conveyed in discrete samples, these too must be genlocked to a reference for most production purposes.

AES11–1991[19] documented standards for digital audio synchronization and requires professional equipment to be able to genlock either to a separate reference input or to the sampling rate of an AES/EBU input.

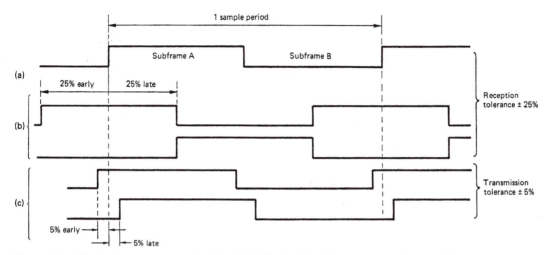

Figure 8.20 The timing accuracy required in AES/EBU signals with respect to a reference (a). Inputs over the range shown at (b) must be accepted, whereas outputs must be closer in timing to the reference as shown at (c).

As the interface uses serial transmission, a shift register is required in order to return the samples to parallel format within equipment. The shift register is generally buffered with a parallel loading latch which allows some freedom in the exact time at which the latch is read with respect to the serial input timing. Accordingly the standard defines synchronism as an identical sampling rate, but with no requirement for a precise phase relationship. Figure 8.20 shows the timing tolerances allowed. The beginning of a frame (the frame edge) is defined as the leading edge of the X preamble. A device which is genlocked must correctly decode an input whose frame edges are within ± 25 per cent of the sample period. This is quite a generous margin, and corresponds to the timing shift due to putting about a kilometre of cable in series with a signal. In order to prevent tolerance build-up when passing through several devices in series, the output timing must be held within ± 5 per cent of the sample period.

The reference signal may be an AES/EBU signal carrying program material, or it may carry muted audio samples; the so-called digital audio silence signal. Alternatively it may just contain the sync patterns. The accuracy of the reference is specified in bits 0 and 1 of byte 4 of channel status (see Figure 8.8). Two zeros indicates the signal is not reference grade (but some equipment may still be able to lock to it). 01 indicates a Grade 1 reference signal which is ±1 ppm accurate, whereas 10 indicates a Grade 2 reference signal which is ±10 ppm accurate. Clearly devices which are intended to lock to one of these references must have an appropriate phase-locked-loop capture range.

In addition to the AES/EBU synchronization approach, some older equipment carries a word clock input which accepts a TTL level square wave at the sampling frequency. This is the reference clock of the old Sony SDIF-2 interface.

Modern digital audio devices may also have a video input for synchronizing purposes. Video syncs (with or without picture) may be input, and a phase-locked loop will multiply the video frequency by an appropriate factor to produce a synchronous audio sampling clock.

8.15   Asynchronous operation

In practical situations, genlocking is not always possible. In a satellite transmission, it is not really practicable to genlock a studio complex half-way around the world to another. Outside broadcasts may be required to generate their own master timing for the same reason. When genlock is not achieved, there will be a slow slippage of sample phase between source and destination due to such factors as drift in timing generators. This phase slippage will be corrected by a synchronizer, which is intended to work with frequencies that are nominally the same. It should be contrasted with the sampling-rate convertor which can work at arbitrary but generally greater frequency relationships. Although a sampling-rate convertor can act as a synchronizer, it is a very expensive way of doing the job. A synchronizer can be thought of as a lower-cost version of a sampling-rate convertor which is constrained in the rate difference it can accept.

In one implementation of a digital audio synchronizer,[20] memory is used as a timebase corrector. Samples are written into the memory with the frequency and phase of the source and, when the memory is half-full, samples are read out with the frequency and phase of the destination. Clearly if there is a net rate difference, the memory will either fill up or empty over a period of time, and in order to recentre the address relationship, it will be necessary to jump the read address. This will cause samples to be omitted or repeated, depending on the relationship of source rate to destination rate, and would be audible on program material. The solution is to detect pauses or low-level passages and permit jumping only at such times. The process is illustrated in Figure 8.21. An alternative to address jumping is to undertake sampling-rate conversion for a short period (Figure 8.22) in order to slip the input/output relationship by one sample.[21] If this is done when the signal level is low, short wordlength logic can be used. However, now that sampling rate convertors are available as a low-cost single chip, these solutions are found less often in hardware, although they may be used in software-controlled processes.

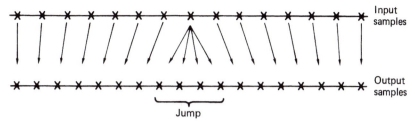

Figure 8.21 In jump synchronizing, input samples are subjected to a varying delay to align them with output timing. Eventually the sample relationship is forced to jump to prevent delay building up. As shown here, this results in several samples being repeated, and can only be undertaken during program pauses, or at very low audio levels. If the input rate exceeds the output rate, some samples will be lost.

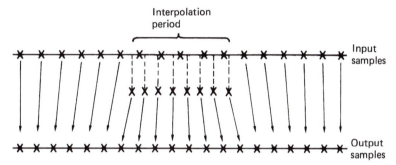

Figure 8.22 An alternative synchronizing process is to use a short period of interpolation in order to regulate the delay in the synchronizer.

The difficulty of synchronizing unlocked sources is eased when the frequency difference is small. This is one reason behind the clock accuracy standards for AES/EBU timing generators.[22]

8.16   Routing

Routing is the process of directing signals between a large number of devices so that any one can be connected to any other. The principle of a router is not dissimilar to that of a telephone exchange. In analog routers, there is the potential for quality loss due to the switching element. Digital routers are attractive because they need introduce no loss whatsoever. In addition, the switching is performed on a binary signal and therefore the cost can be lower. Routers can be either cross-point or time-division multiplexed.

In a BBC proposal[23] the 28 data-bit structure of the AES/EBU subframe has been turned sideways, with one conductor allocated to each bit. Since the maximum transition rate of the AES/EBU interface is 64 times the

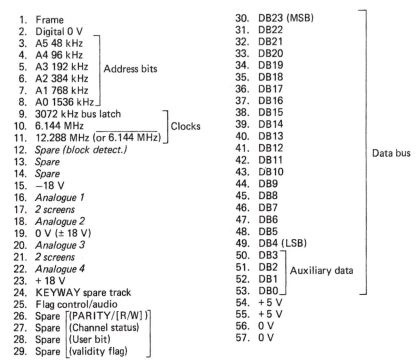

Figure 8.23 Time-division multiplexed 64 channel audio bus proposed by BBC.

sampling rate it follows that, in the parallel implementation, 64 channels could be time-multiplexed into one sample period within the same bandwidth. The necessary signals are illustrated in Figure 8.23. In order to separate the channels on reception, there are six address lines which convey a binary pattern corresponding to the audio channel number of the sample in that time slot. The receiver simply routes the samples according to the attached address. Such a point-to-point system does, however, neglect the potential of the system for more complex use. The bus cable can loop through several different items of equipment, each of which is programmed so that it transmits samples during a different set of time slots from the others. Since all transmissions are available at all receivers, it is only necessary to detect a given address to latch samples from any channel. If two devices decode the same address, the same audio channel will be available at two destinations.

In such a system, channel reassignment is easy. If the audio channels are transmitted in address sequence, it is only necessary to change the addresses which the receiving channels recognize, and a given input channel will emerge from a different output channel. Since the address recognition circuitry is already present in a TDM system, the functionality of a 64×64 point channel-assignment patchboard has been achieved with no extra hardware. The only constraint in the use of TDM systems is that all channels must have synchronized sampling rates. In multitrack recorders this occurs naturally because all the channels are locked by the tape format. With analog inputs it is a simple matter to drive all convertors from a common clock.

Given that the MADI interface uses TDM, it is also possible to perform routing functions using MADI-based hardware.

For asynchronous systems, or where several sampling rates are found simultaneously, a cross-point type of channel-assignment matrix will be necessary, using AES/EBU signals. In such a device, the switching can be performed by logic gates at low cost, and in the digital domain there is, of course, no quality degradation.

8.17   Networks

In the most general sense a network is means of communication between a large number of places. According to this definition the Post Office is a network, as are parcel and courier companies. This type of network delivers physical objects. If, however, we restrict the delivery to information only the result is a telecommunications network. The telephone system is a good example of a telecommunications network because it displays most of the characteristics of later networks.

It is fundamental in a network that any port can communicate with any other port. Figure 8.24 shows a primitive three-port network. Clearly each port must select one or other of the remaining ports in a trivial switching system. However, if it were attempted to redraw Figure 8.24 with one hundred ports, each one would need a 99-way switch and the number of wires needed would be phenomenal. Another approach is needed.

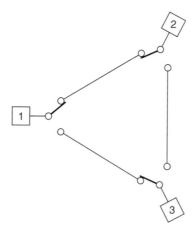

Figure 8.24 Switching is simple with a small number of ports.

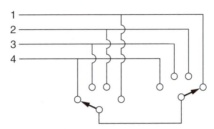

Figure 8.25 An exchange or switch can connect any input to any output, but extra switching is needed to support more than one connection.

Figure 8.25 shows that the common solution is to have an exchange, also known as a router, hub or switch, which is connected to every port by a single cable. In this case when a port wishes to communicate with another, it instructs the switch to make the connection. The complexity of the switch varies with its performance. The minimal case may be to install a single input selector and a single output selector. This allows any port to communicate with any other, but only one at a time. If more simultaneous communications are needed, further switching is needed. The extreme case is where every possible pair of ports can communicate simultaneously.

The amount of switching logic needed to implement the extreme case is phenomenal and in practice it is unlikely to be needed. One fundamental property of networks is that they are seldom implemented with the extreme case supported. There will be an economic decision made balancing the number of simultaneous communications with the equipment cost. Most of the time the user will be unaware that this limit exists, until there is a statistically abnormal condition which causes more than the usual number of nodes to attempt communication.

The phrase 'the switchboard was jammed' has passed into the language and stayed there despite the fact that manual switchboards are only seen in museums. This is a characteristic of networks. They generally only work up to a certain throughput and then there are problems. This doesn't mean that networks aren't useful, far from it. What it means is that with care, networks can be very useful, but without care they can be a nightmare.

There are two key factors to get right in a network. The first is that it must have enough throughput, bandwidth or connectivity to handle the anticipated usage and the second is that a priority system or algorithm is chosen which has appropriate behaviour during overload. These two characteristics are quite different, but often come as a pair in a network corresponding to a particular standard.

Where each device is individually cabled, the result is a radial network shown in Figure 8.26(a). It is not necessary to have one cable per device

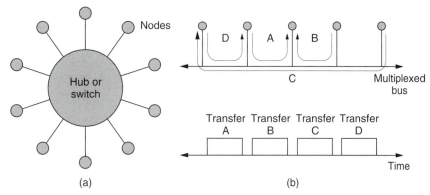

Figure 8.26 (a) Radial installations need a lot of cabling. Time-division multiplexing, where transfers occur during different time frames, reduces this requirement (b).

and several devices can co-exist on a single cable if some form of multiplexing is used. This might be time-division multiplexing (TDM) or frequency division multiplexing (FDM). In TDM, shown in Figure 8.26(b), the time axis is divided into steps which may or may not be equal in length. In Ethernet, for example, these are called frames. During each time step or frame a pair of nodes have exclusive use of the cable. At the end of the time step another pair of nodes can communicate. Rapidly switching between steps gives the illusion of simultaneous transfer between several pairs of nodes. In FDM, simultaneous transfer is possible because each message occupies a different band of frequencies in the cable. Each node has to 'tune' to the correct signal. In practice it is possible to combine FDM and TDM. Each frequency band can be time multiplexed in some applications.

Data networks originated to serve the requirements of computers and it is a simple fact that most computer processes don't need to be performed in real time or indeed at a particular time at all. Networks tend to reflect that background as many of them, particularly the older ones, are asynchronous.

Asynchronous means that the time taken to deliver a given quantity of data is unknown. A TDM system may chop the data into several different transfers and each transfer may experience delay according to what other transfers the system is engaged in. Ethernet and most storage system buses are asynchronous. For broadcasting purposes an asynchronous delivery system is no use at all, but for copying a video data file between two storage devices an asynchronous system is perfectly adequate.

The opposite extreme is the synchronous system in which the network can guarantee a constant delivery rate and a fixed and minor delay. An AES/EBU router is a synchronous network.

In between asynchronous and synchronous networks reside the isochronous approaches. These can be thought of as sloppy synchronous networks or more rigidly controlled asynchronous networks. Both descriptions are valid. In the isochronous network there will be maximum delivery time which is not normally exceeded. The data transmission rate may vary, but if the rate has been low for any reason, it will accelerate to prevent the maximum delay being reached. Isochronous networks can deliver near-real-time performance. If a data buffer is provided at both ends, synchronous data such as AES/EBU audio can be fed through an isochronous network. The magnitude of the maximum delay determines the size of the buffer and the length of the fixed overall delay through the system. This delay is responsible for the term 'near-real time'. ATM is an isochronous network.

These three different approaches are needed for economic reasons. Asynchronous systems are very efficient because as soon as one transfer completes, another can begin. This can only be achieved by making every device wait with its data in a buffer so that transfer can start immediately. Asynchronous sytems also make it possible for low bit rate devices to share a network with high bit rate devices. The low bit rate device will only need a small buffer and will therefore send short data blocks, whereas the high bit rate device will send long blocks. Asynchronous systems have no dificulty in handling blocks of varying size, whereas in a synchronous system this is very difficult.

Isochronous systems try to give the best of both worlds, generally by sacrificing some flexibility in block size. FireWire is an example of a network which is part isochronous and part asynchronous so that the advantages of both are available.

8.18   Introduction to NICAM 728

This system was developed by the BBC to allow the two additional high-quality digital sound channels to be carried on terrestrial television broadcasts. Performance was such that the system was adopted as the UK standard, and was recommended by the EBU to be adopted by its members, many of whom put it into service.[24]

The introduction of stereo sound with television cannot be at the expense of incompatibility with the existing monophonic analog sound channel. In NICAM 728 an additional low-power subcarrier is positioned just above the analog sound carrier, which is retained. The relationship is shown in Figure 8.27. The power of the digital subcarrier is about one hundredth that of the main vision carrier, and so existing monophonic receivers will reject it.

Since the digital carrier is effectively shoe-horned into the gap between TV channels, it is necessary to ensure that the spectral width of the intruder is minimized to prevent interference. As a further measure, the

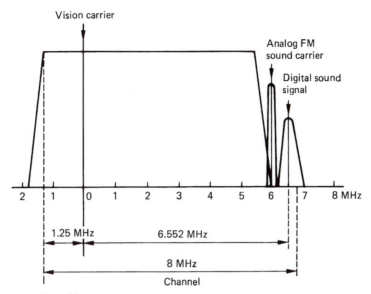

Vision carrier

Analog FM
sound carrier

Digital sound
signal

2   1   0   1   2   3   4   5   6   7   8 MHz

1.25 MHz

6.552 MHz

8 MHz

Channel

Figure 8.27 The additional carrier needed for digital stereo sound is squeezed in between television channels as shown here. The digital carrier is of much lower power than the analog signals, and is randomized prior to transmission so that it has a broad, low-level spectrum which is less visible on the picture.

power of the existing audio carrier is halved when the digital carrier is present.

Figure 8.28 shows the stages through which the audio must pass. The audio sampling rate used is 32 kHz which offers similar bandwidth to that of an FM stereo radio broadcast. Samples are originally quantized to fourteen-bit resolution in two's complement code. From an analog source this causes no problem, but from a professional digital source having longer wordlength and higher sampling rate it would be necessary to pass through a rate convertor, a digital equalizer to provide pre-emphasis, an optional digital compressor in the case of wide dynamic range signals and then through a truncation circuit incorporating digital dither as explained in Chapter 4.

The fourteen-bit samples are block companded to reduce data rate. During each one millisecond block, 32 samples are input from each audio channel. The magnitude of the largest sample in each channel is independently assessed, and used to determine the gain range or scale factor to be used. Every sample in each channel in a given block will then be scaled by the same amount and truncated to ten bits. An eleventh bit present on each sample combines the scale factor of the channel with parity bits for error detection. The encoding process is described as a Near Instantaneously Companded Audio Multiplex, NICAM for short. The resultant data now consists of $2 \times 32 \times 11 = 704$ bits per block. Bit interleaving is employed to reduce the effect of burst errors.

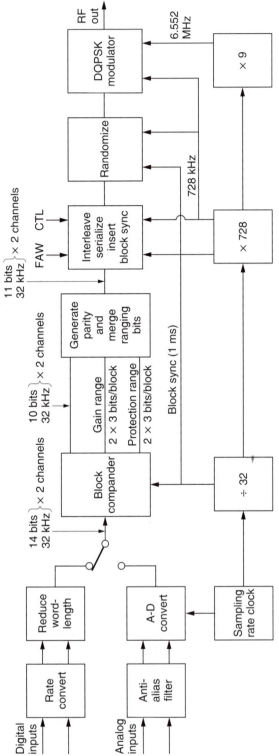

Figure 8.28 The stage necessary to generate the digital subcarrier in NICAM 728. Audio samples are block companded to reduce the bandwidth needed.

At the beginning of each block a synchronizing byte, known as a Frame Alignment Word, is followed by five control bits and eleven additional data bits, making a total of 728 bits per frame, hence the number in the system name. As there are 1000 frames per second, the bit rate is 728 kbits/s. In the UK this is multiplied by 9 to obtain the digital carrier frequency of 6.552 MHz but some other countries use a different subcarrier spacing.

The digital carrier is phase modulated. It has four states which are 90° apart. Information is carried in the magnitude of a phase change which takes place every 18 cycles, or 2.74 μs. As there are four possible phase changes, two bits are conveyed in every change. The absolute phase has no meaning, only the changes are interpreted by the receiver. This type of modulation is known as differentially encoded quadrature phase shift keying (DQPSK), sometimes called four-phase DPSK. In order to provide consistent timing and to spread the carrier energy throughout the band irrespective of audio content, randomizing is used, except during the frame alignment word. On reception, the FAW is detected and used to synchronize the pseudo-random generator to restore the original data.

Figure 8.29 shows the general structure of a frame. Following the sync pattern or FAW is the application control field. The application control bits determine the significance of following data, which can be stereo audio, two independent mono signals, mono audio and data or data only. Control bits C_1, C_2 and C_3 have eight combinations, of which only four are currently standardized. Receivers are designed to mute audio if C_3 becomes 1.

The frame flag bit C_0 spends eight frames high then eight frames low in an endless sixteen-frame sequence which is used to synchronize changes in channel usage. In the last sixteen-frame sequence of the old application, the application control bits change to herald the new application, whereas the actual data change to the new application on the next sixteen frame sequence.

The reserve sound switching flag, C_4, is set to 1 if the analog sound being broadcast is derived from the digital stereo. This fact can be stored by the receiver and used to initiate automatic switching to analog sound in the case of loss of the digital channels.

The additional data bits AD_0 to AD_{10} are as yet undefined, and reserved for future applications.

The remaining 704 bits in each frame may be either audio samples or data. The two channels of stereo audio are multiplexed into each frame, but multiplexing does not occur in any other case. If two mono audio channels are sent, they occupy alternate frames. Figure 8.29(a) shows a stereo frame, where the A channel is carried in odd-numbered samples, whereas Figure 8.29(b) shows a mono frame, where the M1 channel is carried in odd-numbered frames. The format for data has yet to be defined.

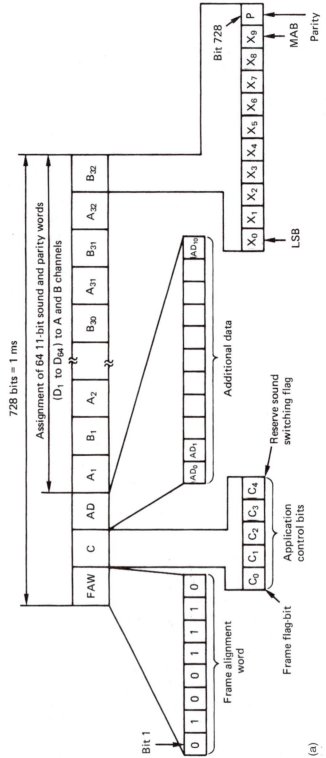

Figure 8.29   In (a) the block structure of a stereo signal multiplexes samples from both channels (A and B) into one block.

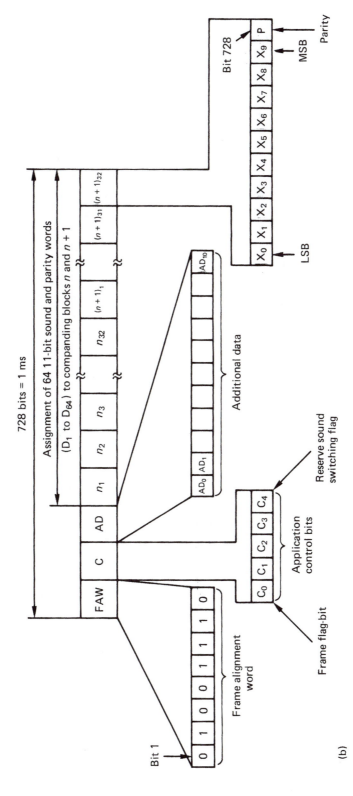

(b)

Figure 8.29 In mono, shown in (b), samples from one channel only occupy a given block. The diagrams here show the data before interleaving. Adjacent bits shown here actually appear at least sixteen bits apart in the data stream.

The sound/data block of NICAM 728 is in fact identical in structure to the first-level protected companded sound signal block of the MAC/packet systems.[25]

8.19   Audio in digital television broadcasting

Digital television broadcasting relies on the combination of a number of fundamental technologies. These are: MPEG-2 compression to reduce the bit rate, multiplexing to combine picture and sound data into a common bitstream, digital modulation schemes to reduce the RF bandwidth needed by a given bit rate and error correction to reduce the error statistics of the channel down to a value acceptable to MPEG data.

MPEG compressed video and audio are both highly sensitive to bit errors, primarily because they confuse the recognition of variable-length codes so that the decoder loses synchronization. However, MPEG is a compression and multiplexing standard and does not specify how error correction should be performed. Consequently a transmission standard must define a system which has to correct essentially all errors such that the delivery mechanism is transparent.

Essentially a transmission standard specifies all the additional steps needed to deliver an MPEG transport stream from one place to another. This transport stream will consist of a number of elementary streams of video and audio, where the audio may be coded according to MPEG audio standard or AC-3. In a system working within its capabilities, the picture and sound quality will be determined only by the performance of the compression system and not by the RF transmission channel. This is the fundamental difference between analog and digital broadcasting. In analog television broadcasting, the picture quality may be limited by composite video encoding artifacts as well as transmission artifacts such as noise and ghosting. In digital television broadcasting the picture quality is determined instead by the compression artifacts and interlace artifacts if interlace has been retained.

If the received error rate increases for any reason, once the correcting power is used up, the system will degrade rapidly as uncorrected errors enter the MPEG decoder. In practice decoders will be programmed to recognize the condition and blank or mute to avoid outputting garbage. As a result, digital receivers tend either to work well or not at all.

It is important to realize that the signal strength in a digital system does not translate directly to picture quality. A poor signal will increase the number of bit errors. Provided that this is within the capability of the error-correction system, there is no visible loss of quality. In contrast, a very powerful signal may be unusable because of similarly powerful reflections due to multipath propagation.

Whilst in one sense an MPEG transport stream is only data, it differs from generic data in that it must be presented to the viewer at a particular rate. Generic data are usually asynchronous, whereas baseband video and audio are synchronous. However, after compression and multiplexing audio and video are no longer precisely synchronous and so the term *isochronous* is used. This means a signal which was at one time synchronous and will be displayed synchronously, but which uses buffering at transmitter and receiver to accommodate moderate timing errors in the transmission.

Clearly another mechanism is needed so that the time axis of the original signal can be recreated on reception. The time stamp and program clock reference system of MPEG does this.

Figure 8.30 shows that the concepts involved in digital television broadcasting exist at various levels which have an independence not found in analog technology. In a given configuration a transmitter can radiate a given payload data bit rate. This represents the useful bit rate and does not include the necessary overheads needed by error correction, multiplexing or synchronizing. It is fundamental that the transmission system does not care what this payload bit rate is used for. The entire capacity may be used up by one high-definition channel, or a large number of heavily compressed channels may be carried. The details of this data usage are the domain of the *transport stream*. The multiplexing of

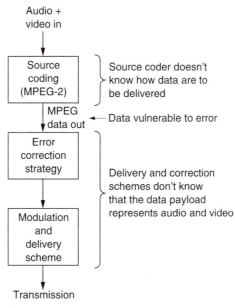

Figure 8.30 Source coder doesn't know delivery mechanism and delivery mechanism doesn't need to know what the data mean.

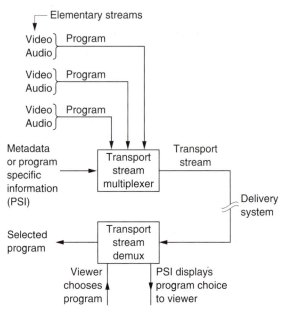

Figure 8.31 Program Specific Information helps the demultiplexer to select the required program.

transport streams is defined by the MPEG standards, but these do not define any error correction or transmission technique.

At the lowest level in Figure 8.31 the source coding scheme, in this case MPEG compression, results in one or more elementary streams, each of which carries a video or audio channel. Elementary streams are multiplexed into a transport stream. The viewer then selects the desired elementary stream from the transport stream. Metadata in the transport stream ensures that when a video elementary stream is chosen, the appropriate audio elementary stream will automatically be selected.

8.20   Packets and time stamps

The video elementary stream is an endless bitstream representing pictures which take a variable length of time to transmit. Bidirection coding means that pictures are not necessarily in the correct order. Storage and transmission systems prefer discrete blocks of data and so elementary streams are packetized to form a PES (packetized elementary stream). Audio elementary streams are also packetized. A packet is shown in Figure 8.32. It begins with a header containing an unique packet start code and a code which identifies the type of data stream. Optionally the packet header also may contain one or more *time stamps* which are used for synchronizing the video decoder to real time and for obtaining lip-sync.

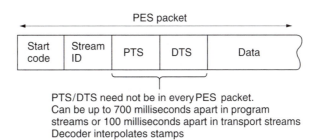

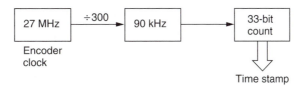

Figure 8.32 A PES packet structure is used to break up the continuous elementary stream.

Figure 8.33 Time stamps are the result of sampling a counter driven by the encoder clock.

Figure 8.33 shows that a time stamp is a sample of the state of a counter which is driven by a 90 kHz clock. This is obtained by dividing down the master 27 MHz clock of MPEG-2. This 27 MHz clock must be locked to the video frame rate and the audio sampling rate of the program concerned. There are two types of time stamp: PTS and DTS. These are abbreviations for presentation time stamp and decode time stamp. A presentation time stamp determines when the associated picture should be displayed on the screen, whereas a decode time stamp determines when it should be decoded. In bidirectional coding these times can be quite different.

Audio packets only have presentation time stamps. Clearly if lip-sync is to be obtained, the audio sampling rate of a given program must have been locked to the same master 27 MHz clock as the video and the time stamps must have come from the same counter driven by that clock.

In practice the time between input pictures is constant and so there is a certain amount of redundancy in the time stamps. Consequently PTS/DTS need not appear in every PES packet. Time stamps can be up to 100 ms apart in transport streams. As each picture type (*I*, *P* or *B*) is flagged in the bitstream, the decoder can infer the PTS/DTS for every picture from the ones actually transmitted.

8.21   MPEG transport streams

The MPEG-2 transport stream is intended to be a multiplex of many TV programs with their associated sound and data channels, although a

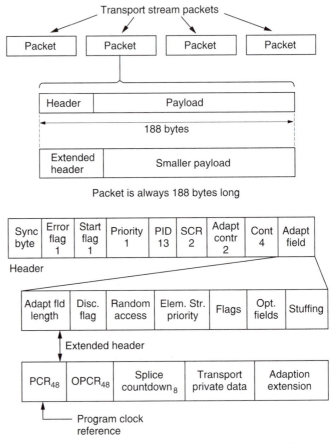

Figure 8.34 Transport stream packets are always 188 bytes long to facilitate multiplexing and error correction.

single program transport stream (SPTS) is possible. The transport stream is based upon packets of constant size so that multiplexing, adding error-correction codes and interleaving in a higher layer is eased. Figure 8.34 shows that these are always 188 bytes long.

Transport stream packets always begin with a header. The remainder of the packet carries data known as the payload. For efficiency, the normal header is relatively small, but for special purposes the header may be extended. In this case the payload gets smaller so that the overall size of the packet is unchanged. Transport stream packets should not be confused with PES packets which are larger and which vary in size. PES packets are broken up to form the payload of the transport stream packets.

The header begins with a sync byte which is an unique pattern detected by a demultiplexer. A transport stream may contain many different elementary streams and these are identified by giving each an unique thirteen-bit packet identification code or PID which is included in the header. A multiplexer seeking a particular elementary stream simply checks the PID of every packet and accepts only those which match.

In a multiplex there may be many packets from other programs in between packets of a given PID. To help the demultiplexer, the packet header contains a continuity count. This is a four-bit value which increments at each new packet having a given PID.

This approach allows statistical multiplexing as it does not matter how many or how few packets have a given PID; the demux will still find them. Statistical multiplexing has the problem that it is virtually impossible to make the sum of the input bit rates constant. Instead the multiplexer aims to make the average data bit rate slightly less than the maximum and the overall bit rate is kept constant by adding 'stuffing' or null packets. These packets have no meaning, but simply keep the bit rate constant. Null packets always have a PID of 8191 (all ones) and the demultiplexer discards them.

8.22   Clock references

A transport stream is a multiplex of several TV programs and these may have originated from widely different locations. It is impractical to expect all the programs in a transport stream to be genlocked and so the stream is designed from the outset to allow unlocked programs. A decoder running from a transport stream has to genlock to the encoder and the transport stream has to have a mechanism to allow this to be done independently for each program. The synchronizing mechanism is called program clock reference (PCR).

Figure 8.35 shows how the PCR system works. The goal is to re-create at the decoder a 27 MHz clock which is synchronous with that at the encoder. The encoder clock drives a 48-bit counter which continuously counts up to the maximum value before overflowing and beginning again.

A transport stream multiplexer will periodically sample the counter and place the state of the count in an extended packet header as a PCR (see Figure 8.34). The demultiplexer selects only the PIDs of the required program, and it will extract the PCRs from the packets in which they were inserted.

The PCR codes are used to control a numerically locked loop (NLL) described in section 3.16. The NLL contains a 27 MHz VCXO (voltage-controlled crystal oscillator). This is a variable-frequency oscillator based on a crystal which has a relatively small frequency range.

The VCXO drives a 48-bit counter in the same way as in the encoder. The state of the counter is compared with the contents of the PCR and the

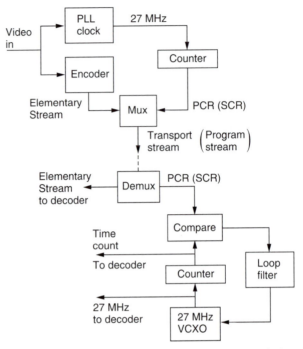

Figure 8.35 Program or System Clock Reference codes regenerate a clock at the decoder. See text for details.

difference is used to modify the VCXO frequency. When the loop reaches lock, the decoder counter would arrive at the same value as is contained in the PCR and no change in the VCXO would then occur. In practice the transport stream packets will suffer from transmission jitter and this will create phase noise in the loop. This is removed by the loop filter so that the VCXO effectively averages a large number of phase errors.

A heavily damped loop will reject jitter well, but will take a long time to lock. Lock-up time can be reduced when switching to a new program if the decoder counter is jammed to the value of the first PCR received in the new program. The loop filter may also have its time constants shortened during lock-up.

Once a synchronous 27 MHz clock is available at the decoder, this can be divided down to provide the 90 kHz clock which drives the time stamp mechanism.

The entire timebase stability of the decoder is no better than the stability of the clock derived from PCR. MPEG-2 sets standards for the maximum amount of jitter which can be present in PCRs in a real transport stream.

Clearly if the 27 MHz clock in the receiver is locked to one encoder it can only receive elementary streams encoded with that clock. If it is attempted to decode, for example, an audio stream generated from a different clock, the result will be periodic buffer overflows or underflows in the decoder. Thus MPEG defines a program in a manner which relates to timing. A program is a set of elementary streams which have been encoded with the same master clock.

8.23   Program Specific Information (PSI)

In a real transport stream, each elementary stream has a different PID, but the demultiplexer has to be told what these PIDs are and what audio belongs with what video before it can operate. This is the function of PSI which is a form of metadata. Figure 8.36 shows the structure of PSI. When a decoder powers up, it knows nothing about the incoming transport stream except that it must search for all packets with a PID of zero. PID zero is reserved for the Program Association Table (PAT). The PAT is transmitted at regular intervals and contains a list of all the programs in this transport stream. Each program is further described by its own Program Map Table (PMT) and the PIDs of of the PMTs are contained in the PAT.

Figure 8.36 also shows that the PMTs fully describe each program. The PID of the video elementary stream is defined, along with the PID(s) of the associated audio and data streams. Consequently when the viewer

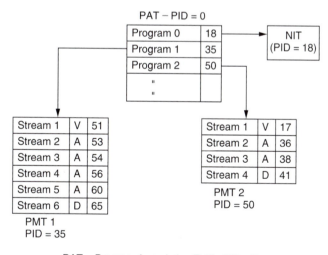

Figure 8.36   MPEG-2 Program Specific Information (PSI) is used to tell a demultiplexer what the transport stream contains.

selects a particular program, the demultiplexer looks up the program number in the PAT, finds the right PMT and reads the audio, video and data PIDs. It then selects elementary streams having these PIDs from the transport stream and routes them to the decoders.

Program 0 of the PAT contains the PID of the Network Information Table (NIT). This contains information about what other transport streams are available. For example, in the case of a satellite broadcast, the NIT would detail the orbital position, the polarization, carrier frequency and modulation scheme. Using the NIT a set-top box could automatically switch between transport streams.

Apart from 0 and 8191, a PID of 1 is also reserved for the Conditional Access Table (CAT). This is part of the access control mechanism needed to support pay per view or subscription viewing.

8.24   Multiplexing

A transport stream multiplexer is a complex device because of the number of functions it must perform. A fixed multiplexer will be considered first. In a fixed multiplexer, the bit rate of each of the programs must be specified so that the sum does not exceed the payload bit rate of the transport stream. The payload bit rate is the overall bit rate less the packet headers and PSI rate.

In practice the programs will not be synchronous to one another, but the transport stream must produce a constant packet rate given by the bit rate divided by 188 bytes, the packet length. Figure 8.37 shows how this is handled. Each elementary stream entering the multiplexer passes

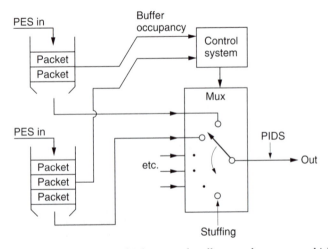

Figure 8.37 A transport stream multiplexer can handle several programs which are asynchronous to one another and to the transport stream clock. See text for details.

through a buffer which is divided into payload-sized areas. Note that periodically the payload area is made smaller because of the requirement to insert PCR.

MPEG-2 decoders also have a quantity of buffer memory. The challenge to the multiplexer is to take packets from each program in such a way that neither its own buffers nor the buffers in any decoder either overflow or underflow. This requirement is met by sending packets from all programs as evenly as possible rather than bunching together a lot of packets from one program. When the bit rates of the programs are different, the only way this can be handled is to use the buffer contents indicators. The fuller a buffer is, the more likely it should be that a packet will be read from it. Thus a buffer content arbitrator can decide which program should have a packet allocated next.

If the sum of the input bit rates is correct, the buffers should all slowly empty because the overall input bit rate has to be less than the payload bit rate. This allows for the insertion of Program Specific Information. Whilst PATs and PMTs are being transmitted, the program buffers will fill up again. The multiplexer can also fill the buffers by sending more PCRs as this reduces the payload of each packet. In the event that the multiplexer has sent enough of everything but still can't fill a packet then it will send a null packet with a PID of 8191. Decoders will discard null packets and as they convey no useful data, the multiplexer buffers will all fill whilst null packets are being transmitted.

The use of null packets means that the bit rates of the elementary streams do not need to be synchronous with one another or with the transport stream bit rate. As each elementary stream can have its own PCR, it is not necessary for the different programs in a transport stream to be genlocked to one another; in fact they don't even need to have the same frame rate.

This approach allows the transport stream bit rate to be accurately defined and independent of the timing of the data carried. This is important because the transport stream bit rate determines the spectrum of the transmitter and this must not vary.

In a statistical multiplexer or statmux, the bit rate allocated to each program can vary dynamically. Figure 8.38 shows that there must be tight connection between the statmux and the associated compressors. Each compressor has a buffer memory which is emptied by a demand clock from the statmux. In a normal, fixed bit rate, coder the buffer content feeds back and controls the requantizer. In statmuxing this process is less severe and only takes place if the buffer is very close to full, because the degree of coding difficulty is also fed to the statmux.

The statmux contains an arbitrator which allocates more packets to the program with the greatest coding difficulty. Thus if a particular program encounters difficult material it will produce large prediction errors and

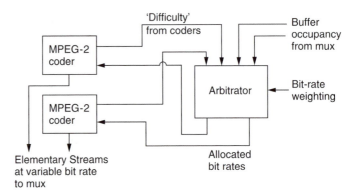

Figure 8.38 A statistical multiplexer contains an arbitrator which allocates bit rate to each program as a function of program difficulty.

begin to fill its output buffer. As the statmux has allocated more packets to that program, more data will be read out of that buffer, preventing overflow. Of course this is only possible if the other programs in the transport stream are handling typical video.

In the event that several programs encounter difficult material at once, clearly the buffer contents will rise and the requantizing mechanism will have to operate.

8.25   Introduction to DAB

Until the advent of NICAM and MAC, all sound broadcasting had been analog. The AM system is now very old indeed, and is not high fidelity by any standards, having a restricted bandwidth and suffering from noise, particularly at night. In theory, the FM system allows high quality and stereo, but in practice things are not so good. Most FM broadcast networks were planned when a radio set was a sizeable unit which usually needed an antenna for AM reception. Signal strengths were based on the assumption that a fixed FM antenna in an elevated position would be used. If such an antenna is used, reception quality is generally excellent. The forward gain of a directional antenna raises the signal above the front-end noise of the receiver and noise-free stereo is obtained. Such an antenna also rejects unwanted signal reflections.

Unfortunately, most FM receivers today are portable radios with whip antennae which have to be carefully oriented to give best reception. It is a characteristic of FM that slight mistuning causes gross distortion so an FM set is harder to tune than an AM set. In many places, nationally broadcast channels can be received on several adjacent frequencies at different strengths. Non-technical listeners tend to find all of this too

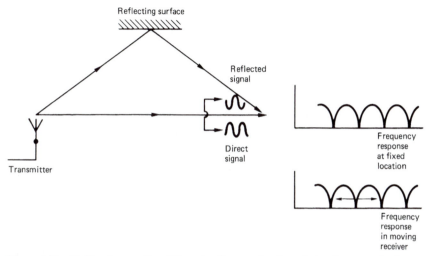

Figure 8.39 Multipath reception. When the direct and reflected signals are received with equal strength, nulling occurs at any frequency where the path difference results in a 180° phase shift.

much and surveys reveal that AM listening is still commonplace despite the same program being available on FM.

Reception on car radios is at a greater disadvantage as directional antennae cannot be used. This makes reception prone to multipath problems. Figure 8.39 shows that when the direct and reflected signals are received with equal strength, nulling occurs at any frequency where the path difference results in a 180° phase shift. Effectively a comb filter is placed in series with the signal. In a moving vehicle, the path lengths change, and the comb response slides up and down the band. When a null passes through the station tuned in, a burst of noise is created. Reflections from aircraft can cause the same problem in fixed receivers.

Digital audio broadcasting (DAB), also known as digital radio, is designed to overcome the problems which beset FM radio, particularly in vehicles. Not only does it do that, it does so using less bandwidth. With increasing pressure for spectrum allocation from other services, a system using less bandwidth to give better quality is likely to be favourably received.

8.26   DAB principles

DAB relies on a number of fundamental technologies which are combined into an elegant system. Compression is employed to cut the required bandwidth. Transmission of digital data is inherently robust as the receiver has only to decide between a small number of possible states.

Sophisticated modulation techniques help to eliminate multipath reception problems whilst further economizing on bandwidth. Error correction and concealment allow residual data corruption to be handled before conversion to analog at the receiver.

The system can only be realized with extremely complex logic in both transmitter and receiver, but with modern VLSI technology this can be inexpensive and reliable. In DAB, the concept of one-carrier-one-program is not used. Several programs share the same band of frequencies. Receivers will be easier to use since conventional tuning will be unnecessary. 'Tuning' consists of controlling the decoding process to select the desired program. Mobile receivers will automatically switch between transmitters as a journey proceeds.

Figure 8.40 shows the block diagram of a DAB transmitter. Incoming digital audio at 32 kHz is passed into the compression unit which uses the techniques described in Chapter 5 to cut the data rate to some fraction of the original. The compression unit could be at the studio end of the line

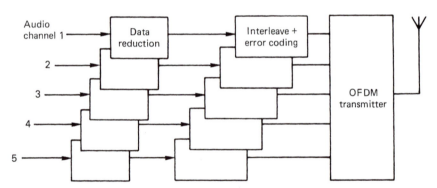

Figure 8.40 Block diagram of a DAB transmitter. See text for details.

to cut the cost of the link. The data for each channel are then protected against errors by the addition of redundancy. Convolutional codes described in Chapter 7 are attractive in the broadcast environment. Several such data-reduced sources are interleaved together and fed to the modulator, which may employ techniques such as randomizing which were introduced in Chapter 6.

Figure 8.41 shows how the multiple carriers in a DAB band are allocated to different program channels on an interleaved basis. Using this technique, it will be evident that when a notch in the received spectrum occurs due to multipath cancellation this will damage a small proportion of all programs rather than a large part of one program. This is the spectral equivalent of physical interleaving on a recording medium. The result is the same in that error bursts are broken up according to the

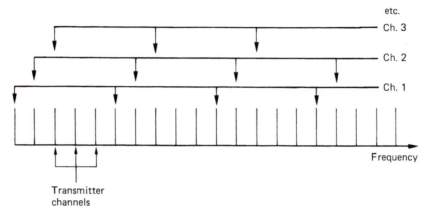

Figure 8.41 Channel interleaving is used in DAB to reduce the effect of multipath notches on a given program.

interleave structure into more manageable sizes which can be corrected with less redundancy.

A serial digital waveform has a $\sin x/x$ spectrum and when this waveform is used to phase modulate a carrier the result is a symmetrical $\sin x/x$ spectrum centred on the carrier frequency. Nulls in the spectrum appear at multiples of the phase switching rate away from the carrier. This distance is equal to 90° or one quadrant of $\sin x$. Further carriers can be placed at spacings such that each is centred at the nulls of the others. Owing to the quadrant spacing, these carries are mutually orthogonal, hence the term orthogonal frequency division.[26,27] A number of such carriers will interleave to produce an overall spectrum which is almost rectangular as shown in Figure 8.42(a). The mathematics describing this process is exactly the same as that of the reconstruction of samples in a low-pass filter and reference should be made to Figure 4.6. Effectively sampling theory has been transformed into the frequency domain.

In practice, perfect spectral interleaving does not give sufficient immunity from multipath reception. In the time domain, a typical reflective environment turns a transmitted pulse into a pulse train extending over several microseconds.[28] If the bit rate is too high, the reflections from a given bit coincide with later bits, destroying the orthogonality between carriers. Reflections are opposed by the use of guard intervals in which the phase of the carrier returns to an unmodulated state for a period which is greater than the period of the reflections. Then the reflections from one transmitted phase decay during the guard interval before the next phase is transmitted.[29] The principle is not dissimilar to the technique of spacing transitions in a recording further apart than the expected jitter. As expected, the use of guard intervals reduces the bit rate of the carrier because for some of the time it is radiating carrier not data. A typical reduction is to around 80 per cent

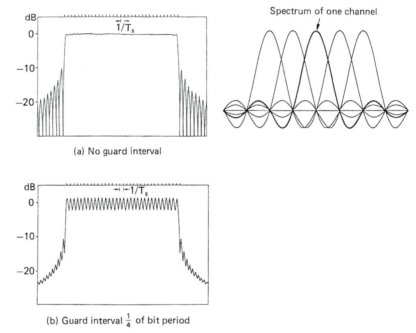

(a) No guard interval

(b) Guard interval $\frac{1}{4}$ of bit period

Figure 8.42 (a) When mutually orthogonal carriers are stacked in a band, the resultant spectrum is virtually flat. (b) When guard intervals are used, the spectrum contains a peak at each channel centre.

of the capacity without guard intervals. This capacity reduction does, however, improve the error statistics dramatically, such that much less redundancy is required in the error correction system. Thus the effective transmission rate is improved. The use of guard intervals also moves more energy from the sidebands back to the carrier. The frequency spectrum of a set of carriers is no longer perfectly flat but contains a small peak at the centre of each carrier as shown in Figure 8.42(b).

A DAB receiver must receive the set of carriers corresponding to the required program channel. Owing to the close spacing of carriers, it can only do this by performing fast Fourier transforms (FFTs) on the DAB band. If the carriers of a given program are evenly spaced, a partial FFT can be used which only detects energy at spaced frequencies and requires much less computation. This is the DAB equivalent of tuning. The selected carriers are then demodulated and combined into a single bitstream. The error-correction codes will then be de-interleaved so that correction is possible. Corrected data then pass through the expansion part of the data reduction coder, resulting in conventional PCM audio which drives DACs.

It should be noted that in the European DVB standard, COFDM transmission is also used. In some respects, DAB is simply a form of DVB without the picture data.

References

1. Audio Engineering Society, AES recommended practice for digital audio engineering – serial transmission format for linearly represented digital audio data. *J. Audio Eng. Soc.*, **33**, 975–984 (1985)
2. EIAJ CP-340, *A Digital Audio Interface*, Tokyo: EIAJ (1987)
3. EIAJ CP-1201, *Digital Audio Interface (revised)*, Tokyo: EIAJ (1992)
4. IEC 958, *Digital Audio Interface*, 1st edn, Geneva: IEC (1989)
5. Finger, R., AES3–1992: the revised two channel digital audio interface. *J. Audio.Eng. Soc.*, **40**, 107–116 (1992)
6. EIA RS-422A. Electronic Industries Association, 2001 Eye St NW, Washington, DC 20006, USA
7. Smart, D.L., Transmission performance of digital audio serial interface on audio tie lines. *BBC Designs Dept Technical Memorandum*, 3.296/84
8. European Broadcasting Union, Specification of the digital audio interface. *EBU Doc. Tech.*, 3250
9. Rorden, B. and Graham, M., A proposal for integrating digital audio distribution into TV production. *J. SMPTE*, 606–608 (Sept.1992)
10. Gilchrist, N., Co-ordination signals in the professional digital audio interface. In *Proc. AES/EBU Interface Conf.*, 13–15. Burnham: Audio Engineering Society (1989)
11. Digital audio taperecorder system (RDAT). Recommended design standard. DAT Conference, Part V (1986)
12. AES18–1992, Format for the user data channel of the AES digital audio interface. *J. Audio Eng. Soc.*, **40** 167–183 (1992)
13. Nunn, J.P., Ancillary data in the AES/EBU digital audio interface. In *Proc. 1st NAB Radio Montreux Symp.*, 29–41 (1992)
14. Komly, A and Viallevieille, A., Programme labelling in the user channel. In *Proc. AES/EBU Interface Conf.*, 28–51. Burnham: Audio Engineering Society (1989)
15. ISO 3309, *Information processing systems – data communications – high level data link frame structure* (1984)
16. AES10–1991, Serial multi-channel audio digital interface (MADI). *J. Audio Eng. Soc.*, **39**, 369–377 (1991)
17. Ajemian, R.G. and Grundy, A.B., Fiber-optics – the new medium for audio: a tutorial. *J.Audio Eng. Soc.*, **38** 160–175 (1990)
18. Lidbetter, P.S. and Douglas, S., A fibre-optic multichannel communication link developed for remote interconnection in a digital audio console. Presented at the 80th Audio Engineering Society Convention (Montreux, 1986), Preprint 2330
19. Dunn, J., Considerations for interfacing digital audio equipment to the standards AES3, AES5 and AES11. In *Proc. AES 10th International Conf.*, 122, New York: Audio Engineering Society (1991)
20. Gilchrist, N.H.C., Digital sound: sampling-rate synchronization by variable delay. *BBC Research Dept Report*, 1979/17
21. Lagadec, R., A new approach to sampling rate synchronisation. Presented at the 76th Audio Engineering Society Convention (New York, 1984), Preprint 2168
22. Shelton, W.T., Progress towards a system of synchronization in a digital studio. Presented at the 82nd Audio Engineering Society Convention (London, 1986), Preprint 2484(K7)
23. Shelton, W.T., Interfaces for digital audio engineering. Presented at the 6th International Conference on Video Audio and Data Recording, Brighton. IERE Publ. No. 67, 49–59 1986
24. Anon., NICAM 728: specification for two additional digital sound channels with System I television. BBC Engineering Information Dept (London, 1988)
25. Anon. Specification of the system of the MAC/packet family. EBU Tech. Doc. 3258 (1986)
26. Cimini, L.J., Analysis and simulation of a digital mobile channel using orthogonal frequency division multiplexing. *IEEE Trans. Commun.*, **COM-33**, No.7 (1985)

27. Pommier, D. and Wu. Y., Interleaving or spectrum spreading in digital radio intended for vehicles. *EBU Tech. Review*, No. 217, 128–142 (1986)

28. Cox, D.C., Multipath delay spread and path loss correlation for 910 MHz urban mobile radio propagation. *IEEE Trans. Vehic. Tech.*, **VT-26** (1977)

29. Alard, M. and Lasalle, R., Principles of modulation and channel coding for digital broadcasting for mobile receivers. *EBU Tech. Review*, No. 224, 168–190 (1987)

9

Digital audio tape recorders

Tape recording has played a large part in the history of digital audio, and continues to be important although the rapid adoption of recorders based on hard disks and optical disks is having a significant effect. Tape recording using the rotary-head principle pioneered in video recorders is used in digital audio alongside the more conventional stationary head approach. Both of these will be considered here. The reader is referred to Chapters 6 and 7 for an explanation of coding and error-correction principles.

9.1   Types of recorder

Digital audio become economic with the development of high-density recorders in the 1970s. The necessary rate of almost two megabits per second for a stereo signal can today be recorded with a very low tape consumption. It is not so long ago, however, that the data rate itself was a problem. When head and tape technology were less advanced than they are today, wavelengths on tape were long, and the only way that high frequencies could be accommodated was to use high speeds. High speed can be achieved in two ways. The head can remain fixed, and the tape can be transported rapidly, with obvious consequences, or the tape can travel relatively slowly, and the head can be moved. The latter is the principle of the rotary-head recorder. Figure 9.1 shows the general arrangement of the two major categories of rotary-head recorder. In transverse-scan recorders, relatively short tracks are recorded almost at right angles to the direction of tape motion by a rotating headwheel containing, typically, four heads. In helical-scan recorders, the tape is wrapped around the drum in such a way that it enters and leaves in two different planes. This

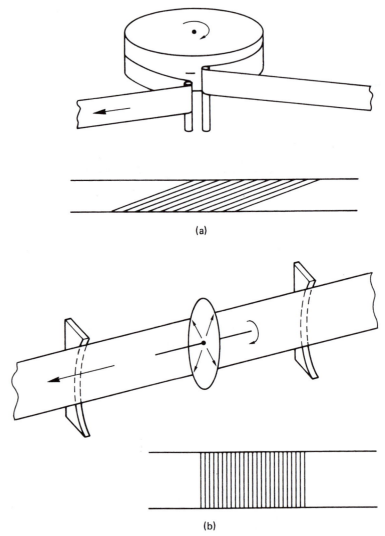

Figure 9.1 Types of rotary-head recorder. (a) Helical scan records long diagonal tracks. (b) Transverse scan records short tracks across the tape.

causes the rotating heads to record long slanting tracks. In both approaches, the width of the space between tracks is determined by the linear tape speed. The track pitch can easily be made much smaller than in stationary-head recorders.

The use of rotary heads was instrumental in the development of the first video recorders. As video signals consist of discrete lines and frames, it was possible to conceal the interruptions in the tracks of a rotary-head machine by making them coincident with the time when the CRT was blanked during flyback.

If digital sample data are encoded to resemble a video waveform, which is known as pseudo-video or composite digital, they can be recorded on a fairly standard video recorder. Digital audio recorders have been made using quadruplex video recorders, one-inch video recorders, U-matic cassette recorders, and the smaller consumer formats. The device needed to format the samples in this way is called a PCM adaptor.

Digital audio recorders have also been made which use only the transport of a video recorder, with specially designed digital signal electronics. Instead of using analog FM, it is possible to use digital recording, as described in Chapter 6, to make a direct digital recorder. The machines built by Decca fall into this category.

The final category of digital audio recorder using rotary heads is one in which direct digital recording is used with a transport specially designed for audio use with no compromises due to a video-based ancestry. DAT is such a machine.

9.2   PCM adaptors

Figure 9.2 shows a block diagram of a PCM adaptor. The unit has five main sections. Central to operation is the sync and timing generation,

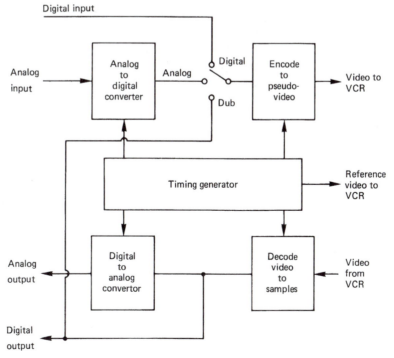

Figure 9.2   Block diagram of PCM adaptor. Note the dub connection needed for producing a digital copy between two VCRs.

which produces sync pulses for control of the video waveform generator and locking the video recorder, in addition to producing sampling-rate clocks and timecode. An ADC allows a conventional analog audio signal to be recorded, but this can be bypassed if a suitable digital input is available. Similarly a DAC is provided to monitor recordings, and this too can be bypassed by using the direct digital output. Also visible in Figure 9.2 are the encoder and decoder stages which convert between digital sample data and the pseudo-video signal.

An example of this type of unit is the PCM-1610/1630 which was designed by Sony for use with a U-matic Video Cassette Recorder (VCR) specifically for Compact Disc mastering. Chapter 4 showed how many audio sampling rates were derived from video frequencies. The Compact Disc format is an international standard, and it was desirable for the mastering recorder to adhere to a single format. Thus the PCM-1610 only worked in conjunction with a 525/60 monochrome VCR. There was no 625/50 version. Thus even in PAL countries Compact Discs were mastered on 60 Hz VCRs to allow the traditional international interchange of recordings. The PCM-1610 was intended for professional use, and thus was not intended to be produced in volume. For this reason the format is simple, even crude, because the LSI technology needed to implement more complex formats was not available.

A typical line of pseudo-video is shown in Figure 9.3. The line is divided into bit cells and, within them, black level represents a binary zero, and about 60 per cent of peak white represents binary one. The reason for the restriction to 60 per cent is that most VCRs use non-linear pre-emphasis and this operating level prevents any distortion due to the pre-emphasis causing misinterpretation of the pseudo-video. The use of a two-level input to a frequency modulator means that the recording is essentially frequency-shift keyed (FSK).

As the video recorder is designed to switch heads during the vertical interval, no samples can be recorded there. In all rotary-head recorders,

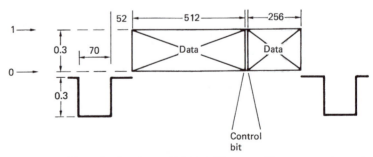

Figure 9.3 Typical line of video from PCM-1610. The control bit conveys the setting of the pre-emphasis switch or the sampling rate depending on position in the frame. The bits are separated using only the timing information in the sync pulses.

some form of time compression is used to squeeze the samples into the active parts of unblanked lines. This is simply done by reading the samples from a memory at an instantaneous rate which is higher than the sampling rate. Owing to the interruptions of sync pulses, the average rate achieved will be the same as the sampling rate. The samples read from the memory must be serialized so that each bit is sent in turn.

It was shown in Chapter 7 that digital audio recorders use extensive interleaving to combat tape dropout. The PCM-1610 subdivides each video field into seven blocks of 35 lines each, and interleaves samples within the blocks. A simple crossword error-correction scheme is used. Some VCRs have dropout compensators built in, which repeat a section of the previous line to conceal the missing picture information. Such circuits must be disabled when used with PCM adaptors because they interfere with the error-correction mechanism.

A PCM adaptor for Compact Disc mastering was also developed by JVC.[1] This format had a more powerful error-correction system, and could be used with VHS recorders; again only 525/60 machines were supported.

For consumer use, a PCM adapter format was specified by the EIAJ[2] which would record stereo with fourteen-bit linear quantizing. These units would be used with a domestic VCR. Since the consumer would expect to be able to use the VCR for conventional TV recording as well, the EIAJ format is in fact two incompatible formats. One uses a sampling rate of 44.0559 kHz in conjunction with 525/59.94 NTSC timing, and one uses 44.1 kHz sampling with 625/50 PAL timing. In the popular PCM-F1, Sony produced a variation on the format which allowed sixteen-bit linear quantizing.

The PCM-F1 was built with LSI technology for low mass-production cost. Owing to the low cost of the product, it found application in professional circles, and indeed served as the introduction to digital audio for many people. Being a consumer product, only one convertor was used between digital and analog domains. This was multiplexed between the two audio channels, resulting in a timeshift between samples of half the sample period, or about 11 μs. This was not a problem in normal use, since the opposite shift was introduced by the multiplexed convertor used for replay. The standard PCM-F1 was not equipped with digital outputs or inputs, and accordingly not too much trouble was taken in controlling DC offsets due to convertor drift. When enthusiasts began to modify the unit to fit digital connections, these problems became significant.

Several companies manufactured adaptor units incorporating digital filters to remove DC offsets and the 11 μs shift to produce a standard AES/EBU output.

In contrast to the formats described above, which used the signal circuitry of video recorders virtually unmodified, the digital recorders developed by Decca[3] for vinyl and Compact Disc mastering use only the transport and servomechanisms of a 625/50 one-inch open-reel video recorder, and make a direct digital recording on the diagonal tracks using MFM channel code.

9.3 Introduction to DAT

When an existing video recorder is used as a basis for a digital audio recorder, the video bandwidth is already defined, and in most cases is much greater than necessary. Furthermore, the signal-to-noise ratio of video recorders is much too high for the purposes of storing binary. The result of these factors is that the tape consumption of such a machine will be far higher than necessary.

As digital audio became established, and markets opened up for large numbers of machines, it was no longer necessary to borrow technology from other disciplines, because it was economically viable to design a purpose-built product. The first of this generation of machines is DAT (digital audio tape). By designing for a specific purpose, the tape consumption can be made very much smaller than that of a converted video machine. In fact the DAT format achieved more bits per square inch than any other form of magnetic recorder at the time of its introduction. The origins of DAT are in an experimental machine built by Sony,[4] but the DAT format has grown out of that through a process of standardization involving some eighty companies.

The general appearance of the DAT cassette is shown in Figure 9.4. The overall dimensions are only 73 mm × 54 mm × 10.5 mm which is rather smaller than the Compact Cassette. The design of the cassette incorporates some improvements over its analog ancestor.[5] As shown in Figure 9.5, the apertures through which the heads access the tape are closed by a hinged door, and the hub drive openings are covered by a sliding panel which also locks the door when the cassette is not in the transport. The act of closing the door operates brakes which act on the reel hubs. This results in a cassette which is well sealed against contamination due to handling or storage. The short wavelengths used in digital recording make it more sensitive to spacing loss caused by contamination.

As in the Compact Cassette, the tape hubs are flangeless, and the edge guidance of the tape pack is achieved by liner sheets. The flangeless approach allows the hub centres to be closer together for a given length of tape. The cassette has recognition holes in four standard places so that players can automatically determine what type of cassette has been inserted. In addition there is a write-protect (record-lockout) mechanism

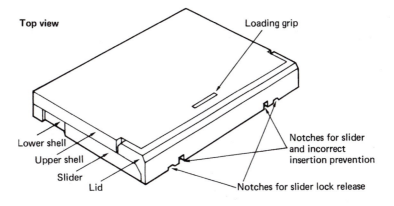

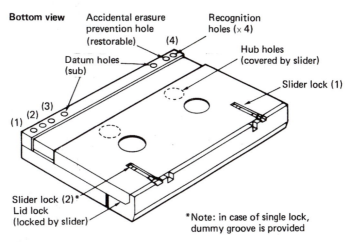

Figure 9.4 Appearance of DAT cassette. Access to the tape is via a hinged lid, and the hub-drive holes are covered by a sliding panel, affording maximum protection to the tape. Further details of the recognition holes are given in Table 9.1. (Courtesy TDK)

which is actuated by a small plastic plug sliding between the cassette halves. The end-of-tape condition is detected optically and the leader tape is transparent. There is some freedom in the design of the EOT sensor. As can be seen in Figure 9.6, transmitted-light sensing can be used across the corner of the cassette, or reflected-light sensing can be used, because the cassette incorporates a prism which reflects light around the back of the tape. Study of Figure 9.6 will reveal that the prisms are moulded integrally with the corners of the transparent insert used for the cassette window.

The high coercivity (typically 1480 oersteds) metal powder tape is 3.81 mm wide, the same width as Compact Cassette tape. The standard overall thickness is 13 μm. A striking feature of the metal tape is that the

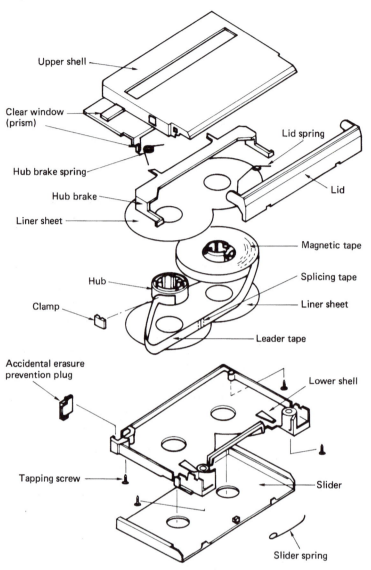

Figure 9.5 Exploded view of DAT cassette showing intricate construction. When the lid opens, it pulls the ears on the brake plate, releasing the hubs. Note the EOT/BOT sensor prism moulded into the corners of the clear window. (Courtesy TDK)

magnetic coating is so thin, at about $3\,\mu m$, that the tape appears translucent. The maximum capacity of the cassette is about $60\,m$.

When the cassette is placed in the transport, the slider is moved back as it engages. This releases the lid lock. Continued movement into the transport pushes the slider right back, revealing the hub openings. The cassette is then lowered onto the hub drive spindles and tape guides, and the door is fully opened to allow access to the tape.

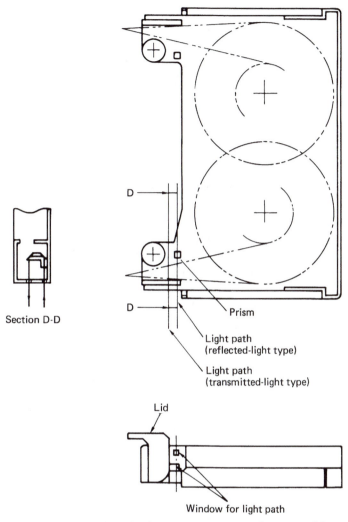

Section D-D

Prism

Light path
(reflected-light type)

Light path
(transmitted-light type)

Lid

Window for light path

Figure 9.6 Tape sensing can be either by transmission across the corner of the cassette, or by reflection through an integral prism. In both cases, the apertures are sealed when the lid closes. (Courtesy TDK)

In DAT, threading is simplified because the digital recording does not need to be continuous. DAT extends the technique of time compression used to squeeze continuous samples into intermittent video lines. Blocks of samples to be recorded are written into a memory at the sampling rate, and are read out at a much faster rate when they are to be recorded. In this way the memory contents can be recorded in less time. Figure 9.7 shows that when the samples are time-compressed, recording is no longer continuous, but is interrupted by long pauses. During the pauses in recording, it is not actually necessary for the head to be in contact with the tape, and so the angle of wrap of the tape around the drum can be

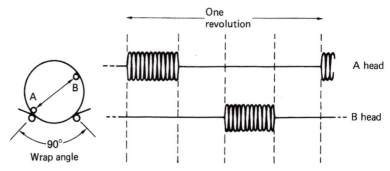

Figure 9.7 The use of time compression reduces the wrap angle necessary, at the expense of raising the frequencies in the channel.

reduced, which makes threading easier. In DAT the wrap angle is only 90° on the commonest drum size. As the heads are 180° apart, this means that for half the time neither head is in contact with the tape. Figure 9.8 shows that the partial-wrap concept allows the threading mechanism to be very simple indeed. As the cassette is lowered into the transport, the pinch roller and several guide pins pass behind the tape. These then simply move toward the capstan and drum and threading is complete. A further advantage of partial wrap is that the friction between the tape and drum is reduced, allowing power saving in portable applications, and allowing the tape to be shuttled at high speed without the partial unthreading needed by videocassettes. In this way the player can read subcode during shuttle to facilitate rapid track access.

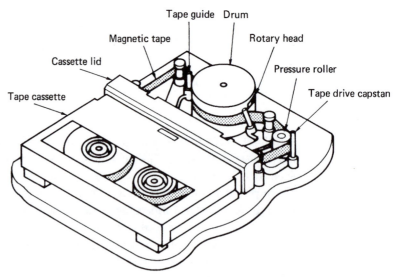

Figure 9.8 The simple mechanism of DAT. The guides and pressure roller move towards the drum and capstan and threading is complete.

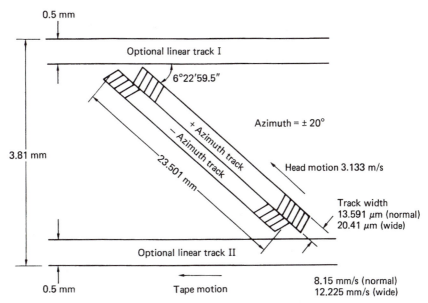

0.5 mm

Optional linear track I

6°22'59.5"

Azimuth = ± 20°

3.81 mm

+ Azimuth track
– Azimuth track

23.501 mm

Head motion 3.133 m/s

Track width
13.591 μm (normal)
20.41 μm (wide)

Optional linear track II

0.5 mm

Tape motion

8.15 mm/s (normal)
12.225 mm/s (wide)

Figure 9.9 The two heads of opposite azimuth angles lay down the above track format. Tape linear speed determines track pitch.

The track pattern laid down by the rotary heads is shown in Figure 9.9. The heads rotate at 2000 rev/min in the same direction as tape motion, but because the drum axis is tilted, diagonal tracks 23.5 mm long result, at an angle of just over six degrees to the edge. The diameter of the scanner needed is not specified, because it is the track pattern geometry which ensures interchange compatibility. For portable machines, a small scanner is desirable, whereas for professional use, a larger scanner allows additional heads to be fitted for confidence replay and editing. It will be seen from Figure 9.9 that azimuth recording is employed as was described in Chapter 6. This requires no spaces or guard bands between the tracks. The chosen azimuth angle of ± 20° reduces crosstalk to the same order as the noise, with a loss of only 1 dB due to the apparent reduction in writing speed.

In addition to the diagonal tracks, there are two linear tracks, one at each edge of the tape, where they act as protection for the diagonal tracks against edge damage. Owing to the low linear tape speed the use of these edge tracks is somewhat limited.

Several related modes of operation are available, some of which are mandatory whereas the remainder are optional. These are compared in Table 9.1. The most important modes use a sampling rate of 48 kHz or 44.1 kHz, with sixteen-bit two's complement uniform quantization. Alongside the audio samples can be carried 273 kbits/s of subcode (about four times that of Compact Disc) and 68.3 kbits/s of ID coding, whose

Table 9.1 The significance of the recognition holes on the DAT cassette. Holes 1, 2 and 3 form a coded pattern; whereas hole 4 is independent.

Hole 1	Hole 2	Hole 3	Function
0	0	0	Metal powder tape or equivalent/13 μm thick
0	1	0	MP tape or equivalent/thin tape
0	0	1	1.5 TP/13 μm thick
0	1	1	1.5 TP/thin tape
1	×	×	(Reserved)

Hole 4		1 = Hole present 0 = Hole blanked off
0	Non-prerecorded tape	
1	Prerecorded tape	

purpose will be explained in due course. With a linear tape speed of 8.15 mm/s, the standard cassette offers 120 min unbroken playing time. Initially it was proposed that all DAT machines would be able to record and play at 48 kHz, whereas only professional machines would be able to record at 44.1 kHz. For consumer machines, playback only of prerecorded media was proposed at 44.1 kHz, so that the same software could be released on CD or prerecorded DAT tape. Now that a SCMS (serial copying management system) is incorporated into consumer machines, they too can record at 44.1 kHz. For reasons which will be explained later, contact duplicated tapes run at 12.225 mm/s to offer a playing time of 80 min. The same subcode and ID rate is offered. The above modes are mandatory if a machine is to be considered to meet the format.

Option 1 is identical to 48 kHz mode except that the sampling rate is 32 kHz. Option 2 is an extra-long-play mode. In order to reduce the data rate, the sampling rate is 32 kHz and the samples change to twelve-bit two's complement with non-linear quantizing. Halving the subcode rate allows the overall data rate necessary to be halved. The linear tape speed and the drum speed are both halved to give a playing time of four hours. All the above modes are stereo, but option 3 uses the sampling parameters of option 2 with four audio channels. This doubles the data rate with respect to option 2, so the standard tape speed of 8.15 mm/s is used.

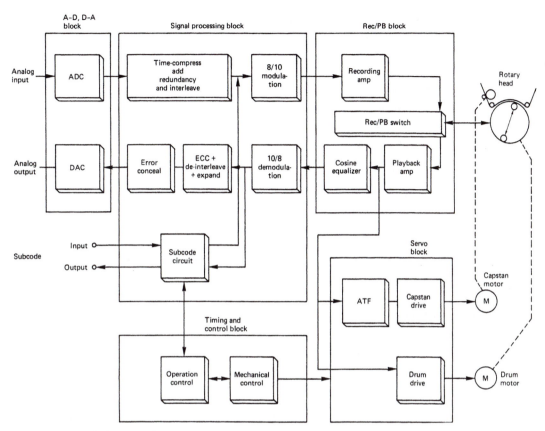

Figure 9.10   Block diagram of DAT.

Figure 9.10 shows a block diagram of a typical DAT recorder, which
will be used to introduce the basic concept of the machine and the major
topics to be described. In order to make a recording, an analog signal is
fed to an input ADC, or a direct digital input is taken from an AES/EBU
interface. The incoming samples are subject to interleaving to reduce the
effects of error bursts. Reading the memory at a higher rate than it was
written performs the necessary time compression. Additional bytes of
redundancy computed from the samples are added to the data stream to
permit subsequent error correction. Subcode information such as the
content of the AES/EBU channel status message is added, and the
parallel byte structure is fed to the channel encoder, which combines a bit
clock with the data, and produces a recording signal according to the
8/10 code which is free of DC (see Chapter 6). This signal is fed to the
heads via a rotary transformer to make the binary recording, which leaves
the tape track with a pattern of transitions between the two magnetic
states.

On replay, the transitions on the tape track induce pulses in the head, which are used to re-create the record current waveform. This is fed to the 10/8 decoder which converts it to the original data stream and a separate clock. The subcode data are routed to the subcode output, and the audio samples are fed into a de-interleave memory which, in addition to time-expanding the recording, functions to remove any wow or flutter due to head-to-tape speed variations. Error correction is performed partially before and partially after de-interleave. The corrected output samples can be fed to DACs or to a direct digital output.

In order to keep the rotary heads following the very narrow slant tracks, alignment patterns are recorded as well as the data. The automatic track-following system processes the playback signals from these patterns to control the drum and capstan motors. The subcode and ID information can be used by the control logic to drive the tape to any desired location specified by the user.

9.4   Track following in DAT

As with any recorder intended for consumer use, economy of tape consumption is paramount, and this involves numerous steps to use the tape area as efficiently as possible. As magnetic tape is flexible and is manufactured to finite tolerances, there will always be some error between the path of the replay head and the recorded track. In the relatively wide tracks of analog audio recorders this is seldom a problem. The high-output metal tape used in DAT allows an adequate signal-to-noise ratio to be obtained with very narrow tracks on the tape. This reduces tape consumption and allows a small cassette, but it becomes necessary actively to control the relative position of the head and the track in order to maximize the replay signal and minimize the error rate.

The track width and the coercivity of the tape largely define the signal-to-noise ratio. A track width has been chosen which makes the signal-to-crosstalk ratio dominant in cassettes which are intended for user recording.

Prerecorded tapes are made by contact duplication, and this process only works if the coercivity of the copy is less than that of the master. The output from prerecorded tapes at the track width of 13.59 μm would be too low, and would be noise-dominated, which would cause the error rate to rise. The solution to this problem is that in prerecorded tapes the track width is increased to be the same as the head pole. The noise and crosstalk are both reduced in proportion to the reduced output of the medium, and the same error rate is achieved as for normal high-coercivity tape.

The 50 per cent increase in track width is achieved by raising the linear tape speed from 8.15 to 12.225 mm/s, and so the playing time of a prerecorded cassette falls to 80 min as opposed to the 120 min of the normal tape.

The track-following principles are the same for prerecorded and normal cassettes, but there are detail differences which will be noted. Tracking is achieved in conventional video recorders by the use of a linear control track which contains one pulse for every diagonal track. The phase of the pulses picked up by a fixed head is compared with the phase of pulses generated by the drum, and the error is used to drive the capstan. This method is adequate for the wide tracks of analog video recorders, but errors in the mounting of the fixed head and variations in tape tension rule it out for high-density use. In any case the control-track head adds undesirable mechanical complexity. In DAT, the tracking is achieved by reading special alignment patterns on the tape tracks themselves, and using the information contained in them to control the capstan.

DAT uses a technique called area-divided track following (ATF) in which separate parts of the track are set aside for track-following purposes. Figure 9.11 shows the basic way in which a tracking error is derived. The tracks at each side of the home track have bursts of pilot tone recorded in two different places. The frequency of the pilot tone is

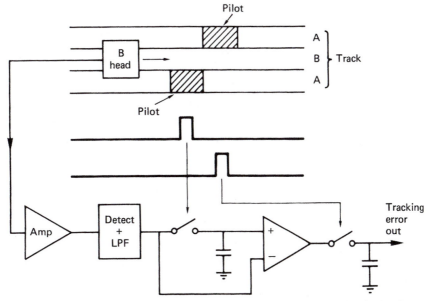

Figure 9.11 In the track-following system of DAT, the signal picked up by the head comes from pilot tones recorded in adjacent tracks at different positions. These pilot tones have low frequency, and are unaffected by azimuth error. The system samples the amplitude of the pilot tones, and subtracts them.

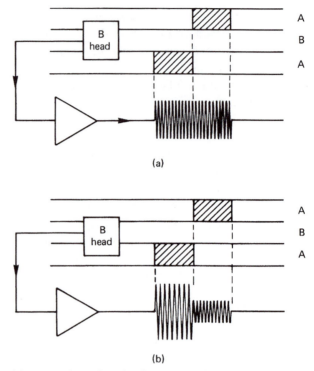

Figure 9.12 (a) A correctly tracking head produces pilot-tones bursts of identical amplitude. (b) The head is off track, and the first pilot burst becomes larger, whereas the second becomes smaller. This produces the tracking error in the circuit of Figure 9.11.

130 kHz, which has been chosen to be relatively low so that it is not affected by azimuth loss. In this way an A head following an A track will be able to detect the pilot tone from the adjacent B tracks.

In Figure 9.12(a) the case of a correctly tracking head is shown. The amount of side-reading pilot tone from the two adjacent B tracks is identical. If the head is off track for some reason, as shown in Figure 9.12(b), the amplitude of the pilot tone from one of the adjacent tracks will increase, and the other will decrease. The tracking error is derived by sampling the amplitude of each pilot-tone burst as it occurs, and holding the result so the relative amplitudes can be compared.

There are some practical considerations to be overcome in implementing this simple system, which result in some added complication. The pattern of pilot tones must be such that they occur at different times on each side of every track. To achieve this there must be a burst of pilot tone in every track, although the pilot tone in the home track does not contribute to the development of the tracking error. Additionally there must be some timing signals in the tracks to determine when the samples of pilot tone should be made. The final issue is to prevent the false locking which could occur if the tape happened to run at twice normal speed.

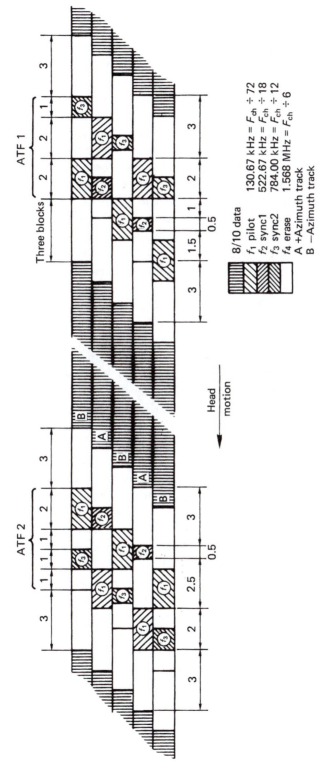

Figure 9.13 The area-divided track-following (ATF) patterns of DAT. To ease generation of patterns on recording, the pattern lengths are related to the data-block dimensions and the frequencies used are obtained by dividing down the channel bit clock F_{ch}. The sync signals are used to control the timing with which the pilot amplitude is sampled.

Figure 9.13 shows how the actual track-following pattern of DAT is laid out.[6] The pilot burst is early on A tracks and late on B tracks. Although the pilot bursts have a two-track cycle, the pattern is made to repeat over four tracks by changing the period of the sync patterns which control the pilot sampling. This can be used to prevent false locking. When an A head enters the track, it finds the home pilot-burst first, followed by pilot from the B track above, then pilot from the B track below. The tracking error is derived from the latter two. When a B head enters the track, it sees pilot from the A track above first, A track below next, and finally home pilot. The tracking error in this case is derived from the former two. The machine can easily tell which processing mode to use because the sync signals have a different frequency depending on whether they are in A tracks (522 kHz) or B tracks (784 kHz). The remaining areas are recorded with the interblock gap frequency of 1.56 MHz which serves no purpose except to erase earlier recordings. Although these pilot and synchronizing frequencies appear strange, they are chosen so that they can be simply obtained by dividing down the master channel-bit-rate clock by simple factors. The channel-bit-rate clock, F_{ch}, is 9.408 MHz; pilot, the two sync frequencies and erase are obtained by dividing it by 72, 18, 12 and 6 respectively. The time at which the pilot amplitude in adjacent tracks should be sampled is determined by the detection of the synchronizing frequencies. As the head sees part of three tracks at all times, the sync detection in the home track has to take place in the presence of unwanted signals. On one side of the home sync signal will be the interblock gap frequency, which is high enough to be attenuated by azimuth. On the other side is pilot, which is unaffected by azimuth. This means that sync detection is easier in the tracking-error direction away from pilot than in the direction towards it. There is an effective working range of about +4 and −5 μm due to this asymmetry, with a dead band of 4 μm between tracks. Since the track-following servo is designed to minimize the tracking error, once lock is achieved the presence of the dead zone becomes academic. The differential amplitude of the pilot tones produces the tracking error, and so the gain of the servo loop is proportional to the playback gain, which can fluctuate due to head contact variations and head tolerance. This problem is overcome by using AGC in the servo system. In addition to subtracting the pilot amplitudes to develop the tracking error, the circuitry also adds them to develop an AGC voltage. Two sample-and-hold stages are provided which store the AGC parameter for each head separately. The heads can thus be of different sensitivities without upsetting the servo. This condition could arise from manufacturing tolerances, or if one of the heads became contaminated.

9.5   Aligning for interchange

One of the most important aspects of DAT maintenance is to ensure that tapes made on a particular machine meet the specifications laid down in the format. If they do, then it will be possible to play those tapes on any other properly aligned machine. In this section the important steps necessary to achieve interchange between transports will be outlined.

When the cassette is lowered into the transport it seats on pillars which hold it level. The tape within the cassette is guided by liner sheets which determine the height of the tape pack above the transport baseplate. The first step in aligning the transport is to ensure that all the guides the tape runs past on its way to and from the scanner are at the same height as the tape. Figure 9.14 shows that the guides are threaded so that they can be screwed up and down. In the correct position, the tape will stay in the cassette plane and distortion will be avoided.

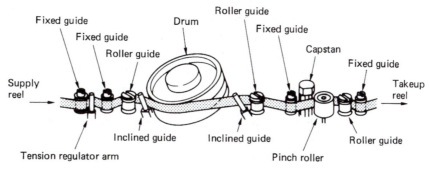

Figure 9.14 Conceptual view of DAT transport with tape path straightened for simplicity. Note the slotted guides which can be turned by screwdriver on threaded posts in order to adjust the height of the tape above the deck. This must be set to feed the tape cleanly from the cassette to the scanner. (Courtesy Sony Broadcast and Communications)

Once the tape can be passed through the machine without damage, the basic transport functions can be checked. Since tape tension affects the track angle, obtaining the correct tension is essential before attempting any adjustments at the scanner. As the DAT mechanism is so small, it is not possible to fit a conventional tape tension gauge. Instead, special test cassettes are made which incorporate torque meters into the reel hubs. Use of these test cassettes will allow the tension to be checked in various transport and shuttle modes. Since the scanner friction is in the opposite sense when the tape is reversed, the back tension must be higher than for forward mode to keep the average scanner tension constant. In some transports the tension-sensing arm is not statically balanced, and the tape tension becomes a function of the orientation of the machine. In this case the adjustment must be made with the machine in the attitude in which it is to be used.

The track spacing on record is determined by the capstan speed, which must be checked. As the capstan speed will be controlled by a frequency-generating wheel on the capstan shaft, it is generally only necessary to check that the capstan FG frequency is correct in record mode. A scratch tape will be used for this check.

Helical interchange can now be considered. Tape passing around the scanner is guided in three ways. On the approach, the tape is steered by the entrance guide, which continues to affect the first part of the scanner wrap. The centre part of the scanner wrap is guided by the machined step on the scanner base. Finally the last part of the wrap is steered by the exit guide. Helical interchange is obtained by adjusting the entrance and exit guide heights so that the tape passes smoothly between the three regions. In video recorders, which use wider tape, it will often also be found necessary to adjust the angle of the guides. In DAT the tape is so narrow that it will flex to accommodate angular errors, and the guides only need a height adjustment.

As tape is flexible, it will distort as it passes round the scanner if the entrance and exit guides are not correctly set. The state of alignment can be assessed by working out the effect of misalignments on the ability of the replay head to follow tape tracks. Figure 9.15(a) shows an example of the entrance guide being too low. The tape is forced to climb up to reach the scanner step, and then it has to bend down again to run along the step. If straight tracks were originally recorded on the tape, they will no longer be straight when the tape is distorted in this way. Figure 9.15(b) shows what happens to the track. Since in an azimuth recording machine the head is larger than the track, small distortions of this kind will be undetectable. It is necessary to offset the tracking deliberately so that the effect of the misalignment can be seen. If the head is offset upwards, then the effect will be that the RF signal grows in level briefly at the beginning of the track, giving the envelope an onion-like appearance on an oscilloscope. If the head is offset downwards, the distortion will take the track away from the head path and the RF envelope will be waisted. If the misalignment is in the exit guide, then the envelope disturbances will appear at the right-hand end of the RF envelope. The height of both the entrance and exit guides is adjusted until no disturbance of the RF envelope is apparent whatever tracking error is applied. One simple check is to disable the track-following system so that the capstan runs at approximately playback speed with no feedback. The tracking will slowly drift in and out of registration. Under these conditions the RF envelope should remain rectangular, so that the amplitude rises and falls equally over the entire head sweep. A rough alignment can be performed with a

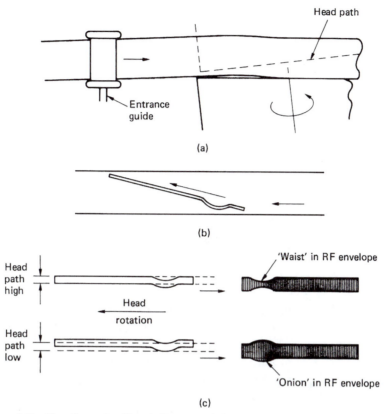

Figure 9.15 The effects of guide misalignment shown exaggerated for clarity. In (a) the entrance guide is too low and forces the tape to flex as it enters the scanner. In (b) the tape relaxes to produce a bent track. In (c) an alignment tape is being played and the transport has bent the tracks. The RF envelope will show different disturbances as the tracking is offset above or below optimum. A correctly aligned transport has an envelope which collapses uniformly as the tracking is offset.

tape previously recorded on a trustworthy machine, but final align-ment requires the use of a reference tape.

Once the mechanical geometry of the transport is set up, straight tracks on tape will appear straight to the scanner, and it is then possible to set up the automatic track-following system so that optimum tracking occurs when the tracking error is zero. This is done by observing the RF level on playback and offsetting the tracking adjustment to each side until two points are found where the level begins to fall. The adjustment is then placed half-way between these points. If the machine has a front panel tracking adjustment, it should be set to zero whilst the internal adjustment is made.

The final interchange adjustment is to ensure that the scanner timing is correct. Even with correct geometry, the tracks can be laid down at the correct angle and spacing, but at the wrong height on the tape, as Figure

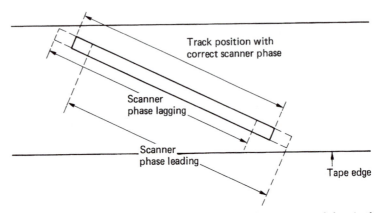

Figure 9.16 Even with correct tape path geometry, a machine can record data in the wrong place along the track if the scanner phase is misadjusted. The scanner pulse generator signal is rephased or delayed in order to make the adjustment.

9.16 shows. The point where recording commences is determined by the sensor which generates a pulse once per revolution of the scanner. The correct timing can be obtained either by physically moving the sensor around the scanner axis, or by adjusting a variable delay in series with an artificially early fixed sensor. A timing reference tape is necessary that has an observable event in the RF waveform. The tape is played, and the sensor or delay is adjusted to give the specified relative timing between the event on the reference tape and the sensor pulse. When this is correct, the machine will record tracks in the right place along the helical sweep.

9.6 DAT data channel

The channel code used in DAT is designed to function well in the presence of crosstalk, to have zero DC component to allow the use of a rotary transformer, and to have a small ratio of maximum and minimum run lengths to ease overwrite erasure. The code used is a group code where eight data bits are represented by ten channel bits, hence the name 8/10. The details of the code are given in Chapter 6.

The basic unit of recording is the sync block shown in Figure 9.17. This consists of the sync pattern, a three-byte header and 32 bytes of data, making 36 bytes in total, or 360 channel bits. The subcode areas each consist of eight of these blocks, and the PCM audio area consists of 128 of them. Note that a preamble is only necessary at the beginning of each area to allow the data separator to phase-lock before the first sync block arrives. Synchronism should be maintained throughout the area, but the

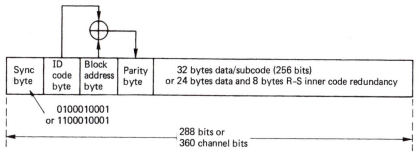

Figure 9.17 The sync block of DAT begins with a sync pattern of ten channel bits, which does not correspond to eight data bits. The header consists of an ID code byte and a block address. Parity is formed on the header bytes. The sync blocks alternate between 32 data (or outer code) bytes and 24 data bytes and 8 bytes of R–S redundancy for the inner codes.

sync pattern is repeated at the beginning of each sync block in case sync is lost due to dropout.

The first byte of the header contains an ID code which in the PCM audio blocks specifies the sampling rate in use, the number of audio channels, and whether there is a copy-prohibit in the recording. The second byte of the header specifies whether the block is subcode or PCM audio with the first bit. If set, the least significant four bits specify the subcode block address in the track, whereas if it is reset, the remaining seven bits specify the PCM audio block address in the track. The final header byte is a parity check and is the exclusive-OR sum of header bytes one and two.

The data format within the tracks can now be explained. The information on the track has three main purposes, PCM audio, subcode data and ATF patterns. It is necessary to be able to record subcode at a different time from PCM audio in professional machines in order to update or post-stripe the timecode. The subcode is placed in separate areas at the beginning and end of the tracks. When subcode is recorded on a tape with an existing PCM audio recording, the heads have to go into record at just the right time to drop a new subcode area onto the track. This timing is subject to some tolerance, and so some leeway is provided by the margin area which precedes the subcode area and the interblock gap (IBG) which follows. Each area has its own preamble and sync pattern so the data separator can lock to each area individually even though they were recorded at different times or on different machines.

The track-following system will control the capstan so that the heads pass precisely through the centre of the ATF area. Figure 9.18 shows that, in the presence of track curvature, the tracking error will be smaller overall if the ATF pattern is placed part-way down the tracks. This explains why the ATF patterns are between the subcode areas and the

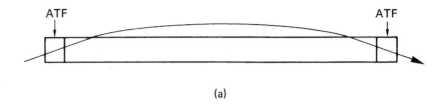

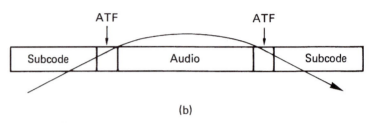

Figure 9.18 (a) The ATF patterns are at the ends of the track, and in the presence of track curvature the tracking error is exaggerated. (b) The ATF patterns are part-way down the track, minimizing mistracking due to curvature, and allowing a neat separation between subcode and audio blocks.

central PCM audio area. The data interleave is block-structured. One pair of tape tracks (one + azimuth and one – azimuth), corresponding to one drum revolution, make up an interleave block. Since the drum turns at 2000 rev/min, one revolution takes 30 ms and, in this time, 1440 samples must be stored for each channel for 48 kHz working.

The first interleave performed is to separate both left- and right-channel samples into odd and even. The right-channel odd samples followed by the left even samples are recorded in the + azimuth track, and the left odd samples followed by the right even samples are recorded in the – azimuth track. Figure 9.19 shows that this interleave allows uncorrectable errors to be concealed by interpolation. At (b) a head becomes clogged and results in every other track having severe errors. The split between right and left samples means that half of the samples in each channel are destroyed instead of every sample in one channel. The missing right even samples can be interpolated from the right odd samples, and the missing left odd samples are interpolated from the left even samples. Figure 9.19(c) shows the effect of a longitudinal tape scratch. A large error burst occurs at the same place in each head sweep. As the positions of left- and right-channel samples are reversed from one track to the next, the errors are again spread between the two channels and interpolation can be used in this case also.

The error-correction system of DAT uses product codes and was treated in detail in Chapter 7.

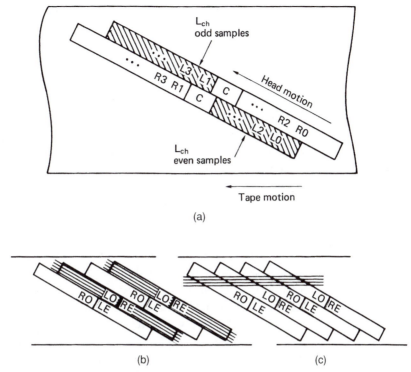

Figure 9.19 (a) Interleave of odd and even samples and left and right channels to permit concealment in case of gross errors. (b) Clogged head loses every other track. Half of the samples of each channel are still available, and interpolation is possible. (c) A linear tape scratch destroys odd samples in both channels. Interpolation is again possible.

9.7 Sound and subcode in shuttle

In all DAT applications it is important to be able to read subcode in shuttle so that wanted areas of the recording can be reached rapidly. In audio recorders, it is also useful to be able to hear at least some sound in shuttle so that the desired part of the recording can be located by ear.

When helical scan recordings are made, the geometry of the tape tracks results from the ratio of the scanner and tape speeds. If it is desired to follow the tracks properly at other than normal speed, then both scanner speed and tape speed must change in the same proportion. Since the scanner speed is locked directly to the sampling clock, it follows that some speed variation can be had simply by changing the clock frequency on replay. If a reference sampling rate fed to a professional DAT machine is reduced in frequency slightly, this will have the effect of slowing down the scanner and all signal processing logic. The slower scanner will now find that the tape tracks are passing through the machine too quickly, and the ATF system will build up a tracking error which in turn causes the capstan to slow down.

This mechanism will be adequate over a small range, perhaps half a semitone, but clearly cannot be used for shuttle. Even if it were possible to turn the scanner at 200× normal speed, it is doubtful whether any useful head contact would be achieved.

When the tape is shuttled, the track-following process breaks down and the heads cross tracks randomly. The head-to-tape speed is the vector sum of the scanner peripheral speed and the tape linear speed. In most formats this is dominated by the head speed, and so the angle at which the heads cross the tracks is relatively shallow. Figure 9.20 shows that using a replay head which is wider than the track allows a reasonable length of track to be correctly recovered even at shuttle speeds. Provided the sync blocks are made shorter than the minimum distance shown, it is possible to recover some data. DAT takes advantage of this effect to allow some sound to be heard in shuttle, to allow the subcode to be read for track searching, and to pick up timecode. The heads in a two-headed machine will typically be 20 μm wide, which is a full 50 per cent wider than the track. This wider replay head is also necessary to replay the wider tracks which are used on contact-duplicated tapes owing to the reduced coercivity needed by the duplication process.

The shuttle readout process is aided by modifying the scanner speed so that the head-to-tape speed remains the same whatever the linear tape speed. Since the scanner turns with the tape direction in DAT, this means speeding up the scanner in forward shuttle and slowing it down in reverse. The effect is that offtape signals have a constant frequency, and so

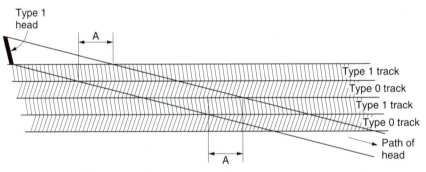

Figure 9.20 During shuttle, the heads cross tracks randomly as shown here (exaggerated). Owing to the use of azimuth recording, a head can only play a track of its own type (0 or 1) but as the head is typically 50 per cent wider than the track it is possible to recover normal signal for the periods above marked A. If sync blocks are shorter than this period, they can be picked up intact. In fact a slightly longer pickup will be possible because the replay system may tolerate a less-than-perfect signal. In any case the final decision is made by checking that the sync block recovered contains valid or correctable codewords.

the filters and phase-locked loops in the replay circuits will be able to stay in lock and recover data whenever the head is sufficiently close to the track centreline.

The data areas of the track consist of numerous short sync blocks, and each one of these is self-contained in that the data separator can resynchronize at the beginning of each, and each contains a Reed–Solomon codeword. If the head crosses the tracks at a shallow angle, then it is highly probable that one or more sync blocks will be recovered correctly. Clearly it is not possible to predict which blocks these will be. In fact total recovery is not necessary, because the goal is to produce a simulation of the recording at much more than normal speed, and this can readily be done by sending to the output convertors only every *n*th sample from the recording. Provided the tape format is designed with this in mind, the track crossings in shuttle will automatically reduce the offtape data rate.

Figure 9.21 shows the principle in simplified form. A head crosses a number of tracks in one rotation and picks up a sync block from each. The interleave used on recording means that the memory which is filled from the successfully recovered sync blocks now contains samples which are more or less evenly spaced throughout a number of tracks. Since each sync block must be independent in this mode, it is important that both bytes of a given sample are in the same sync block. Reference to Figure 7.38 will show that the use of alternate symbols in the columns when forming inner codewords has this effect.

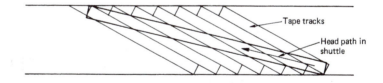

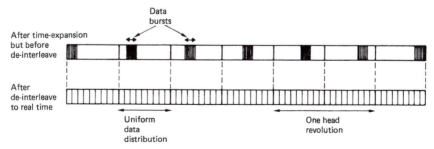

Figure 9.21 In shuttle the tracking process breaks down and bursts of data are recovered as the heads cross tracks. After de-interleave the bursts of data are uniformly spread throughout the frame. (Azimuth is neglected here for simplicity.)

Each subcode track consists of eight sync blocks, but the subcode data rate which can be supported is much less than this would indicate because it is necessary to repeat the subcode information many times in successive tracks to guarantee that it can be read when the heads cross tracks in shuttle. There are a number of incompatible timecodes which have been designed for the various television standards, but it was not appropriate to adopt them because it was desired to have a world standard for DAT timecode which would be independent of television standards. Since the scanner speed in DAT is locked to the sampling rate, it is possible to deduce the exact time by counting head revolutions. There are exactly 100 revolutions in 3 seconds, so it is possible to have a timecode for DAT which counts in hours, minutes, seconds and DAT frames. This timecode is recorded on tape, but real machines will have gearbox software which allows them to convert the tape timecode into any of the television or film timecode formats where necessary. For synchronizing two or more DAT machines, the DAT timecode can be used directly.

9.8   Timecode in DAT

The subcode of DAT is recorded in areas outside the ATF patterns, physically distinct from the PCM area. As a result, the subcode can be independently edited after an audio recording has been made. The DAT subcode performs the functions of program access in much the same way as in the Compact Disc, but it also has a subset of codes for professional use which allows the recording of timecode for synchronizing and edit control purposes.

The PCM audio data are primarily intended to be played at normal speed, with a reduced quality at other speeds. In contrast, the subcode must function well over a wide speed range so that it can be used for high-speed searching to cues. For this reason the structure of the subcode is repetitive to increase the chance of pickup, but it has no outer redundancy, as outer codes could not be assembled in shuttle.

Figure 9.22 shows the general arrangement of the subcode sync blocks. Like the PCM sync blocks, the subcode blocks have eight bytes of C1 redundancy in every other block, so there is a two-block sequence. The subcode data are assembled into standard-sized messages known as packs which contain eight bytes. In the first block of the pair, up to four packs can be accommodated, whereas in the second, only three are present because of the presence of the C1 redundancy.

The header structure of the subcode is identical to that of the PCM data. Figure 9.23 shows that following the sync byte there are two header bytes, followed by a parity byte generated on the first two. The MSB of the block

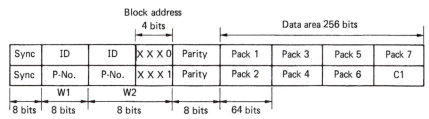

Figure 9.22 Subcode blocks are used in pairs, owing to the inner code interleave. The first block contains up to four packs, whereas the second block contains only three packs and 8 bytes of C1 (inner) redundancy.

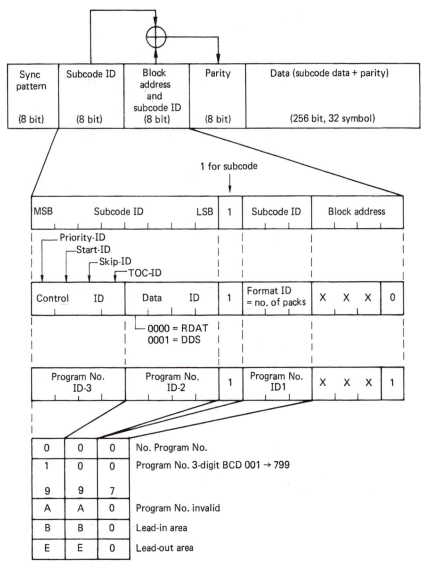

Figure 9.23 The headers in an adjacent pair of subcode sync blocks are interpreted as shown here.

address byte is always 1 in the subcode, to distinguish subcode from PCM data. There are eight subcode blocks at each end of the track, so the four LSBs of the block address byte convey the block number 0–15. The LSB of the block number allows the player to determine whether the first or second block of the subcode sequence has been found.

The smaller range of block addresses leaves three bits in the second header byte for other purposes. In the first block of the pair, these three bits form the Format ID which specifies the number of packs which have been recorded in the pair of blocks.

The first byte of the header in an even-numbered block is split into the Control ID and the Data ID. The Control ID consists of four individual flags. TOC-ID is set if the block contains Table of Contents packs. Skip-ID causes the machine to fast forward to the next Start-ID, which serves a similar function to the P-flag in Compact Disc. Finally, Priority-ID is set if the Program Number (P-No.) in the odd-numbered subcode header has been edited, so that the subcode P-No. has priority over any P-No. in the PCM-ID which cannot be edited independently of the audio. Data-ID serves the same purpose as ID-0 in PCM-ID; all zeros indicate subcode should be interpreted as digital audio standard, 1000 indicates DDS format (Digital Data Storage) subcode.

The majority of subcode data are stored in eight-byte packs in the subdata area. Figure 9.24(a) shows the basic layout of a pack. The four MSBs of the first Pack Contents (PC) symbol contain the Item code which defines the meaning of the rest of the pack. The last PC is a simple XOR parity symbol calculated by adding PC 1 through PC 7 in modulo-2.

Figure 9.24(b) shows the current valid Item codes. Program Time, Absolute Time and Running Time all have the same basic pack layout which is shown in Figure 9.25. As stated, the first four bits of PC 1 are the Item code. Bit 3 of PC 1 must be zero, leaving three bits in PC 1 and the whole of PC 2 to form an eleven-bit Program Number (P-No.). In this context the word 'Program' corresponds to a band on a vinyl LP; it is one song or movement. PC 3 contains the index code which optionally allows a Program to be subdivided. The remaining symbols PC 4–7 carry the time information in hours, minutes, seconds and DAT scanner frames (33.33 . . . Hz).

When DAT is to be used for professional applications, timecode recording is often essential. DAT timecode is carried in a pack known as Professional Running Time, abbreviated to Pro R time.

There are many forms of timecode arising from the variety of frame rates used in television and film. As DAT is in international use, the adoption of a single timecode standard to the exclusion of others is not acceptable. The solution is to record a universal form of timecode on the tape, and to use conversion circuitry appropriate to the frame rate of the system with which it is proposed to work. Internally DAT Pro R time

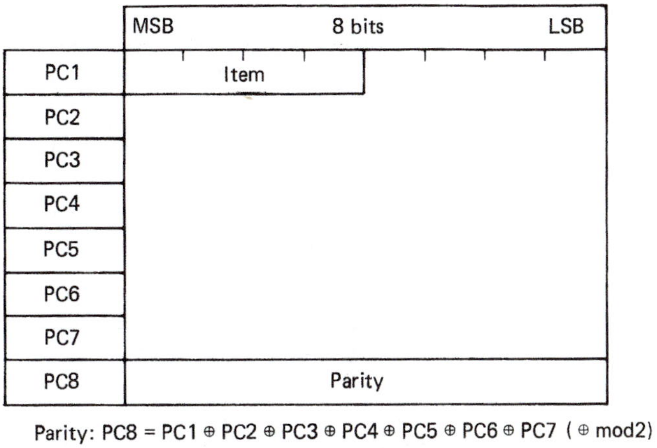
MSB
8 bits
LSB
PC1
Item
PC2
PC3
PC4
PC5
PC6
PC7
PC8
Parity
Parity: PC8 = PC1 ⊕ PC2 ⊕ PC3 ⊕ PC4 ⊕ PC5 ⊕ PC6 ⊕ PC7 (⊕ mod2)

Figure 9.24 (a) General structure of a pack.

Item	*Mode*	*Description*
0000	No information	PC1 through PC8 are all zero
0001	Program time	P-No. Index and continuous time code within a program
0010	Absolute time	P-No. Index and continuous time code on a tape
0011	R time/Pro R time	P-No. Index/TCM continuous time code within one recording
0100	TOC	Table of contents
0101	Calendar	Year, month, day, day of the week, hours, minutes, seconds
0110	Catalog	Catalog number of the cassette
0111	ISRC	International Standard Recording Code
1000	Pro-binary	Time code user bits or AES/EBU static data
1001	reserved	
1010	reserved	
1011	reserved	
1100	reserved	
1101	reserved	
1110	reserved	
1111	Soft mode	Defined by a pre-recorded tape manufacturer

Figure 9.24 (b) Table of item codes which identify the type of pack which has been recovered. Item 0011 carries running time in consumer format and Pro R time in professional timecode version.

	B7	B6	B5	B4	B3	B2	B1	B0
PC1	0	0	Item: 0 (P) 1 1 (A) 0 1 (R) 1		0	Program number 1		
PC2	Program number 2				Program number 3			
PC3	Index number							
PC4	Hours (PH) (AH) (RH)							
PC5	Minutes (PM (AM) (RM)							
PC6	Seconds (PS) (AS) (RS)							
PC7	Frames (PF) (AF) (RF)							
PC8	Pack parity							

Figure 9.25 Program, absolute and running time all share the same pack structure, but with different item codes. For example, with the item code 0001, PC4 contains program hours (PH). With item code 0011, PC7 contains running time frames (RF).

records hours, minutes, seconds and DAT frames (33.33 . . . Hz) which relate simply to the scanner speed. The relationship of DAT frames to frames in one of the standard timecodes produces a variety of phase relationships as shown in Figure 9.26.

Figure 9.26(a) shows the example of EBU 25 Hz television timecode being fed into a DAT recorder. The phase relationship between the frame boundaries changes from frame to frame. The phase relationship measured in samples is known as the Timecode Marker. It is recorded in the Pro R time pack along with the DAT frame number. The pack is also recorded with the sampling rate in use and the type of timecode being input. On replay, there is sufficient information in the pack to allow a suitable processor to compute from the DAT timecode and marker the position and content of EBU timecode frames which will have the same relationship to the audio samples as they originally had. The timecode marker consists of a binary number which can vary from zero up to the number of sample periods in a DAT frame (959, 1322 or 1439 according to the sampling rate in use). Figure 9.26(b) shows the situation with 24 Hz film timecode.

When a DAT recorder contains a built-in timecode generator, it will be simple to synchronize it to the sampling rate, and this is the preferred mode of operation. In this case the next timecode marker can be calculated by subtracting a constant from the previous one and

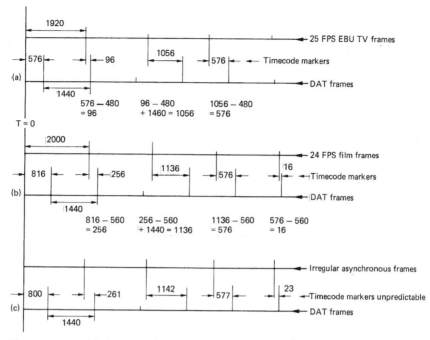

Figure 9.26 At (a) the timecode marker (TCM) can be predicted from the previous TCM in a synchronous system, as shown here for sample-rate-locked 25 Hz timecode. At (b) the TCM can also be predicted if the sampling rate is synchronous to 24 Hz film. At (c) if the source frame rate is not synchronous or unstable, TCMs cannot be predicted but must be individually measured.

expressing the result modulo the number of sample periods in a DAT frame.

Synchronous timecode and sampling rate are essential if a tape is to be played into a digital system via a timecode synchronizer. The system cannot lock to two things at once, so if a tape has asynchronous timecode and sampling rate, the synchronizer will make the replay sampling rate drift, or the sampling-rate reference will make the timecode drift. However, if the replay is to be done in the analog domain, the sampling-rate drift is of no consequence, and asynchronous working is acceptable. The DAT timecode system still works without synchronism between the external timecode signal and the scanner speed. The only difference is that the Timecode Marker parameter cannot be predicted, but will have to be measured at each scanner rotation. Figure 9.26(c) shows an example of asynchronous working.

Pro R time is recorded using a modification of the R time pack (Item 0011) as shown in Figure 9.24. Normally bit 3 of PC 1 in this pack is set to zero; for Pro R time it is set to 1 and the interpretation of the pack changes.

Bits F0 and F1 in PC 2 reflect the audio sampling rate. The Sub-Pack bits SPI-0 and SPI-1 determine whether the timecode recorded is one of the film/television timecodes, or whether it is the embedded timecode of the AES/EBU digital audio interface. When these bits are both zero, the pack is in film/television mode, and bits T0, T1 and T2 specify the frame rate in use. The remaining three bits of PC 2 and the whole of PC 3 form the eleven-bit Timecode Marker.

	B7	B6	B5	B4	B3	B2	B1	B0
		Pack item						
PC1	0	0	1	1	1	0	SPI0	SPI1
PC2	F1	F0	T2	T1	T0	(MSB)		
PC3	11 bit timecode marker							(LSB)
PC4	Hours (RH)							
PC5	Minutes (RM)							
PC6	Seconds (RS)							
PC7	Frames (RF)							
PC8	Pack parity							

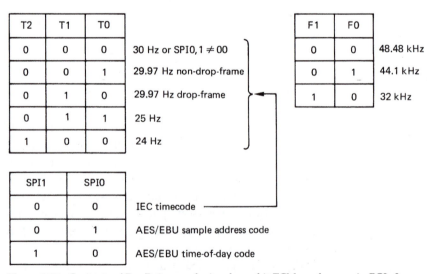

T2	T1	T0	
0	0	0	30 Hz or SPI0,1 ≠ 00
0	0	1	29.97 Hz non-drop-frame
0	1	0	29.97 Hz drop-frame
0	1	1	25 Hz
1	0	0	24 Hz

F1	F0	
0	0	48.48 kHz
0	1	44.1 kHz
1	0	32 kHz

SPI1	SPI0	
0	0	IEC timecode
0	1	AES/EBU sample address code
1	0	AES/EBU time-of-day code

Figure 9.27 Contents of Pro R time pack. An eleven-bit TCM can be seen in PC2–3, when SP10, 1 are 00, T0–2 specify the timecode rate in use when the recording was made. As TCM is in sample periods, F0, F1 are necessary to decode correctly at the sampling rate used.

DAT timecode is measured in the usual hours, minutes, seconds and frames, with the prefix R. RH, RM, RS and RF are all two-digit BCD numbers. Since there are not a whole number of DAT frames in a second, two of the seconds contain 33 frames and the third contains 34 frames. This results in exactly 100 frames in 3 seconds. As in all packs, PC 8 is the modulo-2 sum of all of the other PCs.

The sample address form of timecode conveyed in the AES/EBU digital audio interface (see Chapter 8) can be carried in the pack with the Sub-Pack bits set to 01. The AES/EBU channel-status data frame repeats every 192 sample periods (4 ms at 48 kHz) and contains (among other data) a 32-bit code which is a binary count of the number of sample clocks since midnight at the beginning of the frame. Since there will be several AES/EBU frames in one DAT frame, the DAT pack records the sample address of the AES/EBU frame during which a DAT frame began, converted to DAT timecode, and the Timecode Marker parameter is the number of sample periods from the beginning of the AES/EBU frame to the beginning of the DAT frame. The principle is shown in Figure 9.27.

Conversion from the various forms of input timecode into DAT timecode is based on the fact that all forms of timecode begin from midnight with all parameters at zero. Knowledge of the basic frame rate of the standard concerned allows any actual timecode values to be converted to real time. This can then be converted back to DAT timecode. As a result, the timecode recorded by DAT is truly international. A recording made with EBU 25 Hz timecode as an input results in DAT timecode on the tape. This could, with a suitable player, generate SMPTE timecode when the tape is played. The timecode conversion equations are given in Appendix 9.1.

9.9 Non-tracking replay

For replay only, it is possible to dispense with the scanner and ATF servos in some applications. The scanner free-runs at approximately twice normal speed, whilst the capstan continues to run at the correct speed. The rotary heads cross tracks randomly, but because of the increased speed, virtually every sync block is recovered, many of them twice. The increased scanner speed requires a higher clock frequency in the data separator.

Each pair of sync blocks contains two inner codewords, and those which are found to be error-free or which contain correctable random errors can be used. Each sync block contains an ID pattern and this is used to put the data in the correct place in the product block. If a second copy of any sync block is recovered it is discarded at this stage.

Once the product code memory is full, the de-interleave and error-correction process can occur as normal. Any blocks which are not recovered due to track crossing will be treated as dropouts by the error-correction system, as will genuine dropouts.

In personal portable machines and car-dashboard players the above approach allows a cost saving since two servo systems are eliminated. A further advantage is that alignment of the scanner is not necessary during manufacture, and tapes which are recorded on misaligned machines can still be played. Mistracking resulting from shock and vibration has no effect since the system is mistracking all the time.

The Sony NT (Non-Tracking) Format uses this approach. The rotary-head format uses a postage stamp-sized cassette and has no scanner servo in replay. The non-tracking approach means that interchange alignment is unnecessary. The slant guides on each side of the scanner are actually moulded into the cassette reducing mechanical complexity and cost. A 32 kHz sampling rate and data reduction allow a realistic playing time despite the minute cassette.

9.10   Quarter-inch rotary

Following work which suggests that a rotary-head machine can accept spliced tape, Kudelski[7] proposed a format for 1/4-inch tape using a rotary head which became that of the NAGRA D. This machine offers four independently recordable channels of up to 20-bit wordlength and timecode faciliies. The block structure is basically that of the audio channels of the D-1 DVTR. The format is restricted to low-density recording because of the potential for contamination with open reels. Whilst the recording density is not as great as in DAT, it is still competitive with professional analog machines and as the NAGRA D is a professional only product, tape consumption is of less consequence than reliability. Manual splicing of a helical scan tape causes a serious tracking and data loss problem at the splice. The principle of jump editing (see Chapter 11) is used so that the area of the splice is not played.

9.11   Half-inch and 8 mm rotary formats

A number of manufacturers have developed low-cost digital multitrack recorders for the home studio market. These are based on either VHS or Video-8 rotary-head cassette tape decks and generally offer eight channels of audio. Recording of individual audio channels is possible because the slant tape tracks are divided up into separate blocks for each channel with edit gaps between them. Some models have timecode and include synchronizers so that several machines can be locked together to offer more tracks. These machines represent the future of multitrack recording as their purchase and running costs are considerably lower than that of stationary head machines. It is only a matter of time before a low-cost 24-track is offered.

9.12 Digital audio in VTRs

The audio samples in a DVTR are binary numbers just like the video samples, and although there is an obvious difference in sampling rate and wordlength, this only affects the relative areas of tape devoted to the audio and video samples. The most important difference between audio and video samples is the tolerance to errors. The acuity of the ear means that uncorrected audio samples must not occur more than once every few hours. There is little redundancy in sound, and concealment of errors is not desirable on a routine basis. In video, the samples are highly redundant, and concealment can be effected using samples from previous or subsequent lines or, with care, from the previous frame. Major differences can be expected between the ways that audio and video samples are handled in a DVTR. One such difference is that the audio samples have 100 per cent redundancy: every one is recorded using about twice as much space on tape as the same amount of video data.

In DVTR formats the audio samples are carried by the same channel as the video samples. Using separate heads would have increased tape consumption and machine complexity. The use of the same rotary heads for video and audio reduces the number of preamplifiers and data separators needed in the system, whilst increasing the bandwidth requirement by only a few per cent even with double recording. In order to permit independent audio and video editing, the tape tracks are given a block structure. Editing will require the heads momentarily to go into record as the appropriate audio block is reached. Accurate synchronization is necessary if the other parts of the recording are to remain uncorrupted. The concept of a head which momentarily records in the centre of a track which it is reading is the normal operating procedure for all computer disk drives, as will be seen in Chapter 10. There are in fact many parallels between digital helical recorders and disk drives. Perhaps the only major difference is that in one the heads move slowly and the medium revolves, whereas in the other, the medium moves slowly and the heads revolve. Disk drives support their heads on an air bearing, achieving indefinite head life at the expense of linear density. Helical digital machines must use high-density recording and so there will be head contact and a wear mechanism. With these exceptions, the principles of disk recording apply to DVTRs, and some of the terminology has migrated.

One of these terms is the sector. In moving-head disk drives, the sector address is a measure of the angle through which the disk has rotated. This translates to the phase of the scanner in a rotary-head machine. The part of a track which is in one sector is called a block. The word 'sector' is often used instead of 'block' in casual parlance when it is clear that only one head is involved. However, as DVTRs have two heads in action at any one time, the word 'sector' means the two side-by-side blocks in the segment. As there are four independently recordable audio channels, there are four audio sectors. In D-1 (Figure 9.28), the audio is in the centre of the track, so there must be two video sectors and four audio sectors in one head sweep, and since there are two active heads, in one sweep there will be four video blocks written and eight audio blocks. In D-2 and D-3 there are also two active heads in each sweep, but the audio blocks are at the ends of the tracks, so that there are only two video blocks in the centre.

There is a requirement for the DVTR to produce pictures in shuttle. In this case, the heads cross tracks randomly, and it is most unlikely that complete video blocks can be recovered. To provide pictures in shuttle, each block is broken down into smaller components called sync blocks in the same way as is done in DAT. These contain their own error checking and an address, which in disk terminology would be called a header, which specifies where in the picture the samples in the sync block belong.

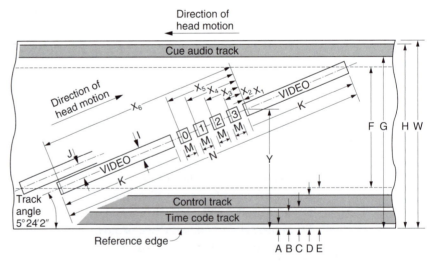

Figure 9.28 The track arrangements of D-1. Note that the segment begins at X_1 after the audio blocks. The arrangement of the D-1 control track is such that servo pulses coincide with the beginning of an even segment after the audio blocks. The next segment has no control track pulse, and so the pulses repeat at drum rotation rate for a four-headed machine. Additional pulses locate the top segment in a frame and record optional colour framing.

In shuttle, if a sync block is read properly, the address can be used to update a frame store. Thus it can be said that a sector is the smallest amount of data which can be written and is that part of a track pair within the same sector address, whereas a sync block is the smallest amount of data which can be read. Clearly there are many sync blocks in a sector.

The sync block structure continues in the audio because the same read/ write circuitry is almost always used for audio and video data. Clearly the address structure must also continue through the audio. In order to prevent audio samples from arriving in the video frame store in shuttle, the audio addresses are different from the video addresses. In all formats, the arrangement of the audio blocks is designed to maximize data integrity in the presence of tape defects and head clogs. The allocation of the audio channels to the sectors is often changed from one segment to the next. If a linear tape scratch damages the data in a given audio channel in one segment, it will damage a different audio channel in the next. Thus the scratch damage is shared between all four audio channels, each of which need correct only one quarter of the damage. It will also be seen that the relationship of the audio channels to the physical tracks rotates by one track against the direction of tape movement from one audio sector to the next. The effect of this is that, if a head becomes clogged, the errors will be distributed through all audio channels, instead of causing severe damage in one channel. In the D-2 format the audio blocks are at the ends of the head sweeps; the audio information is split so that half is recorded at each edge of the tape, and each half will be played with a different head.

In each sector, the track commences with a preamble to synchronize the phase-locked loop in the data separator on replay. Each of the sync blocks begins, as the name suggests, with a synchronizing pattern which allows the read sequencer to deserialize the block correctly. At the end of a sector, it is not possible simply to turn off the write current after the last bit, as the turnoff transient would cause data corruption. It is necessary to provide a postamble such that current can be turned off away from the data. It should now be evident that any editing has to take place a sector at a time. Any attempt to rewrite one sync block would result in damage to the previous block owing to the physical inaccuracy of replacement, damage to the next block due to the turnoff transient, and inability to synchronize to the replaced block because of the random phase jump at the point where it began. The sector in a DVTR is analogous to the cluster in a disk drive. Owing to the difficulty of writing in exactly the same place as a previous recording, it is necessary to leave tolerance gaps between sectors where the write current can turn on and off to edit individual write blocks. For convenience, the tolerance gaps are made the same length as a whole number of sync blocks. The first half of the tolerance gap is the postamble of the previous block, and the second half of the tolerance gap acts as the preamble for the next block. The tolerance gap following editing will contain, somewhere in the centre, an arbitrary jump in bit phase, and a certain amount of corruption due to turnoff transients. Provided that the postamble and preamble remain intact, this is of no consequence.

The number of audio sync blocks in a given time is determined by the number of video fields in that time. It is only possible to have a fixed tape structure if the audio sampling rate is locked to video. With 625/50 machines, the sampling rate of 48 kHz results in exactly 960 audio samples in every field.

For use on 525/60, it must be recalled that the 60 Hz is actually 59.94 Hz. As this is slightly slow, it will be found that in sixty fields, exactly 48 048 audio samples will be necessary. Unfortunately 60 will not divide into 48 048 without a remainder. The largest number which will divide 60 and 48 048 is 12; thus in 60/12 = 5 fields there will be 48 048/12 = 4004 samples. Over a five-field sequence the product blocks contain 801, 801, 801, 801 and 800 samples respectively, adding up to 4004 samples.

In order to comply with the AES/EBU digital audio interconnect, wordlengths between sixteen and twenty bits can be supported, but it is necessary to record a code in the sync block to specify the wordlength in use. Pre-emphasis may have been used prior to conversion, and this status is also to be conveyed, along with the four channel-use bits. The AES/EBU digital interconnect (see Chapter 8) uses a block-sync pattern which repeats after 192 sample periods corresponding to 4 ms at 48 kHz. He who confuses block sync with sync block is lost. Since the block size is different from that of the DVTR interleave block, there can be any phase relationship between interleave-block boundaries and the AES/EBU block-sync pattern. In order to re-create the same phase relationship between block sync and sample data on replay, it is necessary to record the position of block sync within the interleave block. It is the function of the interface control word in the audio data to convey these parameters. There is no guarantee that the 192-sample block-sync sequence will remain intact after audio editing; most likely there will be an arbitrary jump in block-sync, phase. Strictly speaking, a DVTR playing back an edited tape would have to ignore the block-sync positions on the tape, and create new block sync at the standard 192-sample spacing. Unfortunately the DVTR formats are not totally transparent to the whole of the AES/EBU data stream, as certain information is not recorded.

9.13 Stationary-head recorders

Stationary-head digital audio recorders have fixed heads like an analog recorder and often resemble their analog ancestors closely. Stationary head multi-track recorders were developed in preference to rotary head because of the perceived need to support splicing and because the electronic circuitry required was simpler. Stationary head recording is not as efficient as rotary and in the long term the familiar digital multi-track will give way to rotary cassette-based formats with electronic editing. The stereo stationary head PCM recorder has already succumbed to DAT and hard disks in professional use. The use of compression allows the efficiency problem to be overcome for consumer products and this resulted in the digital compact cassette (DCC).

Professional stationary-head recorders were specifically designed for record production and mastering, and had to be able to offer all the features of an analog multitrack. Digital multitracks mimicked analog machines so exactly that they could be installed in otherwise analog studios with the minimum of fuss. When the stationary head formats were first developed, the necessary functions of a professional machine were: independent control of which tracks record and play, synchronous recording, punch-in/punch-out editing, tape-cut editing, variable-speed playback, offtape monitoring in record, various tape speeds and bandwidths, autolocation and the facilities to synchronize several machines.

In both theory and practice a rotary-head recorder can achieve a higher storage density than a stationary-head recorder, thus using less tape. When multitrack digital audio recorders were first proposed, the adaptation of an analog video-recorder transport had to be ruled out because it lacked the necessary bandwidth. For example, a 24-track machine requires about 20 megabits per second. A further difficulty is that helical-scan recorders were not designed to handle tape-cut edits which were then considered necessary. Accordingly, multitrack digital audio recorders evolved with stationary heads and open reels; they look like analog recorders, but offer sufficient bandwidth and support splicing.

A stationary-head digital recorder is basically quite simple, as the block diagram of Figure 9.29 shows. The transport is not dissimilar to that of an analog recorder. The tape substrate used in professional analog recording is quite thick to reduce print-through, whereas in digital recording, the tape is very thin, rather like videotape, to allow it to conform closely to the heads for short-wavelength working. Print-through is not an issue in digital recording. The roughness of the backcoat has to be restricted in digital tape to prevent it embossing the magnetic layer of the adjacent turn when on the reel, since this would nullify the efforts made to provide a smooth surface finish for good head contact. The roughness of the backcoat allows the boundary layer to bleed away between turns when the tape is spooled, and so digital recorders do not spool as quickly as analog recorders. They cannot afford to risk the edge damage which results from storing a poor tape pack. The digital transport has rather

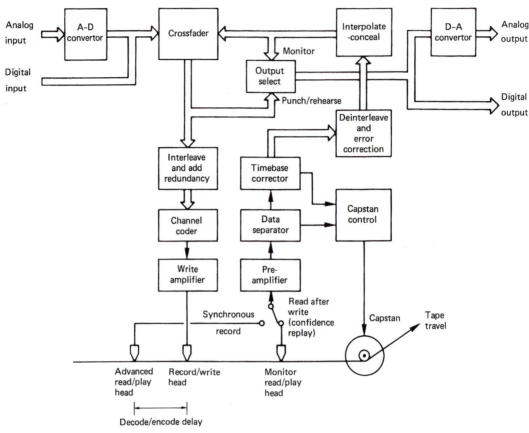

Figure 9.29 Block diagram of typical open-reel digital audio recorder. Note advanced head for synchronous recording, and capstan controlled by replay circuits.

better tension and reel-speed control than an analog machine. Some transports offer a slow-wind mode to achieve an excellent pack on a tape prior to storage.

Control of the capstan is rather different too, being more like that of a video recorder. The capstan turns at constant speed when a virgin tape is being recorded, but for replay, it will be controlled to run at whatever speed is necessary to make the offtape sample rate equal to the reference rate. In this way, several machines can be kept in exact synchronism by feeding them with a common reference. Variable-speed replay can be achieved by changing the reference frequency. It should be emphasized that, when variable speed is used, the output sampling rate changes. This may not be of any consequence if the samples are returned to the analog domain, but it prevents direct connection to a digital mixer, since these usually have fixed sampling rates.

The major items in the block diagram have been discussed in the relevant chapters. Samples are interleaved, redundancy is added, and the bits are converted into a suitable channel code. In stationary-head recorders, the frequencies in each head are low, and complex coding is not difficult. The lack of the rotary transformer of the rotary-head machine means that DC content is less of a problem. The codes used generally try to emphasize density ratio, which keeps down the linear tape speed, and the jitter window, since this helps to reject the inevitable crosstalk between the closely spaced heads. DC content in the code is handled using adaptive slicers as detailed in Chapter 6. On replay there are the usual data separators, timebase correctors and error-correction circuits.

9.14 DASH format

The DASH[8] format was the most successful of the stationary-head formats. It was not one format as such, but a family of like formats, supporting a number of different track layouts. With ferrite-head technology, it was possible to obtain adequate channel SNR with 24 tracks on half-inch tape (H) and eight tracks on quarter-inch tape (Q). The reason that these numbers are not pro-rata is that the same number of analog and control tracks are necessary for both, and take up proportionately more space on the narrower tape. This gave rise to the single-density family of formats known as DASH I. The most successful member of this family was the Sony PCM-3324.

The dimensions of the 24-track tape layout are shown in Figure 9.30. The analog tracks are placed at the edges where they act as guard bands for the digital tracks, protecting them from edge lifting. Additionally there is a large separation between the analog tracks and the digital tracks. This prevents the bias from the analog heads from having an excessive erasing effect on the adjacent digital tracks. For the same reason AC erase may have to be ruled out. One alternative mechanism for erasure of the analog tracks is to use two DC heads in tandem. The first erases the tape by saturating it, and the second is wound in the opposite sense, and carries less current, to return the tape to a near-demagnetized state.

In the half-inch format, the timecode and control tracks are placed at the centre of the tape, where they suffer no more skew with respect to the digital tracks than those at the edge of quarter-inch tape in the presence of tape weave.

The construction of a bulk ferrite multitrack head is shown in Figure 9.31, where it will be seen that space must be left between the magnetic circuits to accommodate the windings. Track spacing is improved by putting the windings on alternate sides of the gap. The parallel close-spaced magnetic circuits have considerable mutual inductance, and suffer from crosstalk. This can be compensated when several adjacent tracks record together by cross-connecting antiphase feeds to the record amplifiers.

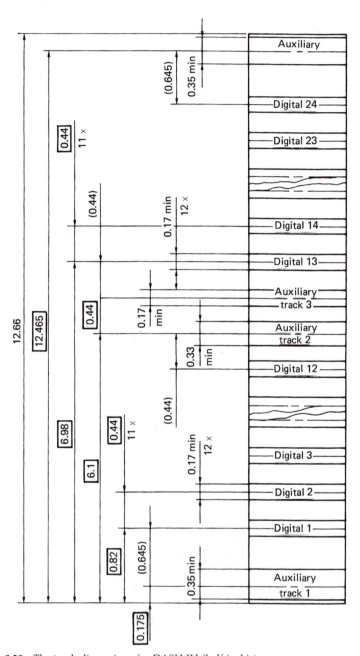

Figure 9.30 The track dimensions for DASH IH (half-inch) tape.

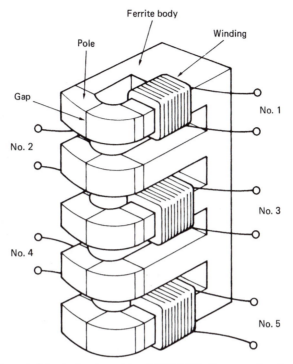

Figure 9.31 A typical ferrite head used for DASH I. Windings are placed on alternate sides to save space, but parallel magnetic circuits have high crosstalk.

Using thin-film heads, the magnetic circuits and windings are produced by deposition on a substrate at right angles to the tape plane, and as seen in Figure 9.32 they can be made very accurately at small track spacings. Perhaps more importantly, because the magnetic circuits do not have such large parallel areas, mutual inductance and crosstalk are smaller, allowing a higher practical track density.

The so-called double-density version, known as DASH II, uses such thin-film heads to obtain 48 digital tracks on half-inch tape and sixteen tracks on quarter-inch tape. The 48-track version of DASH II is shown in Figure 9.33 where it will be seen that the dimensions allow 24 of the replay head gaps on a DASH II machine to align with and play tapes recorded on a DASH I machine. In fact the PCM-3348 could take 24-track tapes and record a further 24 tracks on them.

The DASH format supported three sampling rates and the tape speed is normalized to 30 in./s at the highest rate. The three rates are 32 kHz, 44.1 kHz and 48 kHz. This last frequency was originally 50.4 kHz, which had a simple fractional relationship to 44.1 kHz, but this was dropped in favour of 48 kHz when arbitrary sampling rate conversion was shown to be feasible. In fact most stationary-head recorders will record at any reasonable sampling rate just by supplying them with an external

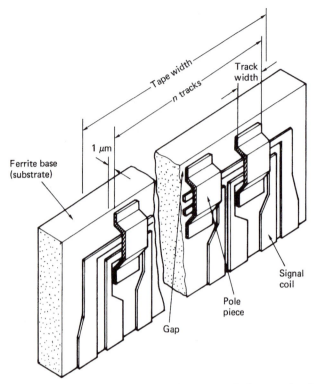

Figure 9.32 The thin-film head shown here can be produced photographically with very small dimensions. Flat structure reduces crosstalk. This type of head is suitable for DASH II which has twice as many tracks as DASH I.

reference, or word clock, at the appropriate frequency. Under these conditions, the sampling-rate switch on the machine only controls the status bits in the recording which set the default playback rate.

In the digital domain it is quite easy to distribute samples from one audio channel over a number of tape tracks. In DASH-F, the fast version, one audio track requires one tape track, and the tape moves at its greatest speed. In DASH-M, the medium version, one audio channel is spread over two tape tracks, and the tape runs at half speed. In DASH-S, the slow version, one audio channel is spread over four tape tracks, and the tape runs at one quarter speed. In twin DASH, the data corresponding to one audio channel are recorded twice, giving advantages in splice tolerance. Clearly the number of audio channels must be halved in twin DASH-F, but in DASH-M and DASH-S, the tape speed could be doubled instead.

By way of example, the well-known PCM-3324 is a DASH-FIH machine:

F = Fast format, one channel per track
I = Single density
H = Half-inch tape, hence 24 tape tracks and 24 audio channels

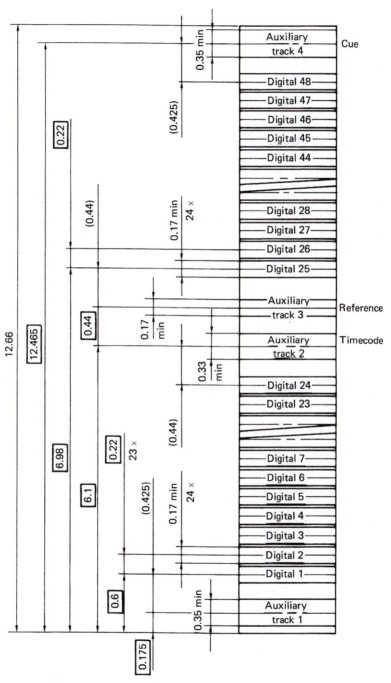

Figure 9.33 The track dimensions for DASH IIH (half inch). Comparison with Figure 9.30 will show that half of the tracks align with the single-density format allowing backwards compatibility.

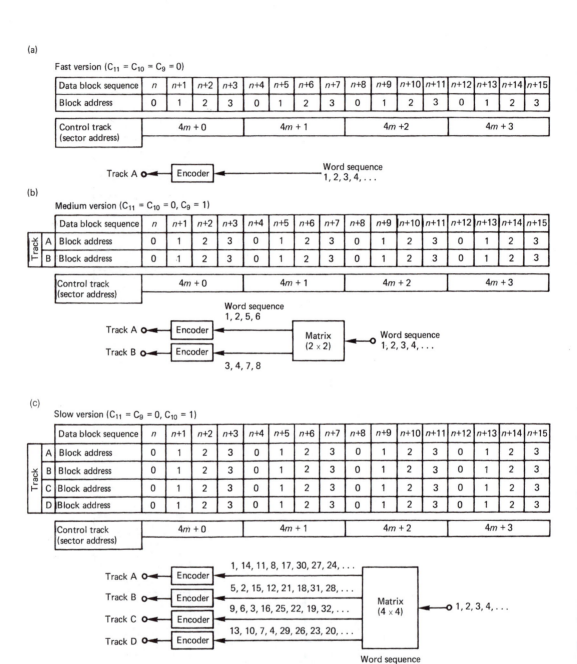

Figure 9.34 Relationships of blocks to control-track sectors. In (a), there are four blocks in one track, representing one audio channel (fast version). In (b), there are eight blocks in two tracks, representing one audio channel. The tape speed can be halved to give the medium version. In (c) there are 16 blocks in four tracks representing one audio channel. In this, the slow version, speed can be 0.25 of fast version.

The track-allocation mechanisms for S, M and F are shown in Figure 9.34 which also depicts the relationship with the control track.

The error-correction strategy of DASH is to form codewords which are confined to single-tape tracks. DASH uses cross-interleaving, which was described in principle in Chapter 7. In all practical recorders measures have to be taken for the rare cases when the error correction is overwhelmed by gross corruption. In open-reel stationary-head recorders, one obvious mechanism is the act of splicing the tape and the resultant contamination due to fingerprints.

The use of interleaving is essential to handle burst errors; unfortunately it conflicts with the requirements of tape-cut editing. Figure 9.35 shows that a splice in cross-interleave destroys codewords for the entire constraint length of the interleave. The longer the constraint length, the greater the resistance to burst errors, but the more damage is done by a splice.

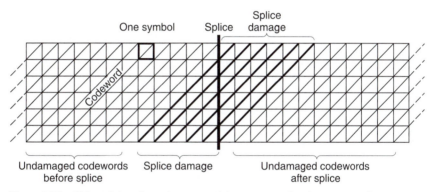

Figure 9.35 Although interleave is a powerful weapon against burst errors, it causes greater data loss when tape is spliced because many codewords are replayed in two unrelated halves.

In order to handle dropouts or splices, samples from the convertor or direct digital input are first sorted into odd and even. The odd/even distance has to be greater than the cross-interleave constraint length. In DASH, the constraint length is 119 blocks, or 1428 samples, and the odd/even delay is 204 blocks, or 2448 samples. In the case of a severe dropout, after the replay de-interleave process, the effect will be to cause two separate error bursts, first in the odd samples, then in the even samples. The odd samples can be interpolated from the even and vice versa in order to conceal the dropout. In the case of a splice, samples are destroyed for the constraint length, but Figure 9.36 shows that this occurs at different times for the odd and even samples. Using interpolation, it is possible simultaneously to obtain the end of the old recording and the

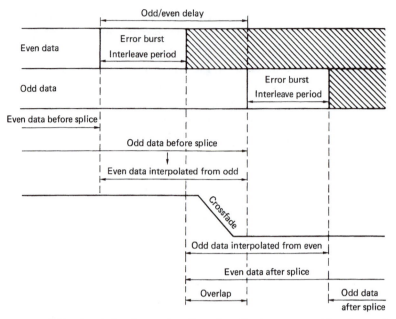

Figure 9.36 Following de-interleave, the effect of a splice is to cause odd and even data to be lost at different times. Interpolation is used to provide the missing samples, and a crossfade is made when both recordings are available in the central overlap.

beginning of the new one. A digital crossfade is made between the old and new recordings.

The interpolation during concealment and splices causes a momentary reduction in frequency response which may result in aliasing if there is significant audio energy above one quarter of the sampling rate. This was overcome in twin DASH machines in the following way. All incoming samples will be recorded twice, which means twice as many tape tracks or twice the linear speed is necessary. The interleave structure of one of the tracks will be identical to the interleave already described, whereas on the second version of the recording, the odd/even sample shuffle is reversed. When a gross error occurs in twin DASH, it will be seen from Figure 9.37 that the result after de-interleave is that when odd samples are destroyed in one channel, even samples are destroyed in the other. By selecting valid data from both channels, a full bandwidth signal can be obtained and no interpolation is necessary. In the presence of a splice, when odd samples are destroyed in one track, even samples will be destroyed in the other track. Thus at all times, all samples will be available without interpolation, and full bandwidth can be maintained across splices. Figure 9.38 shows the results of a splice in twin DASH. The status bits in the control track of twin DASH reflect the use of twin recording.

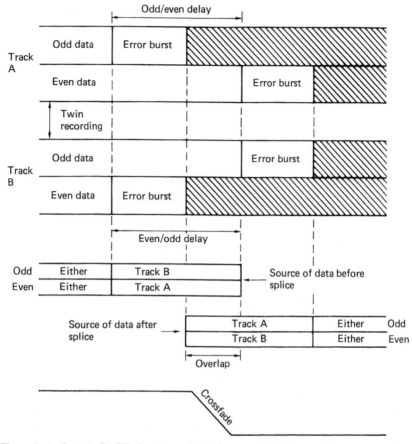

Figure 9.37 In twin DASH, the reversed interleave on the twin recordings means that correct data are always available. This makes twin DASH much more resistant to mishandling.

9.15 DCC – digital compact cassette

DCC is a stationary-head format in which the tape transport is designed to play existing analog Compact Cassettes in addition to making and playing digital recordings. This backward compatibility means that an existing Compact Cassette collection can still be enjoyed whilst newly made or purchased recordings will be digital.[9] To achieve this compatibility, DCC tape is the same width as analog Compact Cassette tape (3.81 mm) and travels at the same speed ($1\frac{7}{8}$ in./s or 4.76 cm/s). The formulation of the DCC tape is different; it resembles conventional chrome video tape, but the principle of playing one 'side' of the tape in one direction and then playing the other side in the opposite direction is retained.

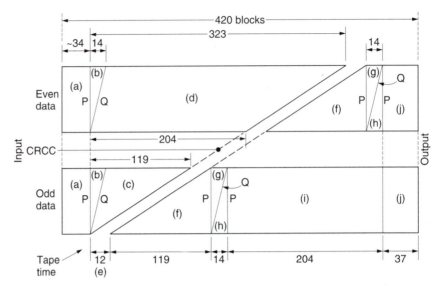

Figure 9.38 In twin DASH, two recordings are made of the same data, but with a reversed interleave. In the area of a splice, when one recording loses odd samples, the other will lose even samples, and vice versa. By selecting samples as above, full bandwidth is maintained through the splice, since no interpolation is necessary.

Although the DCC cassette has similar dimensions to the Compact Cassette so that both can be loaded in the same transport, the DCC cassette is of radically different construction. The DCC cassette only fits in the machine one way, it cannot be physically turned over as it only has hub drive apertures on one side. The head access bulge has gone and the cassette has a uniform rectangular cross-section, taking up less space in storage. The transparent windows have also been deleted as the amount of tape remaining is displayed on the panel of the player. This approach has the advantage that labelling artwork can cover almost the entire top surface. The same approach has been used in pre-recorded MiniDiscs (see Chapter 12). As the cassette cannot be turned over, all transports must be capable of playing in both directions. Thus DCC is an auto-reverse format. In addition to a record lockout plug, the cassette body carries identification holes. Combinations of these specify six different playing times from 45 min to 120 min as in Table 9.1.

The apertures for hub drive, capstans, pinch rollers and heads are covered by a sliding cover formed from metal plate. The cover plate is automatically slid aside when the cassette enters the transport. The cover plate also operates hub brakes when it closes and so the cassette can be left out of its container. The container fits the cassette like a sleeve and has space for an information booklet.

DCC uses a form of data reduction which Philips call Precision Adaptive Sub-band Coding (PASC). PASC is based on MPEG audio compression as described in Chapter 5 and its use allows the recorded data rate to be about one quarter that of the original PCM audio. This allows for conventional chromium tape to be used with a minimum wavelength of about one micrometre instead of the more expensive high-coercivity tapes normally required for use with shorter wavelengths. The advantage of the conventional approach with linear tracks is that tape duplication can be carried out at high speed. This makes DCC attractive to record companies. Even with data reduction, the only way in which the bit rate can be accommodated is to use many tracks in parallel.

Figure 9.39 shows that in DCC audio data are distributed over eight parallel tracks along with a subcode track which together occupy half the width of the tape. At the end of the tape the head rotates about an axis perpendicular to the tape and plays the remaining tracks in reverse. The other half of the head is fitted with magnetic circuits sized for analog tracks and so the head rotation can also select the head type which is in use for a given tape direction.

However, reducing the data rate to one quarter and then distributing it over eight tracks means that the frequency recorded on each track is only 96 kbits/s or about $\frac{1}{16}$ that of a PCM machine recording a single audio channel with a single head. The linear tape speed is incredibly low by stationary-head digital standards in order to obtain the desired playing time. The rate of change of flux in the replay head is very small due to the low tape speed, and conventional inductive heads are at a severe disadvantage because their self-noise drowns the signal. Magneto-resistive heads are necessary because they do not have a derivative

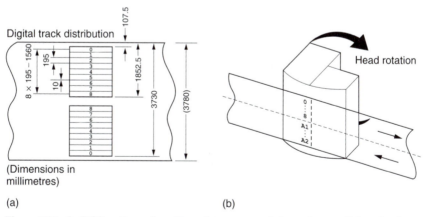

(a) (b)

Figure 9.39 In DCC audio and auxiliary data are recorded on nine parallel tracks along each side of the tape as shown in (a). The replay head shown in (b) carries magnetic poles which register with one set of nine tracks. At the end of the tape, the replay head rotates 180° and plays a further nine tracks on the other side of the tape. The replay head also contains a pair of analog audio magnetic circuits which will be swung into place if an analog cassette is to be played.

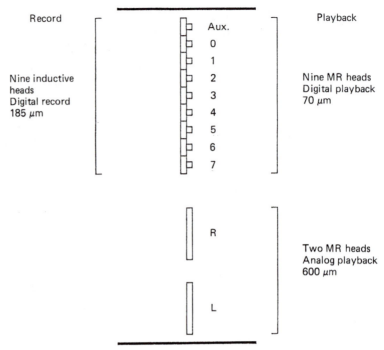

Record

Aux.

0

1

2

3

4

5

6

7

Nine inductive
heads
Digital record
185 μm

Playback

Nine MR heads
Digital playback
70 μm

R

Two MR heads
Analog playback
600 μm

L

Figure 9.40 The head arrangement used in DCC. There are nine record heads which leave tracks wider than the MR replay heads to allow for misregistration. Two MR analog heads allow compact cassette replay.

action, and so the signal is independent of speed. A magnetoresistive head uses an element whose resistance is influenced by the strength of flux from the tape and its operation was discussed in Chapter 6. Magneto-resistive heads are unable to record, and so separate record heads are necessary. Figure 9.40 shows a schematic outline of a DCC head. There are nine inductive record heads for the digital tracks, and these are recorded with a width of 185 μm and a pitch of 195 μm. Alongside the record head are nine MR replay gaps. These operate on a 70 μm band of the tape which is nominally in the centre of the recorded track. There are two reasons for this large disparity between the record and replay track widths. First, replay signal quality is unaffected by a lateral alignment error of ±57 μm and this ensures tracking compatibility between machines. Second, the loss due to incorrect azimuth is proportional to track width and the narrower replay track is thus less sensitive to the state of azimuth adjustment. In addition to the digital replay gaps, a further two analog MR head gaps are present in the replay stack. These are aligned with the two tracks of a stereo pair in a Compact Cassette.

The twenty-gap head could not be made economically by conventional techniques. Instead it is made lithographically using thin film technology.

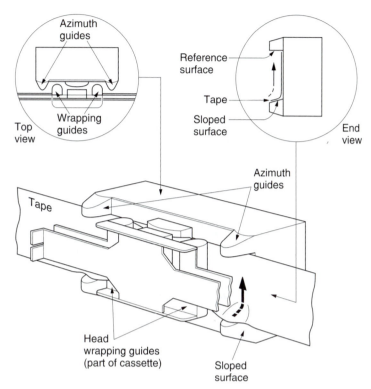

Figure 9.41 The tape guidance of DCC uses a pair of shaped guides on both sides of the head. See text for details.

Tape guidance is achieved by a combination of guides on the head block and pins in the cassette. Figure 9.41 shows that at each side of the head is fitted a C-shaped tape guide. This guide is slightly narrower than the nominal tape width. The reference edge of the runs against a surface which is at right angles to the guide, whereas the non-reference edge runs against a sloping surface. Tape tension tends to force the tape towards the reference edge. As there is such a guide at both sides of the head, the tape cannot wander in the azimuth plane. The tape wrap around the head stack and around the azimuth guides is achieved by a pair of pins behind the tape which are part of the cassette. Between the pins is a conventional sprung pressure pad and screen.

Figure 9.42 shows a block diagram of a DCC machine. The audio interface contains convertors which allow use in analog systems. The digital interface may be used as an alternative. DCC supports 48, 44.1 and 32 kHz sampling rates, offering audio bandwidths of 22, 20 and 14.5 kHz respectively with eighteen-bit dynamic range. Between the interface and the tape subsystem is the PASC coder. The tape subsystem requires error-

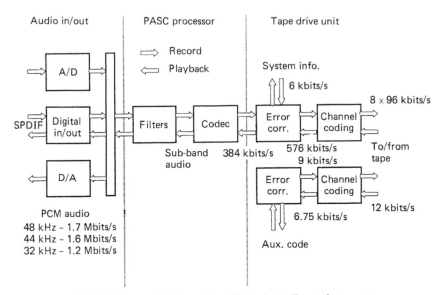

Audio in/out   PASC processor   Tape drive unit

⇒ Record
⇐ Playback

System info.

6 kbits/s

8 × 96 kbits/s

A/D

SPDIF Digital in/out — Filters — Codec — Error corr. — Channel coding

Sub-band audio   384 kbits/s   576 kbits/s
9 kbits/s   To/from tape

D/A

Error corr. — Channel coding

12 kbits/s

6.75 kbits/s

PCM audio
48 kHz – 1.7 Mbits/s
44 kHz – 1.6 Mbits/s
32 kHz – 1.2 Mbits/s

Aux. code

Figure 9.42 Block diagram of DCC machine. This is basically similar to any stationary-head recorder except for the compression (PASC) unit between the convertors and the transport.

correction and channel coding systems not only for the audio data but also for the auxiliary data on the ninth track.

Appendix 9.1   Timecode to Pro R time conversion

As explained in the text, conversion from one timecode standard to Pro R time consists of finding the number of the last timecode frame completed before the beginning of the current Pro R time frame. The beginning of both frames is then expressed in real time, and the timecode marker (TCM) measures the difference between them, in sample periods T_s.

The upper part of Figure 9A.1 shows EBU timecode frames of period T_C beginning from time zero. The number of *complete* timecode frames before the DAT frame in question begins is the Timecode Frame Count, F_C, which is an integer.

The lower section of the diagram shows DAT frames of period T_D, which did not necessarily begin at time zero. The DAT Offset D_O, which is a constant, measures the relationship between the beginning of the first DAT frame and time zero.

The DAT Frame Count F_D is the number of completed DAT frames before the one in question. The time difference between the beginnings of the respective frames is $TCM \times T_s$.

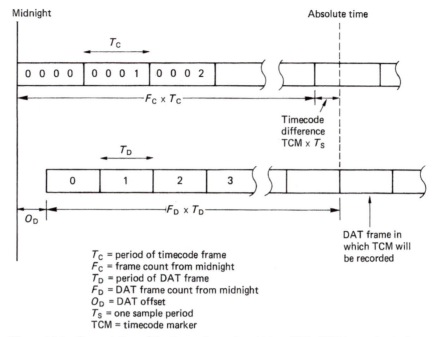

Figure 9A.1 Computation of the timecode marker. At top, 25 Hz EBU frames begin from midnight, but DAT frames (below) are not necessarily aligned with midnight, and the DAT offset O_D takes this into account. The absolute time of the beginning of the DAT frame in question is expressed modulo-T_C which gives the timecode difference. The timecode marker is the timecode difference measured in sample periods T_S.

The absolute time at the beginning of a given DAT frame is:

$$F_D \times T_D + O_D$$

The timecode difference $TCM \times T_S$ is simply the absolute time expressed Modulo-T_C. Thus:

$$TCM = \frac{(F_D \times T_D + O_D)\, \mathrm{Mod}T_C}{T_S}$$

References

1. Yamada, Y., Fujii, Y., Moriyama, M. and Saitoh, S., Professional use PCM audio processor with a high efficiency error-correction system. Presented at the 66th Audio Engineering Society Convention, (Los Angeles, 1980), Preprint 1628(G7)
2. Ishida, Y., Nishi, S., Kunii, S., Satoh, T. and Uetake, K., A PCM digital audio processor for home use VTRs. Presented at the 64th Audio Engineering Society Convention (New York, 1979), Preprint 1528
3. Griffiths, F.A., A digital audio recording system. Presented at the 65th Audio Engineering Society Convention (London, 1980), Preprint 1580(C1)
4. Nakajima, H. and Odaka, K., A rotary-head high-density digital audio tape recorder. *IEEE Trans. Consum. Electron.*, **CE-29**, 430–437 (1983)

5. Itoh, F., Shiba, H., Hayama, M. and Satoh, T., Magnetic tape and cartridge of R-DAT. *IEEE Trans. Consum. Electron.*, **CE-32**, 442–452 (1986)

6. Hitomi, A. and Taki, T., Servo technology of R-DAT. *IEEE Trans. Consum. Electron.*, **CE-32**, 425–432 (1986)

7. Kudelski, S., *et al.*, Digital audio recording format offering extensive editing capabilities. Presented at the 82nd Audio Engineering Society Convention (London, 1987), Preprint 2481(H-7)

8. Doi, T.T., Tsuchiya, Y., Tanaka, M. and Watanabe, N., A format of stationary-head digital audio recorder covering wide range of applications. Presented at the 67th Audio Engineering Society Convention, (New York, 1980), Preprint 1677(H6)

9. Lokhoff, G.C.P., DCC: Digital compact cassette. *IEEE Trans. Consum. Electron.*, **CE-37**, 702–706 (1991)

10

Magnetic disk drives

Disk drives came into being as random-access file-storage devices for digital computers. They were prominent in early experiments with digital audio, but their cost at that time was too great in comparison to emerging tape technology. However, the explosion in personal computers has fuelled demand for low-cost high-density magnetic disk drives and the rapid access offered is increasingly finding applications in digital audio. Optical and magneto-optic (MO) disks are considered in Chapter 12.

10.1   Types of disk drive

Once the operating speed of computers began to take strides forward, it became evident that a single processor could be made to jump between several different programs so fast that they all appeared to be executing simultaneously, a process known as multiprogramming. Computer memory is designed for speed and remains more expensive than other types of mass storage, and so it has never been practicable to store every program or data file necessary within the computer memory. In practice some kind of storage medium is necessary where only programs which are running or are about to run are in the memory, and the remainder are stored on the medium. Punched cards, paper tape and magnetic tape are all computer media, but suffer from the same disadvantage of slow access. The disk drive was developed specifically to offer rapid random access to stored data. Figure 10.1 shows that, in a magnetic disk drive, the data are recorded on a circular track. In floppy disks, the magnetic medium is flexible, and the head touches it. This restricts the rotational speed. In hard-disk drives,

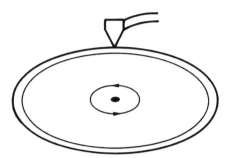

Figure 10.1 The rotating store concept. Data on the rotating circular track are repeatedly presented to the head.

the disk rotates at several thousand rev/min so that the head-to-disk speed is of the order of one hundred miles per hour. At this speed no contact can be tolerated, and the head flies on a boundary layer of air turning with the disk at a height measured in microinches. The longest time it is necessary to wait to access a given data block is a few milliseconds. To increase the storage capacity of the drive without a proportional increase in cost, many concentric tracks are recorded on the disk surface, and the head is mounted on a positioner which can rapidly bring the head to any desired track. Such a machine is termed a moving-head disk drive. The positioner was usually designed so that it could remove the heads away from the disk completely, which could thus be exchanged. The exchangeable-pack moving-head disk drive became the standard for mainframe and minicomputers for a long time, and usually at least two were furnished so that important data could be 'backed up' or copied to a second disk for safe keeping.

Later came the so-called Winchester technology disks, where the disk and positioner formed a sealed unit which allowed increased storage capacity but precluded exchange of the disk pack. This led to the development of high-speed tape drives which could be used as security backup storage.

Disk drive development has been phenomenally rapid. The first flying head disks were about 3 feet across. Subsequently disk sizes of 14, 8, $5\frac{1}{4}$, $3\frac{1}{2}$ and $1\frac{3}{4}$ inches were developed. Despite the reduction in size, the storage capacity is not compromised because the recording density has increased and continues to increase. In fact there is an advantage in making a drive smaller because the moving parts are then lighter and travel a shorter distance, improving access time.

Disk drives are devices which have distinctive characteristics distinguishing them from other media. These will all be explained in this chapter, after which it will become clear how these useful devices can be used in digital audio applications.

10.2   Disk terminology

In all technologies there are specialist terms, and those relating to disks will be explained here. Figure 10.2 shows a typical multiplatter disk pack in conceptual form. Given a particular set of coordinates (cylinder, head, sector), known as a disk physical address, one unique data block is defined. A common block capacity is 512 bytes. The subdivision into sectors is sometimes omitted for special applications. Figure 10.3 introduces the essential subsystems of a disk drive which will be discussed.

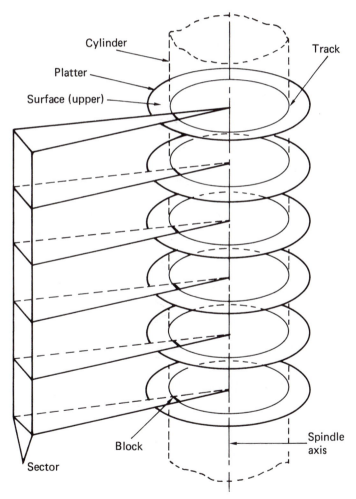

Figure 10.2   Disk terminology. Surface: one side of a platter. Track: path described on a surface by a fixed head. Cylinder: imaginary shape intersecting all surfaces at tracks of the same radius. Sector: angular subdivision of pack. Block: that part of a track within one sector. Each block has a unique cylinder, head and sector address.

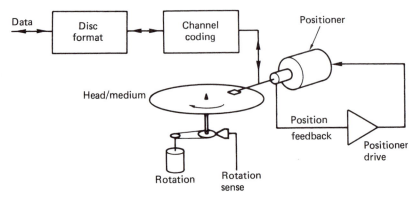

Figure 10.3 The main subsystems of a typical disk drive.

10.3 Structure of disk

The floppy disk is actually made using tape technology, and will be discussed later. Rigid disks are made from aluminium alloy. Magnetic-oxide types use an aluminium oxide substrate, or undercoat, giving a flat surface to which the oxide binder can adhere. Later metallic disks are electroplated with the magnetic medium. In both cases the surface finish must be extremely good owing to the very small flying height of the head. Figure 10.4 shows a cross-section of a typical multiplatter disk pack. As the head-to-disk speed and recording density are functions of track radius, the data are confined to the outer areas of the disks to minimize the change in these parameters. As a result, the centre of the pack is often an empty well. In fixed (i.e. non-interchangeable) disks the drive motor is often installed in the centre well. Removable packs usually seat on a taper to ensure concentricity and elaborate fixing mechanisms are needed on large packs to prevent the pack from working loose in operation. Smaller packs are held to the spindle by a permanent magnet, and a lever mechanism is incorporated into the cartridge to assist their removal.

10.4 Principle of flying head

Disk drives permanently sacrifice storage density in order to offer rapid access. The use of a flying head with a deliberate air gap between it and the medium is necessary because of the high medium speed, but this causes a severe separation loss which restricts the linear density available. The air gap must be accurately maintained, and consequently the head is of low mass and is mounted flexibly.

The aerohydrodynamic part of the head is known as the slipper; it is designed to provide lift from the boundary layer which changes rapidly

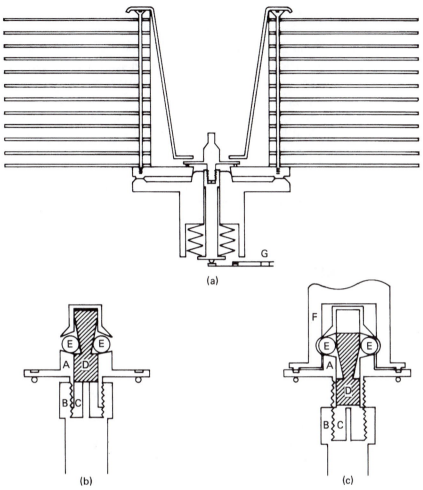

Figure 10.4 (a) Typical construction of multiplatter exchangeable pack. (b) the hold-down screw A is fully engaged with the lockshaft B, and the pin C lifts the ramp D, retracting the balls E. In (c) the hold-down screw is withdrawn from the lockshaft, which retracts, causing the ramp to force the balls into engagement with the cover F. The lockshaft often operates a switch to inform the logic that a pack is present (G).

with changes in flying height. It is not initially obvious that the difficulty with disk heads is not making them fly, but making them fly close enough to the disk surface. The boundary layer travelling at the disk surface has the same speed as the disk, but as height increases, it slows down due to drag from the surrounding air. As the lift is a function of relative air speed, the closer the slipper comes to the disk, the greater the lift will be. The slipper is therefore mounted at the end of a rigid cantilever sprung towards the medium. The force with which the head is pressed towards the disk by the spring is equal to the lift at the designed flying height. Because of the spring, the head may rise and fall over small warps in the disk. It would be virtually impossible to manufacture disks flat enough to dispensed with this feature. As the slipper negotiates a warp it will pitch and roll in addition to rising and falling, but it must be prevented from yawing, as this would cause an azimuth error. Downthrust is applied to the aerodynamic centre by a spherical thrust button, and the required degrees of freedom are supplied by a thin flexible gimbal. The slipper has to bleed away surplus air in order to approach close enough to the disk, and holes or grooves are usually provided for this purpose in the same way that pinch rollers on some tape decks have grooves to prevent tape slip.

In exchangeable-pack drives, there will be a ramp on the side of the cantilever which engages a fixed block when the heads are retracted in order to lift them away from the disk surface.

10.5   Reading and writing

Figure 10.5 shows how disk heads are made. The magnetic circuit of disk heads was originally assembled from discrete magnetic elements. As the gap and flying height became smaller to increase linear recording density, the slipper was made from ferrite, and became part of the magnetic circuit. This was completed by a small C-shaped ferrite piece which carried the coil. In thin-film heads, the magnetic circuit and coil are both formed by deposition on a substrate which becomes the rear of the slipper.

In a moving-head device it is not practicable to position separate erase, record and playback heads accurately. Erase is by overwriting, and reading and writing are carried out by the same head. The presence of the air film causes severe separation loss, and peak shift distortion is a major problem. The flying height of the head varies with the radius of the disk track, and it is difficult to provide accurate equalization of the replay channel because of this. The write current is often controlled as a function of track radius so that the changing reluctance of the air gap does not change the resulting record flux. Automatic gain control (AGC) is used on replay to compensate for changes in signal amplitude from the head.

Equalization may be used on recording in the form of precompensation, which moves recorded transitions in such a way as to oppose the effects of peak shift in addition to any replay equalization used. This was discussed in Chapter 6, which also introduced digital channel coding.

Early disks used FM coding, which was easy to decode but had a poor density ratio. The invention of MFM revolutionized hard disks, and was at one time universal. Further progress led to run-length-limited codes such

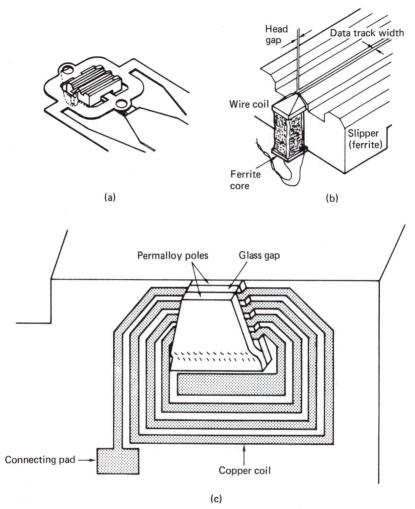

Figure 10.5 (a) Winchester head construction showing large air bleed grooves.
(b) Close-up of slipper showing magnetic circuit on trailing edge. (c) Thin-film head is
fabricated on the end of the slipper using microcircuit technology.

as 2/3 and 2/7 which had a high density ratio without sacrificing the large
jitter window necessary to reject peak-shift distortion. Partial response is
also suited to disks, but is not yet in common use.

Typical drives have several heads, but with the exception of special-
purpose parallel-transfer machines for digital video or instrumentation
work, only one head will be active at any one time, which means that the
read and write circuitry can be shared between the heads. Figure 10.6
shows that in one approach the centre-tapped heads are isolated by
connecting the centre tap to a negative voltage, which reverse-biases the
matrix diodes. The centre tap of the selected head is made positive. When
reading, a small current flows through both halves of the head winding, as

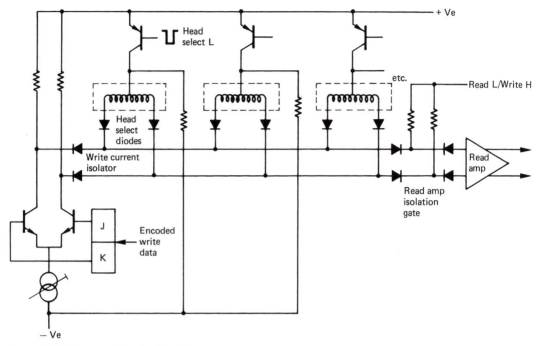

Figure 10.6 Representative head matrix.

the diodes are forward-biased. Opposing currents in the head cancel, but read signals due to transitions on the medium can pass through the forward-biased diodes to become differential signals on the matrix bus. During writing, the current from the write generator passes alternately through the two halves of the head coil. Further isolation is necessary to prevent the write-current-induced voltages from destroying the read preamplifier input. Alternatively, FET analog switches may be used for head selection.

The read channel usually incorporates AGC, which will be overridden by the control logic between data blocks in order to search for address marks, which are short unmodulated areas of track. As a block preamble is entered, the AGC will be enabled to allow a rapid gain adjustment.

The high bit rates of disk drives, due to the speed of the medium, mean that peak detection in the replay channel is usually by differentiation. The detected peaks are then fed to the data separator.

10.6   Moving the heads

The servo system required to move the heads rapidly between tracks, and yet hold them in place accurately for data transfer, is a fascinating and complex piece of engineering.

In exchangeable pack drives, the disk positioner moves on a straight axis which passes through the spindle. The head carriage will usually have preloaded ball races which run on rails mounted on the bed of the machine, although some drives use plain sintered bushes sliding on polished rods.

Motive power on early disk drives was hydraulic, but this soon gave way to moving-coil drive, because of the small moving mass which this technique permits. Another possibility is a coarse-threaded shaft or leadscrew which engages with a nut on the carriage. In very low-cost drives, the motor will be a stepping motor, and the positions of the tracks will be determined by the natural detents of the stepping motor. This has an advantage for portable drives, because a stepping motor will remain detented without power whereas moving-coil actuators require power to stay on track.

When a drive is track-following, it is said to be detented, in fine mode or in linear mode depending on the manufacturer. When a drive is seeking from one track to another, it can be described as being in coarse mode or velocity mode. These are the two major operating modes of the servo.

With the exception of stepping-motor-driven carriages, the servo system needs positional feedback from a transducer of some kind. The purpose of the transducer will be one or more of the following:

1 To count the number of cylinders crossed during a seek
2 To generate a signal proportional to carriage velocity
3 To generate a position error proportional to the distance from the centre of the desired track

Sometimes the same transducer is used for all of these, and so transducers are best classified by their operating principle rather than by their function in a particular drive.

The simplest transducer is the magnetic moving-coil type, with its complementary equivalent, the moving-magnet type. Both generate a voltage proportional to velocity, and can give no positional information, but no precise alignment other than a working clearance is necessary. These devices are usually called tachos and should not be confused with burritos.

Optical transducers have also been used. These consist of gratings, one fixed on the machine base, and one on the carriage. The relative position of the two controls the amount of light which can shine through onto a sensor.

10.7   Controlling a seek

A seek is a process where the positioner moves from one cylinder to another. The speed with which a seek can be completed is a major factor in determining the access time of the drive. The main parameter

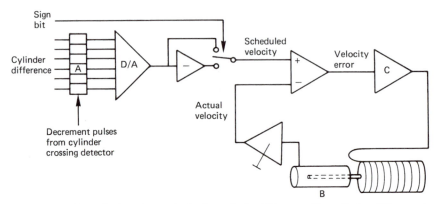

Figure 10.7 Control of carriage velocity by cylinder difference. The cylinder difference is loaded into the difference counter A. A digital-to-analog convertor generates an analog voltage from the cylinder difference, known as the scheduled velocity. This is compared with the actual velocity from the transducer B in order to generate the velocity error which drives the servo amplifier C.

controlling the carriage during a seek is the cylinder difference, which is obtained by subtracting the current cylinder address from the desired cylinder address. The cylinder difference will be a signed binary number representing the number of cylinders to be crossed to reach the target, direction being indicated by the sign. The cylinder difference is loaded into a counter which is decremented each time a cylinder is crossed. The counter drives a DAC which generates an analog voltage proportional to the cylinder difference. As Figure 10.7 shows, this voltage, known as the scheduled velocity, is compared with the output of the carriage-velocity tacho. Any difference between the two results in a velocity error which drives the carriage to cancel the error. As the carriage approaches the target cylinder, the cylinder difference becomes smaller, with the result that the run-in to the target is critically damped to eliminate overshoot.

Figure 10.8(a) shows graphs of scheduled velocity, actual velocity and motor current with respect to cylinder difference during a seek. In the first half of the seek, the actual velocity is less than the scheduled velocity, causing a large velocity error which saturates the amplifier and provides maximum carriage acceleration. In the second half of the graphs, the scheduled velocity is falling below the actual velocity, generating a negative velocity error which drives a reverse current through the motor to slow the carriage down. The scheduled deceleration slope can clearly not be steeper than the saturated acceleration slope. Areas A and B on the graph will be about equal, as the kinetic energy put into the carriage has to be taken out. The current through the motor is continuous, and would result in a heating problem, so to counter this, the DAC is made non-linear so that above a certain cylinder difference no increase in scheduled velocity will occur. This results in the graph of Figure 10.8(b). The actual

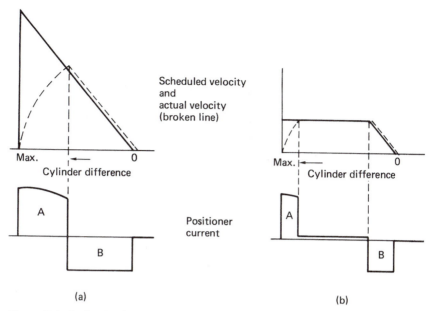

Figure 10.8 In the simple arrangement in (a) the dissipation in the positioner is continuous, causing a heating problem. The effect of limiting the scheduled velocity above a certain cylinder difference is apparent in (b) where heavy positioner current only flows during acceleration and deceleration. During the plateau of the velocity profile, only enough current to overcome friction is necessary. The curvature of the acceleration slope is due to the back EMF of the positioner motor.

velocity graph is called a velocity profile. It consists of three regions: acceleration, where the system is saturated; a constant velocity plateau, where the only power needed is to overcome friction; and the scheduled run-in to the desired cylinder. Dissipation is only significant in the first and last regions.

A consequence of the critically damped run-in to the target cylinder is that short seeks are slow. Sometimes further non-linearity is introduced into the velocity scheduler to speed up short seeks. The velocity profile becomes a piecewise linear approximation to a curve by using non-linear feedback. Figure 10.9 shows the principle of an early analog shaper or profile generator. Later machines compute the curve in microprocessor sofware or use a PROM look-up table.

In small disk drives the amplifier may be linear in all modes of operation, resembling an audio power amplifier. Larger units may employ pulse-width-modulated drive to reduce dissipation, or even switched-mode amplifiers with inductive flywheel circuits. These switching systems can generate appreciable electromagnetic radiation, but this is of no consequence as they are only active during a seek. In track-following mode, the amplifier reverts to linear mode; hence the use of the term linear to mean track-following mode.

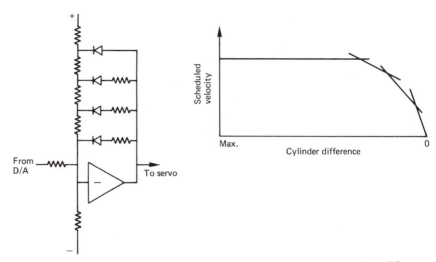

Figure 10.9 The use of voltage-dependent feedback around an operational amplifier permits a piecewise linear approximation to a curved velocity profile. This has the effect of speeding up short seeks without causing a dissipation problem on long seeks. The circuit is referred to as a shaper.

The input of the servo amplifier normally has a number of analog switches which select the appropriate signals according to the mode of the servo. As the output of the position transducer is a triangle or sine wave, the sense of the position feedback has to be inverted on odd-numbered cylinders, to allow detenting on the negative slope. Sometimes a separate transducer is used for head retraction only. A typical system is shown in Figure 10.10.

10.8   Rotation

The rotation subsystems of disk drives will now be covered. The track-following accuracy of a drive positioner will be impaired if there is bearing run-out, and so the spindle bearings are made to a high degree of precision. On early drives, squirrel-cage induction motors were used, driving the spindle through a belt. As recording density has increased, the size of drives has come down, and today brushless DC motors with integral speed control are universal.

In order to control reading and writing, the drive control circuitry needs to know which cylinder the heads are on, and which sector is currently under the head. Sector information in early drives was obtained from a sensor which detects slots cut in the hub of the disk. These can be optical, variable reluctance or eddy-current devices. Pulses from the transducer increment the sector counter, which is reset by a double slot

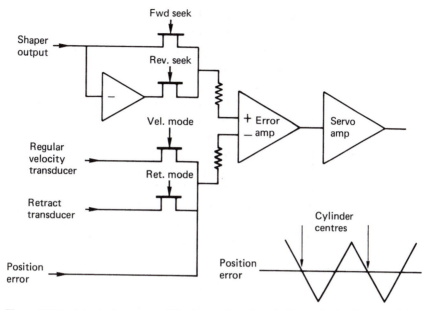

Figure 10.10 A typical servo amplifier input stage. In velocity mode the shaper and the velocity transducer drive the error amp. In track-following mode the position error is the only input.

once per revolution. The desired sector address is loaded into a register, which is compared with the sector counter. When the two match, the desired sector has been found. This process is referred to as a search, and usually takes place after a seek. Having found the correct physical place on the disk, the next step is to read the header associated with the data block to confirm that the disk address contained there is the same as the desired address.

Rotation of a disk pack at speed results in heat build-up through air resistance. This heat must be carried away. A further important factor with exchangeable pack drives is to keep the disk area free from contaminants which might lodge between the head and the disk and cause the destructive phenomenon known as a head crash, where debris builds up on the head until it ploughs the disk surface.

10.9   Servo-surface disks

One of the major problems to be overcome in the development of high-density disk drives was that of keeping the heads on track despite changes of temperature. The very narrow tracks used in digital recording have similar dimensions to the amount a disk will expand as it warms up. The cantilevers and the drive base all expand and contract, conspiring with thermal drift in the cylinder transducer to limit track pitch. The breakthrough in disk density came with the introduction of the servo-surface drive. The position error in a servo-surface drive is derived from a head reading the disk itself. This virtually eliminates thermal effects on head positioning and allows great increases in storage density.

In a multiplatter drive, one surface of the pack holds servo information which is read by the servo head. In a ten-platter pack this means that 5 per cent of the medium area is lost, but this is unimportant since the increase in density allowed is enormous. Using one side of a single-platter cartridge for servo information would be unacceptable as it represents 50 per cent of the medium area, so in this case the servo information can be interleaved with sectors on the data surfaces. This is known as an embedded-servo technique. These two approaches are contrasted in Figure 10.11.

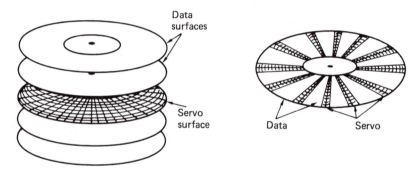

Figure 10.11 In a multiplatter disk pack, one surface is dedicated to servo information. In a single platter, the servo information is embedded in the data on the same surfaces.

The servo surface is written at the time of disk pack manufacture, and the disk drive can only read it. Writing the servo surface has nothing to do with disk formatting, which affects the data storage areas only.

The key to the operation of the servo surface is the special magnetic pattern recorded on it. In a typical servo surface, recorded pairs of transitions, known as dibits, are separated by a space. Figure 10.12 shows that there are two kinds of track. On an A track, the first transition of the pair will cause a positive pulse on reading, whereas on a B track, the first pulse will be negative. In addition the A-track dibits are shifted by one half cycle with respect to the B-track dibits. The width of the magnetic circuit in the servo head is equal to the width of a servo track. During track following, the correct position for the servo head is with half of each type of track beneath it. The read/write heads will then be centred on their respective data tracks. Figure 10.13 illustrates this relationship.

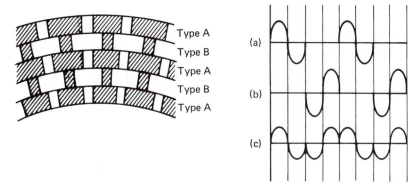

Figure 10.12 The servo surface is divided into two types of track, A and B, which are out of phase by 180° and are recorded with reverse polarity with respect to one another. Waveform (a) results when the servo head is entirely above a type A track, and waveform (b) results from reading solely a type B track. When the servo head is correctly positioned with one-half of its magnetic circuit over each track, waveform (c) results.

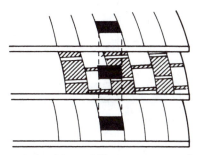

Figure 10.13 When the servo head is straddling two servo tracks, the data heads are correctly aligned with their respective tracks.

The amplitude of dibits from A tracks with respect to the amplitude of dibits from B tracks depends on the relative areas of the servo head which are exposed to the respective tracks. As the servo head has only one magnetic circuit, it will generate a composite signal whose components will change differentially as the position of the servo head changes. Figure 10.14 shows several composite waveforms obtained at different positions of the servo head. The composite waveform is processed by using the first positive and negative pulses to generate a clock. From this clock are derived sampling signals which permit only the second positive and second negative pulses to pass. The resultant waveform has a DC component which after filtering gives a voltage proportional to the distance from the centre of the data tracks. The position error reaches a maximum when the servo head is entirely above one type of servo track, and further movement causes it to fall. The next time the position error falls to zero will be at the centreline of the adjacent cylinder.

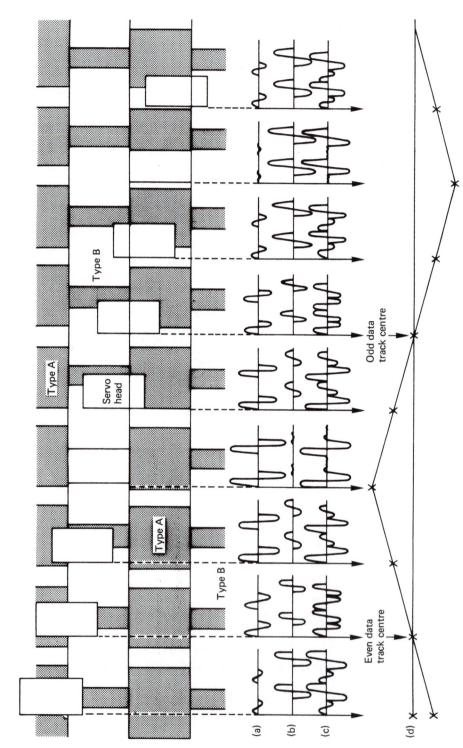

Figure 10.14 Waveforms resulting from several positions of the servo head with respect to the disk. At (a) and (b) are the two components of the waveforms, whose relative amplitudes are controlled by the relative areas of the servo head exposed to the two types of servo track. Because the servo head has only one magnetic circuit, these waveforms are not observed in practice, but are summed together, resulting in the composite waveforms shown at (c). By comparing the magnitudes of the second positive and second negative peaks in the composite waveforms, a position error signal is generated, as shown at (d).

Cylinders with even addresses (LSB = 0) will be those where the servo head is detented between an A track and a B track. Cylinders with odd addresses will be those where the head is between a B track and an A track. It can be seen from Figure 10.14 that the sense of the position error becomes reversed on every other cylinder. Accordingly, an invertor has to be switched into the track-following feedback loop in order to detent on odd cylinders. This inversion is controlled by the LSB of the desired cylinder address supplied at the beginning of a seek, such that the sense of the feedback will be correct when the heads arrive at the target cylinder.

Seeking across the servo surface results in the position-error signal rising and falling in a sawtooth. This waveform can be used to count down the cylinder difference counter which controls the seek. As with any cyclic transducer there is the problem of finding the absolute position. This difficulty is overcome by making all servo tracks outside cylinder 0 type A, and all servo tracks inside the innermost cylinder type B. These areas of identical track are called guard bands, and Figure 10.15 shows the relationship between the position error and the guard bands. During a head load, the servo head generates a constant maximum positive position error in the outer guard band. This drives the carriage forward until the position error first falls to zero. This, by definition, is cylinder zero. Some drives, however, load by driving the heads across the surface until the inner guard band is found, and then perform a full-length reverse seek to cylinder zero.

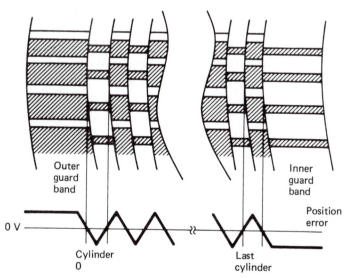

Figure 10.15 The working area of the servo surface is defined by the inner and outer guard bands, in which the position error reaches its maximum value.

10.10   Soft sectoring

It has been seen that a position error and a cylinder count can be derived from the servo surface, eliminating the cylinder transducer. The carriage velocity could also be derived from the slope of the position error, but there would then be no velocity feedback in the guard bands or during retraction, and so some form of velocity transducer is still necessary.

As there are exactly the same number of dibits or tribits on every track, it is possible to describe the rotational position of the disk simply by counting them. All that is needed is a unique pattern of missing dibits once per revolution to act as an index point, and the sector transducer can also be eliminated.

Unlike the read-data circuits, the servo-head circuits are active during a seek as well as when track-following, and have to be protected against interference from switched-mode positioner drivers. The main problem is detecting index, where noise could cause a 'missing' dibit to be masked. There are two solutions available: a preamplifier can be built into the servo-head cantilever, or driver switching can be inhibited when index is expected.

The advantage of deriving the sector count from the servo surface is that the number of sectors on the disk can be varied. Any number of sectors can be accommodated by feeding the dibit-rate signal through a programmable divider, so the same disk and drive can be used in numerous different applications.

In a non-servo-surface disk, the write clock is usually derived from a crystal oscillator. As the disk speed can vary owing to supply fluctuations, a tolerance gap has to be left at the end of each block to cater for the highest anticipated speed, to prevent overrun into the next block on a write. In a servo-surface drive, the write clock is obtained by multiplying the dibit-rate signal with a phase-locked loop. The write clock is then always proportional to disk speed, and recording density will be constant.

Most servo-surface drives have an offset facility, where a register written by the controller drives a DAC which injects a small voltage into the track-following loop. The action of the servo is such that the heads move off-track until the position error is equal and opposite to the injected voltage. The position of the heads above the track can thus be program-controlled. Offset is only employed on reading if it is suspected that the pack in the drive has been written by a different drive with non-standard alignment. A write function will cancel the offset.

10.11   Winchester technology

In order to offer extremely high capacity per spindle, which reduces the cost per bit, a disk drive must have very narrow tracks placed close together, and must use very short recorded wavelengths, which implies that the flying height of the heads must be small. The so-called Winchester technology is one approach to high storage density. The technology was developed by IBM, and the name came about because the model number of the development drive was the same as that of the famous rifle.

Reduction in flying height magnifies the problem of providing a contaminant-free environment. A conventional disk is well protected whilst inside the drive, but outside the drive the effects of contamination become intolerable.

In exchangeable-pack drives, there is a real limit to the track pitch that can be achieved because of the difficulty or cost of engineering head-alignment mechanisms to make the necessary minute adjustments to give interchange compatibility.

The essence of Winchester technology is that each disk pack has its own set of read/write and servo heads, with an integral positioner. The whole is protected by a dust-free enclosure, and the unit is referred to as a head disk assembly, or HDA.

As the HDA contains its own heads, compatibility problems do not exist, and no head alignment is necessary or provided for. It is thus possible to reduce track pitch considerably compared with exchangeable pack drives. The sealed environment ensures complete cleanliness which permits a reduction in flying height without loss of reliability, and hence leads to an increased linear density. If the rotational speed is maintained, this can also result in an increase in data transfer rate.

The HDA is completely sealed, but some have a small filtered port to equalize pressure. Into this sealed volume of air, the drive motor delivers the majority of its power output. The resulting heat is dissipated by fins on the HDA casing. Some HDAs are filled with helium which significantly reduces drag and heat build-up.

An exchangeable-pack drive must retract the heads to facilitate pack removal. With Winchester technology this is not necessary. An area of the disk surface is reserved as a landing strip for the heads. The disk surface is lubricated, and the heads are designed to withstand landing and take-off without damage. Winchester heads have very large air-bleed grooves to allow low flying height with a much smaller downthrust from the cantilever, and so they exert less force on the disk surface during contact. When the term *parking* is used in the context of Winchester technology, it refers to the positioning of the heads over the landing area.

Disk rotation must be started and stopped quickly to minimize the length of time the heads slide over the medium. A powerful motor will accelerate the pack quickly. Eddy-current braking cannot be used, since a power failure would allow the unbraked disk to stop only after a prolonged head-contact period. A failsafe mechanical brake is used, which is applied by a spring and released with a solenoid.

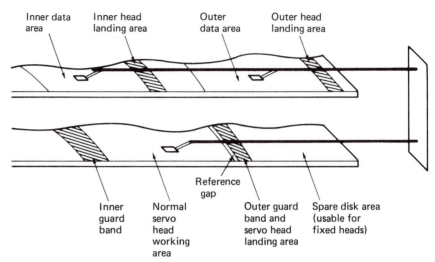

Inner data area   Inner head landing area   Outer data area   Outer head landing area

Reference gap

Inner guard band   Normal servo head working area   Outer guard band and servo head landing area   Spare disk area (usable for fixed heads)

Figure 10.16 When more than one head is used per surface, the positioner still only requires one servo head. This is often arranged to be equidistant from the read/write heads for thermal stability.

A major advantage of contact start/stop is that more than one head can be used on each surface if retraction is not needed. This leads to two gains: first, the travel of the positioner is reduced in proportion to the number of heads per surface, reducing access time; and, second, more data can be transferred at a given detented carriage position before a seek to the next cylinder becomes necessary. This increases the speed of long transfers. Figure 10.16 illustrates the relationships of the heads in such a system.

10.12 Servo-surface Winchester drives

With contact start/stop, the servo head is always on the servo surface, and it can be used for all the transducer functions needed by the drive. Figure 10.17 shows the position-error signal during a seek. The signal rises and falls as servo tracks are crossed, and the slope of the signal is proportional to positioner velocity. The position-error signal is differentiated and rectified to give a velocity feedback signal. Owing to the cyclic nature of the position-error signal, the velocity signal derived from it has troughs where the derivative becomes zero at the peaks. These cannot be filtered out, as the signal is in a servo loop, and the filter would introduce an additional lag. The troughs would, however, be interpreted by the servo driver as massive momentary velocity errors which might overload the amplifier. The solution which can be adopted is to use a signal obtained by integrating the positioner-motor current which is

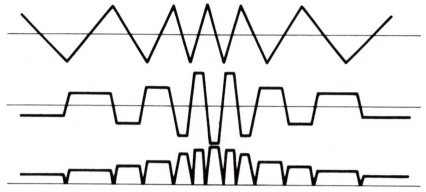

Figure 10.17 To generate a velocity signal, the position error from the servo head is differentiated and rectified.

selected when there is a trough in the differentiated position-error signal.

In order to make velocity feedback available over the entire servo surface, the conventional guard-band approach cannot be used since it results in steady position errors in the guard bands. In contact start/stop drives, the servo head must be capable of detenting in a guard band for the purpose of landing on shutdown.

A modification to the usual servo surface is used in Winchester drives, one implementation of which is shown in Figure 10.18, where it will be seen that there are extra transitions, identical in both types of track, along with the familiar dibits. The repeating set of transitions is known as a frame, in which the first dibit is used for synchronization, and a phase-locked oscillator is made to run at a multiple of the sync signal rate. The PLO is used as a reference for the write clock, as well as to generate sampling pulses to extract a position error from the composite waveform and to provide a window for the second dibit in the frame, which may or may not be present. Each frame thus contains one data bit, and succesive frames are read to build up a pattern in a shift register. The parallel output of the shift register is examined by a decoder which recognizes a number of unique patterns. In the guard bands, the decoder will repeatedly recognize the guard band code as the disk revolves. An index is generated in the same way, by recognizing a different pattern. In a contact start/stop drive, the frequency of index detection is used to monitor pack speed in order to dispense with a separate transducer. This does mean, however, that it must be possible to detect index everywhere, and for this reason, index is still recorded in the guard bands by replacing the guard-band code with index code once per revolution.

A consequence of deriving velocity information from the servo surface is that the location of cylinder zero is made more difficult, as there is no

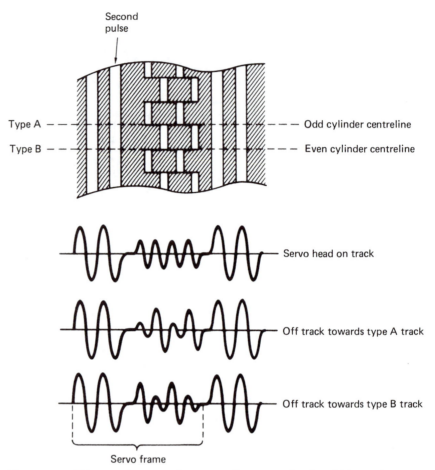

Figure 10.18 This type of servo surface pattern has a second pulse which may be omitted to act as a data bit. This is used to detect the guard bands and index.

longer a continuous maximum position error in the guard band. A common solution is to adopt a much smaller area of continuous position error known as a reference gap; this is typically three servo tracks wide. In the reference gap and for several tracks outside it, there is an unique reference gap code recorded in the frame-data bits. Figure 10.19 shows the position error which is generated as the positioner crosses this area of the disk, and shows the plateau in the position-error signal due to the reference gap. During head loading, which in this context means positioning from the parking area to cylinder zero, the heads move slowly inwards. When the reference code is detected, positioner velocity is reduced, and the position error is sampled. When successive position-error samples are the same, the head must be on the position-error plateau, and if the servo is put into track-following mode, it will

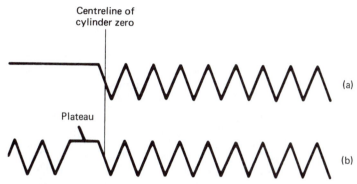

Figure 10.19 (a) Conventional guard band. (b) Winchester guard band, showing the plateau in the position error, known as the reference gap, which is used to locate cylinder 0.

automatically detent on cylinder zero, since this is the first place that the position error falls to zero.

10.13   Rotary positioners

Figure 10.20 shows that rotary positioners are feasible in Winchester drives; they cannot be used in exchangeable-pack drives because of interchange problems. There are some advantages to a rotary positioner.

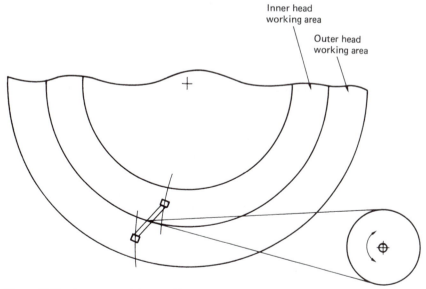

Figure 10.20 A rotary positioner with two heads per surface. The tolerances involved in the spacing between the heads and the axis of rotation mean that each arm records data in a unique position. Those data can only be read back by the same heads, which rules out the use of a rotary positioner in exchangeable-pack drives. In a head disk assembly the problem of compatibility does not arise.

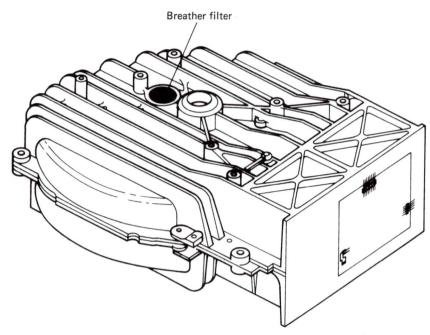

Figure 10.21 Head disk assembly with a rotary positioner. The adoption of this technique allows a very compact structure.

It can be placed in the corner of a compact HDA allowing smaller overall size. The manufacturing cost will be less than a linear positioner because fewer bearings and precision bars are needed. Significantly, a rotary positioner can be made faster since its inertia is smaller. With a linear positioner all parts move at the same speed. In a rotary positioner, only the heads move at full speed, as the parts closer to the shaft must move more slowly. Figure 10.21 shows a typical HDA with a rotary positioner. The principle of many rotary positioners is exactly that of a moving-coil ammeter, where current is converted directly into torque. Alternatively various configurations of electric motor or stepping motor can be used with band or wire drive.

One disadvantage of rotary positioners is that there is a component of windage on the heads which tends to pull the positioner in towards the spindle. In linear positioners windage is at right angles to motion and can be neglected. Windage can be overcome in rotary positioners by feeding the current cylinder address to a ROM which sends a code to a DAC. This produces an offset voltage which is fed to the positioner driver to generate a torque which balances the windage whatever the position of the heads.

When extremely small track spacing is contemplated, it cannot be assumed that all the heads will track the servo head due to temperature gradients. In this case the embedded-servo approach must be used, where each head has its own alignment patterns. The servo surface is often retained in such drives to allow coarse positioning, velocity feedback and index and write-clock generation, in addition to locating the guard bands for landing the heads.

Winchester drives have been made with massive capacity, but the problem of backup is then magnified, and the general trend has been for the physical size of the drive to come down as the storage density increases in order to improve access time. Very small Winchester disk drives are now available which plug into standard integrated circuit sockets. These are competing with RAM for memory applications where non-volatility is important.

10.14   Floppy disks

Floppy disks are the result of a search for a fast yet cheap non-volatile memory for the programmable control store of a processor under development at IBM in the late 1960s. Both magnetic tape and hard disk were ruled out on grounds of cost since only intermittent duty was required. The device designed to fulfil these requirements – the floppy disk drive – incorporated both magnetic-tape and disk technologies.

The floppy concept was so cost-effective that it transcended its original application to become a standard in industry as an online data-storage device. The original floppy disk, or diskette as it is sometimes called, was 8 inches in diameter, but a $5\frac{1}{4}$-inch diameter disk was launched to suit more compact applications. More recently Sony introduced the $3\frac{1}{2}$-inch floppy disk which has a rigid shell with sliding covers over the head access holes to reduce the likelihood of contamination.

Strictly speaking, the floppy is a disk, since it rotates and repeatedly presents the data on any track to the heads, and it has a positioner to give fast two-dimensional access, but it also resembles a tape drive in that the magnetic medium is carried on a flexible substrate which deforms when the read/write head is pressed against it.

Floppy disks are stamped from wide, thick tape, and are anisotropic, because the oxide becomes oriented during manufacture. On many disks this can be seen by the eye as parallel striations on the disk surface. A more serious symptom is the presence of sinusoidal amplitude modulation of the replay signal at twice the rotational frequency of the disk, as illustrated in Figure 10.22.

Floppy disks have radial apertures in their protective envelopes to allow access by the head. A further aperture allows a photoelectric index sensor to detect a small hole in the disk once per revolution.

Figure 10.23 shows that the disk is inserted into the drive edge-first, and slides between an upper and a lower hub assembly. One of these has

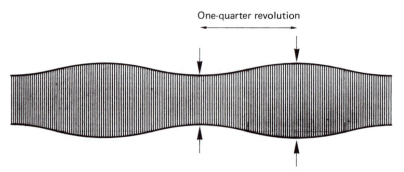

Figure 10.22 Sinusoidal amplitude modulation of floppy disk output due to anisotropy of medium.

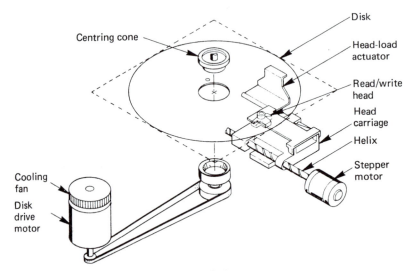

Figure 10.23 The mechanism of a floppy disk drive.

a fixed bearing which transmits the drive; the other is spring-loaded and mates with the drive hub when the door is closed, causing the disk to be centred and gripped firmly. The moving hub is usually tapered to assist centring. To avoid frictional heating and prolong life, the spindle speed is restricted when compared with that of hard disks. Recent drives almost univerally use direct-drive brushless DC motors. Since the rotational latency is so great, there is little point in providing a fast positioner, and the use of leadscrews driven by a stepping motor is universal. The permanent magnets in the stepping motor provide the necessary detenting, and to seek it is only necessary to provide a suitable number of drive pulses to the motor. As the drive is incremental, some form of reference is needed to determine the position of cylinder zero. At the rearward limit of carriage travel, a light beam is interrupted which resets the cylinder count. Upon power-up, the drive has to reverse-seek until this limit is found in order to calibrate the positioner. The grinding noise this makes is a characteristic of most PCs on power-up.

One of the less endearing features of plastics materials is a lack of dimensional stability. Temperature and humidity changes affect plastics much more than metals. The effect on the anisotropic disk substrate is to distort the circular tracks into a shape resembling a dog bone. For this reason, the track width and pitch have to be generous.

The read/write head of a single-sided floppy disk operates on the lower surface only, and is rigidly fixed to the carriage. Contact with the medium is achieved with the help of a spring-loaded pressure pad applied to the top surface of the disk opposite the head. Early drives retracted the pressure pad with a solenoid when not actually transferring data; later drives simply stop the disk. In double-sided drives, the pressure pad is replaced by a second sprung head.

Because of the indifferent stability of the medium, side trim or tunnel erasing is used, because it can withstand considerable misregistration. Figure 10.24 shows the construction of a typical side-trimming head, which has erase poles at each side of the magnetic circuit. When such a head writes, the erase poles are energized, and erase a narrow strip of the disk either side of the new data track. If the recording is made with misregistration, the side-trim prevents traces of the previous recording from being played back as well (Figure 10.25).

As the floppy-disk drive was originally intended to be a low-cost item, simple channel codes were used. Early drives used FM and double-density drives used MFM, but RLL codes were adopted later to further increase performance. As the recording density becomes higher at the inner tracks, the write current is sometimes programmed to reduce with inward positioner travel.

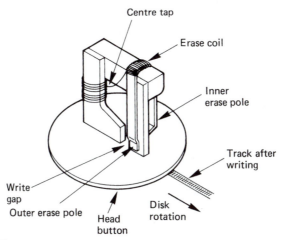

Figure 10.24 The poor dimensional stability of the plastic diskette means that tunnel erase or side trim has to be used. The extra erase poles can be seen here.

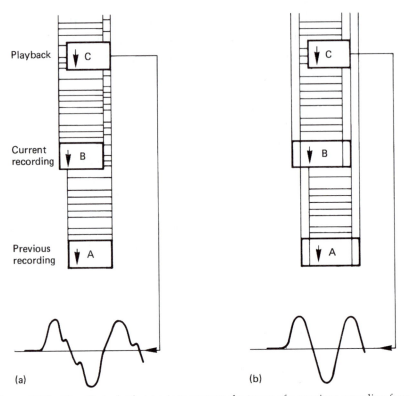

Figure 10.25 The effect of side trim is to prevent the traces of a previous recording from interfering with the latest recording: (a) without side trim; (b) with side trim.

The capacity of floppy disks is of the order of a few megabytes. This virtually precludes their use for digital audio recording except in conjunction with samplers. However, floppy disks find almost universal application in edit-list storage, console set-up storage, and as a software-loading medium for computer-based equipment where the sheer capacity of CDROM is not necessary.

10.15   Error handling

The protection of data recorded on disks differs considerably from the approach used on other media in digital audio. This has much to do with the traditional intolerance of data processors to errors when compared with audio. In particular, it is not possible to interpolate to conceal errors in a computer program or a data file.

In the same way that magnetic tape is subject to dropouts, magnetic disks suffer from surface defects whose effect is to corrupt data. The shorter wavelengths employed as disk densities increase are affected more by a given size of defect. Attempting to make a perfect disk is subject to a law of diminishing returns, and eventually a state is reached where it becomes more cost-effective to invest in a defect-handling system.

There are four main methods of handling media defects in magnetic media, and further techniques needed in WORM laser disks (see Chapter 12), whose common goal is to make their presence transparent to the data. These methods vary in complexity and cost of implementation, and can often be combined in a particular system.

In the construction of bad-block files, a brand new disk is tested by the operating system. Known patterns are written everywhere on the disk, and these are read back and verified. Following this the system gives the disk a volume name, and creates on it a directory structure which keeps records of the position and size of every file subsequently written. The physical disk address of every block which fails to verify is allocated to a file which has an entry in the disk directory. In this way, when genuine data files come to be written, the bad blocks appear to the system to be in use storing a fictitious file, and no attempt will be made to write there. Some disks have dedicated tracks where defect information can be written during manufacture or by subsequent verification programs, and these permit a speedy construction of the system bad-block file.

In association with the bad-block file, many drives allocate bits in each header to indicate that the associated block is bad. If a data transfer is attempted at such a block, the presence of these bits causes the function to be aborted. The bad-block file system gives very reliable protection against defects, but can result in a lot of disk space being wasted. Systems often use several disk blocks to store convenient units of data called clusters, which will all be written or read together. Figure 10.26 shows how a bit map is searched to find free space, and illustrates how the presence of one bad block can write off a whole cluster.

In sector skipping, space is made at the end of every track for a spare data block, which is not normally accessible to the system. Where a track

	1	1	1	1	1	1	1	1	1	1	1	0	0	0	0	0	A
A	0	0	0	0	0	0	1	1	1	1	1	1	1	1	1	1	
	1	1	1	1	1	1	1	1	1	1	1	1	1	1	1	1	
	1	1	1	1	1	1	1	0	0	0	0	1	0	0	0	0	B
	0	0	1	1	1	1	1	1	1	1	1	1	1	0	0	0	
	0	0	0	0	0	0	0	0	e	tc.							

Figure 10.26 A disk-block-usage bit map in 16 bit memory for a cluster size of 11 blocks. Before writing on the disk, the system searches the bit map for contiguous free space equal to or larger than the cluster size. The first available space is unusable because the presence of a bad block B destroys the contiguity of the cluster. Thus one bad block causes the loss of a cluster.

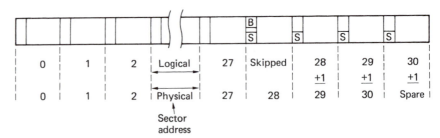

Figure 10.27 Skip sectoring. The bad block in this example has a physical sector address of 28. By setting the skip-sector flags in the header, this and subsequent logical blocks have one added to their sector addresses, and the spare block is brought into use.

is found to contain a defect, the affected block becomes a skip sector. In this block, the regular defect flags will be set, but in addition, a bit known as the skip-sector flag is set in this and every subsequent block in the track. When the skip-sector flag is encountered, the effect is to add one to the desired sector address for the rest of the track, as in Figure 10.27. In this way the bad block is unused, and the track format following the bad block is effectively slid along by one block to bring into use the spare block at the end of the track. Using this approach, the presence of single bad blocks does not cause the loss of clusters, but requires slightly greater control complexity. If two bad blocks exist in a track, the second will be added to the bad-block file as usual.

The two techniques described so far have treated the block as the smallest element. In practice, the effect of a typical defect is to corrupt only a few bytes. The principle of defect skipping is that media defects can be skipped over within the block so that a block containing a defect is made usable. The header of each block contains the location of the first defect in bytes away from the end of the header, and the number of bytes from the first defect to the second defect, and so on up to the maximum of four shown in the example of Figure 10.28. Each defect is overwritten with a fixed number of bytes of preamble code and a sync pattern. The skip is positioned so that there is sufficient undamaged preamble after the defect for the data separator to regain lock. Each defect lengthens the block, causing the format of the track to slip round. A space is left at the end of each track to allow a reasonable number of skips to be accommodated. Often a track descriptor is written at the beginning of each track which contains the physical position of defects relative to index. The disk format needed for a particular system can then be rapidly arrived at by reading the descriptor, and translating the physical defect locations into locations relative to the chosen sector format. Figure 10.29 shows how a soft-sectoring drive can have two different formats around the same defects using this principle.

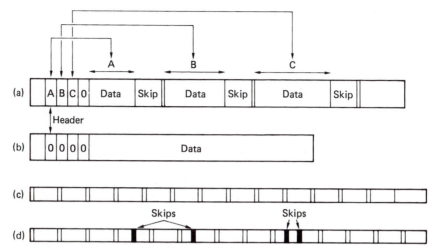

Figure 10.28 Defect skipping. (a) A block containing three defects. The header contains up to four parameters which specify how much data is to be written before each skip. In this example only three entries are needed. (b) An error-free block for comparison with (a); the presence of the skips lengthens the block. To allow for this lengthening, the track contains spare space at the end, as shown in (c), which is an error-free track. (d) A track containing the maximum of four skips, which have caused the spare space to be used up.

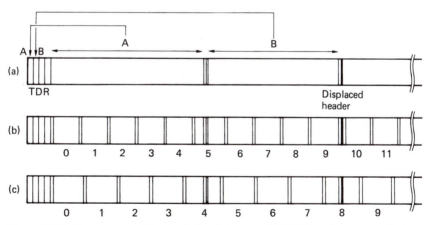

Figure 10.29 The purpose of the track descriptor record (TDR) is to keep a record of defects independent of disk format. The positions of the defects stored in the TDR (a) are used by the formatter to establish the positions relative to the format used. With the format (b), the first defect appears in sector 5, but the same defect would be in sector 4 for format (c). The second defect falls where a header would be written in (b) so the header is displaced for sector 10. The same defect falls in the data area of sector 8 in (c).

In the case where there are too many defects in a track for the skipping to handle, the system bad-block file will be used. This is rarely necessary in practice, and the disk appears to be contiguous error-free logical and physical space. Defect skipping requires fast processing to deal with events in real time as the disk rotates. Bit-slice microsequencers are one approach, as a typical microprocessor would be too slow.

A refinement of sector skipping which permits the handling of more than one bad block per track without the loss of a cluster is revectoring. A bad block caused by a surface defect may only have a few defective bytes, so it is possible to record highly redundant information in the bad block. On a revectored disk, a bad block will contain in the data area repeated records pointing to the address where data displaced by the defect can be found. The spare block at the end of the track will be the first such place, and can be read within the same disk revolution, but out of sequence, which puts extra demands on the controller. In the less frequent case of more than one defect in a track, the second and subsequent bad blocks revector to spare blocks available in an area dedicated to that purpose. The principle is illustrated in Figure 10.30. In this case a seek will be necessary to locate the replacement block. The low probability of this means that access time is not significantly affected.

These steps are the first line of defence against errors in disk drives, and serve to ensure that, by and large, the errors due to obvious surface defects are eliminated. There are other error mechanisms in action, such as noise and jitter, which can result in random errors, and it is necessary to protect disk data against these also. The error-correction mechanisms described in Chapter 7 will be employed. In general each data block is made into a codeword by the addition of redundancy at the end. The error-correcting code used in disks was, for a long time, Fire code, because it allowed correction with the minimum circuit complexity. It could, however, only

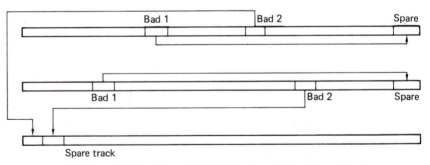

Figure 10.30 Revectoring. The first bad block in each track is revectored to the spare block at the end of the track. Unlike skip sectoring, subsequent good blocks are unaffected, and the replacement block is read out of sequence. The second bad block on any one track is revectored to one of a number of spare tracks kept on the disk for this purpose.

correct one error burst per block, and it had a probability of miscorrection which was marginal for some applications. The advances in complex logic chips meant that the adoption of a Reed–Solomon code was a logical step, since these have the ability to correct multiple error bursts. As the larger burst errors in disk drives are taken care of by verifying the medium, interleaving in the error-correction sense is not generally needed. When interleaving is used in the context of disks, it usually means that the sectors along a track are interleaved so that reading them in numerical order requires two revolutions. This will slow down the data transfer rate where the drive is too fast for the associated circuitry.

In some systems, the occurrence of errors is monitored to see if they are truly random, or if an error persistently occurs in the same physical block. If this is the case, and the error is small, and well within the correction power of the code, the block will continue in use. If, however, the error is larger than some threshold, the data will be read, corrected and rewritten elsewhere, and the block will then be added to the bad-block file so that it will not be used again.

10.16   RAID arrays

Whilst the MTBF of a disk drive is very high, it is a simple matter of statistics that when a large number of drives is assembled in a system the time between failures becomes shorter. Winchester drives are sealed units and the disks cannot be removed if there is an electronic failure. Even if this were possible the system cannot usually afford downtime whilst such a data recovery takes place.

Consequently any system in which the data are valuable must take steps to ensure data integrity. This is commonly done using RAID (redundant array of inexpensive disks) technology. Figure 10.31 shows that in a RAID array data blocks are spread across a number of drives.

An error-correcting check symbol (typically Reed–Solomon) is stored on a redundant drive. The error correction is powerful enough to fully correct any error in the block due to a single failed drive. In RAID arrays the drives are designed to be hot-plugged (replaced without removing power) so if a drive fails it is simply physically replaced with a new one. The error-correction system will rewrite the drive with the data which was lost with the failed unit.

When a large number of disk drives are arrayed together, it is necessary and desirable to spread files across all the drives in a RAID array. Whilst this ensures data integrity, it also means that the data transfer rate is multiplied by the number of drives sharing the data. This means that the data transfer rate can be extremely high and new approaches are necessary to move the data in and out of the disk system.

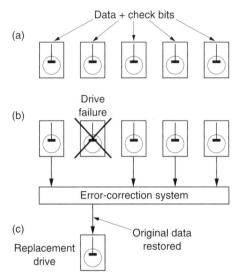

Figure 10.31 In RAID technology, data and redundancy are spread over a number of drives (a). In the case of a drive failure (b) the error-correction system can correct for the loss and continue operation. When the drive is replaced (c) the data can be rewritten so that the system can then survive a further failure.

10.17   The disk controller

A disk controller is a unit which is interposed between the drives and the rest of the system. It consists of two main parts; that which issues control signals to and obtains status from the drives, and that which handles the data to be stored and retrieved. Both parts are synchronized by the control sequencer. The essentials of a disk controller are determined by the characteristics of drives and the functions needed, and so they do not vary greatly. For digital audio use, it is desirable for economic reasons to use a commercially available disk controller intended for computers. Disk drives are generally built to interface to a standard controller interface, such as the SCSI bus.

The execution of a function by a disk subsystem requires a complex series of steps, and decisions must be made between the steps to decide what the next will be. There is a parallel with computation, where the function is the equivalent of an instruction, and the sequencer steps needed are the equivalent of the microinstructions needed to execute the instruction. The major failing in this analogy is that the sequence in a disk drive must be accurately synchronized to the rotation of the disk.

Most disk controllers use direct memory access, which means that they have the ability to transfer disk data in and out of the associated memory without the assistance of the processor. In order to cause an audio-file transfer, the disk controller must be told the physical disk address (cylinder, sector, track), the physical memory address where the audio file begins, the size of the file and the direction of transfer (read or write). The controller will then position the disk heads, address the memory, and transfer the samples. One disk transfer may consist of many contiguous disk blocks, and the controller will automatically increment the disk-address registers as each block is completed. As the disk turns, the sector address increases until the end of the track is reached. The track or head address will then be incremented and the sector address reset so that transfer continues at the beginning of the next track. This process continues until all the heads have been used in turn. In this case both the head address and sector address will be reset, and the cylinder address will be incremented, which causes a seek. A seek which takes place because of a data transfer is called an implied seek, because it is not necessary formally to instruct the system to perform it. As disk drives are block-structured devices, and the error correction is codeword-based, the controller will always complete a block even if the size of the file is less than a whole number of blocks. This is done by packing the last block with zeros.

The status system allows the controller to find out about the operation of the drive, both as a feedback mechanism for the control process and to handle any errors. Upon completion of a function, it is the status system which interrupts the control processor to tell it that another function can be undertaken.

In a system where there are several drives connected to the controller via a common bus, it is possible for non-data-transfer functions such as seeks to take place in some drives simultaneously with a data transfer in another.

Before a data transfer can take place, the selected drive must physically access the desired block, and confirm this by reading the block header. Following a seek to the required cylinder, the positioner will confirm that the heads are on-track and settled. The desired head will be selected, and then a search for the correct sector begins. This is done by comparing the desired sector with the current sector register, which is typically incremented by dividing down servo-surface pulses. When the two counts are equal, the head is about to enter the desired block. Figure 10.32 shows the structure of a typical disk track. In between blocks are placed address marks, which are areas without transitions which the read circuits can detect. Following detection of the address mark, the sequencer is roughly synchronized to begin handling the block. As the block is entered, the data separator locks to the preamble, and in due course the sync pattern will be found. This sets to zero a counter which divides the data-bit rate by eight, allowing the serial recording to be correctly assembled into bytes, and also allowing the sequencer to count the position of the head through the block in order to perform all the necessary steps at the right time.

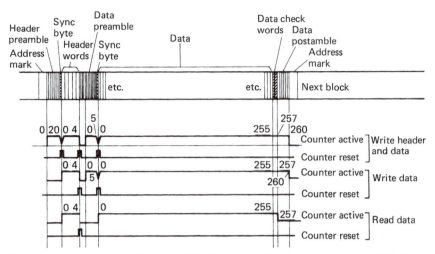

Figure 10.32 The format of a typical disk block related to the count process which is used to establish where in the block the head is at any time. During a read the count is derived from the actual data read, but during a write, the count is derived from the write clock.

The first header word is usually the cylinder address, and this is compared with the contents of the desired cylinder register. The second header word will contain the sector and track address of the block, and these will also be compared with the desired addresses. There may also be bad-block flags and/or defect-skipping information. At the end of the header is a CRCC which will be used to ensure that the header was read correctly. Figure 10.33 shows a flowchart of the position verification, after which a data transfer can proceed. The header reading is completely automatic. The only time it is necessary formally to command a header to be read is when checking that a disk has been formatted correctly.

During the read of a data block, the sequencer is employed again. The sync pattern at the beginning of the data is detected as before, following which the actual data arrive. These bits are converted to byte or sample parallel, and sent to the memory by DMA. When the sequencer has counted the last data-byte off the track, the redundancy for the error-correction system will be following.

During a write function, the header-check function will also take place as it is perhaps even more important not to write in the wrong place on a disk. Once the header has been checked and found to be correct, the write process for the associated data block can begin. The preambles, sync pattern, data block, redundancy and postamble have all to be written contiguously. This is taken care of by the sequencer, which is obtaining timing information from the servo surface to lock the block structure to the angular position of the disk. This should be contrasted with the read function, where the timing comes directly from the data.

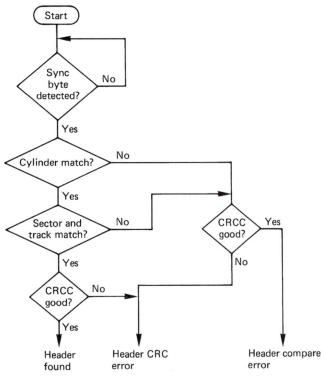

Figure 10.33 The vital process of position confirmation is carried out in accordance with the above flowchart. The appropriate words from the header are compared in turn with the contents of the disk-address registers in the subsystem. Only if the correct header has been found and read properly will the data transfer take place.

10.18   Digital audio disk systems

In order to use disk drives for the storage of audio samples, a system like the one shown in Figure 10.34 is needed. The control computer determines where and when samples will be stored and retrieved, and sends instructions to the disk controller which causes the drives to read or write, and transfers samples between them and the memory. The instantaneous data rate of a typical drive is roughly ten times higher than the sampling rate, and this may result in the system data bus becoming choked by the disk transfers so other transactions are locked out. This is avoided by giving the disk controller DMA system a lower priority for bus access so that other devices can use the bus. A rapidly spinning disk cannot wait, and in order to prevent data loss, a silo or FIFO memory is necessary in the disk controller.[1] The operation of these devices was described in Chapter 3. A silo will be interposed in the disk controller data stream in the fashion shown in Figure 10.35 so that it can buffer data both to and from the disk. When reading the disk, the silo starts empty,

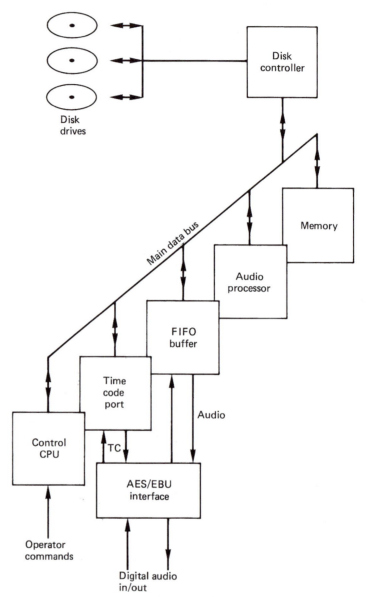

Figure 10.34 The main parts of a digital audio disk system. Memory and FIFO allow continuous audio despite the movement of disk heads between blocks.

and if there is bus contention, the silo will start to fill as shown in Figure 10.36(a). Where the bus is free, the disk controller will attempt to empty the silo into the memory. The system can take advantage of the interblock gaps on the disk, containing headers, preambles and redundancy, for in these areas there are no data to transfer, and there is some breathing space to empty the silo before the next block. In practice the silo need not be empty at the start of every block, provided it never becomes full before

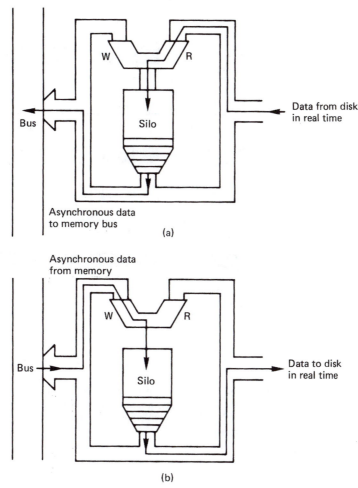

Figure 10.35 In order to guarantee that the drive can transfer data in real time at regular intervals (determined by disk speed and density) the silo provides buffering to the asynchronous operation of the memory access process. In (a) the silo is configured for a disk read. The same silo is used in (b) for a disk write.

the end of the transfer. If this happens some data are lost and the function must be aborted. The block containing the silo overflow will generally be reread on the next revolution. In sophisticated systems, the silo has a kind of dipstick which indicates how far up the V bit is set, and can interrupt the CPU if the data get too deep. The CPU can then suspend some bus activity to allow the disk controller more time to empty the silo.

When the disk is to be written, a continuous data stream must be provided during each block, as the disk cannot stop. The silo will be prefilled before the disk attempts to write as shown in Figure 10.36(b), and the disk controller attempts to keep it full. In this case all will be well if the silo does not become empty before the end of the transfer.

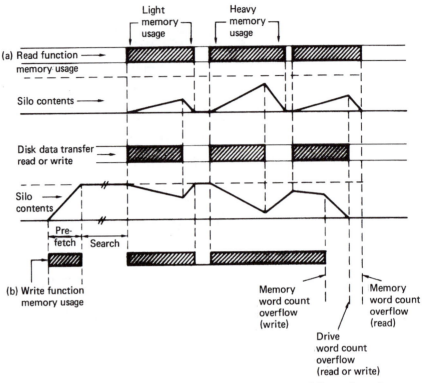

Figure 10.36 The silo contents during read functions (a) appear different from those during write functions (b). In (a), the control logic attempts to keep the silo as empty as possible; in (b) the logic prefills the silo and attempts to keep it full until the memory word count overflows.

The disk controller cannot supply samples at a constant rate, because of gaps between blocks, defective blocks and the need to move the heads from one track to another and because of system bus contention. In order to accept a steady audio sample stream for storage, and to return it in the same way on replay, hard disk-based audio recorders must have a quantity of RAM for buffering. Then there is time for the positioner to move whilst the audio output is supplied from the RAM. In replay, the drive controller attempts to keep the RAM as full as possible by issuing a read command as soon as one block space appears in the RAM. This allows the maximum time for a seek to take place before reading must resume. Figure 10.37 shows the action of the RAM during reading. Whilst recording, the drive controller attempts to keep the RAM as empty as possible by issuing write commands as soon as a block of data is present, as in Figure 10.38. In this way the amount of time available to seek is maximized in the presence of a continuous audio sample input.

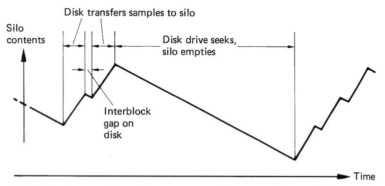

Figure 10.37 During an audio replay sequence, silo is constantly emptied to provide samples, and is refilled in blocks by the drive.

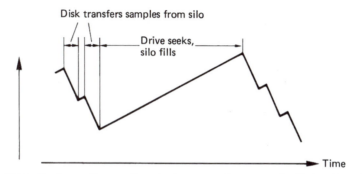

Figure 10.38 During audio recording, the input samples constantly fill the silo, and the drive attempts to keep it empty by reading from it.

10.19   Arranging the audio data on disk

When playing a tape recording or a disk having a spiral track, it is only necessary to start in the right place and the data are automatically retrieved in the right order. Such media are also driven at a speed which is proportional to the sampling rate. In contrast, a hard disk has a discontinuous recording and acts more like a RAM in that it must be addressed before data can be retrieved. The rotational speed of the disk is constant and not locked to anything. A vital step in converting a disk drive into an audio recorder is to establish a link between the time through the recording and the location of the data on the disk.

When audio samples are fed into a disk-based system, from an AES/EBU interface or from a convertor, they will be placed initially in RAM, from which the disk controller will read them by DMA. The continuous-input sample stream will be split up into disk blocks for disk storage. The AES/EBU interface carries a timecode in the channel status data, and this timecode, or that from a local generator, will be used to assemble a table which contains a conversion from real time in the recording to the physical disk address of the corresponding audio files. As an alternative, an interface may be supplied which allows conventional SMPTE or EBU timecode to be input. Wherever possible, the disk controller will allocate incoming audio samples to contiguous disk addresses, since this eases the conversion from timecode to physical address.[2] This is not, however, always possible in the presence of defective blocks, or if the disk has become chequerboarded from repeated rerecording.

The table of disk addresses will also be made into a named disk file and stored in an index which will be in a different area of the disk from the audio files. Several recordings may be fed into the system in this way, and each will have an entry in the index.

If it is desired to play back one or more of the recordings, then it is only necessary to specify the starting timecode and the filename. The system will look up the index file in order to locate the physical address of the first and subsequent sample blocks in the desired recording, and will begin to read them from disk and write them into the RAM. Once the RAM is full, the real-time replay can begin by sending samples from RAM to the output or to local convertors. The sampling rate clock increments the RAM address and the timecode counter. Whenever a new timecode frame is reached, the corresponding disk address can be obtained from the index table, and the disk drive will read a block in order to keep the RAM topped up.

The disk transfers must by definition take varying times to complete because of the rotational latency of the disk. Once all the sectors on a particular cylinder have been read, it will be necessary to seek to the next cylinder, which will cause a further extension of the reading sequence. If a bad block is encountered, the sequence will be interrupted until it has passed. The RAM buffering is sufficient to absorb all these access time variations. Thus the RAM acts as a delay between the disk transfers and the sound which is heard. A corresponding advance is arranged in timecodes fed to the disk controller. In effect the actual timecode has a constant added to it so that the disk is told to obtain blocks of samples in advance of real time. The disk takes a varying time to obtain the samples, and the RAM then further delays them to the correct timing. Effectively the disk/RAM subsystem is a timecode-controlled memory. One need only put in the time and out comes the audio corresponding to that time. This is the characteristic of an audio synchronizer. In most audio equipment the synchronizer is extra; the hard disk needs one to work at all, and so every hard disk comes with a free synchronizer. This makes disk-based systems very flexible as they can be made to lock to almost any reference and care little what sampling rate is used or if it varies. They perform well locked to videotape or film via timecode because no matter how the pictures are shuttled or edited, the timecode link always produces the correct sound to go with the pictures.

A multitrack recording can be stored on a single disk and, for replay, the drive will access the files for each track faster than real time so that they all become present in the memory simultaneously. It is not, however, compulsory to play back the tracks in their original time relationship. For the purpose of synchronization,[3] or other effects, the tracks can be played with any time relationship desired, a feature not possible with multitrack tape drives.

In order to edit the raw audio files fed into the system, it is necessary to listen to them in order to locate the edit points. This can be done by playback of the whole file at normal speed if time is no object, but this neglects the random access capability of a disk-based system. If an event list has been made at the time of the recordings, it can be used to access any part of them within a few tens of milliseconds, which is the time taken for the heads to traverse the entire disk surface. This is far superior to the slow spooling speed of tape recorders.

10.20   Spooling files

If an event list is not available, it will be necesary to run through the recording at a raised speed in order rapidly to locate the area of the desired edit. If the disk can access fast enough, an increase of up to ten times normal speed can be achieved simply by raising the sampling-rate clock, so that the timecode advances more rapidly, and new data blocks are requested from the disk more rapidly. If a constant sampling-rate output is needed, then rate reduction via a digital filter will be necessary.[4,5] Some systems have sophisticated signal processors which allow pitch changing, so that files can be played at non-standard speed but with normal pitch or vice versa.[6] If higher speeds are required, an alternative approach to processing on playback only is to record spooling files[7] at the same time as an audio file is made. A spooling file block contains a sampling-rate-reduced version of several contiguous audio blocks. When played at standard sampling rate, it will sound as if it is playing faster by the factor of rate reduction employed. The spooling files can be accessed less often for a given playback speed, or higher speed is possible within a given access-rate constraint.

Once the rough area of the edit has been located by spooling, the audio files from that area can be played to locate the edit point more accurately. It is often not sufficiently accurate to mark edit points on the fly by listening to the sound at normal speed. In order to simulate the rock-and-roll action of edit-point location in an analog tape recorder, audio blocks in the area of the edit point can be transferred to memory and accessed at variable speed and in either direction by deriving the memory addresses from a hand-turned rotor. A description of this process is included in the next chapter where digital audio editing is discussed.

10.21 Broadcast applications

In a radio broadcast environment it is possible to contain all the commercials and jingles in daily use on a disk system thus eliminating the doubtful quality of analog cartridge machines.[8] Disk files can be cued almost instantly by specifying the file name of the wanted piece, and once RAM resident play instantly they are required. Adding extra output modules means that several audio files can be played back simultaneously if a station broadcasts on more than one channel. If a commercial break contains several different spots, these can be chosen at short notice just by producing a new edit list.

10.22 Sampling rate and playing time

The bit rate of a digital audio system is such that high-density recording is mandatory for long playing time. A disk drive can never reach the density of a rotary-head tape machine because it is optimized for fast random access, but the performance of all types of data recorders has advanced so rapidly that disk drive capacity is no longer an issue in digital audio applications.

One high-quality digital audio channel requires nearly a megabit per second, which means that a gigabyte of storage (the usual unit for disk measurement) offers about three hours of monophonic audio. There is, however, no compulsion to devote the whole disk to one audio channel, and so in one gigabyte of storage, two channels could be recorded for 90 min, or four channels for 45 min and so on. For broadcast applications, where an audio bandwidth of 15 kHz is imposed by the FM stereo transmission standard, the alternative sampling rate of 32 kHz can be used, which allows about four hours of monophonic digital audio per gigabyte. Where only speech is required, an even lower rate can be employed. Compression can also be used to obtain greater playing time, but with the penalty of loss of quality.

In practice, multitrack working with disks is better than these calculations would indicate, because on a typical multitrack master tape, all tracks are not recorded continuously. Some tracks will contain only short recordings in a much longer overall session. A tape machine has no option but to leave these tracks unrecorded either side of the wanted recording, whereas a disk system will only store the actual wanted samples. The playing time in a real application will thus be greater than expected.

A further consideration is that hard disks systems do not need to edit the actual data files on disk. The editing is performed in the memory of the control system and is repeated dynamically under the control of an EDL (edit decision list) each time the edited work is required. Thus a lengthy editing session on a hard disk system does not result in the disk becoming fuller as only a few bytes of EDL are generated.

References

1. Ingbretsen, R.B. and Stockham, T.G., Random access editing of digital audio. *J. Audio Eng. Soc.*, **32**, 114–122 (1982)
2. McNally, G.W., Gaskell, P.S. and Stirling, A.J., Digital audio editing. *BBC Research Dept Report*, RD 1985/10
3. McNally, G.W., Bloom, P.J. and Rose, N.J., A digital signal processing system for automatic dialogue post-synchronisation. Presented at the 82nd Audio Engineering Society Convention (London, 1987), Preprint 2476(K-6)
4. McNally, G.W., Varispeed replay of digital audio with constant output sampling rate. Presented at the 76th Audio Engineering Society Convention (New York, 1984), Preprint 2137(A-9)
5. Gaskell, P.S., A hybrid approach to the variable speed replay of audio. Presented at the 77th Audio Engineering Society Convention (Hamburg, 1985), Preprint 2202(B-1)
6. Gray, E., The Synclavier digital audio system: recent developments in audio post production. *Int. Broadcast Eng.*, **18**, 55 (March 1987)
7. McNally, G.W., Fast edit-point location and cueing in disk-based digital audio editing. Presented at the 78th Audio Engineering Society Convention (Anaheim, 1985), Preprint 2232(D-10)
8. Itoh, T., Ohta, T. and Sohma, Y., Real time transmission system of commercial messages in radio broadcasting. Presented at the 67th Audio Engineering Society Convention (New York, 1980), Preprint 1682(H-1)

11

Digital audio editing

Unlike analog, digital audio can take advantage of the freedom to store data in random access media and the signal processing techniques developed in computation. This has had an enormous impact in the way audio is edited, completely displacing traditional methods. This chapter shows how the digital edit process is achieved using combinations of storage media, processing and control systems.

11.1   Introduction

At its most basic, editing may be no more than a punch-in on a multi-track recorder, or the removal of hesitations from an interview. At a higher level, it includes assembling myriad sound effects and mixing them with timecode-locked dialogue in order to create a film soundtrack.

Mastering is a further form of editing where various tracks are put together to create the master recording from which an album will be made. Unlike vinyl disk cutting, where the operator controls the cutter parameters, the CD and MD 'cutting' process is just a data transfer and is independent of musical content, so responsibility for the subjective quality of the final disk falls entirely on those who make the master recording. The duration of each musical piece, the length of any pauses between pieces and the relative levels of the pieces on the disk have to be determined at the time of mastering. The master recording will be compiled from source media which may each contain only some of the pieces required on the final CD, in any order. The recordings will vary in level, and may contain several retakes of a passage which was unsatisfactory.

The purpose of the digital mastering editor is to take each piece, and insert sections from retakes to correct errors, and then to assemble the pieces in the correct order, with appropriate pauses between and with the correct relative levels to create the master tape. All this is done by copying in the digital domain. The source recordings need not be changed in any way, and degradation of quality is minimal. The master recording will also have contiguous timecode, and with the addition of the subcode information, it is ready for cutting the CD.

At the other end of the spectrum, editors may be used for audio post-production of film or video sound tracks. The acoustic of most film sets precludes the use of live recording and the recording made on the set is used only as a guide. In ADR (automatic dialogue replacement), dialogue recorded later under better conditions is fitted to lip movement on the pictures. Sound effects from an effects bank will need to be triggered so that they coincide with visible events. Foreign-language dubs may require the time axis of the audio to be stretched or compressed without pitch change in order to achieve more convincing lip-sync. Each source recording may need equalization, compression or some effect such as reverberation before being added to the mix.

Digital audio editors work in two basic ways, by assembling or by inserting sections of audio waveform to build the finished waveform. Both terms have the same meaning as in the context of video recording. Assembly begins with a blank master file or recording. The beginning of the work is copied from the source, and new material is successively appended to the end of the previous material. Figure 11.1 shows how a

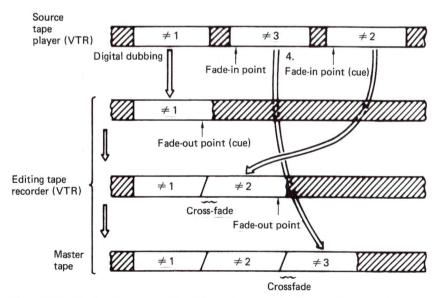

Figure 11.1 The function of an editor is to perform a series of assembles to produce a master tape from source tapes.

master recording is made up by assembly from source recordings. Insert editing begins with an existing recording in which a section is replaced by the edit process. Punch-in in multi-track recorders is a form of insert editing.

11.2   Editing with random access media

In all types of audio editing the goal is the appropriate sequence of sounds at the appropriate times. In analog audio equipment, editing was almost always performed using tape or magnetically striped film. These media have the characteristic that the time through the recording is proportional to the distance along the track. Editing consisted of physically cutting and splicing the medium, in order mechanically to assemble the finished work, or of copying lengths of source medium to the master.

When this was the only way of editing, it did not need a qualifying name. Now that audio is stored as data, alternative storage media have become available which allow editors to reach the same goal but using different techniques. Whilst early open-reel digital audio tape formats were designed to support splice editing, this was an evolutionary dead end. In all other digital audio editing samples from various sources are brought from the storage media to various pages of RAM. The edit is performed by crossfading between sample streams retrieved from RAM and subsequently rewriting on the output medium. Thus the nature of the storage medium does not affect the form of the edit in any way except the amount of time needed to execute it.

Tapes only allow serial access to data, whereas disks and RAM allow random access and so can be much faster. Editing using random access storage devices is very powerful as the shuttling of tape reels is avoided. The technique is generally called non-linear editing because the time axis of the storage medium is non-linear.

11.3   Editing on recording media

Audio editing requires the modification of source material in the correct real-time sequence to sample accuracy. However, all real digital storage media are block-based. Blocks are needed to allow the inclusion of preambles so that synchronized replay can begin and to allow an addressing mechanism. Most media have some form of error correction requiring an interleave, or reordering, of samples to reduce the impact of large errors. As a result, editing to sample accuracy simply cannot be performed directly on real media. Even if an individual sample could be

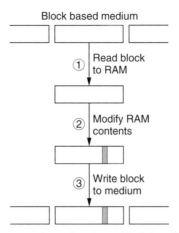

Block based medium

① Read block to RAM

② Modify RAM contents

③ Write block to medium

Figure 11.2 A single byte cannot be updated on a block-based medium. Instead the whole block is transferred to RAM, which is byte addressable. Following the change, the entire block is written back to the medium.

located in a block, replacing the samples after it would destroy the codeword structure and render the block uncorrectable.

The only solution is to ensure that the medium itself is only edited at block boundaries so that entire error-correction codewords are written down. Figure 11.2 shows that in order to edit to sample accuracy, entire blocks must be read from the medium and de-interleaved into RAM. The de-interleaved data are then modified in RAM by the edit process and re-interleaved for writing back on the medium. This technique is called *read–modify–write* and is only an extension of the technique used in word processors to correct a spelling mistake in a text file. The block containing the error is read from disk into RAM, the error (which may be a single byte) is corrected in RAM and the whole block is written back to disk.

In disks, blocks are often associated into clusters which consist of a fixed number of blocks in order to increase data throughput. When clustering is used, editing on the disk can only take place by rewriting entire clusters.

11.4   The structure of an editor

The digital audio editor consists of three main areas. First, the various contributory recordings must enter the processing stage at the right time with respect to the master recording. This will be achieved using a combination of timecode, transport synchronization and RAM timebase correction. The synchronizer will take control of the various transports during an edit so that one section reaches its out-point just as another reaches its in-point.

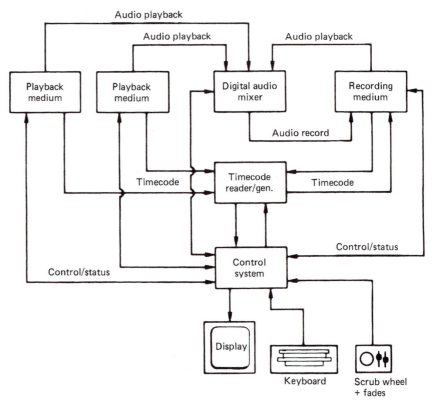

Figure 11.3 A digital audio editor requires an audio path to process the samples, and a timing and synchronizing section to control the time alignment of signals from the various sources. A supervisory control system acts as the interface between the operator and the hardware.

Second, the audio signal path of the editor must take the appropriate action, such as a crossfade, at the edit point. This requires some digital processing circuitry.

Third, the editing operation must be supervised by a control system which coordinates the operation of the transports and the signal processing to achieve the desired result.

Figure 11.3 shows a simple block diagram of an editor. Each source device, be it disk or tape or some other medium, must have timecode locked to the audio samples in some way. The synchronizer section of the control system uses the timecode to determine the relative timing of sources and sends remote control signals to the transport to make the timing correct. The master recorder is also fed with timecode in such a way that it can make a contiguous timecode track when performing assembly edits. The control system also generates a master sampling rate clock to which contributing devices must lock in order to feed samples into the edit process. The audio signal processor takes contributing sources and mixes them as instructed by the control system. The mix is then routed to the recorder.

11.5   Timecode

Timecode is essential to editing, but the standardization of timecode for digital audio recorders has been hampered by the diversity of standards in video. Synchronization between timecode and the sampling rate is essential, otherwise there will be a conflict between the need to lock the various sampling rates in the system with the need to lock the timecodes. This can only be resolved with synchronous timecode. The EBU timecode format relates easily to digital audio sampling rates of 48 kHz, 44.1 kHz and 32 kHz, but it is not so easy with the dropframe SMPTE timecode necessary for NTSC recording due to the 0.1 per cent slip between the actual field rate and 60 Hz.

The timecode used in a great deal of equipment follows the SMPTE standard for 525/60 and is shown in Figure 11.4. EBU timecode is basically similar to the SMPTE code except that it is designed for 50 Hz frame rate systems. These timecode systems encode hours, minutes, seconds and frames as binary-coded decimal (BCD) numbers. In tape media, the timecode data may be serially encoded along with user bits into an FM channel code (see Chapter 6) which is recorded on a dedicated linear track. The user bits are not specified in the standard, but a common use is to record the take or session number.

Disks also use timecode for audio synchronization, but the timecode is not recorded on the disk as such. Instead timecode forms part of the access mechanism so that samples are retrieved by specifying the required timecode which the disk subsystem converts into a physical block address. This mechanism was detailed in Chapter 10.

A further problem with the use of video-based timecode is that the accuracy to which the edit must be made in audio is much greater than the frame boundary accuracy needed in video. A video frame lasts 33 or 40 ms and a DAT frame lasts 30 ms, whereas audio needs to be edited to an accuracy of a few samples. When the exact edit point is chosen in an audio editor, it will be described to great accuracy and is stored as hours, minutes, seconds, frames and the number of the sample within the frame.

11.6   Locating the edit point

Digital audio editors must simulate the traditional 'rock and roll' process of edit-point location in analog tape recorders where the tape reels were

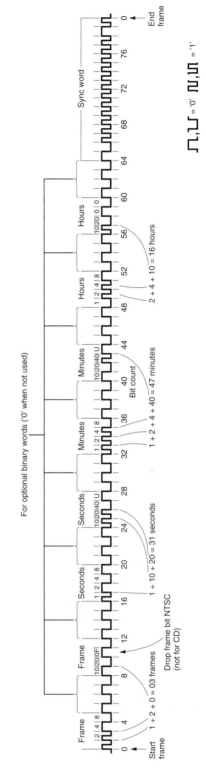

Figure 11.4 In SMPTE standard timecode, the frame number and time are stored as eight BCD symbols. There is also space for 32 user-defined bits. The code repeats every frame. Note the asymmetrical sync word which allows the direction of tape movement to be determined.

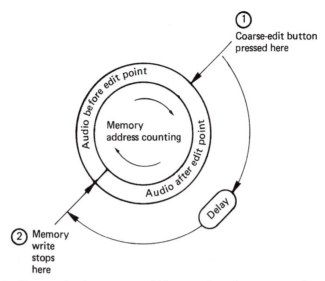

Figure 11.5 The use of a ring memory which overwrites allows storage of samples before and after the coarse edit point.

moved to and fro by hand. Digital media cannot do this directly as they can only play at one bit rate in one direction. As with editing, the solution is to transfer the recording in the area of the edit point to RAM in the editor. Samples can be read from RAM at any speed in either direction and the precise edit point can then be conveniently found by monitoring audio from the RAM.

Figure 11.5 shows how the area of the edit point is transferred to the memory. The source device is commanded to play, and the operator listens to replay samples via a DAC in the monitoring system. The same samples are continuously written into a memory within the editor. This memory is addressed by a counter which repeatedly overflows to give the memory a ring-like structure rather like that of a timebase corrector, but somewhat larger. When the operator hears the rough area in which the edit is required, he will press a button. This action stops the memory writing, not immediately, but one half of the memory contents later. The effect of this deliberate overrun is that the memory contains an equal number of samples before and after the rough edit point.

Typically an operator needs to be able to hear about 30 seconds of audio to be able mentally to synchronize to the rhythm and anticipate the edit point. This requires, in a stereo PCM system, a storage requirement of around five megabytes. In early digital audio editors this represented a significant cost, and to reduce the size of memory needed, many early editors used some form of compression, sampling rate reduction or mixing stereo material down to mono. With today's RAM prices this is no longer an issue. Samples which will be used to make the master recording need never pass through these processes; they are solely to assist in the location of the edit points. The sound quality in edit-point location mode can be impaired, but this does not affect the finished work.

Once the recording is in the memory, it can be accessed at leisure, and the constraints of the source device play no further part in the edit-point location. There are a number of ways in which the the memory can be read. If the memory address is supplied by a counter which is clocked at the appropriate rate, the edit area can be replayed at normal speed, or at some fraction of normal speed repeatedly. In order to simulate the analog method of finding an edit point, the operator is usually provided with a *scrub wheel* or rotor, and the memory address will change at a rate proportional to the speed with which the rotor is turned, and in the same direction. Thus the sound can be heard forward or backward at any speed, and the effect is exactly that of manually rocking an analog tape past the heads of an ATR.

The operation of a scrub wheel encoder was shown in section 3.13. Although a simple device, there are some difficulties to overcome. There are not enough pulses per revolution to create a clock directly and the human hand cannot turn the rotor smoothly enough to address the memory directly without flutter. A phase-locked loop is generally employed to damp fluctuations in rotor speed and multiply the frequency. A standard sampling rate must be recreated to feed the monitor DAC and a rate convertor, or interpolator, is necessary to restore the sampling rate to normal. These items can be seen in Figure 11.6. In low-cost editors the function of the scrub wheel will be performed by the mouse or a trackball.

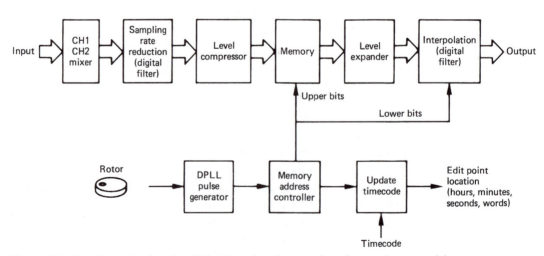

Figure 11.6 In order to simulate the edit location of analog recorders, the samples are read from memory under the control of a hand-operated rotor.

The act of pressing the coarse edit-point button stores the timecode of the source at that point, which is frame-accurate. As the rotor is turned, the memory address is monitored, and used to update the timecode to sample accuracy within the frame.

Before assembly can be performed, two edit points must be determined, the out-point at the end of the previously recorded signal and the in-point at the beginning of the new signal. The editor's microprocessor stores these in an edit decision list (EDL) in order to control the automatic assemble process.

Edit-point location can also be done on the fly by reading the musical score as the recording plays, and pressing the edit button at the right instant.

However the edit point is established, the subjective effect can be assessed in a preview process and if the outcome is unsatisfactory the in- or out-point can be trimmed any number of times until the desired result is obtained.

11.7   Editing with disk drives

Using one or other of the above methods, an edit list can be made which contains an in-point, an out-point and an audio filename for each of the segments of audio which need to be assembled to make the final work, along with a crossfade period and a gain parameter. This edit list will also be stored on disk. When a preview of the edited work is performed, the edit list is used to determine what files will be necessary and when, and this information drives the disk controller.

Figure 11.7 shows the events during an edit between two files. The edit list causes the relevant audio blocks from the first file to be transferred from disk to memory, and these will be read by the signal processor to produce the preview output. As the edit point approaches, the disk controller will also place blocks from the incoming file into the memory. It can do this because the rapid data-transfer rate of the drive allows blocks to be transferred to memory much faster than real time, leaving time for the positioner to seek from one file to another. In different areas of the memory there will be simultaneously the end of the outgoing recording and the beginning of the incoming recording.

Using timecode alone, the editor can only change the relative timing of the two recordings in frame increments. However, timing to sample accuracy can be obtained by using an area of RAM as a variable delay. The signal processor will use the fine edit-point parameters to work out the relationship between the actual edit points and the cluster or block boundaries. The relationship between the cluster on disk and the RAM address to which it was transferred is known, and this allows the memory

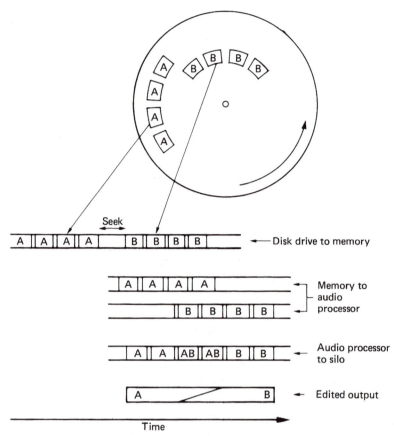

Seek

A A A A   B B B B   ← Disk drive to memory

A A A A   ← Memory to
             audio
B B B B   ← processor

A A AB AB B B   ← Audio processor
                    to silo

A   B   ← Edited output

Time

Figure 11.7 In order to edit together two audio files, they are brought to memory sequentially. The audio processor accesses file pages from both together, and performs a crossfade between them. The silo produces the final output at constant steady-sampling rate.

read addresses to be computed in order to obtain samples with the correct timing.

Prior to the edit point, only samples from the outgoing recording are accessed, but as the crossfade begins, samples from the incoming recording are also accessed, multiplied by the gain parameter and then mixed with samples from the outgoing recording according to the crossfade period required. The output of the signal processor becomes the edited preview material, which can be checked for the required subjective effect. If necessary the in- or out-points can be trimmed, or the crossfade period changed, simply by modifying the edit-list file. The preview can be repeated as often as needed, until the desired effect is obtained. At this stage the edited work does not exist as a file, but is re-created each time by a further execution of the EDL. Thus a lengthy editing session need not fill up the disk.

It is important to realize that at no time during the edit process were the original audio files modified in any way. The editing was done solely by reading the audio files. The power of this approach is that if an edit list is created wrongly, the original recording is not damaged, and the problem can be put right simply by correcting the edit list. The advantage of a disk-based system for such work is that location of edit points, previews and reviews are all performed almost instantaneously, because of the random access of the disk. This can reduce the time taken to edit a program to a quarter of that needed with a linear tape machine.[1]

During an edit, the disk drive has to provide audio files from two different places on the disk simultaneously, and so it has to work much harder than for a simple playback. If there are many close-spaced edits, the drive may be hard-pressed to keep ahead of real time, especially if there are long crossfades, because during a crossfade the source data rate is twice as great as during replay. A large buffer memory helps this situation because the drive can fill the memory with files before the edit actually begins, and thus the instantaneous sample rate can be met by the memory's emptying during disk-intensive periods. In practice crossfades measured in seconds can be achieved easily.

Disk formats which handle defects dynamically, using techniques such as defect skipping, will also be superior to those using bad-block files when throughput is important. Some drives rotate the sector addressing from one cylinder to the next so that the drive does not lose a revolution when it moves to the next cylinder. Disk-editor perform-ance is usually specified in terms of peak editing activity which can be achieved, but with a recovery period between edits. If an unusually severe editing task is necessary where the drive just cannot access files fast enough, it will be necessary to rearrange the files on the disk surface so that files which will be needed at the same time are on nearby cylinders.[2] An alternative is to spread the material between two or more drives so that overlapped seeks are possible.

Once the editing is finished, it will generally be necessary to transfer the edited material to form a contiguous recording so that the source files can make way for new work. If the source files already exist on tape the disk files can simply be erased. If the disks hold original recordings they will need to be backed up to tape if they will be required again. If editing is not complete, and the editor is required for another purpose, the disk files can be transferred in their entirety to tape so that the disk data can be exactly restored at a later time. In large broadcast systems, the edited work can be broadcast directly from the disk file. In smaller systems it will be necessary to output to some removable medium, since the Winchester drives in the editor have fixed media. It is only necessary to connect the AES/EBU output of the signal processor to any type of digital recorder, and then the edit list is executed once more. The edit sequence will be performed again, exactly as it was during the last preview, and the results will be recorded on the external device.

11.8   CD mastering

At the time of the introduction of the Compact Disc, mastering was carried out using a PCM adaptor and U-matic rotary-head VCRs. Each frame on tape was treated as a data block addressed by timecode, and Figure 11.8 shows that editing to sample accuracy was achieved by setting the VCR into record at a frame boundary, and rerecording what was already on the tape up to the edit point, where the new recording will appear to commence. A crossfade of appropriate length is carried out in the digital domain. The basic operation of the editor was exactly as described above for disk-based editing. As U-matic VCRs had a habit of locking up one frame early or late, a page of RAM was used to resynchronize the data by inserting or removing frame delays so that the edit would not have to be aborted.

The first CD cutters could only operate at normal playing speed and a U-matic master tape which could only play at normal speed was acceptable. However, CD cutting is only a data transfer process, and the most economic data transfer rate is as fast as possible so that one CD cutter can do more work. Today, CD cutters can operate many times faster than real time. Master recordings can be supplied on any economic medium, typically a computer-type data tape. The tape is read into a disk store and checked for data integrity and the disk store then delivers data to the cutter using a RAM TBC to deliver an exactly constant bit rate. Instead of delivering a physical medium, the master recording can also be transferred to the cutter as a data file over a communications network.

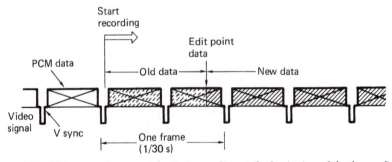

Figure 11.8 Video recorders can only start recording at the beginning of the frame; fine position of the edit point is determined by rerecording the old data up to the edit point.

11.9   Editing in DAT

In order to edit a DAT tape, many of the constraints of pseudo-video editing apply. Editing can only take place at the the beginning of an interleave block, known as a frame, which is contained in two diagonal tracks. The transport would need to perform a preroll, starting before the edit point, so that the drum and capstan servos would be synchronized to the tape tracks before the edit was reached. Fortunately, the very small drum means that mechanical inertia is minute by the standards of video recorders, and lock-up can be very rapid. One way in which a read–modify–write edit could be performed would be to use an editor of the type designed for PCM adaptors. This would permit editing on a DAT machine which could only record or play.

A better solution used in professional machines is to fit two sets of heads in the drum. The standard permits the drum size to be increased and the wrap angle to be reduced provided that the tape tracks are recorded to the same dimensions. In normal recording, the first heads to reach the tape tracks would make the recording, and the second set of heads would be able to replay the recording immediately afterwards for confidence monitoring. For editing, the situation would be reversed. The first heads to meet a given tape track would play back the existing recording, and this would be de-interleaved and corrected, and presented as a sample stream to the record circuitry. The record circuitry would then interleave the samples ready for recording. If the heads are mounted a suitable distance apart in the scanner along the axis of rotation, the time taken for tape to travel from the first set of heads to the second will be equal to the decode/encode delay. If this process goes on for a few blocks, the signal going to the record head will be exactly the same as the pattern already on the tape, so the record head can be switched on at the beginning of an interleave block. Once this has been done, new material can be crossfaded into the sample stream from the advanced replay head, and an edit will be performed.

If insert editing is contemplated, following the above process, it will be necessary to crossfade back to the advanced replay samples before ceasing rerecording at an interleave block boundary. The use of overwrite to produce narrow tracks causes a problem at the end of such an insert. Figure 11.9 shows that this produces a track which is half the width it should be. Normally the error-correction system would take care of the consequences, but if a series of inserts were made at the same point in an attempt to make fine changes to an edit, the result could be an extremely weak signal for one track duration. One solution is to incorporate an algorithm into the editor so that the points at which the tape begins and ends recording change on every attempt. This does not affect the audible result as this is governed by the times at which the crossfader operates.

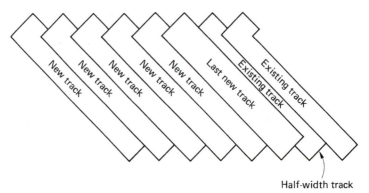

Half-width track

Figure 11.9 When editing a small track-pitch recording, the last track written will be 1.5 times the normal track width, since that is the width of the head. This erases half of the next track of the existing recording.

11.10   Editing in open-reel digital recorders

On many occasions in studio multitrack recording it is necessary to replace a short section of a long recording, because a wrong note was played or something fell over and made a noise. The tape is played back to the musicians before the bad section, and they play along with it. At a musically acceptable point prior to the error, the tape machine passes into record, a process known as punch-in, and the offending section is rerecorded. At another suitable time, the machine ceases recording at the punch-out point, and the musicians can subsequently stop playing.

The problem with an interleaved recording is that it is not possible to point to a specific place on the tape and say that it represents the recording at a particular instant. It is not possible just to begin recording at some arbitrary place, as the interleave structure would be destroyed. Once more, a read–modify–write approach is necessary, using a record head positioned *after* the replay head. The mechanism necessary is shown in Figure 11.10. Prior to the punch-in point, the replay-head signal is de-interleaved, and this signal is fed to the record channel. The record channel re-interleaves the samples, and after some time will produce a signal which is identical to what is already on the tape. At a block boundary the record current can be turned on, when the existing recording will be rerecorded. At the punch-in point, the samples fed to the record encoder will be crossfaded to samples from the ADC. The crossfade takes place in the non-interleaved domain. The new recording is made to replace the unsatisfactory section, and at the end, punch-out is commenced by returning the crossfader to the samples from the replay head. After some time, the record head will once more be rerecording what is already on the tape, and at a block boundary the record current can be switched off. The crossfade

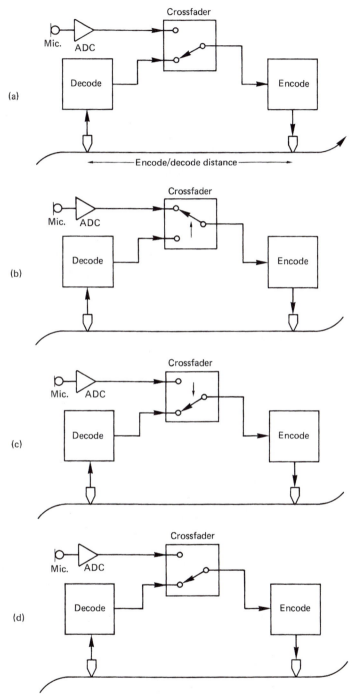

Figure 11.10 The four stages of an insert (punch-in/out) with interleaving: (a) rerecord existing samples for at least one constraint length; (b) crossfade to incoming samples (punch-in point); (c) crossfade to existing replay samples (punch-out point); (d) rerecord existing samples for at least one constraint length. An assemble edit consists of steps (a) and (b) only.

duration can be chosen according to the nature of the recorded material. If a genuine silence appears between notes played in a dead acoustic, a rapid crossfade may be optimum. With a large chorus in reverberant surroundings, a long crossfade might go unnoticed. It is possible to rehearse the punch-in process and monitor what it would sound like by feeding headphones from the crossfader, and doing everything described except that the record head is disabled. The punch-in and punch-out points can then be moved to give the best subjective result. The machine can learn the sector addresses at which the punches take place, so the final punch is fully automatic.

Assemble editing, where parts of one or more source tapes are dubbed from one machine to another to produce a continuous recording, is performed in the same way as a punch-in, except that the punch-out never comes. After the new recording from the source machine is faded in, the two machines continue to dub until one of them is stopped. This will be done some time after the next assembly point is reached.

11.11   Jump editing

Conventional splice handling in stationary-head recorders was detailed in Chapter 9. An extension to the principle has been suggested by Lagadec[3] in which the samples from the area of the splice are not heard. Instead an electronic edit is made between the samples before the splice and those after.

In this system, a tape splice is made physically with excess tape adjacent to the intended edit points. The timebase corrector has two read-address generators which can access the memory independently. It will be seen in Figure 11.11 that when the machine plays the tape, the capstan is phase-advanced so that the timebase corrector is causing a long delay to compensate. As the splice is detected, the corruption due to the splice enters the TBC memory and travels towards the output. As the splice nears the end of the memory, the machine output crossfades to a signal from the second TBC output which has been delayed much less. The data in the area of the tape splice are thus omitted. The capstan will now effectively be lagging because the delay has been shortened, and it will speed up slightly for a short period until the lead condition is re-established. This can be done without ill effect since the sample rate from the memory remains constant throughout. Although the splice is an irrevocable mechanical act, the precise edit timing can be changed at will by controlling the sector address at which the TBC jumps, which determines the out-point, and the address difference, which determines the length of tape omitted,

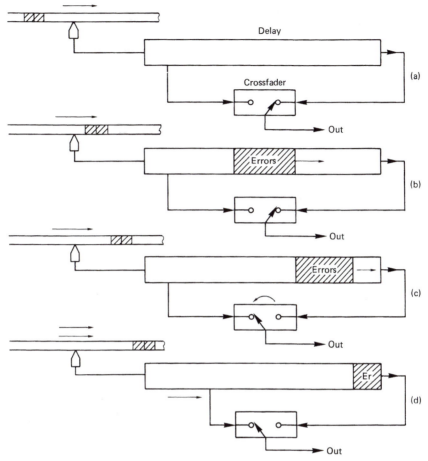

Figure 11.11 Jump editing. (a) Splice approaches, capstan is advanced, and audio is delayed. (b) Splice passes head, and error burst travels down delay. (c) Crossfader fades to signal after splice. (d) Capstan accelerates, and delay increases. When the delay tap reaches the end, the crossfader can switch back ready for the next splice.

and thus controls the in-point. The size of the jump is limited by the available memory.

If only a short section of audio is to be removed, no splice is necessary at all as a memory jump can be used to omit a short length of the recording. Such a system would be excellent for news broadcasts where it is often necessary to remove many short sections of tape to eliminate hesitations and unwanted pauses from interviews. Control of the jumping could be by programming a CPU to recognize timecode or sector addresses and insert the commands, or, as suggested by Lagadec, inserting the jump distance in the reference track prior to the splice. In either case machines not equipped to jump would handle any splices with mechanically determined timing.

Jump editing can also be used in rotary-head recorders such as DAT and the Nagra-D. Rotary-head machines have a low linear tape speed and so can accelerate the tape to omit quite long sections whilst replay continues from memory.

References

1. Todoroki, S., *et al.*, New PCM editing system and configuration of total professional digital audio system in near future. Presented at the 80th Audio Engineering Society Convention (Montreux, 1986), Preprint 2319(A8)
2. McNally, G.W., Gaskell, P.S. and Stirling, A.J. Digital Audio Editing. *BBC Research Dept. Report*, RD 1985/10
3. Lagadec, R., Current status in digital audio. Presented at the IERE Video and Data Recording Conference (Southampton, 1984)

12

Digital audio in optical disks

Optical disks are particularly important to digital audio, not least because of the success of the Compact Disc and subsequent devlopments such as MiniDisc, magneto-optical production recorders and DVD. CD, DVD and MiniDisc are worthy of detailed consideration, as they are simultaneously consumer products available in large numbers at low cost, and yet are technically advanced devices. Optical disks result from the marriage of many disciplines, including laser optics, servomechanisms, error correction and both analog and digital circuitry in VLSI form.

12.1   Types of optical disk

There are numerous types of optical disk, which have different characteristics.[1] There are, however, three broad groups, shown in Figure 12.1, which can be usefully compared:

1 The Compact Disc, the Digital Video Disc and the prerecorded MiniDisc are read-only laser disks, which are designed for mass duplication by stamping. They cannot be recorded.
2 Some laser disks can be recorded, but once a recording has been made, it cannot be changed or erased. These are usually referred to as write-once-read-mostly (WORM) disks. Recordable CDs and DVDs work on this principle.
3 Erasable optical disks have essentially the same characteristic as magnetic disks, in that new and different recordings can be made in the same track indefinitely. Recordable MiniDisc is in this category. Sometimes a separate erase process is necessary before rewriting.

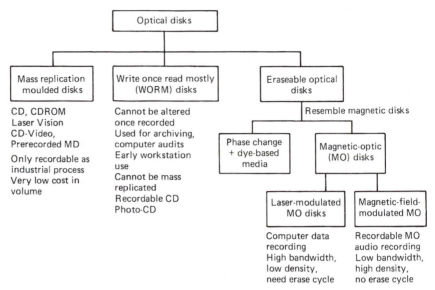

Figure 12.1 The various types of optical disk. See text for details.

The Compact Disc, generally abbreviated to CD, is a consumer digital audio recording which is intended for mass replication. When optical recording was in its infancy, many companies were experimenting with a variety of optical media. In most cases the goal was to make an optical recorder where the same piece of apparatus could record and immediately reproduce information. This would be essential for most computer applications and for use in audio or video production. This was not, however, the case in the consumer music industry, where the majority of listening was, and still is, to prerecorded music. The vinyl disk could not be recorded in the home, yet it sold by the million. Individual vinyl disks were not recorded as such by the manufacturer as they were replicated by pressing, or moulding, molten plastic between two surfaces known as stampers which were themselves made from a master disk produced on a cutting lathe. This master disk was the recording.

Philips' approach was to invent an optical medium which would have the same characteristics as the vinyl disk in that it could be mass replicated by moulding or stamping with no requirement for it to be recordable by the user. The information on it is carried in the shape of flat-topped physical deformities in a layer of plastic, and as a result the medium has no photographic, magnetic or electronic properties, but is simply a relief structure. Such relief structures lack contrast and are notoriously difficult to study with conventional optics, but in 1934 Zernike[2] described a Nobel Prize-winning technique called phase contrast microscopy which allowed an apparent contrast to be obtained

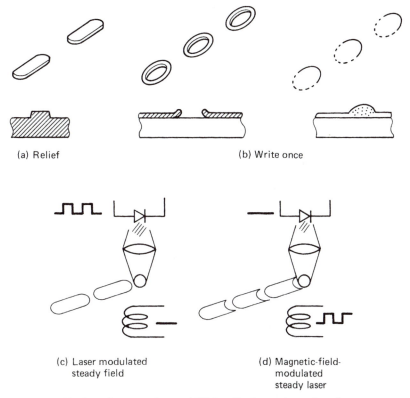

(a) Relief   (b) Write once

(c) Laser modulated
steady field

(d) Magnetic-field-
modulated
steady laser

Figure 12.2 (a) The information layer of CD is reflective and uses interference. (b) Write-once disks may burn holes or raise blisters in the information layer. (c) High data rate MO disks modulate the laser and use a constant magnetic field. (d) At low data rates the laser can run continuously and the magnetic field is modulated.

from such a structure using optical interference. This principle is used to read relief recordings.

Figure 12.2(a) shows that the information layer of CD and the prerecorded MiniDisc is an optically flat mirror upon which microscopic bumps are raised. A thin coating of aluminium renders the layer reflective. When a small spot of light is focused on the information layer, the presence of the bumps affects the way in which the light is reflected back, and variations in the reflected light are detected in order to read the disk. Figure 12.2 also illustrates the very small dimensions which are common to both disks. For comparison, some sixty CD/MD tracks can be accommodated in the groove pitch of a vinyl LP. These dimensions demand the utmost cleanliness in manufacture.

Figure 12.2(b) shows that there are several types of WORM disks. The disk may contain a thin layer of metal; on recording, a powerful laser melts spots on the layer. Surface tension causes a hole to form in the metal, with a thickened rim around the hole. Subsequently a low-power laser can read the disk because the metal reflects light, but the hole passes it through. Computer WORM disks work on this principle. As an alternative, the layer of metal may be extremely thin, and the heat from the laser heats the material below it to the point of decomposition. This causes gassing which raises a blister or bubble in the metal layer. Recordable CDs can use this principle as the relief structure can be read like a normal CD. It is also possible to impregnate a disk with a chemical dye which darkens when it is struck by a high level of radiation from a writing laser. Clearly once such a pattern of holes, blisters or dark areas has been made it is permanent.

Rerecordable or erasable optical disks rely on magneto-optics,[3] also known more fully as thermomagneto-optics. Writing in such a device makes use of a thermomagnetic property possessed by all magnetic materials, which is that above a certain temperature, known as the Curie temperature, their coercive force becomes zero. This means that they become magnetically very soft, and take on the flux direction of any externally applied field. On cooling, this field orientation will be frozen in the material, and the coercivity will oppose attempts to change it. Although many materials possess this property, there are relatively few which have a suitably low Curie temperature. Compounds of terbium and gadolinium have been used, and one of the major problems to be overcome is that almost all suitable materials from a magnetic viewpoint corrode very quickly in air.

There are two ways in which magneto-optic (MO) disks can be written. Figure 12.2(c) shows the first system, in which the intensity of laser is modulated with the waveform to be recorded. If the disk is considered to be initially magnetized along its axis of rotation with the north pole upwards, it is rotated in a field of the opposite sense, produced by a steady current flowing in a coil which is weaker than the room-temperature coercivity of the medium. The field will therefore have no effect. A laser beam is focused on the medium as it turns, and a pulse from the laser will momentarily heat a very small area of the medium past its Curie temperature, whereby it will take on a reversed flux due to the presence of the field coils. This reversed-flux direction will be retained indefinitely as the medium cools.

Alternatively the waveform to be recorded modulates the magnetic field from the coils as shown in Figure 12.2(d). In this approach, the laser is operating continuously in order to raise the track beneath the beam above the Curie temperature, but the magnetic field recorded is determined by the current in the coil at the instant the track cools. Magnetic field modulation is used in the recordable MiniDisc.

In both of these cases, the storage medium is clearly magnetic, but the writing mechanism is the heat produced by light from a laser; hence the term thermomagneto-optics. The advantage of this writing mechanism is that there is no physical contact between the writing head and the medium. The distance can be several millimetres, some of which is taken up with a protective layer to prevent corrosion. In prototypes, this layer is glass, but commercially available disks use plastics.

The laser beam will supply a relatively high power for writing, since it is supplying heat energy. For reading, the laser power is reduced, such that it cannot heat the medium past the Curie temperature, and it is left on continuously. Readout depends on the so-called Kerr effect, which describes a rotation of the plane of polarization of light due to a magnetic field. The magnetic areas written on the disk will rotate the plane of polarization of incident polarized light to two different planes, and it is possible to detect the change in rotation with a suitable pickup.

Erasable disks may also be made which rely on phase changes. Certain chemical compounds may exist in two states, crystalline or amorphous. One state appears shiny and the other dull to a readout beam. It is possible to switch from one state to another by using different power levels in the laser.

12.2   CD, DVD and MD contrasted

CD and MD have a great deal in common. Both use a laser of the same wavelength which creates a spot of the same size on the disk. The track pitch and speed are the same and both offer the same playing time. The channel code and error-correction strategy are the same. DVD uses the same principle as CD, but the readout spot is smaller so that the recording density can significantly be raised.

CD carries 44.1 kHz sixteen-bit PCM audio and is intended to be played in a continuous spiral like a vinyl disk. The CD process, from cutting, through pressing and reading, produces no musical degradation whatso-ever, since it simply conveys a series of numbers which are exactly those recorded on the master tape. The only part of a CD player which can cause subjective differences in sound quality in normal operation is the DAC, although in the presence of gross errors some players will correct and/or conceal better than others. Chapter 4 deals with the principles of conversion.

DVD records an MPEG program stream. This is a multiplex of compressed audio and video elementary streams which can be decoded to reproduce television programs or movies.

MD begins with the same PCM data as CD, but uses a form of compression known as ATRAC (see Chapter 5) having a compression factor of 0.2 which introduces a slight but audible loss of quality. After the addition of subcode and housekeeping data MD has an average data rate which is 0.225 that of CD. However, MD has the same recording density and track speed as CD, so the data rate from the disk is greatly in excess of that needed by the audio decoders. The difference is absorbed in RAM exactly as shown in Figure 10.34.

The RAM in a typical MD player is capable of buffering about 3 seconds of audio. When the RAM is full, the disk drive stops transferring data but keeps turning. As the RAM empties into the decoders, the disk drive will top it up in bursts. As the drive need not transfer data for over three quarters of the time, it can reposition between transfers and so it is capable of editing in the same way as a magnetic hard disk. A further advantage of a large RAM buffer is that if the pickup of a CD or MD player is knocked off-track by an external shock the RAM continues to provide data to the audio decoders and provided the pickup can get back to the correct track before the RAM is exhausted there will be no audible effect.

When recording an MO disk, the MiniDisc drive also uses the RAM buffer to allow repositioning so that a continuous recording can be made on a disk which has become chequerboarded through selective erasing. The full total playing time is then always available irrespective of how the disk is divided into different recordings.

CD and MD have a fixed bit rate, whereas DVD does not. It is a characteristic of video material that the degree of compression possible varies with program material. Consequently the greatest efficiency is obtained if the bit rate can increase to handle complex pictures and slow down again to handle simple images. In a multiplex a variable video bit rate can exist alongside a fixed audio bit rate.

12.3   CD and MD – disk construction

Figure 12.3 shows the mechanical specification of CD. Within an overall diameter of 120 mm the program area occupies a 33 mm-wide band between the diameters of 50 and 116 mm. Lead-in and lead-out areas increase the width of this band to 35.5 mm. As the track pitch is a constant 1.6 μm, there will be

$$\frac{35.6 \times 1000}{1.6} = 22\,188$$

tracks crossing a radius of the disk. As the track is a continuous spiral, the track length will be given by the above figure multiplied by the average circumference:

$$\text{Length} = 2 \times \pi \times \frac{58.5 + 23}{2} \times 22\,188 = 5.7\,\text{km}$$

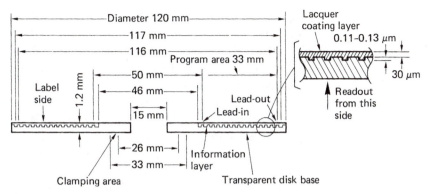

Figure 12.3 Mechanical specification of CD. Between diameters of 46 and 117 mm is a spiral track 5.7 km long.

These figures give a good impression of the precision involved in CD manufacture. The CD case is for protection in storage and the CD has to be taken out of its case and placed in the player. The disk has a plain centre hole and most players clamp the disk onto the spindle from both sides. There are some CD players designed for broadcasters which require the standard CD to be placed into a special cassette. The CD can then be played inside the cassette.

Figure 12.4 shows the mechanical specification of prerecorded Mini-Disc. Within an overall diameter of 64 mm the lead-in area begins at a diameter of 29 mm and the program area begins at 32 mm. The track pitch is exactly the same as in CD, but the MiniDisc can be smaller than CD without any sacrifice of playing time because of the use of compression. For ease of handling, MiniDisc is permanently enclosed in a shuttered plastic cartridge which is 72 × 68 × 5 mm. The cartridge resembles a smaller version of a $3\frac{1}{2}$-inch floppy disk, but unlike a floppy, it is slotted into the drive with the shutter at the side. An arrow is moulded into the cartridge body to indicate this.

In the prerecorded MiniDisc, it was a requirement that the whole of one side of the cartridge should be available for graphics. Thus the disk is

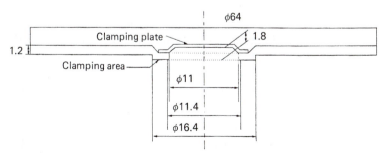

Figure 12.4 The mechanical dimensions of MiniDisc.

designed to be secured to the spindle from one side only. The centre of the disk is fitted with a ferrous clamping plate and the spindle is magnetic. When the disk is lowered into the drive it simply sticks to the spindle. The ferrous disk is only there to provide the clamping force. The disk is still located by the moulded hole in the plastic component. In this way the ferrous component needs no special alignment accuracy when it is fitted in manufacture. The back of the cartridge has a centre opening for the hub and a sliding shutter to allow access by the optical pickup.

The recordable MiniDisc and cartridge has the same dimensions as the prerecorded MiniDisc, but access to both sides of the disk is needed for recording. Thus the recordable MiniDisc has a a shutter which opens on both sides of the cartridge, rather like a double-sided floppy disk. The opening on the front allows access by the magnetic head needed for MO recording, leaving a smaller label area.

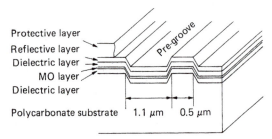

Figure 12.5 The construction of the MO recordable MiniDisc.

Figure 12.5 shows the construction of the MO MiniDisc. The 1.1 μm wide tracks are separated by grooves which can be optically tracked. Once again the track pitch is the same as in CD. The MO layer is sandwiched between protective layers.

12.4   Rejecting surface contamination

A fundamental goal of consumer optical disks is that no special working environment or handling skill is required. The bandwidth needed by PCM audio is such that high-density recording is mandatory if reasonable playing time is to be obtained in CD. Although MiniDisc uses compression, it does so in order to make the disk smaller and the recording density is actually the same as for CD.

High-density recording implies short wavelengths. Using a laser focused on the disk from a distance allows short-wavelength recordings to be played back without physical contact, whereas conventional magnetic recording requires intimate contact and implies a wear

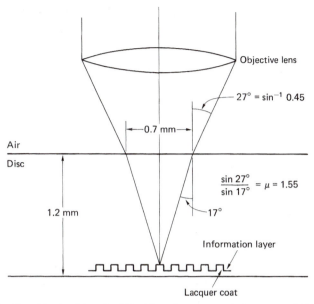

Figure 12.6 The objective lens of a CD pickup has a numerical aperture (NA) of 0.45; thus the outermost rays will be inclined at approximately 27° to the normal. Refraction at the air/disk interface changes this to approximately 17° within the disk. Thus light focused to a spot on the information layer has entered the disk through a 0.7 mm diameter circle, giving good resistance to surface contamination.

mechanism, the need for periodic cleaning, and susceptibility to contamination.

The information layer of CD and MD is read through the thickness of the disk. Figure 12.6 shows that this approach causes the readout beam to enter and leave the disk surface through the largest possible area. The actual dimensions involved are shown in the figure. Despite the minute spot size of about 1.2 μm diameter, light enters and leaves through a 0.7 mm-diameter circle. As a result, surface debris has to be three orders of magnitude larger than the readout spot before the beam is obscured. This approach has the further advantage in MO drives that the magnetic head, on the opposite side to the laser pickup, is then closer to the magnetic layer in the disk.

The bending of light at the disk surface is due to refraction of the wavefronts arriving from the objective lens. Wave theory of light suggests that a wavefront advances because an infinite number of point sources can be considered to emit spherical waves which will only add when they are all in the same phase. This can only occur in the plane of the wavefront. Figure 12.7 shows that at all other angles, interference between spherical waves is destructive.

When such a wavefront arrives at an interface with a denser medium, such as the surface of an optical disk, the velocity of propagation is

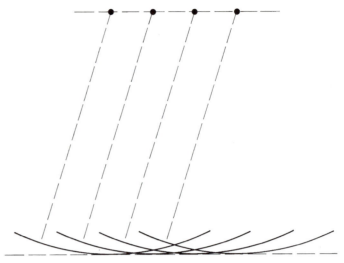

Figure 12.7 Plane-wave propagation considered as infinite numbers of spherical waves.

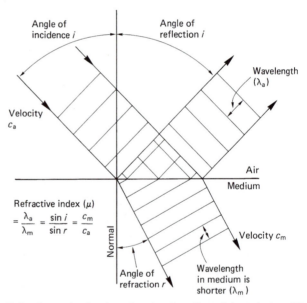

Figure 12.8 Reflection and refraction, showing the effect of the velocity of light in a medium.

reduced; therefore the wavelength in the medium becomes shorter, causing the wavefront to leave the interface at a different angle (Figure 12.8). This is known as refraction. The ratio of velocity *in vacuo* to velocity in the medium is known as the refractive index of that medium; it determines the relationship between the angles of the incident and refracted wavefronts. Reflected light, however, leaves at the same angle to the normal as the incident light. If the speed of light in the medium varies with wavelength, dispersion takes place, where incident white light will be split into a rainbow-like spectrum leaving the interface at different angles. Glass used for chandeliers and cut glass is chosen to be highly dispersive, whereas glass for optical instruments will be chosen to have a refractive index which is as constant as possible with changing wavelength. The use of monochromatic light in optical disks allows low-cost optics to be used as they only need to be corrected for a single wavelength.

The size of the entry circle in Figure 12.6 is a function of the refractive index of the disk material, the numerical aperture of the objective lens and the thickness of the disk. MiniDiscs are permanently enclosed in a cartridge, and scratching is unlikely. This is not so for CD, but fortunately the method of readout through the disk thickness tolerates surface scratches very well. In extreme cases of damage, a scratch can often successfully be removed with metal polish. By way of contrast, the label side is actually more vulnerable than the readout side, since the lacquer coating is only $30\,\mu m$ thick. For this reason, writing on the label side of CD is not recommended. Pressure from a ballpoint pen could distort the information layer, and solvents from marker pens have been known to penetrate the lacquer and cause corruption. The common party-piece of writing on the readout surface of CD with a felt pen to show off the error-correction system is quite harmless, since the disk base material is impervious to most solvents.

The base material is in fact a polycarbonate plastic produced by (among others) Bayer under the trade name of Makrolon. It has excellent mechanical and optical stability over a wide temperature range, and lends itself to precision moulding and metallization. It is often used for automotive indicator clusters for the same reasons. An alternative material is polymethyl methacrylate (PMMA), one of the first optical plastics, known by such trade names as Perspex and Plexiglas, and widely used for illuminated signs and aircraft canopies. Polycarbonate is preferred by some manufacturers since it is less hygroscopic than PMMA. The differential change in dimensions of the lacquer coat and the base material can cause warping in a hygroscopic material. Audio disks are too small for this to be a problem, but the larger analog video disks are actually two disks glued together back-to-back to prevent this warpage.

12.5   Playing optical disks

A typical laser disk drive resembles a magnetic drive in that it has a spindle drive mechanism to revolve the disk, and a positioner to give radial access across the disk surface. The positioner has to carry a collection of lasers, lenses, prisms, gratings and so on, and cannot be accelerated as fast as a magnetic-drive positioner. A penalty of the very small track pitch possible in laser disks, which gives the enormous storage capacity, is that very accurate track following is needed, and it takes some time to lock onto a track. For this reason tracks on laser disks are usually made as a continuous spiral, rather than the concentric rings of magnetic disks. In this way, a continuous data transfer involves no more than track following once the beginning of the file is located.

In order to record MO disks or replay any optical disk, a source of monochromatic light is required. The light source must have low noise otherwise the variations in intensity due to the noise of the source will mask the variations due to reading the disk. The requirement for a low-noise monochromatic light source is economically met using a semiconductor laser.

The semiconductor laser is a relative of the light-emitting diode (LED). Both operate by raising the energy of electrons to move them from one valence band to another conduction band. Electrons which fall back to the valence band emit a quantum of energy as a photon whose frequency is proportional to the energy difference between the bands. The process is described by Planck's Law:

Energy difference $E = H \times f$

where H = Planck's Constant

$$= 6.6262 \times 10^{-34} \text{ joules/Hertz}$$

For gallium arsenide, the energy difference is about 1.6 eV, where 1 eV is 1.6×10^{-19} joules.

Using Planck's Law, the frequency of emission will be:

$$f = \frac{1.6 \times 1.6 \times 10^{-19}}{6.6262 \times 10^{-34}} \text{ Hz}$$

The wavelength will be c/f where

c = the velocity of light = $3 \times 10^8 \text{m/s}$

$$\text{Wavelength} = \frac{3 \times 10^8 \times 6.6262 \times 10^{-34}}{2.56 \times 10^{-19}} \text{m}$$

$$= 780 \text{ nanometres}$$

In the LED, electrons fall back to the valence band randomly, and the light produced is incoherent. In the laser, the ends of the semiconductor are optically flat mirrors, which produce an optically resonant cavity. One photon can bounce to and fro, exciting others in synchronism, to produce coherent light. This is known as light amplification by stimulated emission of radiation, mercifully abbreviated to LASER, and can result in a runaway condition, where all available energy is used up in one flash. In injection lasers, an equilibrium is reached between energy input and light output, allowing continuous operation. The equilibrium is delicate, and such devices are usually fed from a current source. To avoid runaway when temperature change disturbs the equilibrium, a photosensor is often fed back to the current source. Such lasers have a finite life, and become steadily less efficient. The feedback will maintain output, and it is possible to anticipate the failure of the laser by monitoring the drive voltage needed to give the correct output.

Some of the light reflected back from the disk re-enters the aperture of objective lens. The pickup must be capable of separating the reflected light from the incident light. Figure 12.9 shows two systems. In (a) an intensity beamsplitter consisting of a semisilvered mirror is inserted in the optical path and reflects some of the returning light into the photosensor. This is not very efficient, as half of the replay signal is lost by transmission straight on. In the example at (b) separation is by polarization.

In natural light, the electric-field component will be in many planes. Light is said to be polarized when the electric field direction is constrained. The wave can be considered as made up from two orthogonal components. When these are in phase, the polarization is said to be linear. When there is a phase shift between the components, the polarization is said to be elliptical, with a special case at 90° called circular polarization. These types of polarization are contrasted in Figure 12.10.

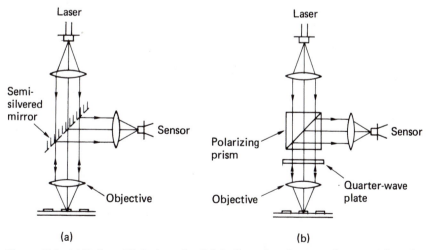

Figure 12.9 (a) Reflected light from the disk is directed to the sensor by a semisilvered mirror. (b) A combination of polarizing prism and quarter-wave plate separates incident and reflected light.

Polarization

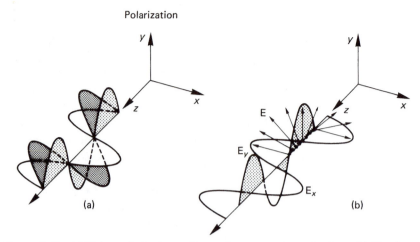

Figure 12.10 (a) Linear polarization: orthogonal components are in phase. (b) Circular polarization: orthogonal components are in phase quadrature.

In order to create polarized light, anisotropic materials are necessary. Polaroid material, invented by Edwin Land, is vinyl which is made anisotropic by stretching it while hot. This causes the long polymer molecules to line up along the axis of stretching. If the material is soaked in iodine, the molecules are rendered conductive, and short out any electric-field component along themselves. Electric fields at right angles are unaffected; thus the transmission plane is at right angles to the stretching axis.

Stretching plastics can also result in anisotropy of refractive index; this effect is known as birefringence. If a linearly polarized wavefront enters such a medium, the two orthogonal components propagate at different velocities, causing a relative phase difference proportional to the distance travelled. The plane of polarization of the light is rotated. Where the thickness of the material is such that a 90° phase change is caused, the device is known as a quarter-wave plate. The action of such a device is shown in Figure 12.11. If the plane of polarization of the incident light is at 45° to the planes of greatest and least refractive index, the two orthogonal components of the light will be of equal magnitude, and this results in circular polarization. Similarly, circular-polarized light can be returned to the linear-polarized state by a further quarter-wave plate. Rotation of the plane of polarization is a useful method of separating incident and reflected light in a laser pickup. Using a quarter-wave plate, the plane of polarization of light leaving the pickup will have been turned 45°, and on return it will be rotated a further 45°, so that it is now at right angles to the plane of polarization of light from the source. The two can easily be separated by a polarizing prism, which acts as a transparent block to light in one

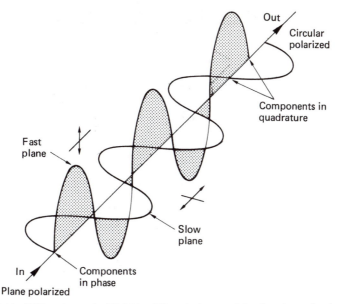

Figure 12.11 Different speed of light in different planes rotates the plane of polarization in a quarter-wave plate to give a circular-polarized output.

plane, but as a prism to light in the other plane, such that reflected light is directed towards the sensor.

In a CD player, the sensor is concerned only with the intensity of the light falling on it. When playing MO disks, the intensity does not change, but the magnetic recording on the disk rotates the plane of polarization one way or the other depending on the direction of the vertical magnetization. MO disks cannot be read with circular-polarized light. Light incident on the medium must be plane polarized and so the quarter-wave plate of the CD pickup cannot be used. Figure 12.12(a) shows that a polarizing prism is still required to linearly polarize the light from the laser on its way to the disk. Light returning from the disk has had its plane of polarization rotated by approximately ± 1°. This is an extremely small rotation. Figure 12.12(b) shows that the returning rotated light can be considered to be composed of two orthogonal components. R_x is the component which is in the same plane as the illumination and is called the *ordinary* component and R_y is the component due to the Kerr effect rotation and is known as the *magneto-optic* component. A polarizing beam splitter mounted squarely would reflect the magneto-optic component R_y very well because it is at right angles to the transmission plane of the prism, but the ordinary component would pass straight on in the direction of the laser. By rotating the prism slightly a small amount of the ordinary component is also reflected. Figure 12(c) shows that when combined with the magneto-optic component, the angle of rotation has

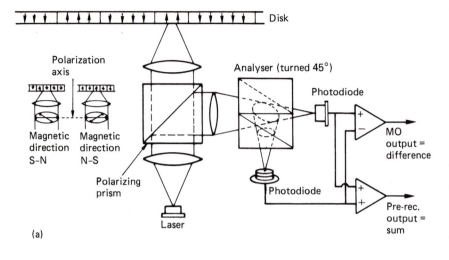

(a)

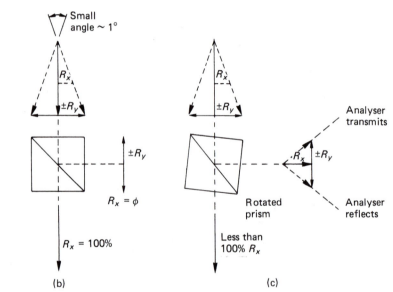

(b)   (c)

Figure 12.12 A pickup suitable for the replay of magneto-optic disks must respond to very small rotations of the plane of polarization.

increased.[4] Detecting this rotation requires a further polarizing prism or analyser as shown in Figure 12.12. The prism is twisted such that the transmission plane is at 45° to the planes of R_x and R_y. Thus with an unmagnetized disk, half of the light is transmitted by the prism and half is reflected. If the magnetic field of the disk turns the plane of polarization towards the transmission plane of the prism, more light is transmitted and less is reflected. Conversely if the plane of polarization is rotated away from the transmission plane, less light is transmitted and more is reflected. If two sensors are used, one for transmitted light and one for reflected light, the difference between the two sensor outputs will be a waveform representing the angle of polarization and thus the recording on the disk. This differential analyser eliminates common-mode noise in the reflected beam.[5] As Figure 12.12 shows, the output of the two sensors is summed as well as subtracted in a MiniDisc player. When playing MO disks, the difference signal is used. When playing prerecorded disks, the sum signal is used and the effect of the second polarizing prism is disabled.

Since the residual stresses set up in moulding plastic tend to align the long-chain polymer molecules, plastics can have different refractive indices in different directions, a phenomenon known as birefringence. It is possible accidentally to rotate the plane of polarization of light in birefringent plastics, and there were initial reservations as to the feasibility of this approach for playing moulded CDs. The necessary quality of disk moulding was achieved by using relatively high temperatures where the material flows more easily. This has meant that birefringence is negligible, and the polarizing beamsplitter is widely used.

12.6 Focus systems

The frequency response of the laser pickup and the amount of crosstalk are both a function of the spot size and care must be taken to keep the beam focused on the information layer. If the spot on the disk becomes too large, it will be unable to discern the smaller features of the track, and can also be affected by the adjacent track. Disk warp and thickness irregularities will cause focal-plane movement beyond the depth of focus of the optical system, and a focus servo system will be needed. The depth of field is related to the numerical aperture, which is defined, and the accuracy of the servo must be sufficient to keep the focal plane within that depth, which is typically ± 1 μm.

The focus servo moves a lens along the optical axis in order to keep the spot in focus. Since dynamic focus-changes are largely due to warps, the focus system must have a frequency response in excess of the rotational speed. A moving-coil actuator is often used owing to the small moving mass which this permits. Figure 12.13 shows that a cylindrical magnet assembly almost identical to that of a loudspeaker can be used, coaxial with the light beam. Alternatively a moving-magnet design can be used. A rare earth magnet allows a sufficiently strong magnetic field without excessive weight.

A focus-error system is necessary to drive the lens. There are a number of ways in which this can be derived, the most common of which will be described here.

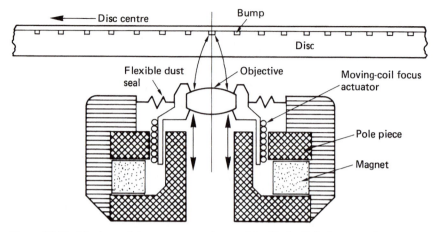

Figure 12.13 Moving-coil-focus servo can be coaxial with the light beam as shown.

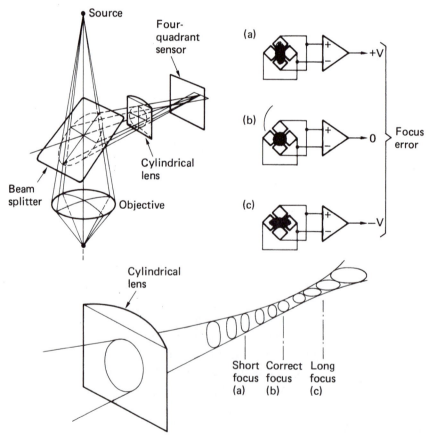

Figure 12.14 The cylindrical lens focus method produces an elliptical spot on the sensor whose aspect ratio is detected by a four-quadrant sensor to produce a focus error.

In Figure 12.14 a cylindrical lens is installed between the beamsplitter and the photosensor. The effect of this lens is that the beam has no focal point on the sensor. In one plane, the cylindrical lens appears parallel-sided, and has negligible effect on the focal length of the main system, whereas in the other plane, the lens shortens the focal length. The image will be an ellipse whose aspect ratio changes as a function of the state of focus. Between the two foci, the image will be circular. The aspect ratio of the ellipse, and hence the focus error, can be found by dividing the sensor

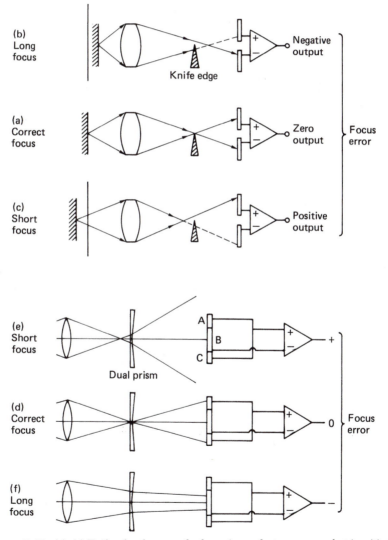

Figure 12.15 (a)–(c) Knife-edge focus method requires only two sensors, but is critically dependent on knife-edge position. (d)–(f) Twin-prism method requires three sensors (A, B, C), where focus error is (A + C) − B. Prism alignment reduces sensitivity without causing focus offset.

into quadrants. When these are connected as shown, the focus-error signal is generated. The data readout signal is the sum of the quadrant outputs.

Figure 12.15 shows the knife-edge method of determining focus. A split sensor is also required. At (a) the focal point is coincident with the knife edge, so it has little effect on the beam. At (b) the focal point is to the right of the knife edge, and rising rays are interrupted, reducing the output of the upper sensor. At (c) the focal point is to the left of the knife edge, and descending rays are interrupted, reducing the output of the lower sensor. The focus error is derived by comparing the outputs of the two halves of the sensor. A drawback of the knife-edge system is that the lateral position of the knife edge is critical, and adjustment is necessary. To overcome this problem, the knife edge can be replaced by a pair of prisms, as shown in Figure 12.15(d)–(f). Mechanical tolerances then only affect the sensitivity, without causing a focus offset.

The cylindrical lens method is compared with the knife-edge/prism method in Figure 12.16, which shows that the cylindrical lens method has a much smaller capture range. A focus-search mechanism will be required, which moves the focus servo over its entire travel, looking for a zero crossing. At this time the feedback loop will be completed, and the sensor will remain on the linear part of its characteristic. The spiral track of CD and MiniDisc starts at the inside and works outwards. This was deliberately arranged because there is less vertical run-out near the hub, and initial focusing will be easier.

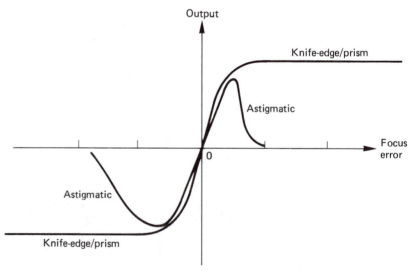

Figure 12.16 Comparison of captive range of knife-edge/prism method and astigmatic (cylindrical lens) system. Knife edge may have a range of 1 mm, whereas astigmatic may only have a range of 40 μm, requiring a focus-search mechanism.

12.7 Tracking systems

The track pitch is only 1.6 μm, and this is much smaller than the accuracy to which the player chuck or the disk centre hole can be made; on a typical player, run-out will swing several tracks past a fixed pickup. The non-contact readout means that there is no inherent mechanical guidance of the pickup. In addition, a warped disk will not present its surface at 90° to the beam, but will constantly change the angle of incidence during two whole cycles per revolution. Owing to the change of refractive index at the disk surface, the tilt will change the apparent position of the track to the pickup, and Figure 12.17 shows that this makes it appear wavy. Warp also results in coma of the readout spot. The disk format specifies a maximum warp amplitude to keep these effects under control. Finally, vibrations induced in the player from outside, particularly in portable and automotive players, will tend to disturb tracking. A track-following servo is necessary to keep the spot centralized on the track in the presence of these difficulties. There are several ways in which a tracking error can be derived.

In the three-spot method, two additional light beams are focused on the disk track, one offset to each side of the track centreline. Figure 12.18 shows that, as one side spot moves away from the track into the mirror

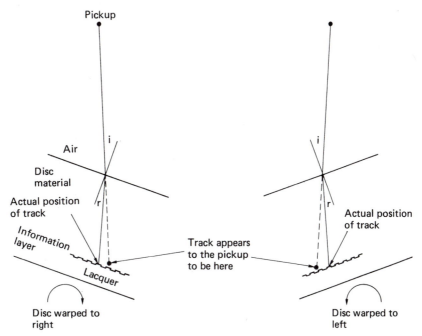

Figure 12.17 Owing to refraction, the angle of incidence (*i*) is greater than the angle of refraction (*r*). Disk warp causes the apparent position of the track (dashed line) to move, requiring the tracking servo to correct.

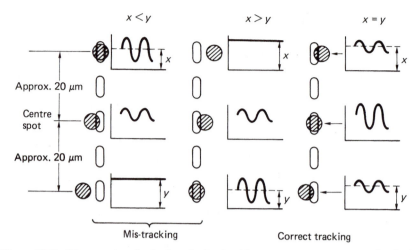

Figure 12.18 Three-spot method of producing tracking error compares average level of side-spot signals. Side spots are produced by a diffraction grating and require their own sensors.

area, there is less destructive interference and more reflection. This causes the average amplitude of the side spots to change differentially with tracking error. The laser head contains a diffraction grating which produces the side spots, and two extra photosensors onto which the reflections of the side spots will fall. The side spots feed a differential amplifier, which has a low-pass filter to reject the channel-code information and retain the average brightness difference. Some players use a delay line in one of the side-spot signals whose period is equal to the time taken for the disk to travel between the side spots. This helps the differential amplifier to cancel the channel code.

The side spots are generated as follows. When a wavefront reaches an aperture which is small compared to the wavelength, the aperture acts as a point source, and the process of diffraction can be observed as a spherical wavefront leaving the aperture as in Figure 12.19. Where the

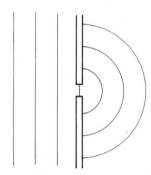

Figure 12.19 Diffraction as a plane wave reaches a small aperture.

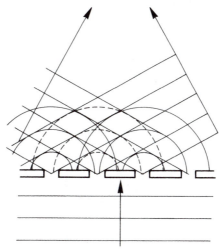

Figure 12.20 In a diffraction grating, constructive interference can take place at more than one angle for a single wavelength.

wavefront passes through a regular structure, known as a diffraction grating, light on the far side will form new wavefronts wherever radiation is in phase, and Figure 12.20 shows that these will be at an angle to the normal depending on the spacing of the structure and the wavelength of the light. A diffraction grating illuminated by white light will produce a dispersed spectrum at each side of the normal. To obtain a fixed angle of diffraction, monochromatic light is necessary.

The alternative approach to tracking-error detection is to analyse the diffraction pattern of the reflected beam. The effect of an off-centre spot is to rotate the radial diffraction pattern about an axis along the track. Figure 12.21 shows that, if a split sensor is used, one half will see greater modulation than the other when off-track. Such a system may be prone to

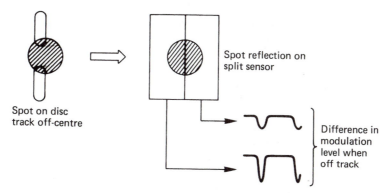

Figure 12.21 Split-sensor method of producing tracking error focuses image of spot onto sensor. One side of spot will have more modulation when off-track.

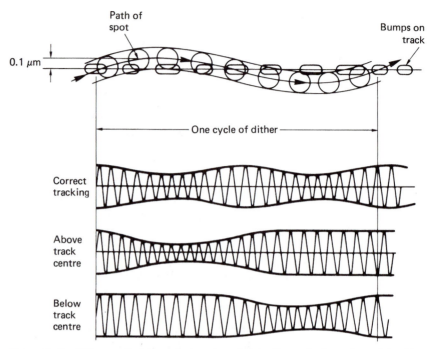

Figure 12.22 Dither applied to readout spot modulates the readout envelope. A tracking error can be derived.

develop an offset due either to drift or to contamination of the optics, although the capture range is large. A further tracking mechanism is often added to obviate the need for periodic adjustment. Figure 12.22 shows this dither-based system, which resembles in many respects the track-following method used in many professional videotape recorders. A sinusoidal drive is fed to the tracking servo, causing a radial oscillation of spot position of about ± 50 nm. This results in modulation of the envelope of the readout signal, which can be synchronously detected to obtain the sense of the error. The dither can be produced by vibrating a mirror in the light path, which enables a high frequency to be used, or by oscillating the whole pickup at a lower frequency.

12.8   Typical pickups

It is interesting to compare different designs of laser pickup. Figure 12.23 shows a Philips laser head.[6] The dual-prism focus method is used, which combines the output of two split sensors to produce a focus error. The focus amplifier drives the objective lens which is mounted on a parallel motion formed by two flexural arms. The capture range of the focus system is sufficient to accommodate normal tolerances without assis-

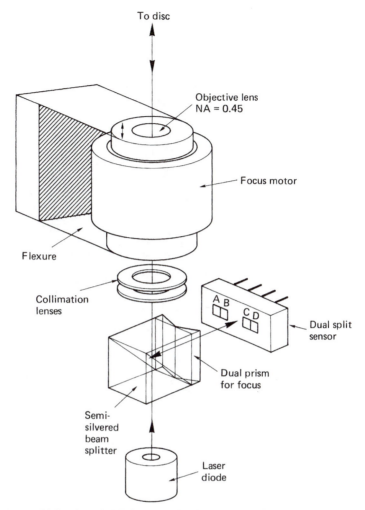

Figure 12.23 Philips laser head showing semisilvered prism for beam splitting. Focus error is derived from dual-prism method using split sensors. Focus error (A + D) – (B + C) is used to drive focus motor which moves objective lens on parallel action flexure. Radial differential tracking error is derived from split sensor (A + B) – (C + D). Tracking error drives entire pickup on radial arm driven by moving coil. Signal output is (A + B + C + D). System includes 600 Hz dither for tracking. (Courtesy *Philips Technical Review*)

tance. A radial differential tracking signal is extracted from the sensors as shown in the figure. Additionally, a dither frequency of 600 Hz produces envelope modulation which is synchronously rectified to produce a drift-free tracking error. Both errors are combined to drive the tracking system. As only a single spot is used, the pickup is relatively insensitive to angular errors, and a rotary positioner can be used, driven by a moving coil. The assembly is statically balanced to give good resistance to lateral shock.

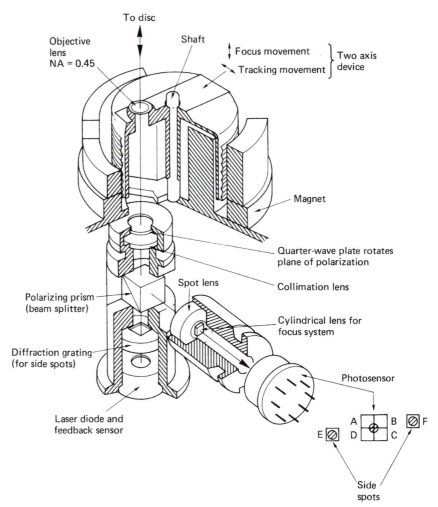

Figure 12.24 Sony laser head showing polarizing prism and quarter-wave plate for beam splitting, and diffraction grating for production of side spots for tracking. The cylindrical lens system is used for focus, with a four-quadrant sensor (A, B, C, D) and two extra sensors E, F for the side spots. Tracking error is E–F; focus error is (A + C) – (B + D). Signal output is (A + B + C + D). The focus and tracking errors drive the two-axis device. (Courtesy *Sony Broadcast*)

Figure 12.24 shows a Sony laser head used in consumer players. The cylindrical-lens focus method is used, requiring a four-quadrant sensor. Since this method has a small capture range, a focus-search mechanism is necessary. When a disk is loaded, the objective lens is ramped up and down looking for a zero crossing in the focus error. The three-spot method is used for tracking. The necessary diffraction grating can be seen adjacent to the laser diode. Tracking error is derived from side-spot sensors (E, F). Since the side-spot system is sensitive to angular error, a

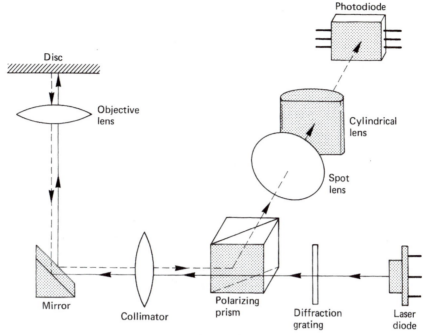

Figure 12.25 For automotive and portable players, the pickup can be made more compact by incorporating a mirror, which allows most of the elements to be parallel to the disk instead of at right angles.

parallel-tracking laser head traversing a disk radius is essential. A cost-effective linear motion is obtained by using a rack-and-pinion drive for slow, coarse movements, and a laterally moving lens in the light path for fine rapid movements. The same lens will be moved up and down for focus by the so-called two-axis device, which is a dual-moving coil mechanism. In some players this device is not statically balanced, making the unit sensitive to shock, but this was overcome on later heads designed for portable players. Figure 12.25 shows a later Sony design having a prism which reduces the height of the pickup above the disk.

12.9   DVD and CD readout in detail

The CD medium is designed to be read with a phase contrast microscope, and so it is correct to describe the deformities on the information layer as a phase structure. The original LaserVision disk patent[7] contains a variety of approaches, but in the embodiment used in CD it consists of two parallel planes separated by a distance which is constant and specifically related to the wavelength of the light which will be used to read it. The phase structure is created with deformities which depart from the first of the planes and whose extremities are in the second plane. These deformities are called pits when the second plane is below the first and bumps when the second plane is above the first.

Whilst a phase structure can be read by transmission or reflection, commercial designs based the Philips medium, such as LaserVision,[8] DVD, CD, CD-Video, CDROM and prerecorded MiniDisc, use reflective readout exclusively.

Optical physicists characterize materials by their reflectivity, transmissivity and absorption. Light energy cannot disappear, so when light is incident on some object, the amounts of light transmitted, absorbed and reflected must add up to the original incident amount. When no light is absorbed, the incident light is divided between that transmitted and that reflected. When light is absorbed, the transmitted and reflected amounts of light are both reduced. A medium such as a photograph contains pigments which absorb light more in the dark areas and less in the light areas. Thus the amount of light reflected varies. A medium such as a transparency also contains such pigments but in this case it is primarily the amount of light transmitted which varies. Such a variation in transmitted or reflected light from place to place is known as contrast.

Figure 12.2(a) showed that in CD, the information layer consists of an optically flat surface above which flat-topped bumps project. The entire surface of the phase structure is metallized to render it reflective. This metallization of the entire information layer means that little light is transmitted or absorbed, and as a result virtually all incident light must be reflected. The information layer of CD does not have conventional contrast. Contrast is in any case unnecessary as interference is used for readout, and this works better with a totally reflecting structure. Referring to Figure 12.26 it will be seen that a spot of light is focused onto the phase structure such that it straddles a bump. Ideally half the light energy should be incident on the top of the bump and half on the surrounding mirror surface. The height of the bump is ideally one quarter the wavelength of the light in a reflective system and as a result light which has reflected from the mirror surface has travelled one half a wavelength further than light which has reflected from the mirror surface. Consequently, along the normal, there are two components of light of almost equal energy, but they are in phase opposition, and destructive interference occurs, such that no wavefront can form in that direction. As light energy cannot disappear, wavefronts will leave the phase structure at any oblique angle at which constructive interference between the components can be achieved, creating a diffraction pattern. In the case of the light beam straddling the centre of a long bump the diffraction pattern will be in a plane which is normal to the disk surface and which intersects a disk radius. It is thus called a radial diffraction pattern. The zeroth-order radiation (that along the normal) will be

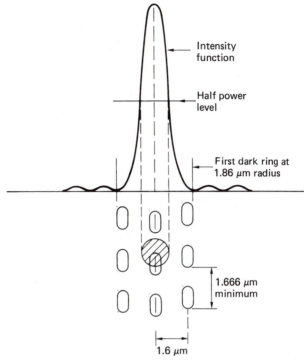

Intensity
function

Half power
level

First dark ring at
1.86 μm radius

1.666 μm
minimum

1.6 μm

Figure 12.26 The structure of a maximum frequency recording is shown here, related to the intensity function of an objective of 0.45NA with 780 μm light. Note that track spacing puts adjacent tracks in the dark rings, reducing crosstalk. Note also that as the spot has an intensity function it is meaningless to specify the spot diameter without some reference such as an intensity level.

heavily attenuated, and most of the incident energy will concentrated in the first- and second-order wavefronts.

Some treatments use the word scattering to describe the effect of the interaction of the readout beam with the relief structure. This is technically incorrect as scattering is a random phenomenon which is independent of wavelength. The diffraction pattern is totally predictable and strongly wavelength dependent.

When the light spot is focused on a plain part of the mirror surface, known as a land, clearly most of the energy is simply reflected back whence it came. Thus when a bump is present, light is diffracted away from the normal, whereas in the absence of a bump, it returns along the normal. Although all incident light is reflected at all times, the effect of diffraction is that the direction in which wavefronts leave the phase structure is changed by the presence of the bump. What then happens is a function of the optical system being used. In a conventional CD player the angle to the normal of the first diffracted order in the radial diffraction pattern due to a long bump will be sufficiently oblique that it passes outside the aperture of the objective and does not return to the photosensor. Thus the bumps appear dark to the photosensor and the lands appear bright. Although all light is reflected at all times and there is no conventional contrast, inside the pickup there are variations in the light falling on the photosensor, a phenomenon called phase contrast.

The phase contrast technique described will only work for a given wavelength and with an appropriate aperture and lens design, and so the CD must be read with monochromatic light. Whilst the ideal case is where the two components of light are equal to give exact cancellation, in practice this ideal is not met but instead there is a substantial reduction in the light returning to the pickup.

Some treatments of CD refer to a 'beam' of light returning from the disk to the pickup, but this is incorrect. What leaves the disk is a hemispherical diffraction pattern certain orders of which enter the aperture of the pickup. The destructive interference effect can be seen with the naked eye by examining any CD under a conventional incandescent lamp. The data surface of a CD has many parallel tracks and works somewhat like a diffraction grating by dispersing the incident white light into a spectrum. However, the resultant spectrum is not at all like that produced by a conventional diffraction grating or by a prism. These latter produce a spectrum in which the relative brightness of the colours is like that of a rainbow, i.e. the green in the centre is brightest, the red at one end is less bright and the blue at the other end is fainter still. This is due to the unequal response of the eye to various colours, where equal red, green and blue stimuli produce responses in approximately the proportions 2:5:1 respectively. In the diffracted spectrum from a CD, however, the blue component appears as strong or stronger than the other colours. This is because the relief structure of CD is designed not to reflect infrared light of 780 nm wavelength. This relief structure will, however, reflect perfectly ultraviolet light of half that wavelength as the zeroth-order light reflected from the top of the bumps will be in phase with light reflected from the land. Thus a CD reflects visible blue light much more strongly than longer-wavelength colours.

It is essential to the commercial success of CD that a useful playing time (75 min max.) should be obtained from a recording of reasonable size (12 cm). The size was determined by the European motor industry as being appropriate for car dashboard-mounted units. It follows that the smaller the spot of light which can be created, the smaller can be the deformities carrying the information, and so more information per unit area (known in the art as the superficial recording density) can be stored. Development of a successful high-density optical recorder requires an intimate knowledge of the behaviour of light focused into small spots. If it is attempted to focus a uniform beam of light to an infinitely small spot on a surface normal to the optical axis, it will be found that it is not possible. This is probably just as well as an infinitely small spot would have infinite intensity and any matter it fell on would not survive. Instead the result of such an attempt is a distribution of light in the area of the focal point which has no sharply defined boundary. This is called the Airy distribution[9] (sometimes pattern or disk) after Lord Airy (1835), the then astronomer royal. If a line is considered to pass across the focal plane, through the theoretical focal point, and the intensity of the light is plotted on a graph as a function of the distance along that line, the result is the intensity function shown in Figure 12.26. It will be seen that this contains a central sloping peak surrounded by alternating dark rings and light rings of diminishing intensity. These rings will in theory reach to infinity before their intensity becomes zero. The intensity distribution or function described by Airy is due to diffraction effects across the finite aperture of the objective. For a given wavelength, as the aperture of the objective is increased, so the diameter of the features of the Airy pattern reduces. The Airy pattern vanishes to a singularity of infinite intensity with a lens of infinite aperture which, of course, cannot be made. The approximation of geometric optics is quite unable to predict the occurrence of the Airy pattern.

An intensity function does not have a diameter, but for practical purposes an effective diameter typically quoted is that at which the intensity has fallen to some convenient fraction of that at the peak. Thus one could state, for example, the half-power diameter.

Since light paths in optical instruments are generally reversible, it is possible to see an interesting corollary which gives a useful insight into the readout principle of CD. Considering light radiating from a phase structure, as in Figure 12.27, the more closely spaced the features of the phase structure, i.e. the higher the spatial frequency, the more oblique the direction of the wavefronts in the diffraction pattern which results and the larger the aperture of the lens needed to collect the light if the

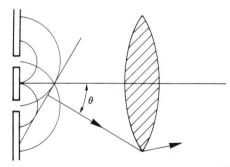

Figure 12.27 Fine detail in an object can only be resolved if the diffracted wavefront due to the highest spatial frequency is collected by the lens. Numerical aperture (NA) = sin θ, and as θ is the diffraction angle it follows that, for a given wavelength, NA determines resolution.

resolution is not to be lost. The corollary of this is that the smaller the Airy distribution it is wished to create, the larger must be the aperture of the lens. Spatial frequency is measured in lines per millimetre and as it increases, the wavefronts of the resultant diffraction pattern become more oblique. In the case of a CD, the smaller the bumps and the spaces between them along the track, the higher the spatial frequency, and the more oblique the diffraction pattern becomes in a plane tangential to the track. With a fixed-objective aperture, as the tangential diffraction pattern becomes more oblique, less light passes the aperture and the depth of modulation transmitted by the lens falls. At some spatial frequency, all the diffracted light falls outside the aperture and the modulation depth transmitted by the lens falls to zero. This is known as the spatial cut-off frequency. Thus a graph of depth of modulation versus spatial frequency can be drawn and which is known as the modulation transfer function (MTF). This is a straight line commencing at unity at zero spatial frequency (no detail) and falling to zero at the cut-off spatial frequency (finest detail). Thus one could describe a lens of finite aperture as a form of spatial low-pass filter. The Airy function is no more than the spatial impulse response of the lens, and the concentric rings of the Airy function are the spatial analog of the symmetrical ringing in a phase-linear electrical filter. The Airy function and the triangular frequency response form a transform pair[10] as shown in Chapter 3.

When an objective lens is used in a conventional microscope, the MTF will allow the resolution to be predicted in lines per millimetre. However, in a scanning microscope the spatial frequency of the detail in the object is multiplied by the scanning velocity to give a temporal frequency measured in Hertz. Thus lines per millimetre multiplied by millimetres per second gives lines per second. Instead of a straight-line MTF falling to the spatial cut-off frequency, a scanning microscope has a temporal frequency response falling to zero at the optical cut-off frequency. Whilst this concept requires a number of idiomatic terms to be assimilated at once, the point can be made clear by a simple analogy. Imagine the evenly spaced iron railings outside a schoolyard. These are permanently fixed, and can have no temporal frequency, yet they have a spatial frequency which is the number of railings per unit distance. A small boy with a stick takes great delight in running along the railings so that his stick hits each one in turn and makes a great noise. The rate at which his stick hits the railings is the temporal frequency which results from their being scanned. This rate would increase if the boy ran faster, but it would also increase if the rails were closer together. As a consequence it can be seen that the temporal frequency is proportional to the spatial frequency multiplied by the scanning speed. Put more technically, the frequency response of an optical recorder is the Fourier transform of the Airy distribution of the readout spot multiplied by the track velocity.

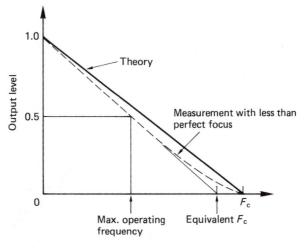

Figure 12.28 Frequency response of laser pickup. Maximum operating frequency is about half of cut-off frequency F_c.

In magnetic recorders and vinyl disk recorders there is at least a frequency band where the response is reasonably flat. CD is basically a phase contrast scanning microscope. Figure 12.28 shows that the frequency response falls progressively from DC to the optical cutoff frequency which is given by:

$$F_c = \frac{2NA}{\text{wavelength}} \times \text{velocity}$$

The minimum linear velocity of CD is 1.2 m/s, giving a cutoff frequency of

$$F_c = \frac{2 \times 0.45 \times 1.2}{780 \times 10^{-9}} = 1.38\,\text{MHz}$$

Actual measurements reveal that the optical response is only a little worse than the theory predicts. This characteristic has a large bearing on the type of modulation schemes which can be successfully employed. Clearly, to obtain any noise immunity, the maximum operating frequency must be rather less than the cutoff frequency. The maximum frequency used in CD is 720 kHz, which represents an absolute minimum wavelength of 1.666 μm, or a bump length of 0.833 μm, for the lowest permissible track speed of 1.2 m/s used on the full-length 75 min-playing disks. One-hour-playing disks have a minimum bump length of 0.972 μm at a track velocity of 1.4 m/s. The maximum frequency is the same in both cases. This maximum frequency should not be confused with the bit rate of CD since this is different owing to the channel code used. Figure 12.26

showed a maximum-frequency recording, and the physical relationship of the intensity function to the track dimensions.

In a CD player, the source of light is a laser, and this does not produce a beam of uniform intensity. It is more intense in the centre than it is at the edges, and this has the effect of slightly increasing the half-power diameter of the intensity function. The effect is analogous to the effect of window functions in FIR filters (see Chapter 3). The intensity function can also be enlarged if the lens used suffers from optical aberrations. This was studied by Maréchal[11] who established criteria for the accuracy to which the optical surfaces of the lens should be made to allow the ideal Airy distribution to be obtained. CD player lenses must meet the Maréchal criterion. With such a lens, the diameter of the distribution function is determined solely by the combination of numerical aperture (NA) and the wavelength. When the size of the spot is as small as the NA and wavelength allow, the optical system is said to be diffraction limited. Figure 12.27 showed how numerical aperture is defined, and illustrates that the smaller the spot needed, the larger must be the NA. Unfortunately the larger the NA, the more obliquely to the normal the light arrives at the focal plane and the smaller the depth of focus will be. This was investigated by Hopkins,[12] who established the depth of focus available for a given NA. CD players have to use an NA of 0.45 which is a compromise between a small spot and an impossibly small depth of focus.[13] The later DVD uses an NA of 0.6 in conjunction with a shorter-wavelength laser. This allows a significant reduction in the size of the recorded bumps and a corresponding increase in storage density. Essentially the information layer of DVD is a scaled-down CD.

The intensity function will also be distorted and grossly enlarged if the optical axis is not normal to the medium. The initial effect is that the energy in the first bright ring increases strongly in one place and results in a secondary peak adjacent to the central peak. This is known as coma and its effect is extremely serious as the enlargement of the spot restricts the recording density. The larger the NA, the smaller becomes the allowable tilt of the optical axis with respect to the medium before coma becomes a problem. With the NA of CD this angle is less than a degree.[13]

Numerical aperture is defined as the cosine of the angle between the optical axis and rays converging from the perimeter of the lens. It will be apparent that there are many combinations of lens diameter and focal length which will have the same NA. As the difficulty of manufacture, and consequently the cost, of a lens meeting the Maréchal criterion increases disproportionately with size, it is advantageous to use a small lens of short focal length, mounted close to the medium and held precisely perpendicular to the medium to prevent coma. As the lens needs to be driven along its axis by a servo to maintain focus, the smaller lens will facilitate the design of the servo by reducing the mass to be driven. It is extremely difficult to make a lens which meets the Maréchal criterion over a range of wavelengths because of dispersion. The use of monochromatic light eases the lens design as it has only to be correct for one wavelength.

At the high recording density of CD, there is literally only one scanning mechanism with which all the optical criteria can be met and this is the approach known from the scanning microscope. The optical pickup is mounted in a carriage which can move it parallel to the medium in such a way that the optical axis remains at all times parallel to the axis of rotation of the medium. The latter rotates as the pickup is driven away from the axis of rotation in such a way that a spiral track on the disk is followed. The pickup contains a short focal length lens of small diameter which must therefore be close to the disk surface to allow a large NA. All high-density optical recorders operate on this principle in which the readout of the carrier is optical but the scanning is actually mechanical.

12.10   How optical disks are made

The steps used in the production of CDs will next be outlined. Prerecorded MiniDiscs are made in an identical fashion except for detail differences which will be noted. MO disks need to be grooved so that the track-following system will work. The grooved substrate is produced in a similar way to a CD master, except that the laser is on continuously instead of being modulated with a signal to be recorded. As stated, CD is replicated by moulding, and the first step is to produce a suitable mould. This mould must carry deformities of the correct depth for the standard wavelength to be used for reading, and as a practical matter these deformities must have slightly sloping sides so that it is possible to release the CD from the mould.

The major steps in CD manufacture are shown in Figure 12.29. The mastering process commences with an optically flat glass disk about 220 mm in diameter and 6 mm thick. The blank is washed first with an alkaline solution, then with a fluorocarbon solvent, and spun dry prior to polishing to optical flatness. A critical cleaning process is then undertaken using a mixture of de-ionized water and isopropyl alcohol in the presence of ultrasonic vibration, with a final fluorocarbon wash. The blank must now be inspected for any surface irregularities which would cause data errors. This is done by using a laser beam and monitoring the reflection as the blank rotates. Rejected blanks return to the polishing process, those which pass move on, and an adhesive layer is applied followed by a coating of positive photoresist. This is a chemical substance which softens when exposed to an appropriate intensity of light of a

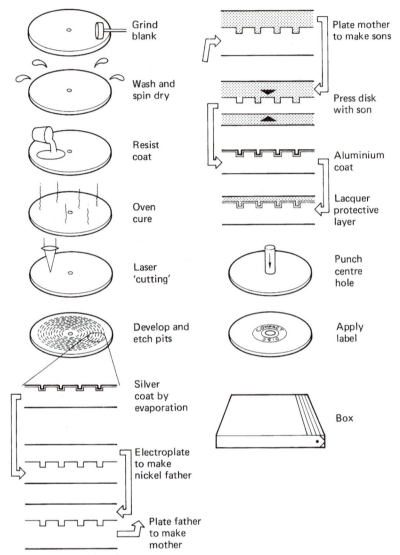

Figure 12.29 The many stages of CD manufacture, most of which require the utmost cleanliness.

certain wavelength, typically ultraviolet. Upon being thus exposed, the softened resist will be washed away by a developing solution down to the glass to form flat-bottomed pits whose depth is equal to the thickness of the undeveloped resist. During development the master is illuminated with laser light of a wavelength to which it is insensitive. The diffraction pattern changes as the pits are formed. Development is arrested when the appropriate diffraction pattern is obtained.[14] The thickness of the resist layer must be accurately controlled, since it affects the height of the bumps on the finished disk, and an optical scanner is used to check that

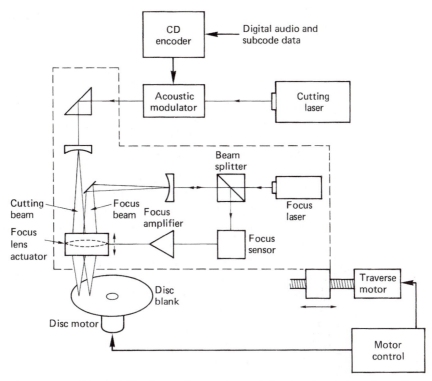

Figure 12.30 CD cutter. The focus subsystem controls the spot size of the main cutting laser on the photosensitive blank. Disc and traverse motors are coordinated to give constant track pitch and velocity. Note that the power of the focus laser is insufficient to expose the photoresist.

there are no resist defects which would cause data errors or tracking problems in the end product. Blanks which pass this test are oven-cured, and are ready for cutting. Failed blanks can be stripped of the resist coating and used again.

The cutting process is shown in simplified form in Figure 12.30. A continuously operating helium cadmium[15] or argon ion[16] laser is focused on the resist coating as the blank revolves. Focus is achieved by a separate helium neon laser sharing the same optics. The resist is insensitive to the wavelength of the He–Ne laser. The laser intensity is controlled by a device known as an acousto-optic modulator which is driven by the encoder. When the device is in a relaxed state, light can pass through it, but when the surface is excited by high-frequency vibrations, light is scattered. Information is carried in the lengths of time for which the modulator remains on or remains off. As a result the deformities in the resist produced as the disk turns when the modulator allows light to pass are separated by areas unaffected by light when the modulator is shut off. Information is carried solely in the variations of the lengths of these two areas.

The laser makes its way from the inside to the outside as the blank revolves. As the radius of the track increases, the rotational speed is proportionately reduced so that the velocity of the beam over the disk remains constant. This constant linear velocity (CLV) results in rather longer playing time than would be obtained with a constant speed of rotation. Owing to the minute dimensions of the track structure, the cutter has to be constructed to extremely high accuracy. Air bearings are used in the spindle and the laser head, and the whole machine is resiliently supported to prevent vibrations from the building from affecting the track pattern.

Early CD cutters worked in real time, but subsequently the operating speed has been increased dramatically to increase the throughput.

As the player is a phase contrast microscope, it must produce an intensity function which straddles the deformities. As a consequence the intensity function which produces the deformities in the photoresist must be smaller in diameter than that in the reader. This is conveniently achieved by using a shorter wavelength of 400–500 nm from a helium–cadmium or argon–ion laser combined with a larger lens aperture of 0.9. These are expensive, but are only needed for the mastering process.

It is a characteristic of photoresist that its development rate is not linearly proportional to the intensity of light. This non-linearity is known as 'gamma'. As a result there are two intensities of importance when scanning photoresist; the lower sensitivity, or threshold, below which no development takes place, and the upper threshold above which there is full development. As the laser light falling on the resist is an intensity function, it follows that the two thresholds will be reached at different diameters of the function. It can be seen in Figure 12.31 that advantage is taken of this effect to produce tapering sides to the pits formed in the resist. In the centre, the light is intense enough to fully develop the resist right down to the glass. This gives the deformity a flat bottom. At the edge, the intensity falls and as some light is absorbed by the resist, the diameter of the resist which can be developed falls with depth in the resist. By controlling the intensity of the laser, and the development time, the slope of the sides of the pits can be controlled.

In summary, the resist thickness controls the depth of the pits, the cutter laser wavelength and the NA of the objective together control the width of the pits in the radial direction, and the laser intensity and sensitivity of the resist together with the development time control the slope. The length of the pits in the tangential direction, i.e. along the track, is controlled by the speed of the disk past the objective and the length of time for which the modulator allows light to pass. The space between the pits along the track is controlled by the speed of the disk past the objective and the length of time for which the modulator blocks the laser

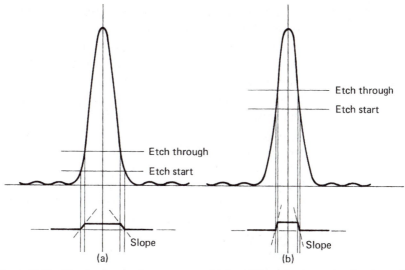

Figure 12.31 The two levels of exposure sensitivity of the resist determine the size and edge slope of the bumps in the CD. (a) Large exposure results in large bump with gentle slope; (b) less exposure results in smaller bump with steeper sloped sides.

light. In practice all these values are constant for a given cutting process except for the times for which the modulator turns on or off. As a result pits of constant depth and cross-section are formed, and only their length and the space between them along the track is changed in order to carry information.

The specified wavelength of 780 nm and the numerical aperture of 0.45 used for playback results in an Airy function where the half-power level is at a diameter of about 1 μm. The first dark ring will be at about 1.9 μm diameter. As the illumination follows an intensity function, it is really meaningless to talk about spot size unless the relative power level is specified. The analogy is quoting frequency response without dB limits. Allowable crosstalk between tracks then determines the track pitch. The first ring outside the central disk carries some 7 per cent of the total power, and limits crosstalk performance. The track spacing is such that with a slightly defocused beam and a slight tracking error, crosstalk due to adjacent tracks is acceptable. Since aberrations in the objective will increase the spot size and crosstalk, the CD specification requires the lens to be within the Maréchal criterion. Clearly the numerical aperture of the lens, the wavelength of the laser, the refractive index and thickness of the disk and the height and size of the bumps must all be simultaneously specified.

The master recording process has produced a phase structure in relatively delicate resist, and this cannot be used for moulding directly. Instead a thin metallic silver layer is sprayed onto the resist to render it electrically conductive so that electroplating can be used to make robust copies of the relief structure. This conductive layer then makes the resist optically reflective and it is possible to 'play' the resist master for testing purposes. However, it cannot be played by the cutter, as the beam in the cutter is too small and it would not straddle the pits. The necessary phase contrast between light energy leaving the lands and pits would not then be achieved. Unfortunately the resist master cannot be played by a normal CD pickup either, because the pits in the resist are full of air, in which the velocity (and therefore the wavelength) of light is different from the value it will have in the finished disk when the pits are filled with plastic. Thus the correct pit depth for a plastic disk is incorrect in air; a third type of optical system is needed to test play a resist master in which the wavelength is shorter than the wavelength used in the normal player. This would produce an Airy pattern which was too small with a conventional lens aperture, and so a lens of smaller aperture is needed to produce a spot of the correct diameter from the 'wrong' wavelength.

The electrically conductive resist master is then used as the cathode of an electroplating process where a first layer of metal is laid down over the resist, conforming in every detail to the relief structure thereon. This metal layer can then be separated from the glass and the resist is dissolved away and the silver is recovered leaving a laterally inverted phase structure on the surface of the metal, in which the pits in the photoresist have become bumps in the metal. From this point on, the production of CD is virtually identical to the replication process used for vinyl disks, save only that a good deal more precision and cleanliness is needed.

This first metal layer could itself be used to mould disks, or it could be used as a robust submaster from which many stampers could be made by pairs of plating steps. The first metal phase structure can itself be used as a cathode in a further electroplating process in which a second metal layer is formed having a mirror image of the first. A third such plating step results in a stamper. The decision to use the master or substampers will be based on the number of disks and the production rate required.

The master is placed in a moulding machine, opposite a flat plate. A suitable quantity of molten plastic is injected between, and the plate and the master are forced together. The flat plate renders one side of the disk smooth, and the bumps in the metal stamper produce pits in the other surface of the disk. The surface containing the pits is next metallized, with any good electrically conductive material, typically aluminium. This metallization is then covered with a lacquer for protection. In the case of CD, the label is printed on the lacquer. In the case of a prerecorded MiniDisc, the ferrous hub needs to be applied prior to fitting the cartridge around the disk.

12.11   Direct metal mastering

An alternative method of CD duplication has been developed by Teldec.[17] As Figure 12.32 shows, the recording process is performed by a diamond stylus which embosses the pit structure into a thin layer of copper. The stylus is driven by a piezoelectric element using motional feedback. The element is supported in an elastic medium so that its own centre of gravity tends to remain stationary. Application of drive voltage makes the element contract, lifting the stylus completely off the copper between pits. Since the channel code of CD is DC-free, the stylus spends exactly half its

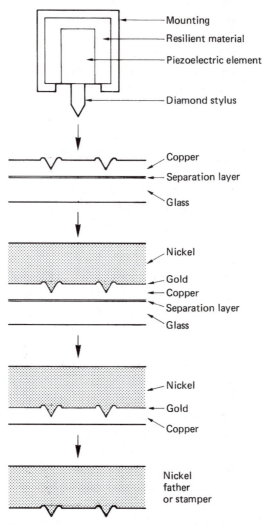

Figure 12.32   In direct metal mastering, a piezoelectric element embosses a copper layer, which is plated over with nickel and subsequently etched away to make a father, or for direct use as a stamper for short runs.

time in contact with the copper, and therefore the embossing force is exactly twice the static force applied. The pits produced are vee-shaped, rather than the flat-bottomed type produced by the photoresist method, but this is not of much consequence, since the diffraction-limited optics of the player cannot determine any more about the pit than its presence or absence. It is claimed that this form of pit is easier to mould.

A glass master disk is prepared as before, and following a thin separation layer, a coating of copper about 300 nm thick is sputtered on. The recording is made on this copper layer, which is then gold-coated, and nickel-plated to about 0.25 mm thick. The resultant metal sandwich can then be peeled off the glass master, which can be re-used. The recording is completely buried in the sandwich, and can be stored or transported in this form.

In order to make a stamper, the copper is etched away with ferric chloride to reveal the gold-coated nickel. This can then be used as an electroplating father, as before. For short production runs, the nickel layer can be used as a stamper directly, but it is recommended that the gold layer be replaced by rhodium for this application.

12.12   MiniDisc read/write in detail

MiniDisc has to operate under a number of constraints which largely determine how the read/write pickup operates. A prerecorded MiniDisc has exactly the same track dimensions as CD so that it can be mastered on similar equipment. When playing a prerecorded disk, the MiniDisc player pickup has to act in the same way as a CD pickup. This determines the laser wavelength, the NA of the objective and the effective spot diameter on the disk. This spot diameter must also be used when the pickup is operating with an MO disk.

Figure 12.33(a) shows to scale a CD track being played by a standard pickup. The readout spot straddles the track so that two antiphase components of reflected light can be obtained. As was explained in section 12.10, the CD mastering cutter must use a shorter wavelength and larger NA than the subsequent player in order to 'cut' the small pits in the resist. Figure 12.33(b) shows that the cutting process convolves the laser-enabling pulse with the spot profile so that the pit is actually longer than the pulse duration by a spot diameter. The effect is relatively small in CD because of the small spot used in the cutter.

When using an MO disk, the tracks recorded will be equal in width to the spot diameter and so will be wider than CD tracks as Figure 12.33(c) shows. MO writing can be performed in two ways. The conventional method used in computer disks is to apply a steady current to the coil and to modulate the laser. This is because the coil has to be some distance

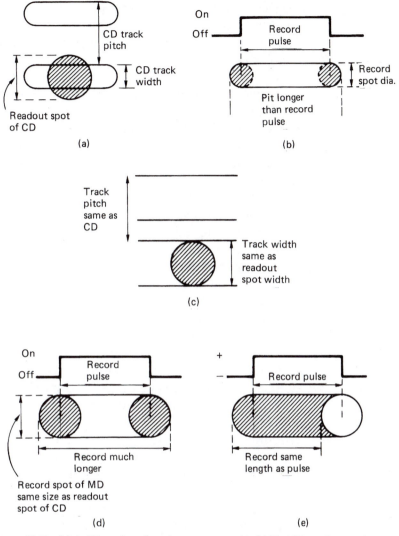

Figure 12.33 (a) A CD track and readout spot to scale. (b) The CD track is cut by a smaller spot, but the process results in pits which are longer than the pulse duration. (c) An MO track and readout spot to scale. If this spot is pulsed for writing, the magnetized areas are much larger than the pulse period and density is compromised as in (d). If, however, the magnetic field is modulated, as in (e), the recording is made at the trailing edge of the spot and short wavelengths can be used.

from the magnetic layer of the disk and must be quite large. The inductance of the coil is too great to allow it to be driven at the data frequency in computer applications.

Figure 12.33(d) shows what would happen if MD used laser modulation. The spot profile is convolved with the modulation pulse as for a CD cutter, but the spot is the same size as a replay spot. As a result the magnetized area is considerably longer than the modulation pulse. The shortest wavelengths of a recording could not be reproduced by this system, and it would be necessary to increase the track speed, reducing the playing time.

The data rate of MiniDisc is considerably lower than is the case for computer disks, and it is possible to use magnetic field modulation instead of laser modulation. The laser is then on continuously, and the spot profile is no longer convolved with the modulation. The recording is actually made at the instant the magnetic layer cools below the Curie temperature of about 180°C just after the spot has passed. The state of the magnetic field at this instant is preserved on the disk. Figure 12.33(e) shows that the recorded wavelength can be much shorter because the recording is effectively made by the trailing edge of the spot. This makes the ends of the recorded flux patterns somewhat crescent-shaped. Thus a spot the same size as a CD readout spot can be made to record flux patterns as short as the pits made by the smaller spot of a cutter. The recordable MiniDisc can thus have the same playing time as a prerecorded disk. The optical pickup is simplified because no laser modulator is needed.

The magnetic layer of MO disks should show a large Kerr rotation angle in order to give an acceptable SNR on replay. A high Curie temperature requires a high recording power, but allows greater readout power to be used without fear of demagnetization. This increases the readout signal with respect to the photodiode noise. As a result the Curie temperature is a compromise. Magnetic layers with practical Curie temperatures are made from proprietary alloys of iron, cobalt, platinum, terbium, gadolinium and various other rare earths. These are all highly susceptible to corrosion in air and are also incompatible with the plastics used for moulded substrates. The magnetic layer must be protected by sandwiching it between layers of material which require to be impervious to corrosive ions but which must be optically transmissive. Thus only dielectrics such as silicon dioxide or aluminium nitride can be used.

The disk pickup is concerned with analysing light which has returned from within the MO layer as only this will have the Kerr rotation. Reflection from the interface between the MO layer and the dielectric overlayer will have no Kerr rotation. The optical characteristics of the dielectric layers can be used to enhance readout by reducing the latter reflection. Figure 12.34 shows that the MO disks have an optically reflective layer behind the sandwiched MO layer. The thickness of the dielectric between the MO layer and the reflector is selected such that light from the reflector is antiphase with light from the overlayer/MO layer interface and instead of being reflected back to the pickup is absorbed in the MO layer. Conversely, light originating in the MO layer and leaving in the direction of the pickup experiences constructive

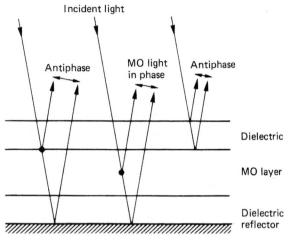

Figure 12.34 In MO disks the dielectric and reflective layers are as important as the magneto-optic layer itself.

interference with reflected components of that light. These components which contain Kerr rotation are readily able to exit the disk. These measures enhance the ratio of the magneto-optic component to ordinary light at the pickup.

12.13   How recordable MiniDiscs are made

Recordable MiniDiscs make the recording as flux patterns in a magnetic layer. However, the disks need to be pre-grooved so that the tracking systems described in section 12.7 can operate. The grooves have the same pitch as CD and the prerecorded MD, but the tracks are the same width as the laser spot: about 1.1 μm. The grooves are not a perfect spiral, but have a sinusoidal waviness at a fixed wavelength. Like CD, MD uses constant track linear velocity, not constant speed of rotation. When recording on a blank disk, the recorder needs to know how fast to turn the spindle to get the track speed correct. The wavy grooves will be followed by the tracking servo and the frequency of the tracking error will be proportional to the disk speed. The recorder simply turns the spindle at a speed which makes the grooves wave at the correct frequency. The groove frequency is 75 Hz; the same as the data sector rate. Thus a zero crossing in the groove signal can also be used to indicate where to start recording. The grooves are particularly important when a chequer-boarded recording is being replayed. On a CLV disk, every seek to a new track radius results in a different track speed. The wavy grooves allow the track velocity to be monitored as soon as a new track is reached.

The pre-grooves are moulded into the plastics body of the disk when it is made. The mould is made in a similar manner to a prerecorded disk master, except that the laser is not modulated and the spot is larger. The track velocity is held constant by slowing down the resist master as the radius increases, and the waviness is created by injecting 75 Hz into the lens radial positioner. The master is developed and electroplated as normal in order to make stampers. The stampers make pre-grooved disks which are then coated by vacuum deposition with the MO layer, sandwiched between dielectric layers. The MO layer can be made less susceptible to corrosion if it is smooth and homogeneous. Layers which contain voids, asperities or residual gases from the coating process present a larger surface area for attack. The life of an MO disk is affected more by the manufacturing process than by the precise composition of the alloy.

Above the sandwich an optically reflective layer is applied, followed by a protective lacquer layer. The ferrous clamping plate is applied to the centre of the disk, which is then fitted in the cartridge. The recordable cartridge has a double-sided shutter to allow the magnetic head access to the back of the disk.

12.14   Channel code of CD and MiniDisc

CD and MiniDisc use the same channel code known as EFM. This was optimized for the optical readout of CD and prerecorded MiniDisc, but is also used for the recordable version of MiniDisc for simplicity. DVD uses a refinement of the CD code called EFM+.

The frequency response falling to the optical cut-off frequency is only one of the constraints within which the modulation scheme has to work. There are a number of others. In all players the tracking and focus servos operate by analysing the average amount of light returning to the pickup. If the average amount of light returning to the pickup is affected by the content of the recorded data, then the recording will interfere with the operation of the servos. Debris on the disk surface affects the light intensity and means must be found to prevent this reducing the signal quality excessively.

Optical disks are serial media which produce on replay only a single voltage varying with time. If it is attempted to simply serialize raw data, a process known as direct recording, it is not difficult to see what will happen in the case where the data are digital audio samples where the audio is muted. Upon serializing the all-zeros code for muting the serial waveform is simply a steady logical low level and in the absence of a separate clock it is impossible to tell how many zeros were present, nor in the case of CD will there be a track to follow. A similar problem would be experienced if all ones occur in the data except that a steady high logic level results in a continuous bump. In digital logic circuits it is common to have signal lines and separate clock lines to overcome this problem, but with a single signal the separate clock is not possible. A further problem with direct optical recording is that the average brightness of the track is a function of the relative proportion of ones and zeros. Focus and tracking servos cannot be used with direct recordings because the data determine the average brightness and confuse the servos. Chapter 6 discussed modulation schemes known as DC-free codes. If such a code is used, the average brightness of the track is constant and independent of the data bits. Figure 12.35(a) shows the replay signal from the pickup being

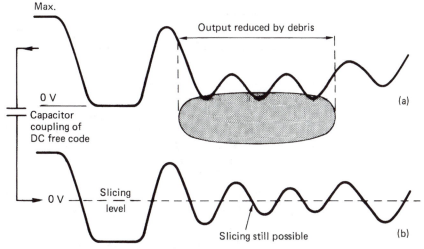

Figure 12.35 A DC-free code allows signal amplitude variations due to debris to be rejected.

compared with a threshold voltage in order to recover a binary waveform from the analog pickup waveform, a process known as slicing. If the light beam is partially obstructed by debris, the pickup signal level falls, and the slicing level is no longer correct and errors occur. If, however, the code is DC-free, the waveform from the pickup can be passed through a high-pass filter (e.g. a series capacitor) and Figure 12.35(b) shows that this rejects the falling level and converts it to a reduction in amplitude about the slicing level so that the slicer still works properly. This step cannot be performed unless a DC-free code is used.

As the frequency response on replay falls linearly to the cut-off frequency determined by the aperture of the lens and the wavelength of light used, the shorter bumps and lands produce less modulation than longer ones. Figure 12.36(a) shows what happens to the replay waveform

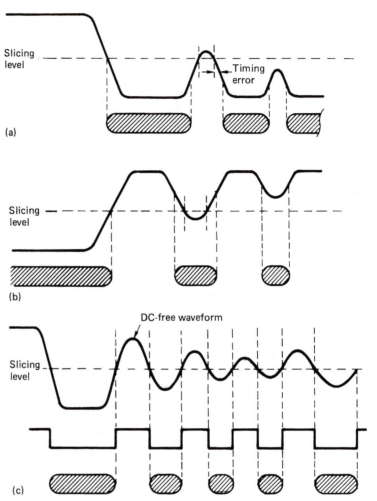

Figure 12.36 If the recorded waveform is not DC-free, timing errors occur until slicing becomes impossible. With a DC-free code, jitter-free slicing is possible in the presence of serious amplitude variation.

as a bump between two long lands is made shorter. At some point the replayed signal no longer crosses the slicing level and readout is impossible. Figure 12.36(b) shows that the same effect occurs as a land between two long bumps is made shorter. In these cases recorded frequencies have to be restricted to those which produce wavelengths long enough for the player to register. Using direct recording where, for example, lands represent a 1 and bumps represent a 0 it is clear that the length of track corresponding to a one or a zero would have to be greater than the limit at which the slicing in the player failed and this would restrict the playing time.

Figure 12.36(c) shows that if the recorded waveform is restricted to one which is DC-free, as the length of bumps and lands falls with rising density, the replay waveform simply falls in amplitude but the average voltage remains the same and so the slicer still operates correctly. It will be clear that by using a DC-free code correct slicing remains possible with much shorter bumps and lands than with direct recording. Thus in practical high-density optical disk players, including CD players, a DC-free code must be used. The output of the pickup passes to two filters. A low-pass filter removes the DC-free modulation and leaves a signal which can be used for tracking, and the high-pass filter removes the effect of debris and allows the slicer to continue to function properly. Clearly direct recording of serial data from a shift register cannot be DC-free and so it cannot be read at high density, it will not be self-clocking and it will not be resistant to errors caused by debris, and it will interfere with the operation of the servos. The solution to all these problems is to use a suitable channel code. The concepts of channel coding were discussed in Chapter 6, in which frequency shift keying (FSK) was described. In FSK it is possible to use a larger number of different discrete frequencies, for example four frequencies allow all combinations of two bits to be conveyed, eight frequencies allow all combinations of three bits to be conveyed and so on. The channel code of CD is similar in that it is the minimal case of multi-tone FSK where only a half-cycle of each of nine different frequencies is used. These frequencies are 196, 216, 240, 270, 308, 360, 430, 540 and 720 kHz and are obtained by dividing a master clock of 2.16 MHz by 11, 10, 9, 8, 7, 6, 5, 4, and 3. There are therefore nine different periods or run lengths in the CD signal, and it does not matter whether the period is the length of a land or the length of a bump. In fact the signal from a CD pickup could be inverted without making the slightest difference to the data recovery as all that is of any consequence is the time between successive zero crossings of the signal. In run-length-limited coding of this kind, the time periods are described in a relative rather than an absolute manner. Thus if half a cycle of the master clock has a period T, then the periods or run lengths of the code can be from $3T$ to $11T$. The run lengths are combined in ways which make the resulting waveform DC-free and so the slicer will function properly as the response falls at higher frequencies. The various frequencies or periods used in CD can be seen by examining the replay waveform from the pickup with an oscilloscope. Figure 12.37 shows the resultant eye pattern. It will be seen that the higher frequencies (period $3T$) have the smallest amplitude on replay. Note that the optical cut-off frequency of CD is only 1.4 MHz, and so it will be evident that the master clock frequency of 2.16 MHz cannot be recorded or reproduced. This is of no consequence in CD as it does not need to be recorded. 1.4 MHz is the frequency at which the depth of modulation has fallen to zero. As stated, the highest frequency which can be reliably recorded is about one half of the optical cut-off frequency. Frequencies above this replay with an amplitude so small that they have

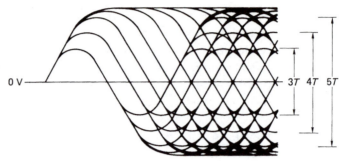

Figure 12.37 The characteristic eye pattern of EFM observed by oscilloscope. Note the reduction in amplitude of the higher-frequency components. The only information of interest is the time when the signal crosses zero.

inadequate signal-to-noise ratio. It will be seen that the highest frequency in CD is 720 kHz which is about half of 1.4 MHz. Although frequencies lower than 196kHz can be replayed easily, the clock content of lower frequencies is considered inadequate.

CD uses a coding scheme where combinations of the data bits to be recorded are represented by unique waveforms. These waveforms are created by combining various run lengths from $3T$ to $11T$ together to give a channel pattern which is $14T$ long.[18] Within the run length limits of $3T$ to $11T$, a waveform $14T$ long can have 267 different patterns. This is slightly more than the 256 combinations of eight data bits and so eight bits are represented by a waveform lasting $14T$. Some of these patterns are shown in Figure 12.38. As stated, these patterns are not polarity conscious and they could be inverted without changing the meaning.

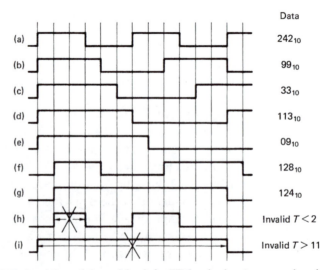

Figure 12.38 (a–g) Part of the codebook for EFM code showing examples of various run lengths from $3T$ to $11T$. (h,i) Invalid patterns which violate the run-length limits.

Not all the 14T patterns used are DC-free, some spend more time in one state than the other. The overall DC content of the recorded waveform is rendered DC-free by inserting an extra portion of waveform, known as a packing period, between the 14T channel patterns. This packing period is 3T long and may or may not contain a transition, which if it is present can be in one of three places. The packing period contains no information, but serves to control the DC content of the overall waveform.[19] The packing waveform is generated in such a way that in the long term the amount of time the channel signal spends in one state is equal to the time it spends in the other state. A packing period is placed between every pair of channel patterns and so the overall length of time needed to record eight bits is 17T. Packing periods were discussed in Chapter 6.

CD is recorded using such patterns where the lengths of bumps and lands are modulated in ideally discrete steps. The simplest way in which such patterns can be generated is to use a look-up table which converts the data bits to a control code for a programmable waveform generator. As stated, the polarity of the CD waveform is irrelevant. What matters on the disk are the lengths of the bumps or lands. The change of state in the signal sent to the cutter laser is called a transition. Clearly if a bump is being cut, it will be terminated by interrupting the light beam. If a land is being recorded, it will be terminated by allowing through the light beam. Both of these are classified as a transition, therefore it is logical for the control code to cause transitions rather than to control the waveform level as it is not concerned with the polarity of the waveform. This is conveniently achieved by controlling the cutter laser with the output waveform of a JK type bistable as shown in Figure 12.39. A bistable of this kind can be configured to have a data input and a clock input. If the data input is 0, there is no effect on the output when the clock edge arrives, whereas if the data input is 1 the output changes state when the clock edge arrives. The change of state causes a transition on the disk. If the clock has a period of T, at each channel

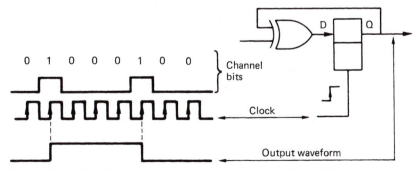

Figure 12.39 A bistable is necessary to convert a stream of channel bits to a channel-coded waveform. It is the waveform which is recorded not the channel bits.

time period or detent the output waveform will contain a transition if the control code is 1 or not if it is 0.

The control code is a binary word having fourteen bits which are known in the art as channel bits or binits. Thus a group of eight data bits is represented by a code of fourteen channel bits, hence the name of eight to fourteen modulation (EFM). The use of groups gives rise to the generic name of group code recording (GCR). It is a common misconception that the channel bits of a group code are recorded; in fact they are simply a convenient but not essential way of synthesizing a coded waveform having uniform time steps. It should be clear that channel bits cannot be recorded as they have a rate of 4.3 Mbits/s whereas the optical cut-off frequency of CD is only 1.4 MHz.

Another common misconception is that channel bits are data. If channel bits were data, all combinations of fourteen bits, or 16 384 different values could be used. In fact only 267 combinations produce waveforms which can be recorded.

In a practical CD modulator, the eight-bit data symbols to be recorded are used as the address of a look-up table which outputs a fourteen-bit channel bit pattern. As the highest frequency which can be used in CD is 720 kHz, transitions cannot be closer together than $3T$ and so successive 1s in the channel bit stream must have two or more zeros between them. Similarly transitions cannot be further apart than $11T$ or there will be insufficient clock content. Thus there cannot be more than 10 zeros between channel 1s. Whilst the look-up table can be pro-grammed to prevent code violations within the $14T$ pattern, they could occur at the junction of two successive patterns. Thus a further function of the packing period is to prevent violation of the run-length limits. If the previous pattern ends with a transition and the next begins with one, there will be no packing transition and so the $3T$ minimum requirement can be met. If the patterns either side have long run lengths, the sum of the two might exceed $11T$ unless the packing period contained a transition. In fact the minimum run-length limit could be met with $2T$ of packing, but the requirement for DC control dictated $3T$ of packing.

The coding of CD may appear complex, but this is because it was designed to offer the required playing time on a disk of restricted size. It does this by reducing the frequency of the recorded signal compared to the data frequency. Eight data bits are represented by a length of track corresponding to $17T$. The shortest run length in a conventional recording code such as MFM would be the length of one bit, and as eight bits require $17T$ of track, the length of one bit would be $17/8T$ or $2.125T$. Using the CD code the shortest run length is $3T$. Thus the highest frequency in the CD code is less than that of an MFM recording, so a density improvement of $3/2.125$ or 1.41 is obtained. Thus CD can record 41 per cent more using EFM than if it used MFM. A CD can play for 75 minutes maximum. Using MFM a CD would only play for 53 minutes.

The high-pass filtered DC-free signal from the CD pickup can be readily sliced back to a binary signal having transitions at the zero crossings. A group-coded waveform needs a suitably designed data separator to decode and deserialize the replay signal. When the disk is initially scanned, the data separator simply sees a single voltage varying with time, and it has no other information to go on whatsoever. The scanning of the disk will not necessarily be at the correct speed, and the transitions recovered will suffer from jitter. The jitter comes from two main sources. The first of these is variations in the thickness of the disk. Everyone is familiar with the illusion that the bottom of a shallow pond is moving when there are ripples in the water. In the same way, ripples in the disk thickness make the track appear to vary in speed. The second source is simply in the production tolerance to which bump edges can be made. The replication process from master to stamper will cause some slight migration of edge position, and stampers can wear in service. In order to interpret the replay waveform in the presence of jitter, use is made of the fact that transitions ideally occur at integer multiples of T. When a real transition occurs at a time other than an exact multiple of T, it can be attributed to the nearest multiple if the jitter is not too serious, and the jitter will be completely rejected. If, however the jitter is too great, the wrongly timed transition will be attributed to the incorrect detent, and the wrong pattern will be identified.

A phase-locked loop is an essential part of a practical high-density data separator. The operation of a phase-locked loop was described in section 5.9. If the input is a group-coded signal, it will contain transitions at certain multiples of the basic time period T, but not at every cycle owing to the run-length limits. The reason for the use of multiples of a basic time period in group codes is simply that a phase-locked loop can lock to such a waveform. When a transition occurs, a phase comparison can be made, but when no transition occurs, there is no phase comparison but the VCO will continue to run at the same frequency like a flywheel. The maximum run-length limit of $11T$ in CD is to ensure that the VCO does not have to run for too long between phase corrections. As a result, the VCO recreates a continuous clock from the intermittent clock content of the channel-coded signal. In a group-coded system, the VCO recreates the channel bit rate. In CD this is the only way in which the channel bit rate can be reproduced, as the disk itself cannot record the channel bit rate.

Jitter in the transition timing is handled by inserting a low-pass or averaging filter between the phase detector and the VCO and/or by increasing the division ratio in the feedback. Both of these steps increase the flywheel inertia. The VCO then runs at the average frequency obtained from many channel transitions and the jitter is substantially removed from the re-created clock. With a jitter-free continuous clock available from the VCO, the actual time at which a transition occurs can differ from the ideal by a considerable amount. When the recording was made, the transitions were intended to be spaced at multiples of the channel bit period, and the run lengths in the code ideally should be discrete. In practice the analog nature of the channel causes the run lengths to vary. A certain amount of variation can be rejected in a properly engineered channel code. The VCO is used to create windows called detents along the time axis of the replay signal. An ideal jitter-free signal would have a transition in the centre of the window, but real transitions may occur before or after the centre. As long as the variation is within the window, it is rejected, but if the jitter were so large that a transition crossed into an adjacent window, an error would occur. It was shown in Chapter 6 that the jitter window of EFM is 8/17 of a data bit. Transitions on a CD replay signal can be up to plus or minus 4/17 of a data bit period out of time before errors are caused. This jitter rejection is a requirement of the CD system because such jitter actually occurs on real disks as has been described. Indeed if it did not, the designers would have used a code with less jitter tolerance and even higher recording density. Thus it is simplistic to regard the surface of a high-density recording as a nice neat set of areas like toy bricks. In practice the manufacturing tolerances are eased so that the recording becomes cheaper even if the transitions become a little jittery. Provided the channel code can reject the jitter, the extra density makes the product more cost-effective. The deformities on real CDs are not exact multiples of the basic unit in practice. If this were a requirement they could never be sold on the consumer market. The jitter-rejection mechanism allows considerable production tolerances to be absorbed so that disks can be mass produced.

The length of a deformity on a CD master is affected not only by the duration of the record pulse, which can be as accurate as necessary, but also by the sensitivity of the resist and the intensity function of the laser. The pit which is formed in the resist is the result of the convolution of the rectangular pulse operating the modulator with the Airy function. Thus the pit will be longer than the period of the pulse would suggest. The pit edge is then subject to further position tolerance as a result of electroplating mothers and sons to create a large number of stampers. The stampers themselves will wear in service. The position of a transition is now subject to the tolerance of the cutting laser intensity function and state of focus, resist sensitivity, electroplating accuracy and wear and so the actual disk will be non-ideal.

The shortest deformity in CD is nominally $3T$ long or $3 \times 8/17$ data bits long. This can suffer nearly plus or minus 4/17 data bit periods of jitter at each end before it cannot be read properly. Thus in the worst case, where the leading edge was early and the trailing edge late, the deformity could be almost 30 per cent longer than the ideal. In typical production disks, the edge position is held a little more accurately than this theoretical limit in order to allow extra jitter in the replay process due to thickness ripple, coma due to warped disks or out-of-focus conditions.

Once the phase-locked loop has reached the lock condition, it outputs a clock whose frequency is proportional to the speed of the track. If the track speed is correct it will have the same frequency as the channel bit clock in the cutter. This clock can then be used to sample the sliced analog signal from the pickup. As can be seen from Figure 12.40, transitions nominally occur in the centre of a T period. If the samples are taken on the edge of every T period, a transition will be reliably detected as the difference between two successive samples even if it has positional jitter approaching plus or minus $T/2$. Thus the output of the sampler is a jitter-free replica of the replay signal, and in the absence of errors it will be identical to the output of the JK bistable in the cutter. The sampling clock runs at the average phase of a large number of transitions from the track. Every transition not only conveys part of the waveform representing data, but also allows the phase of the clock to be updated and so every transition can also be considered to have a synchronizing function. The $11T$ maximum run-length limit is necessary to ensure that synchronizing information for the VCO is regularly available in the replay waveform; a requirement that cannot be met by direct recording.

The information in the CD replay waveform is carried in the timing of the transitions, not in the polarity. It is thus necessary to create a polarity-independent signal from the sliced de-jittered replay waveform. This is done by differentiating the sampler output. Figure 12.40 shows that this can be achieved by a D-type latch and an exclusive-OR gate. The latch is

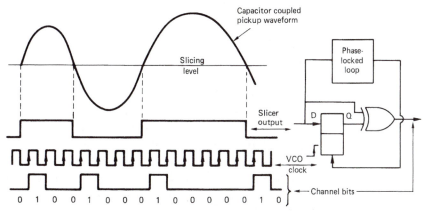

Figure 12.40 The output of the slicer is sampled at the boundary of every T period. Where successive samples differ, a channel bit 1 is generated.

clocked at the channel bit rate, and so acts as a one-bit delay. The gate compares the input and output of the delay. When they are the same, there is no transition and the gate outputs 0. When a transition passes through, the input and output of the latch will be different and the gate outputs 1. Thus some distance through the replay circuitry from the pickup, the channel bits reappear, just as they disappeared before reaching the cutter laser.

12.15   Deserialization

Decoding the stream of channel bits into data requires that the boundaries between successive 17T periods are identified. This is the process of deserialization. On the disk one 17T period runs straight into the next; there are no dividing marks. Symbol separation is performed by counting channel bit periods and dividing them by 17 from a known reference point. The three packing periods are discarded and the remaining 14T symbol is decoded to eight data bits. The reference point is provided by the synchronizing pattern which is given that name because its detection synchronizes the deserialization counter to the replay waveform.

Synchronization has to be as reliable as possible because if it is incorrect all the data will be corrupted up to the next sync pattern. Synchronization is achieved by the detection of an unique waveform periodically recorded on the track at a regular spacing. It must be unique in the strict sense in that nothing else can give rise to it, because the detection of a false sync is just as damaging as failure to detect a correct one. Clearly the sync pattern cannot be a data code value in CD as there would then be a Catch-22 situation. It would not be possible to deserialize the EFM symbols in order to decode them until the sync pattern had been detected, but if the sync pattern were a data code value, it could not be detected until the deserialization of the EFM waveform had been synchronized. Thus in a group code recording a data code value simply cannot be used for synchronizing. In any case it is undesirable and unnecessary to restrict the data code values which can be recorded; CD requires all 256 combinations of the eight-bit symbols recorded.

In practice CD synchronizes deserialization with a waveform which is unique in that it is different from any of the 256 waveforms which represent data. For reliability, the sync pattern should have the best signal-to-noise ratio possible, and this is obtained by making it one complete cycle of the lowest frequency (11T plus 11T) which gives it the largest amplitude and also makes it DC-free. Upon detection of the 2 × T_{max} waveform, the deserialization counter which divides the channel bit count by 17 is reset. This occurs on the next system clock, which is the

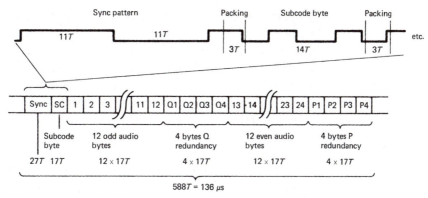

Figure 12.41 One CD data block begins with a unique sync pattern, and one subcode byte, followed by 24 audio bytes and eight redundancy bytes. Note that each byte requires 14T in EFM, with 3T packing between symbols, making 17T.

reason for the 0 in the sync pattern after the third 1 and before the merging bits. CD therefore uses forward synchronization and correctly deserialized data are available immediately after the first sync pattern is detected. The sync pattern is longer than the data symbols, and so clearly no data code value can create it, although it would be possible for certain adjacent data symbols to create a false sync pattern by concatenation were it not for the presence of the packing period. It is a further job of the packing period to prevent false sync patterns being generated at the junction of two channel symbols.

Each data block or frame in CD and MD, shown in Figure 12.41, consists of 33 symbols 17T each following the preamble, making a total of 588T or 136 μs. Each symbol represents eight data bits. The first symbol in the block is used for subcode, and the remaining 32 bytes represent 24 audio sample bytes and 8 bytes of redundancy for the error-correction system. The subcode byte forms part of a subcode block which is built up over 98 successive data frames, and this will be described in detail later in this chapter.

The channel bits which are re-created by sampling and differentiating the sliced replay waveform in time to the restored clock from the VCO are conveniently converted to parallel format for decoding in a shift register which need only have fourteen stages. The bit counter which is synchronized to the serial replay waveform by the detection of the sync pattern will output a pulse every 17T when a complete 14T pattern of channel bits is in the register. This pattern can then be transferred in parallel to the decoder which will identify the channel pattern and output the data code value.

Detection of sync in CD is simply a matter of identifying a complete cycle of the lowest recorded frequency. In practical players the sync pattern will be sliced, sampled and differentiated to channel bits along with the rest of the replay waveform. As a shift register is already present it is a matter of convenience to extend it to 23 stages so that the sync pattern can be detected by continuously examining the parallel output as the patterns from the track shift by. The pattern will be detected by a combination of logic gates which will only output a 'true' value when the shift register contains 10000000000100000000001 in the correct place.

This is not a bit pattern which exists on the disk; the disk merely contains two maximum run-lengths in series and it does not matter whether these are a bump followed by a land or a land followed by a bump. The sliced replay waveform cannot be sampled at the correct frequency until the VCO has locked and this requires the T rate synchronizing information from a prior length of data track. If the VCO were not locked, the sync waveform would be sampled into the wrong number of periods and would not be detected. Following sampling, the replay signal is differentiated so that transitions of either direction produce a channel bit 1.

Figure 12.42 shows an overall block diagram of the record modulation scheme used in CD mastering and the corresponding replay system or data separator. The input to the record channel coder consists of sixteen-bit audio samples which are divided in two to make symbols of eight bits.

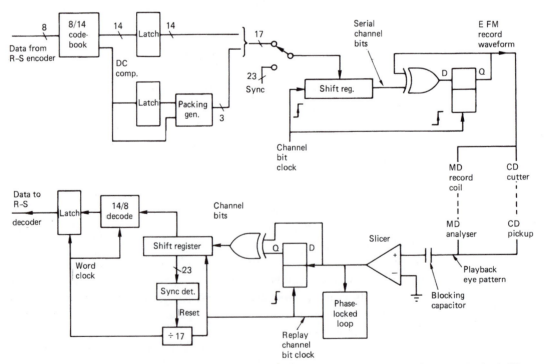

Figure 12.42 Overall block diagram of the EFM encode/decode process. A MiniDisc will contain both. A CD player only has the decoder; the encoding is in the mastering cutter.

These symbols are used in the error-correction system which interleaves them and adds redundant symbols. For every twelve audio symbols, there are four symbols of redundancy, but the channel coder is not concerned with the sequence or significance of the symbols and simply records their binary code values.

Symbols are provided to the coder in eight-bit parallel format, with a symbol clock. The symbol clock is obtained by dividing down the 4.3218 MHz T rate clock by a factor of 17. Each symbol is used to address the look-up table which outputs a corresponding fourteen-channel bit pattern in parallel into a shift register. The T rate clock then shifts the channel bits along the register. The look-up table also outputs data corresponding to the digital sum value (DSV) of the fourteen-bit symbol to the packing generator. The packing generator determines if action is needed between symbols to control DC content. The packing generator checks for run-length violations and potential false sync patterns. As a result of all the criteria, the packing generator loads three channel bits into the space between the symbols, such that the register then contains fourteen-bit symbols with three bits of packing between them. At the beginning of each frame, the sync pattern is loaded into the register just before the first symbol is looked up in such a way that the packing bits are correctly calculated between the sync pattern and the first symbol.

A channel bit one indicates that a transition should be generated, and so the serial output of the shift register is fed to the JK bistable along with the T rate clock. The output of the JK bistable is the ideal channel coded waveform containing transitions separated by $3T$ to $11T$. It is a self-clocking, run-length-limited waveform. The channel bits and the T rate clock have done their job of changing the state of the JK bistable and do not pass further on. At the output of the JK the sync pattern is simply two $11T$ run lengths in series.

At this stage the run-length-limited waveform is used to control the acousto-optic modulator in the cutter. This actually results in pits which are slightly too long and lands which are too short because of the convolution of the record waveform with the Airy function which was mentioned above. As the cutter spot is about 0.4 μm across, the pit edges in the resist are moved slightly. Thus although the ideal waveform is created in the encoding circuitry, having integer multiples of T between transitions, the pit structure is non-ideal and pit edges are not located at exact multiples of a basic distance. The duty cycle of the pits and lands is not exactly 50 per cent and the replay waveform will have a DC offset. This is of no consequence in CD as the channel code is known to be DC-free and an equivalent offset can be generated in the slicing level of the player such that the duty cycle of the slicer output becomes 50 per cent.

The resist master is developed and used to create stampers. The resulting disks can then be replayed. The track velocity of a given CD is constant, but the rotational speed depends upon the radius. In order to get into lock, the disk must be spun at roughly the right track speed. This is done using the run-length limits of the recording. The pick-up is focused and the tracking is enabled. The replay waveform from the pick-up is passed through a high-pass filter to remove level variations due to contamination and sliced to return it to a binary waveform. The slicing level is self-adapting as Figure 12.43 shows, so that a 50 per cent duty cycle is obtained. The slicer output is then sampled by the unlocked VCO running at approximately T rate. If the disk is running too slowly, the longest run length on the disk will appear as more than $11T$, whereas if the disk is running too fast, the shortest run length will appear as less than $3T$. As a result, the disk speed can be brought to approximately the

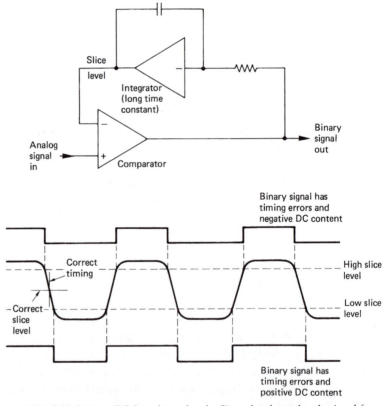

Figure 12.43 Self-slicing a DC-free channel code. Since the channel code signal from the disk is band limited, it has finite rise times, and slicing at the wrong level (as shown here) results in timing errors, which cause the data separator to be less reliable. As the channel code is DC-free, the binary signal when correctly sliced should integrate to zero. An incorrect slice level gives the binary output a DC content and, as shown here, this can be fed back to modify the slice level automatically.

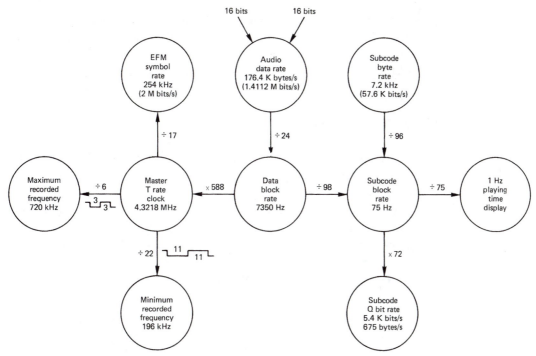

Figure 12.44 CD timing structure.

right speed and the VCO will then be able to lock to the clock content of the EFM waveform from the slicer. Once the VCO is locked, it will be possible to sample the replay waveform at the correct T rate. The output of the sampler is then differentiated and the channel bits reappear and are fed into the shift register. The sync pattern detector will then function to reset the deserialization counter which allows the $14T$ symbols to identified. The $14T$ symbols are then decoded to eight bits in the reverse coding table.

Figure 12.44 reveals the timing relationships of the CD format. The sampling rate of 44.1 kHz with sixteen-bit words in left and right channels results in an audio data rate of 176.4 kb/s (k = 1000 here, not 1024). Since there are 24 audio bytes in a data frame, the frame rate will be:

$$\frac{176.4}{24} \, \text{kHz} \, = \, 7.35 \, \text{kHz}$$

If this frame rate is divided by 98, the number of frames in a subcode block, the subcode block or sector rate of 75 Hz results. This frequency can be divided down to provide a running-time display in the player. Note that this is the frequency of the wavy grooves in recordable MDs.

If the frame rate is multiplied by 588, the number of channel bits in a frame, the master clock-rate of 4.3218 MHz results. From this the maximum and minimum frequencies in the channel, 720 kHz and 196 kHz, can be obtained using the run-length limits of EFM.

12.16   Error-correction strategy

This section discusses the track structure of CD in detail. The track structure of MiniDisc is based on that of CD and the differences will be noted in the next section.

Each sync block was seen in Figure 12.41 to contain 24 audio bytes, but these are non-contiguous owing to the extensive interleave.[20-22] There are a number of interleaves used in CD, each of which has a specific purpose. The full interleave structure is shown in Figure 12.45. The first stage of interleave is to introduce a delay between odd and even samples. The

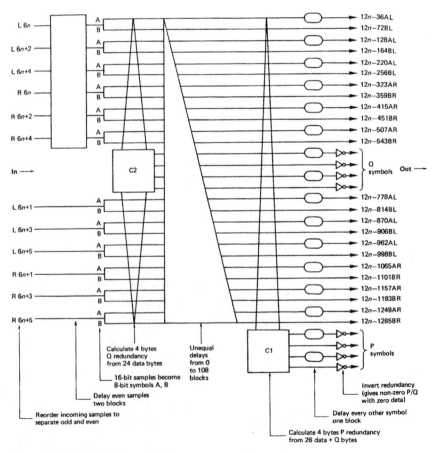

Figure 12.45   CD interleave structure.

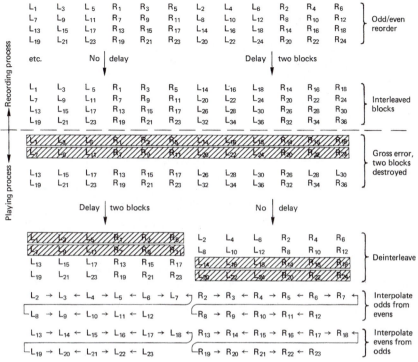

Figure 12.46 Odd/even interleave permits the use of interpolation to conceal uncorrectable errors.

effect is that uncorrectable errors cause odd samples and even samples to be destroyed at different times, so that interpolation can be used to conceal the errors, with a reduction in audio bandwidth and a risk of aliasing. The odd/even interleave is performed first in the encoder, since concealment is the last function in the decoder. Figure 12.46 shows that an odd/even delay of two blocks permits interpolation in the case where two uncorrectable blocks leave the error-correction system.

Left and right samples from the same instant form a sample set. As the samples are sixteen bits, each sample set consists of four bytes, AL, BL, AR, BR. Six sample sets form a 24-byte parallel word, and the C2 encoder produces four bytes of redundancy Q. By placing the Q symbols in the centre of the block, the odd/even distance is increased, permitting interpolation over the largest possible error burst. The 28 bytes are now subjected to differing delays, which are integer multiples of four blocks. This produces a convolutional interleave, where one C2 codeword is stored in 28 different blocks, spread over a distance of 109 blocks.

At one instant, the C2 encoder will be presented with 28 bytes which have come from 28 different codewords. The C1 encoder produces a further four bytes of redundancy P. Thus the C1 and C2 codewords are

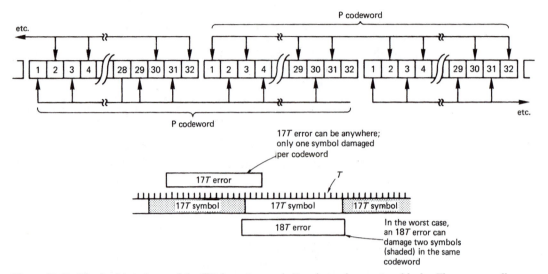

Figure 12.47 The final interleave of the CD format spreads P codewords over two blocks. Thus any small random error can only destroy one symbol in one codeword, even if two adjacent symbols in one block are destroyed. Since the P code is optimized for single-symbol error correction, random errors will always be corrected by the C1 process, maximizing the burst-correcting power of the C2 process after de-interleave.

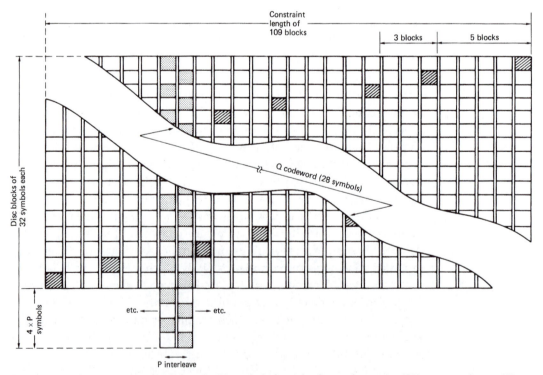

Figure 12.48 Owing to cross-interleave, the 28 symbols from the Q encode process (C2) are spread over 109 blocks, shown hatched. The final interleave of P codewords (as in Figure 12.47) is shown stippled. The result of the latter is that Q codeword has 5, 3, 5, 3 spacing rather than 4, 4.

produced by crossing an array in two directions. This is known as cross-interleaving.

The final interleave is an odd/even output symbol delay, which causes P codewords to be spread over two blocks on the disk as shown in Figure 12.47. This mechanism prevents small random errors destroying more than one symbol in a P codeword. The choice of eight-bit symbols in EFM assists this strategy. The expressions in Figure 12.45 determine how the interleave is calculated. Figure 12.48 shows an example of the use of these expressions to calculate the contents of a block and to demonstrate the cross-interleave.

The calculation of the P and Q redundancy symbols is made using Reed–Solomon polynomial division. The P redundancy symbols are primarily for detecting errors, to act as pointers or error flags for the Q system. The P system can, however, correct single-symbol errors.

12.17   Track layout of MD

MD uses the same channel code and error-correction interleave as CD for simplicity and the sectors are exactly the same size. The interleave of CD is convolutional, which is not a drawback in a continuous recording. However, MD uses random access and the recording is discontinuous. Figure 12.49 shows that the convolutional interleave causes codewords to run between sectors. Rerecording a sector would prevent error correction in the area of the edit. The solution is to use a buffering zone in the area of an edit where the convolution can begin and end. This is the job of the link sectors. Figure 12.50 shows the layout of data on a recordable MD. In each cluster of 36 sectors, 32 are used for encoded audio data. One is used for subcode and the remaining three are link sectors. The cluster is the

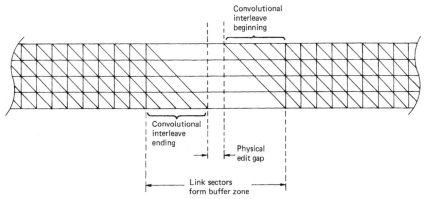

Figure 12.49 The convolutional interleave of CD is retained in MD, but buffer zones are needed to allow the convolution to finish before a new one begins, otherwise editing is impossible.

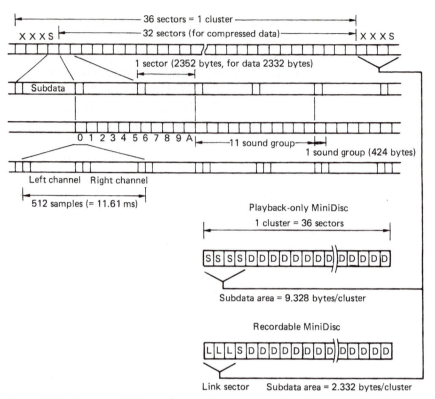

Figure 12.50 Format of MD uses clusters of sectors including link sectors for editing. Prerecorded MDs do not need link sectors, so more subcode capacity is available. The ATRAC coder of MD produces the sound groups shown here.

minimum data quantum which can be recorded and represents just over two seconds of decoded audio. The cluster must be recorded continuously because of the convolutional interleave. Effectively the link sectors form an edit gap which is large enough to absorb both mechanical tolerances and the interleave overrun when a cluster is rewritten. One or more clusters will be assembled in memory before writing to the disk is attempted.

Prerecorded MDs are recorded at one time, and need no link sectors. In order to keep the format consistent between the two types of MiniDisc, three extra subcode sectors are made available. As a result it is not possible to record the entire audio and subcode of a prerecorded MD onto a recordable MD because the link sectors cannot be used to record data.

The ATRAC coder produces what are known as sound groups (see Chapter 5). Figure 12.50 shows that these contain 212 bytes for each of the two audio channels and are the equivalent of 11.6 ms of real-time audio. Eleven of these sound groups will fit into two standard CD sectors with 20 bytes to spare. The 32 audio data sectors in a cluster thus contain a total of $16 \times 11 = 176$ sound groups.

12.18   CD subcode

Subcode is essentially an auxiliary data stream which is merged with the audio samples, and which has numerous functions. One of these is to assist in locating the beginning of the different musical pieces on a disk, and providing a catalogue of their location on the disk and their durations. A further vital function is to convey the status of pre-emphasis in the recording, so that de-emphasis can be automatically selected in the player. The subcode information in CD is conveyed by including an extra byte, which corresponds to one EFM symbol, in the main frame structure. As the format of the disk is standardized, the player is designed to route the subcode byte in the frame to a different destination from that of the audio sample bytes. The separation is based upon the physical position of the subcode byte in the frame. The player uses the sync pattern at the beginning of the frame to reset a byte count so that it always knows how far through the frame it is. As a result, subcode bytes will be separated from the data stream at frame frequency.

It has been shown that there are 98 bytes in a subcode block, since this results in a subcode block rate of exactly 75 Hz. This frequency can be used to run the playing-time display.

It is necessary for the player to know when a new subcode block is beginning. This is the function of the subcode sync patterns which are placed in the subcode byte position of two successive frames. There are more than 256 legal fourteen-bit patterns in EFM, and two of these additional legal channel-bit patterns are used for subcode-block synchronizing. The EFM decoder will be able to distinguish them from the patterns used to represent subcode-data bytes. For this reason it is impossible to describe the subcode sync patterns by a byte, and they have to be specified as fourteen channel bits.

Figure 12.51 shows the subcode sync patterns, and illustrates the contents of the subcode block. After the subcode sync patterns, there are 96 bytes in the block. The block is arranged as eight 96-bit words, labelled P Q R S T U V and W. The choice of labelling is unfortunate because the letters P and Q have already been used to describe the redundancy in the error-correction system. The subcode P and Q data have absolutely nothing to do with that. The eight words are quite independent, and each subcode byte in a disk frame contains one bit from each word. This is a form of interleaving which reduces the damage done to a particular word by an error.

The P data word is used to denote the start of specific bands (having the same meaning as the bands on a vinyl disk) in the sound recorded. The entire word is recorded as data ones during the start-flag period. It can be used even where there is no audible pause in the music, since the start point is defined as where the P data become zeros again. The CD

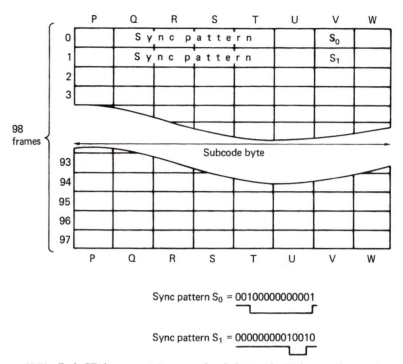

Sync pattern S_0 = 00100000000001

Sync pattern S_1 = 00000000010010

Figure 12.51 Each CD frame contains one subcode byte. Afer 98 frames, the structure above will repeat. Each subcode byte contains 1 bit from eight 96 bit words following the two synchronizing patterns. These patterns cannot be expressed as a byte, because they are 14 bit EFM patterns additional to those which describe the 256 combinations of eight data bits.

standard calls for a minimum of two seconds of start flag to be recorded. This seems wasteful, but it allows a very simple player to recognize the beginning of a piece easily by skipping tracks. The fact that every bit is a one means that it is not necessary to wait for subcode block sync to be found before finding pause status on the disk track. The two-second-flag period means that the status will be seen a few tracks in advance of the actual start-point, helping to prevent the pickup from overshooting. If a genuine pause exists in the music, the start flag may be extended to the length of the pause if it exceeds two seconds. Again, for the benefit of simple players, the start flag alternates on and off at 2 Hz in the lead-out area at the end of the recording.

At the time of writing the only other defined subcode data word is the Q word. This word has numerous modes and uses which can be taken advantage of by CD players with greater processing and display capability.

Figure 12.52 shows the structure of the Q subcode word. In the 96 bits following the sync patterns, there are two four-bit words for control, a

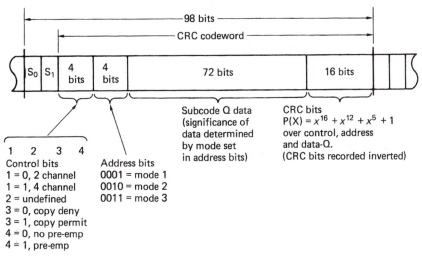

Figure 12.52 The structure of the Q data block. The 72-bit data can be interpreted in three ways determined by the address bits.

72-bit data block, and a sixteen-bit CRC character which makes all 96 bits a codeword.

The first four-bit control word contains flags specifying the number of audio channels encoded, to permit automatic decoding of four channel-disks, the copy-prohibit status and the pre-emphasis status. Since de-emphasis is often controlled by a relay or electronic switch in the analog stages of the player, the pre-emphasis status is only allowed to change during a P code start flag.

The second four-bit word determines the meaning of the subsequent 72-bit block. There can be three meanings: mode 1, which tells the player the number and start times of the bands on the disk; mode 2, which carries the disk catalogue number; and mode 3, which carries the ISRC (International Standard Recording Code) of each band. Of all the subcode blocks on a disk, the mode 1 blocks are by far the most common.

Mode 1 has two major functions. During the lead-in track it contains a table of contents (TOC), listing each piece of music and the absolute playing time when it starts. During the music content of the disk, it contains running time.

Figure 12.53 shows that the 72-bit block is subdivided into nine bytes, one of which is unused and permanently zero. Each byte represents two hexadecimal digits where not all codes are valid. The first byte in the block is the music number (MNR), which specifies the number of the track on the disk; where in this context 'track' corresponds to the bands on a vinyl disk. The tracks are numbered from one upwards, and the track number of 00 indicates that the pickup is in the lead-in area and that the rest of the block contains an entry in the table of contents.

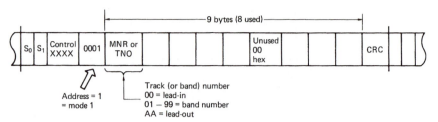

Figure 12.53 General format of Q subcode frame in mode 1. There are eight unused bits, leaving eight active bytes. First byte is music or track number, which determines meaning of remaining bytes.

First disk: five bands 1–5

Frame no.	Point	Point (min,sec,frame)		
n, $n+1$, $n+2$	01	00	02	00
$n+3$, $n+4$, $n+5$	02	12	09	10
$n+6$, $n+7$, $n+8$	03	24	11	20
$n+9$, $n+10$, $n+11$	04	36	59	74
$n+12$, $n+13$, $n+14$	05	48	50	22
$n+15$, $n+16$, $n+17$	A0	01	00	00
$n+18$, $n+19$, $n+20$	A1	05	00	00
$n+21$, $n+22$, $n+23$	A2	59	27	31

Second disk: six bands 6–11

n, $n+1$, $n+2$	06	00	02	10
$n+3$, $n+4$, $n+5$	07	10	03	20
$n+6$, $n+7$, $n+8$	08	20	05	11
$n+9$, $n+10$, $n+11$	09	30	04	03
$n+12$, $n+13$, $n+14$	10	40	02	19
$n+15$, $n+16$, $n+17$	11	50	59	70
$n+18$, $n+19$, $n+20$	A0	06	00	00
$n+21$, $n+22$, $n+23$	A1	11	00	00
$n+24$, $n+25$, $n+26$	A2	59	20	74

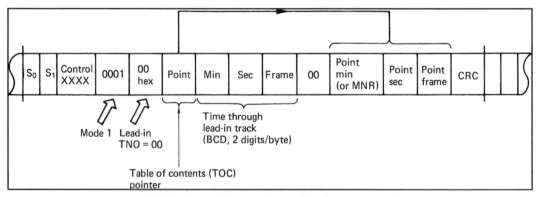

Figure 12.54 During lead-in TNO is zero and Q subcode builds up a table of contents using numbered points with starting times. For multidisk sets, the band numbering can continue from one disk to the next, and there are point-limit codes A0 and A1 which specify the range of bands on a given disk. The example of a two-disk set is given, with five bands on the first disk and six bands on the second. Point = 00–99, point = music number, and point (min, sec, frame) denotes absolute starting time of that music number. This forms an entry in TOC. Point = A0 hex, point min byte = music number of *first* band on this disk, denotes beginning MNR of TOC. Point = A1 hex, point min byte = music number of *last* band on this disk, denotes end MNR of TOC. Point = A2 hex, point (min, sec, frame), denotes absolute starting time of lead-out track.

The table of contents is built up by listing points in time where each track starts. One point can be described in one subcode block. Figure 12.54 shows that the second byte of the block is the point number. The absolute time at which that point will be reached after the start of the first track is contained in the last three bytes as point minutes, point seconds and point frames. These bytes are two BCD digits, where the maximum value of point frame is 74. As there is only error detection in the Q data, the point is repeated in three successive subcode blocks. The number of points allowed is 99, but the track numbering can continue through a set of disks. For example, in a two-disk set, there could be five tracks, 1 to 5, on the first disk, and six tracks, 6 to 11, on the second disk. Clearly the first point on the second disk is going to be point 6, and to prevent the player fruitlessly looking for points that are absent, the point range is specified.

If the point byte has the value A0 hex, the point-minute byte contains the number of the first track on the disk, which in the example given would be 6. If the point byte has the value A1 hex, the point-minute byte contains the number of the last track on the disk, which would here be 11. A further point is specified, which is the absolute running time of the start of the lead-out track, which uses the point code of A2 hex. These three points come after the actual music start points. During the lead-in track, the running time is counted by the minute, second and frame bytes in the block.

If the first byte of the block is between 00 and 99, the block is in a music track, and the meaning shown in Figure 12.55 applies. The running time is given in three ways. Minute, second and frame are the running time from the start of that track, and A(bsolute)min, Asec and Aframe are the

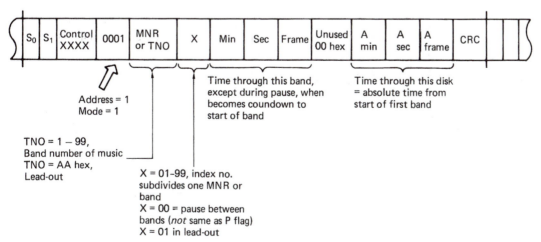

Figure 12.55 During music bands TNO. is 01–99, and subcode shows time through band and time through disk. The former counts down during pause. Each band can be subdivided by index count X.

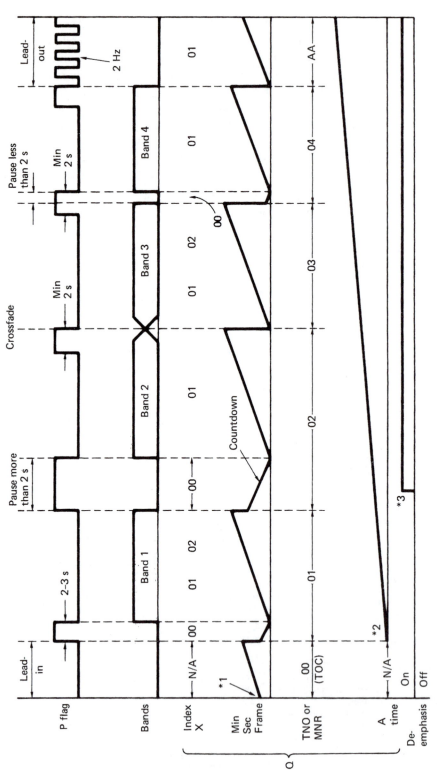

Figure 12.56 The relationship of P and Q subcode timing to the music bands. P flag is never less than 2 s between bands, whereas index reflects actual pause, and vanishes at a crossfade. Time counts down during index 00.
*1: lead-in time does not have to start from zero;
*2: A time must start from zero;
*3: de-emphasis can only change during pause of 2 s or more.

running time from the start of the first track on the disk. The third running-time mode employs the index or X byte. When this is zero, it denotes a pause, which corresponds to the P subcode's being 1. During this pause, which precedes the start of a track, the running time counts down to zero, so that a player can display the time to go before a track starts to play. The absolute time is unaffected by this mode. Non-zero values of X denote a subdivision of the track into shorter sections. This would be useful to locate individual phrases on a language-course disk, or the individual effects on a sound-effects disk. Figure 12.56 shows an example of the use of P and Q subcode and the relationship between them and the music bands.

Mode 2 of the Q subcode allows the recording of the barcode number of the disk, and is denoted by the address code of 2 in the block as shown in Figure 12.57. The 52-bit barcode, along with twelve zeros and a continuation of the absolute frame count, are protected by the CRC character. If this mode is used, it should show up at least once in every 100 subcode blocks and the contents of each block should be identical. The use of the mode is not compulsory.

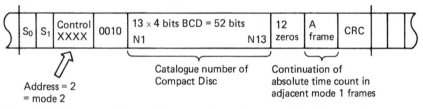

Figure 12.57 In mode 2, the catalogue number can be recorded. This must always be the same throughout the disk, and must appear in at least one out of a hundred successive blocks.

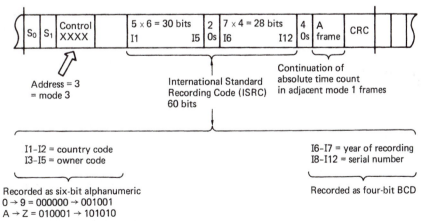

Figure 12.58 ISRC format in mode 3 allows each band to have a different code. All mode 3 frames must be the same within same TNO. Must appear in at least one out of every hundred successive blocks. Not present in lead-in or lead-out tracks.

Mode 3 of Q subcode is similar to mode 2, except that a code number can be allocated to each track on the disk. Figure 12.58 shows that the ISR code requires five alphanumeric characters of six bits each and seven BCD characters of four bits each. Again the mode is optional but, if used, the mode 3 subcode block must occur at least once in every 100 blocks.

The R to W subcode is currently not standardized, but proposed uses for this data include a text display which would enable the words of a song to appear on a monitor in synchronism with the sound played from the disk. A difficulty in this area is the requirement to support not only the kind of alphanumerics in which this book is written but also the complex Kanji characters which would be needed for the Japanese market.

12.19   MD table of contents

The TOC of the pre-recorded MiniDisc is basically similar to the CD TOC as it performs the same function. Recordable MiniDiscs have a different approach. Recordable MD is more like a hard disk than a real-time audio recorder, and the buffer memory allows continuous audio listening from records which are fragmented across the disk surface. Thus the UTOC (user table of contents) of the recordable MD is more like the directory of a data disk (see Chapter 10). UTOC contains one entry for each numbered recorded item which lists the physical cluster addresses at which the data for that item are recorded. When the user selects the number of an item, the player reads the UTOC in order to locate the data addresses. Item numbers are contiguous, so if an item is deleted or if two items are merged, the numbering scheme beyond will move up by one. It is not necessary to actually erase unwanted recordings. Instead the directory entry is deleted and then as far as the system is concerned the recording no longer exists and the clusters it uses are available for overwriting. Figure 12.59 shows some examples of UTOC operations. Despite the internal complexity, the disk mapping is taken care of by a microprocessor and the user simply selects item numbers.

12.20   CD player structure

The physics of the manufacturing process and the readout mechanism have been described, along with the format on the disk. Here, the details of actual CD and MD players will be explained. One of the design constraints of the CD and MD formats was that the construction of players should be straightforward, since they were to be mass-produced.

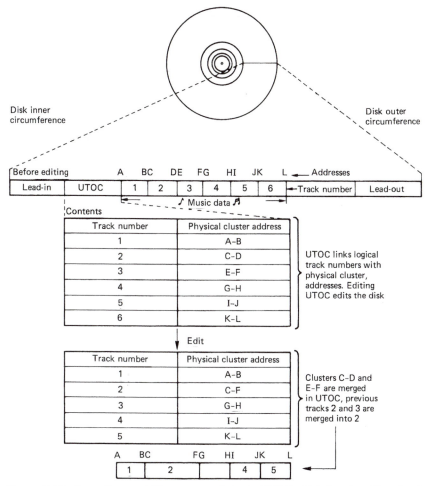

Figure 12.59 Recordings on MD are accessed via the UTOC which functions like a hard disk index. Editing is achieved simply by altering UTOC.

Figure 12.60 shows the block diagram of a typical CD player, and illustrates the essential components. The most natural division within the block diagram is into the control/servo system and the data path. The control system provides the interface between the user and the servo mechanisms, and performs the logical interlocking required for safety and the correct sequence of operation.

The servo systems include any power-operated loading drawer and chucking mechanism, the spindle-drive servo, and the focus and tracking servos already described.

Power loading is usually implemented on players where the disk is placed in a drawer. Once the drawer has been pulled into the machine, the disk is lowered onto the drive spindle, and clamped at the centre, a process known as chucking. In the simpler top-loading machines, the

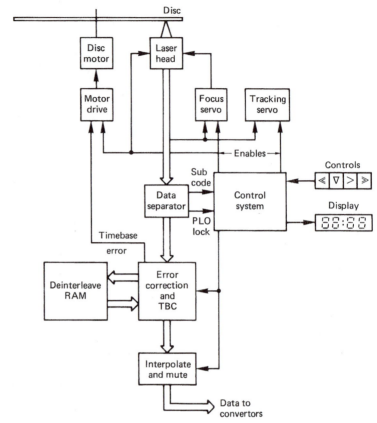

Figure 12.60 Block diagram of CD player showing the data path (broad arrow) and control/servo systems.

disk is placed on the spindle by hand, and the clamp is attached to the lid so that it operates as the lid is closed.

The lid or drawer mechanisms have a safety switch which prevents the laser operating if the machine is open. This is to ensure that there can be no conceivable hazard to the user. In actuality there is very little hazard in a CD pickup. This is because the beam is focused a few millimetres away from the objective lens, and beyond the focal point the beam diverges and the intensity falls rapidly. It is almost impossible to position the eye at the focal point when the pickup is mounted in the player, but it would be foolhardy to attempt to disprove this.

The data path consists of the data separator, timebase correction and the de-interleaving and error-correction process followed by the error-concealment mechanism. This results in a sample stream which is fed to the convertors.

The data separator which converts the readout waveform into data was detailed in the description of the CD channel code. LSI chips have been developed to perform the data-separation function: for example, the Philips SAA 7010 or the Sony CX 7933. The separated output from both of these consists of subcode bytes, audio samples, redundancy and a clock. The data stream and the clock will contain speed variations due to disk run-out and chucking tolerances, and these have to be removed by a timebase corrector.

The timebase corrector is a memory addressed by counters which are arranged to overflow, giving the memory a ring structure as described in Chapter 1. Writing into the memory is done by using clocks from the data separator whose frequency rises and falls with run-out, whereas reading is done using a crystal-controlled clock, which removes speed variations from the samples, and makes wow and flutter unmeasurable. The timebase-corrector will only function properly if the two addresses are kept apart. This implies that the long-term data rate from the disk must equal the crystal-clock rate. The disk speed must be controlled to ensure that this is always true, and there are two contrasting ways in which it can be done.

The data-separator clock counts samples from the disk. By phase-comparing this clock with the crystal reference, the phase error can be used to drive the spindle motor. This system was used in the Sony CDP-101, where the principle was implemented with a CX-193 chip, originally designed for DC turntable motors. The data-separator signal replaces the feedback signal which would originally have come from a toothed wheel on the turntable.

The alternative approach is to analyse the address relationship of the timebase corrector. If the disk is turning too fast, the write address will move towards the read address; if the disk is turning too slowly, the write address moves away from the read address. Subtraction of the two addresses produces an error signal which can be fed to the motor. The TBC RAM controller produces the motor-control signal. In these systems, and in all CD players, the speed of the motor is unimportant. The important factor is that the sample rate is correct, and the system will drive the spindle at whatever speed is necessary to achieve the correct rate. As the disk cutter produces constant bit density along the track by reducing the rate of rotation as the track radius increases, the player will automatically duplicate that speed reduction. The actual linear velocity of the track will be the same as the velocity of the cutter, and although this will be constant for a given disk, it can vary between 1.2 and 1.4 m/s on different disks.

These speed-control systems can only operate when the data separator has phase-locked, and this cannot happen until the disk speed is almost correct. A separate mechanism is necessary to bring the disk up to roughly the right speed. One way of doing this is to make use of the run-length limits of the channel code. Since transitions closer than $3T$ and further apart than $11T$ are not present, it is possible to estimate the disk speed by analysing the run lengths. The period between transitions should be from 694 ns to 2.55 μs. During disk run-up the periods between transitions can be measured, and if the longest period found exceeds 2.55 μs, the disk must be turning too slowly, whereas if the shortest period is less than 694 ns, the disk must be turning too fast. Once the data separator locks up, the coarse speed control becomes redundant. The method relies upon the regular occurrence of maximum and minimum run lengths in the channel. Synchronizing patterns have the maximum run length, and occur regularly. The description of the disk format showed that the C1 and C2 redundancy was inverted. This injects some ones into the channel even when the audio is muted. This is the situation during the lead-in track – the very place that lock must be achieved. The presence of the table of contents in subcode during the lead-in also helps to produce a range of run lengths.

Owing to the use of constant linear velocity, the disk speed will be wrong if the pickup is suddenly made to jump to a different radius using manual search controls. This may force the data separator out of lock, and the player will mute briefly until the correct track speed has been restored, allowing the PLO to lock again. This can be demonstrated with most players, since it follows from the format.

Following data separation and timebase correction, the error-correction and de-interleave processes take place. Because of the cross-interleave system, there are two opportunities for correction, first, using the C1 redundancy prior to deinterleaving, and second, using the C2 redundancy after de-interleaving. In Chapter 6 it was shown that interleaving is designed to spread the effects of burst errors among many different codewords, so that the errors in each are reduced. However, the process can be impaired if a small random error, due perhaps to an imperfection in manufacture, occurs close to a burst error caused by surface contamination. The function of the C1 redundancy is to correct single-symbol errors, so that the power of interleaving to handle bursts is undiminished, and to generate error flags for the C2 system when a gross error is encountered.

The EFM coding is a group code which means that a small defect which changes one channel pattern into another will have corrupted up to eight data bits. In the worst case, if the small defect is on the boundary between two channel patterns, two successive bytes could be corrupted. However, the final odd/even interleave on encoding ensures that the two bytes damaged will be in different C1 codewords; thus a random error can never corrupt two bytes in one C1 codeword, and random errors are therefore always correctable by C1. From this it follows that the maximum size of a defect considered random is $17T$ or 3.9 μs. This corresponds to about a 5 μm length of the track. Errors of greater size are, by definition, burst errors.

The de-interleave process is achieved by writing sequentially into a memory and reading out using a sequencer. The RAM can perform the function of the timebase-corrector as well. The size of memory necessary follows from the format; the amount of interleave used is a compromise between the resistance to burst errors and the cost of the de-interleave memory. The maximum delay is 108 blocks of 28 bytes, and the minimum delay is negligible. It follows that a memory capacity of $54 \times 28 = 1512$ bytes is necessary. Allowing a little extra for timebase error, odd/even interleave and error flags transmitted from C1 to C2, the convenient capacity of 2048 bytes is reached.

The C2 decoder is designed to locate and correct a single-symbol error, or to correct two symbols whose locations are known. The former case occurs very infrequently, as it implies that the C1 decoder has miscorrected. However, the C1 decoder works before de-interleave, and there is no control over the burst-error size that it sees. There is a small but finite probability that random data in a large burst could produce the same syndrome as a single error in good data. This would cause C1 to miscorrect, and no error flag would accompany the miscorrected symbols. Following de-interleave, the C2 decode could detect and correct the miscorrected symbols as they would now be single-symbol errors in many codewords. The overall miscorrection probability of the system is thus quite minute. Where C1 detects burst errors, error flags will be attached to all symbols in the failing C1 codeword. After de-interleave in the memory, these flags will be used by the C2 decoder to correct up to two corrupt symbols in one C2 codeword. Should more than two flags appear in one C2 codeword, the errors are uncorrectable, and C2 flags the entire codeword bad, and the interpolator will have to be used. The final odd/even sample de-interleave makes interpolation possible because it displaces the odd corrupt samples relative to the even corrupt samples.

If the rate of bad C2 codewords is excessive, the correction system is being overwhelmed, and the output must be muted to prevent unpleasant noise. Unfortunately digital audio cannot be muted by simply switching the sample stream to zero, since this would produce a click. It is necessary to fade down to the mute condition gradually by multiplying sample values by descending coefficients, usually in the form of a half-cycle of a cosine wave. This gradual fade-out requires some advance warning, in order to be able to fade out before the errors arrive. This is achieved by feeding the fader through a delay. The mute status bypasses the delay, and allows the fade-out to begin sufficiently in advance of the error. The final output samples of this system will be either correct, interpolated or muted, and these can then be sent to the convertors in the player.

The power of the CD error correction is such that damage to the disk generally results in mistracking before the correction limit is reached.

There is thus no point in making it more powerful. CD players vary tremendously in their ability to track imperfect disks and expensive models are not automatically better. It is generally a good idea when selecting a new player to take along some marginal disks to assess tracking performance.

The control system of a CD player is inevitably microprocessor-based, and as such does not differ greatly in hardware terms from any other microprocessor-controlled device. Operator controls will simply interface to processor input ports and the various servo systems will be enabled or overridden by output ports. Software, or more correctly firmware, connects the two. The necessary controls are Play and Eject, with the addition in most players of at least Pause and some buttons which allow rapid skipping through the program material.

Although machines vary in detail, the flowchart of Figure 12.61 shows the logic flow of a simple player, from start being pressed to sound emerging. At the beginning, the emphasis is on bringing the various servos into operation. Towards the end, the disk subcode is read in order to locate the beginning of the first section of the program material.

When track-following, the tracking-error feedback loop is closed, but for track-crossing, in order to locate a piece of music, the loop is opened, and a microprocessor signal forces the laser head to move. The tracking error becomes an approximate sinusoid as tracks are crossed. The cycles of tracking error can be counted as feedback to determine when the correct number of tracks have been crossed. The 'mirror' signal obtained when the read-out spot is half a track away from target is used to brake pickup motion and re-enable the track-following feedback.

The control system of a professional player for broadcast use will be more complex because of the requirement for accurate cueing. Professional machines will make extensive use of subcode for rapid access, and in addition are fitted with a hand-operated rotor which simulates turning a vinyl disk by hand. In this mode the disk constantly repeats the same track by performing a single track-jump once every revolution. Turning the rotor moves the jump point to allow a cue point to be located. The machine will commence normal play from the cue point when the start button is depressed or from a switch on the audio fader. An interlock is usually fitted to prevent the rather staccato cueing sound from being broadcast.

CD changers running from 12 volts are available for remote installation in cars. These can be fitted out of sight in the luggage trunk and controlled from the dashboard. The RAM buffering principle can be employed to overcome skipping as in MD, but a larger memory is required.

Personal portable CD players are available, but these have not displaced the personal analog cassette in the youth market. This is possibly due to the cost of player and disks relative to the Compact Cassette. The Compact Cassette is also more immune to rough handling. Personal CD players are

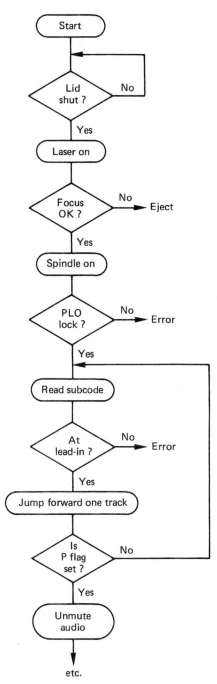

Figure 12.61 Simple flowchart for control system, focuses, starts disk, and reads subcode to locate first item of programme material.

more of a niche market, being popular with professionals who are more likely to have a quality audio system and CD collection. The same CDs can then be enjoyed whilst travelling.

12.21   MD recorder/player structure

Figure 12.62 shows the block diagram of an MD player. There is a great deal of similarity with a conventional CD player in the general

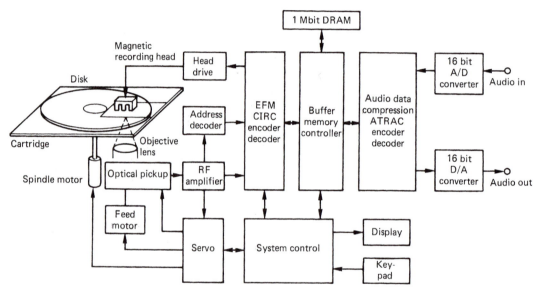

Figure 12.62   MiniDisc block diagram. See text for details.

arrangement. Focus, tracking and spindle servos are basically the same, as is the EFM and Reed–Solomon replay circuitry. The main difference is the presence of recording circuitry connected to the magnetic head, the large buffer memory and the data reduction codec. The figure also shows the VLSI chips developed by Sony for MD. Whilst MD machines are capable of accepting 44.1 kHz PCM or analog audio in real time, there is no reason why a twin-spindle machine should not be made which can dub at four to five times normal speed.

12.22   Structure of a DVD player

Figure 12.63 shows the block diagram of a typical DVD player, and illustrates the essential components. The most natural division within the

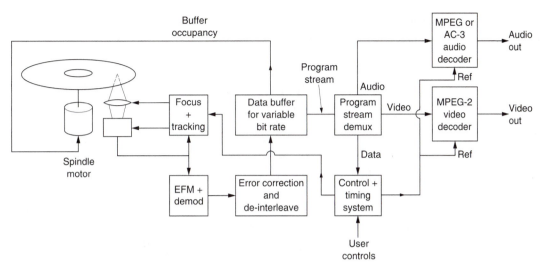

Figure 12.63 A DVD player's essential parts. See text for details.

block diagram is into the control/servo system and the data path. The control system provides the interface between the user and the servo mechanisms, and performs the logical interlocking required for safety and the correct sequence of operation.

The servo systems include any power-operated loading drawer and chucking mechanism, the spindle-drive servo, and the focus and tracking servos already described for CD.

The data path consists of the data separator, the de-interleaving and error-correction process followed by a RAM buffer which supplies the MPEG decoders.

The data separator converts the EFM+ read-out waveform into data. Following data separation the error-correction and de-interleave processes take place. Because of the interleave system, there are two opportunities for correction, first, using the inner code prior to de-interleaving, and second, using the outer code after de-interleaving. As MPEG data are very sensitive to error the correction performance has to be extremely good.

Following the de-interleave and outer error-correction process an MPEG program stream emerges. Some of the program stream data will be video, some will be audio and this will be routed to the appropriate decoder. It is a fundamental concept of DVD that the bit rate of this program stream is not fixed, but can vary with the difficulty of the program material in order to maintain consistent image quality. Although the bit rate allocated to the audio remains constant, the video bit rate doesn't. The bit rate is changed by changing the linear speed of the disk track. However, there is a complication because the disk uses constant linear velocity rather than constant angular velocity. It is not possible to obtain a particular bit rate with a fixed spindle speed.

The solution is to use a RAM buffer between the transport and the MPEG decoders. The amount of data read from the disk over the long term must equal the amount of data used by the MPEG decoders. The speed of the motor is unimportant. The important factor is that the data rate needed by the decoder is correct, and the system will drive the spindle at whatever speed is necessary so that the buffer neither underflows nor overflows.

The MPEG decoder will convert the compressed elementary streams into PCM video and audio and place the pictures and audio blocks into RAM. These will be read out of RAM whenever the time stamps recorded with each picture or audio block match the state of a time stamp counter. If bidirectional coding is used, the RAM readout sequence will convert the recorded picture sequence back to the real-time sequence. The time stamp counter is derived from a crystal oscillator in the player which is divided down to provide the 90 kHz time stamp clock.

As a result the frame rate at which the disk was mastered will be replicated as the pictures are read from RAM. Once a picture buffer is read out, this will trigger the decoder to decode another picture. It will read data from the buffer until this has been completed and thus indirectly influence the disk speed.

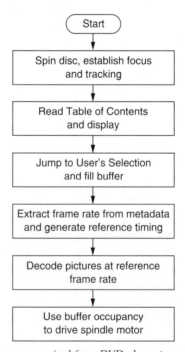

Figure 12.64 Simple processes required for a DVD player to operate.

Owing to the use of constant linear velocity, the disk speed will be wrong if the pickup is suddenly made to jump to a different radius using manual search controls. This may force the data separator out of lock, or cause a buffer overflow and the decoder may freeze briefly until this has been remedied.

Although machines vary in detail, the flowchart of Figure 12.64 shows the logic flow of a simple player, from start being pressed to pictures and sound emerging. At the beginning, the emphasis is on bringing the various servos into operation. Towards the end, the disk subcode is read in order to locate the beginning of the first section of the program material.

References

1. Bouwhuis, G. *et al.*, *Principles of Optical Disc Systems*, Bristol: Adam Hilger (1985)
2. Zernike, F., Beugungstheorie des schneidenverfahrens und seiner verbesserten form, der phasenkontrastmethode. *Physica*, **1**, 689 (1934)
3. Mee, C.D. and Daniel, E.D. (eds) *Magnetic Recording*, Vol. III, New York: McGraw-Hill (1987)
4. Connell, G.A.N., Measurement of the magneto-optical constants of reactive metals. *Appl. Opt.*, **22**, 3155 (1983)
5. Goldberg, N., A high density magneto-optic memory. *IEEE Trans. Magn.*, **MAG-3**, 605 (1967)
6. Various authors, *Philips Tech. Rev.*, **40**, 149–180 (1982).
7. German Patent No. 2,208,379
8. Various authors, Video long-play systems. *Appl. Opt.*, **17**, 1993–2036 (1978)
9. Airy, G.B., *Trans. Camb. Phil. Soc.*, **5**, 283 (1835)
10. Ray, S.F., *Applied Photographic Optics*, Oxford: Focal Press (1988)
11. Maréchal, A., *Rev. d'Optique*, **26**, 257 (1947)
12. Hopkins, H.H., Diffraction theory of laser read-out systems for optical video discs. *J. Opt. Soc. Am.*, **69**, 4 (1979)
13. Bouwhuis *et al.*, *op. cit.*, Chapter 2.
14. Pasman, J.H.T., Optical diffraction methods for analysis and control of pit geometry on optical discs. *J. Audio Eng. Soc.*, **41**, 19–31 (1993)
15. Verkaik, W., Compact Disc (CD) mastering – an industrial process. in *Digital Audio*, edited by B.A. Blesser, B. Locanthi and T.G. Stockham Jr, New York: Audio Engineering Society, 189–195 (1983)
16. Miyaoka, S., Manufacturing technology of the Compact Disc. In *Digital Audio, op. cit.*, 196–201
17. Redlich, H. and Joschko, G., CD direct metal mastering technology: a step toward a more efficient manufacturing process for Compact Discs. *J. Audio Eng. Soc.*, **35**, 130–137 (1987)
18. Ogawa, H., and Schouhamer Immink, K.A., EFM – the modulation system for the Compact Disc digital audio system. In *Digital Audio, op. cit.*, 117–124
19. Schouhamer Immink, K.A. and Gross, U., Optimization of low-frequency properties of eight-to-fourteen modulation. *Radio Electron. Eng.*, **53**, 63–66 (1983)
20. Peek, J.B.H., Communications aspects of the Compact Disc digital audio system. *IEEE Commun. Mag.*, **23**, 7–15 (1985)
21. Vries, L.B. *et al.*, The digital Compact Disc – modulation and error correction. Presented at the 67th Audio Engineering Society Convention (New York, 1980), Preprint 1674
22. Vries, L.B. and Odaka, K., CIRC – the error correcting code for the Compact Disc digital audio system. In *Digital Audio, op. cit.*, 178–186

13

Sound quality considerations

Sound reproduction can only achieve the highest quality if every step is carried out to an adequate standard. This requires a knowledge of what the standards are and the means to test whether they are being met, as well as some understanding of psychology and a degree of tolerance!

13.1 Introduction

In principle, quality can be lost almost anywhere in the audio chain whether by poor interconnections, the correct use of poorly designed equipment or incorrect use of good equipment. The best results will only be obtained when good equipment is used and connected correctly and frequently tested or monitored.

Figure 13.1 shows a representative digital audio system which contains all the components needed to go from original sound to reproduced sound. Like any chain, the system is only as good as its weakest link. It should be clear that a significant number of analog stages remain, especially in the area of the transducers. There is nothing fundamentally wrong with analog equipment, provided that it is engineered to the appropriate quality level. It is regrettable that the enormous advance in signal performance which digital techniques have brought to recording, processing and delivery have largely not been paralleled by advances in subjective or objective testing.

It ought to be possible, even straightforward, to monitor every stage of Figure 13.1 to see if its performance is adequate. If adequate stages are retained and inadequate stages are improved, the entire system can be improved. In practice this is not as easy as it seems. As will be seen in this chapter, the audio industry has not established standards by which the

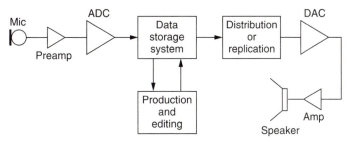

Figure 13.1 The major processes in a digital audio system. Quality can be lost anywhere and can only be assessed by considering every stage.

entire system of Figure 13.1 can be objectively measured in a way which relates to what it will sound like. The measurements techniques which exist are incomplete and the criteria to be met are not established.

Figure 13.2 shows that the human hearing system operates in three domains; time, frequency and space. In real life we know where a sound source is, when it made a sound and the timbral character of the sound. If a high-quality reproduction is to be achieved, it is necessary to test the equipment in all these domains to ensure that it meets or exceeds the accuracy of the human ear. The frequency domain is reasonably well served, but the time and space domains are still suffering serious neglect.

The ultimate criterion for sound reproduction is that the human ear is fooled by the overall system into thinking it has heard the real thing. It follows that even to approach that ideal, the properties of human hearing have to form the basis for design judgements of every part of the audio chain. It is generally assumed that the more accurate some aspect of a reproduction system is, the more realistic it will sound, but this doesn't

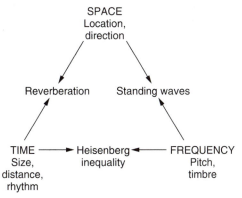

Figure 13.2 The hearing system is part of the mind's mechanism for modelling its surroundings and it works in time, space and frequency. The traditional fixation with the frequency domain leads to neglect of or damage to the other two domains.

follow. In practice this assumption is only true if the accuracy is less than the accuracy of the hearing system. Once some aspect of an audio system exceeds the accuracy of the hearing system, further improvement is not only unnecessary, but it diverts effort away from other areas where it could be more useful.

Figure 13.3(a) shows some criteria by which audio accuracy can be assessed. In each area, it should be possible to measure the requirements of the listener using whatever units are appropriate in order to create a multi-criterion threshold. Figure 13.3(b) shows that these criteria must be met equally despite the sum of all degradations in every stage of the audio chain from microphone to speaker. Figure 13.3(c) shows that a balance must be reached in two dimensions so that each criterion is given equal attention in every stage through which the signal passes. This gives the best value for money whatever standard is achieved. Figure (d) shows a more common approach which is where the excess quality of some parts of the system is wasted because other weak parts dominate the

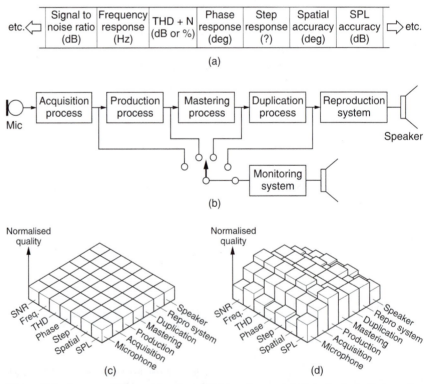

Figure 13.3 Some quality considerations. At (a) an incomplete list of criteria that must be met for accurate reproduction. At (b) in a real-world system, the criteria of (a) must be met in all the components in the chain. The monitoring system should be at least as good as the best anticipated consumer reproduction system. (c) an ideal system has ideally equal performance in all stages. Real system at (d) has better performance in signal processing than in transducers. Excess performance is a waste of money.

impression of the listener. The most commonly found error is that the electronic aspects of audio systems are overspecified whilst the transducers are underspecified.

13.2 Information capacity

The descriptions of time, frequency and space carried in an audio system represent information, and an an analysis of audio system quality falls within the scope of information theory. In the digital domain, the information rate is fixed by the wordlength and the sampling rate. The wordlength determines how many different conditions can be described by each sample. For example a sixteen-bit sample can have 65 536 different values.

In order to reach this performance, every item in the chain needs to have the same information capacity. Recent ADCs and DACs approach this performance, but transducers in general and loudspeakers in particular don't.

Any audio device, analog or digital, can be modelled as an information channel of finite capacity whose equivalent bit rate can be measured or calculated. This can also be done with the loudspeaker. This equivalent bit rate relates to the realism which the speaker can achieve.

When the speaker information capacity is limited, the presence of an earlier restriction in the signal being monitored may go unheard and it may erroneously be assumed that the signal is ideal when in fact it is not.

13.3 Loudspeaker problems

The use of poor loudspeakers simply enables other poor audio devices to enter service. When most loudspeakers have such poor information capacity, how can they be used to assess this capacity in earlier components in the audio chain? ADCs, DACs, pre- and power amplifiers and compression codecs can only meaningfully be assessed on speakers of adequate information capacity. It also follows that the definition of a high-quality speaker is one which readily reveals compression artifacts.

Non-ideal loudspeakers act like compressors in that the distortions, delayed resonances and delayed reradiation they create conceal or mask information in the original audio signal. If a real compressor is tested with non-ideal loudspeakers certain deficiencies of the compressor will not be heard. Others, notably the late Michael Gerzon, have correctly suggested that compression artifacts which are inaudible in mono may be audible in stereo. The spatial compression of non-ideal stereo loudspeakers conceals real spatial compression artifacts.

The ear is a lossy device because it exhibits masking. Not all the presented sound is sensed. If a lossy loudspeaker is designed to a high standard, the losses may be contained to areas which are masked by the ear and then that loudspeaker would be judged transparent. Douglas Self has introduced the term 'blameless' for a device whose imperfections are undetectable; an approach which commands respect. However, the majority of legacy loudspeakers are not in this category. Audible defects are intoduced into the reproduced sound in frequency, time and spatial domains, giving the loudspeaker a kind of character which is best described as a signature or footprint.

13.4   Subjective and objective testing

An audio waveform is simply a voltage changing with time within a limited spectrum, and a digital audio waveform is simply a number changing its value at the sampling rate. As a result any error introduced by a given device can in principle be extracted and analysed. If this approach is used during the design process the performance of any unit can be refined until the error is inaudible.

Naturally the determination of what is meant by inaudible has to be done carefully using valid psychoacoustic experiments. Using such subjective results it is possible to design objective tests which will measure different aspects of the signal to determine if it is acceptable. With care and precision loudspeakers it is equally valid to use listening tests where the reproduced sound is compared with that of other audio systems or of live performances. In fact the combination of careful listening tests with objective technical measurements is the only way to achieve outstanding results. The reason is that the ear can be thought of as making all tests simultaneously. If any aspect of the system has been overlooked the ear will detect it. Then objective tests can be made to find the deficiency the ear detected and remedy it.

If two pieces of equipment consistently measure the same but sound different, the measurement technique must be inadequate. In general, the audio industry survives on inadequate measurement in the misguided belief that the ear has some mysterious power to detect things that can never be measured.

A further difficulty in practice is that the ideal combination of subjective and objective testing is not achieved as often as might be thought. Unfortunately the audio industry represents one of the few remaining opportunities to find employment without qualifications. Given the combination of advanced technologies and marginal technical knowledge, it should be no surprise that the audio industry periodically produces theories which are at variance with scientific knowledge.

People tend to be divided into two camps where audio quality is concerned.

Subjectivists are those who listen at length to both live and recorded music and can detect quite small defects in audio systems. Unfortunately most subjectivists have little technical knowledge and are quite unable to convert their detection of a problem into a solution. Although their perception of a problem may be genuine, their hypothesis or proposed solution may require the laws of physics to be altered. A particular problem with subjective testing is the avoidance of bias. Technical knowledge is essential to understand the importance of experimental design. Good experimental design is important to ensure that only the parameter to be investigated changes so that any difference in the result can only be due to the change. If something else is unwittingly changed the experiment is void. It is also important to avoiding bias. Statistical analysis is essential to determine the *significance* of the results, i.e. the degree of confidence that the results are not due to chance alone. As most subjectivists lack such technical knowledge it is hardly surprising that a lot of deeply held convictions about audio are simply due to unwitting bias where repeatable results are simply not obtained.

A classic example of bias is the understandable tendency of the enthusiast who has just spent a lot of money on a device which has no audible effect whatsoever to 'hear' an improvement.

Another problem with subjectivism is caused by those who don't regularly listen to live music. They become imprinted on the equipment they normally use and subconsciously regard it as 'correct'. Any other equipment to which they are exposed will automatically be judged incorrect, even if it is technically superior. On the introduction of FM radio, with 15 kHz bandwidth, broadcasters received complaints that the audio was too shrill or bright. In comparison with the 7 kHz of AM radio it was!

The author loaned an experimental loudspeaker with a ruler-flat frequency response to an experienced sound engineer. It was returned with the complaint that the response had a peak. The engineer even estimated the frequency of the peak. It was precisely at the crossover frequency of the speakers he normally uses, which have a notorious dip in power response.

It is extremely difficult to make progress when experienced people acting in good faith make statements which are completely incorrect, but exposure to the audio industry will show that this is surprisingly common.

Objectivists are those who make a series of measurements and then pronounce a system to have no audible defect. They frequently have little experience of live performance or critical listening. One of the most frequent mistakes made by objectivists is to assume that because a piece of equipment passes a given test under certain conditions then it is ideal. This is simply untrue for several reasons. The criteria by which the equipment is considered to pass or fail may be incorrect or inappropriate. The equipment might fail other tests or the same test under other conditions. In some cases the tests which equipment might fail have yet to be designed.

Not surprisingly, the same piece of equipment can be received quite differently by the two camps. The introduction of transistor audio amplifiers caused an unfortunate and audible step backwards in audio quality. The problem was that vacuum-tube amplifiers were generally operated in Class A and had low distortion except when delivering maximum power. Consequently valve amplifiers were tested for distortion at maximum power and in a triumph of tradition over reason transistor amplifiers were initially tested in the same way. However, transistor amplifiers generally work in Class B and produce more distortion at very low levels due to crossover between output devices. Naturally this distortion was not detectable on a high-power distortion test. Early transistor amplifiers sounded dreadful at low level and it is astonishing that this was not detected before they reached the market when the subjectivists rightly gave them a hard time.

Whilst the objectivists looked foolish over the crossover distortion issue, subjectivists have no reason to crow with their fetishes for gold-plated AC power plugs, special feet for equipment, exotic cables and mysterious substances to be applied to Compact Discs.

The only solution to the subjectivist/objectivist schism is to arrange them in pairs and bang their heads together.

13.5   Objective testing

Objective testing consists of making measurements which indicate the accuracy to which audio signals are being reproduced. When the measurements are standardized and repeatable they do form a basis for comparison even if *per se* they do not give a complete picture of what a device-under-test (DUT) or system will sound like.

There is only one symptom of quality loss which is where the reproduced waveform differs from the original. For convenience the differences are often categorized.

Any error in an audio waveform can be considered as an unwanted signal which has been linearly added to the wanted signal. Figure 13.4 shows that there are only two classes of waveform error. The first is where the error is not a function of the audio signal, but results from some uncorrelated process. This is the definition of noise. The second is where the error is a direct function of the audio signal which is the definition of distortion.

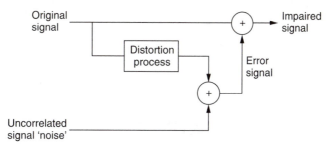

Figure 13.4 All waveform errors can be broken down into two classes: those due to distortions of the signal and those due to an unwanted additional signal.

Noise can be broken into categories according to its characteristics. Noise due to thermal and electronic effects in components or analog tape hiss has essentially stationary statistics and forms a constant background which is subjectively benign in comparison with most other errors. Noise can be periodic or impulsive. Power frequency-related hum is often rich in harmonics. Interference due to electrical switching or lightning is generally impulsive. Crosstalk from other signals is also noise in that it does not correlate with the wanted signal. An exception is crosstalk between the signals in stereo or surround systems.

Distortion is a signal-dependent error and has two main categories. Non-linear distortion arises because the transfer function is not straight. The result in analog parts of audio systems is harmonic distortion where new frequencies which are integer multiples of the signal frequency are added, changing the spectrum of the signal. In digital parts of systems non-linearity can also result in anharmonic distortion because harmonics above half the sampling rate will alias. Non-linear distortions are subjectively the least acceptable as the resulting harmonics are frequently not masked, especially on pure tones.

Linear distortion is a signal-dependent error in which different frequencies propagate at different speeds due to lack of phase linearity. In complex signals this has the effect of altering the waveform but without changing the spectrum. As no harmonics are produced, this form of distortion is more benign than non-linear distortion. Whilst a completely linear phase system is ideal, the finite phase accuracy of the ear means that in practice a *minimum phase system* is probably good enough. Minimum phase implies that phase error changes smoothly and continuously over the audio band without any sudden discontinuities. Loudspeakers often fail to achieve minimum phase, with legacy techniques such as reflex tuning and passive crossovers being particularly unsatisfactory.

Figure 13.5(a) shows that the signal-to-noise ratio (SNR) is the ratio in dB between the largest amplitude undistorted signal the DUT can pass

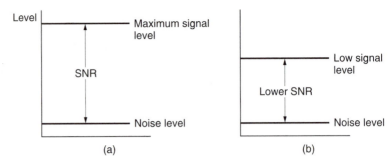

Figure 13.5 (a) SNR is the ratio of the largest undistorted signal to the noise floor. (b) Maximum SNR is not reached if the signal never exercises the highest possible levels.

and the amplitude of the output with no input whatsoever, which is presumed to be due to noise. The spectrum of the noise is as important as the level. When audio signals are present, auditory masking occurs which reduces the audibility of the noise. Consequently the noise floor is most significant during extremely quiet passages or in pauses. Under these conditions the threshold of hearing is extremely dependent on frequency.

A measurement which more closely resembles the effect of noise on the listener is the use of an A-weighting filter prior to the noise level measurement stage. The result is then measured in dB(A).

Just because a DUT measures a given number of dB of SNR does not guarantee that SNR will be obtained in use. The measured SNR is only obtained when the DUT is used with signals of the correct level. Figure 13.5(b) shows that if the DUT is installed in a system where the input level is too low, the SNR of the output will be impaired. Consequently in any system where this is likely to happen the SNR of the equipment must exceed the required output SNR by the amount by which the input level is too low. This is the reason why quality mixing consoles offer apparently phenomenal SNRs.

Often the operating level is deliberately set low to provide *headroom* so that occasional transients are undistorted. The art of quality audio production lies in setting the level to the best compromise between elevating the noise floor and increasing the occurrences of clipping.

Another area in which conventional SNR measurements are meaningless is where gain ranging or floating point coding is used. With no signal the system switches to a different gain range and an apparently high SNR is measured which does not correspond to the subjective result.

The frequency response of audio equipment is measured by the system shown in Figure 13.6(a). The same level is input at a range of frequencies and the output level is measured. The end of the frequency range is

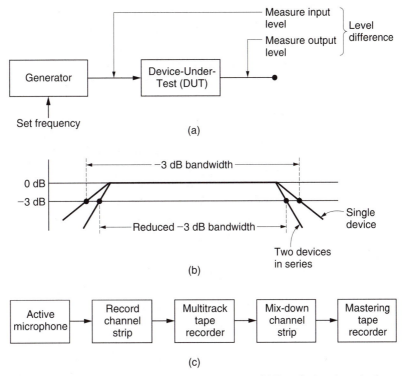

Figure 13.6 (a) Frequency response measuring system. (b) Two devices in series have narrower bandwidth. (c) Typical record production signal chain, showing potential severity of generation loss.

considered to have been reached where the output level has fallen by 3 dB with respect to the maximum level. The correct way of expressing this measurement is:

Frequency response: −3 dB, 20 Hz – 20 kHz

or similar. If the level limit is omitted, as it often is, the figures are meaningless.

There is a seemingly endless debate about how much bandwidth is necessary in analog audio and what sampling rate is needed in digital audio. There is no one right answer as will be seen. In analog systems, there is generation loss. Figure 13.6(b) shows that two identical DUTs in series will cause a loss of 6 dB at the frequency limits. Depending on the shape of the roll-off, the −3 dB limit will be reached over a narrower frequency range. Conversely if the original bandwidth is to be maintained, then the −3 dB range of each DUT must be wider.

In analog production systems, the number of different devices an audio signal must pass through is quite large. Figure 13.6(c) shows the signal chain of a multi-track-produced vinyl disk. The number of stages involved mean that if each stage has a seemingly respectable –3 dB, 20 Hz – 20 kHz response, the overall result will be dreadful with a phenomenal rate of roll-off at the band edge. The only solution is that each item in the chain has to have wider bandwidth making it considerably overspecified in a single-generation application.

Another factor is phase response. At the –3 dB point of a DUT, if the response is limited by a first-order filtering effect, the phase will have shifted by 45°. Clearly in a multi-stage system these phase shifts will add. An eight-stage system, not at all unlikely, will give a complete phase rotation as the band-edge is approached. The phase error begins a long way before the –3 dB point, preventing the system from displaying even a minimum phase characteristic.

Consequently in complex analog audio systems each stage must be enormously overspecified in order to give good results after generation loss. It is not unknown for mixing consoles to respond down to 6 Hz in order to prevent loss of minimum phase in the audible band. Obviously such an extended frequency response on its own is quite inaudible, but when cascaded with other stages, the overall result *will* be audible.

In the digital domain there is no generation loss if the numerical values of the samples are not altered. Consequently digital data can be copied from one tape to another, or to a Compact Disc without any quality loss whatsoever. Simple digital manipulations, such as level control, do not impair the frequency or phase response and, if well engineered, the only loss will be a slight increase in the noise floor. Consequently digital systems do not need overspecified bandwidth. The bandwidth needs only to be sufficient for the application because there is nothing to impair it. However, those brought up on the analog tradition of overspecified bandwidth find this hard to believe.

In a digital system the bandwidth and phase response is defined at the anti-aliasing filter in the ADC. Early anti-aliasing filters had such dreadful phase response that the aliasing might have been preferable, but this has been overcome in modern oversampled convertors which can be highly phase-linear.

One simple but good way of checking frequency response and phase linearity is squarewave testing. A squarewave contains indefinite harmonics of a known amplitude and phase relationship. If a squarewave is input to an audio DUT, the characteristics can be assessed almost instantly. Figure 13.7 shows some of the defects which can be isolated. (a) shows inadequate low-frequency response causing the horizontal sections of the waveform to droop. (b) shows poor high-frequency response in conjunction with poor phase linearity which turns the edges into exponential curves. (c) shows a phase-linear system of finite bandwidth, e.g. a good anti-aliasing filter. Note that the transitions are symmetrical

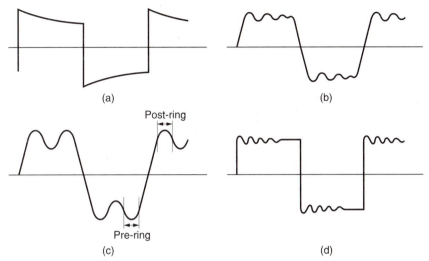

Figure 13.7 Squarewave testing gives a good insight into audio performance. (a) Poor low-frequency response causing droop. (b) Poor high-frequency response with poor phase linearity. (c) Phase-linear bandwidth limited system. (d) Asymmetrical ringing shows lack of phase linearity.

with equal pre- and post-ringing. This is one definition of phase linearity. (d) shows a system with wide bandwidth but poor HF phase response. Note the asymmetrical ringing.

One still hears from time to time that squarewave testing is illogical because squarewaves never occur in real life. The explanation is simple. Few would argue that any sine wave should come out of a DUT with the same amplitude and no phase shift. A linear audio system ought to be able to pass any number of superimposed signals simultaneously. A squarewave is simply one combination of such superimposed sine waves. Consequently if an audio system cannot pass a squarewave as shown in Figure 13.7(c) then it will cause a problem with real audio.

As linearity is extremely important in audio, relevant objective linearity testing is vital. Real sound consists of many different contributions from different sources which all superimpose in the sound waveform reaching the ear. If an audio system is not capable of carrying an indefinite number of superimposed sounds without interaction then it will cause an audible impairment. Interaction between a single waveform and a non-linear system causes distortion. Interaction between waveforms in a non-linear system is called *intermodulation distortion* (*IMD*) whose origin is shown in Figure 13.8(a). As the transfer function is not straight, the low-frequency signal has the effect of moving the high-frequency signal to parts of the transfer function where the slope differs. This results in the high frequency being amplitude modulated by the low. The amplitude

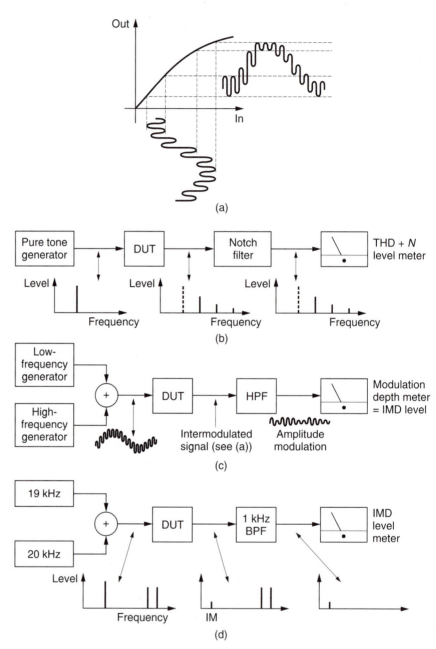

Figure 13.8 (a) Non-linear transfer function allows large low-frequency signal to modulate small high-frequency signal. (b) Harmonic distortion test using notch to remove fundamental measures THD + N. (c) Intermodulation test measuring depth of modulation of high frequency by low-frequency signal. (d) Intermodulation test using a pair of tones and measuring the level of the difference frequency.

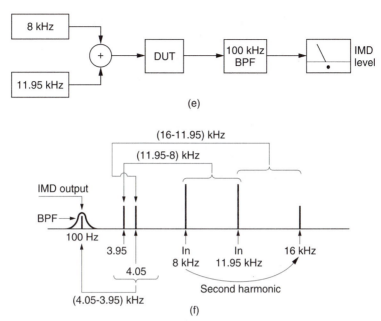

Figure 13.8 (e) Complex test using intermodulation between harmonics. (f) Spectrum analyser output with two-tone input.

modulation will also produce sidebands. Clearly a system which is perfectly linear will be free of both types of distortion.

Figure 13.8(b) shows a simple harmonic distortion test. A low-distortion oscillator is used to inject a clean sine wave into the DUT. The output passes through a switchable sharp 'notch' filter which rejects the fundamental frequency. With the filter bypassed, an AC voltmeter is calibrated to 100 per cent. With the filter in circuit, any remaining output must be harmonic distortion or noise and the measured voltage is expressed as a percentage of the calibration voltage. The correct way of expressing the result is as follows:

$$\text{THD} + N \text{ at } 1\,\text{kHz} = 0.1\%$$

or similar. A stringent test would repeat the test at a range of frequencies. There is not much point in conducting THD + N tests at high frequencies as the harmonics will be beyond the audible range.

The THD + N measurement is not particularly useful in high-quality audio because it only measures the amount of distortion and tells nothing about its distribution. A vacuum-tube amplifier displaying 0.1 per cent distortion may sound very good indeed, whereas a transistor amplifier or a DAC with the same distortion figure will sound awful. This is because the vacuum-tube amplifier produces primarily low-order harmonics, which some listeners even find pleasing, whereas transistor amplifiers and digital devices can produce higher-order harmonics, which are unpleasant.

Very high-quality audio equipment has a characteristic whereby the equipment itself seems to recede, leaving only the sound. This will only happen when the entire reproduction chain is sufficiently free of any characteristic footprint which it impresses on the sound. The term *resolution* is used to describe this ability. Audio equipment which offers high resolution appears to be free of distortion products which are simply not measured by THD + N tests. Figure 13.9(a) shows the spectrum of a sine wave emerging from an ideal audio system. Figure 13.9(b) shows the spectrum of a low-resolution signal. Note the presence of sidebands around the original signal. Analog tape displays this characteristic, which is known as modulation noise. The Compact Cassette is notorious for a high level of modulation noise and poor resolution.

Analog circuitry can have the same characteristic. Figure 13.9(c) shows that signal or power amplifiers using negative feedback to linearize the

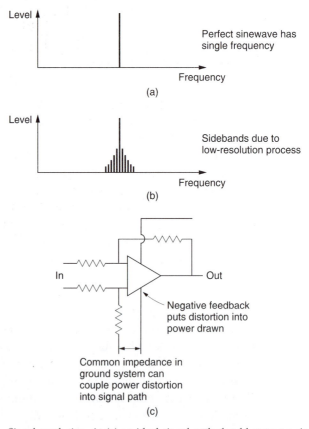

Figure 13.9 Signal resolution. At (a) an ideal signal path should output a sinewave as a single frequency. Low-resolution signal paths produce sidebands around the signal (b). In (c), amplifiers which linearize their output using negative feedback do so by forcing the distortion into the power rails. Poor layout can couple the power rails back into the signal path.

output do so by pushing the non-linearities into the power rails. If the circuit board layout is poor, the power rail distortion can enter the signal path through common impedances.

In Chapter 4 the subject of sampling clock jitter was introduced. The effect of sampling clock jitter is to produce sidebands of the kind shown in Figure 13.9(b). This will be considered further in section 13.7.

Few people realize that loudspeakers can also display the effect of Figure 13.9(b). Traditional loudspeakers use ferrite magnets for economy. However, ferrite is an insulator and so there is nothing to stop the magnetic field moving within the magnet due to the Newtonian reaction to the coil drive force. Figure 13.10(a) shows that when the coil is quiescent, the lines of flux are symmetrically disposed about the coil turns, but when coil current flows, as in (b), the flux must be distorted in order to create a thrust. In magnetic materials the magnetic field can only move by the motion of domain walls and this is a non-linear process. The result in a conductive magnet is flux modulation and Barkhausen noise. The flux modulation and noise make the transfer function of the transducer non-linear and result in intermodulation.

The author did not initially believe the results of estimates of the magnitude of the problem, which showed that ferrite magnets cannot reach the sixteen-bit resolution of CD. Consequently two designs of tweeter were built, identical except for the magnet. The one with the conductive neodymium magnet has audibly higher resolution, approaching that of an electrostatic transducer, which, of course, has no magnet at all.

Given the damaging effect on realism caused by sidebands, a more meaningful approach than THD + N testing is to test for intermodulation distortion. There are a number of ways of conducting intermodulation tests, and, of course, anyone who understands the process can design a test from first principles. An early method standardized by the SMPTE was to exploit the amplitude modulation effect in two widely spaced frequencies, typically 70 Hz with 7 kHz added at one tenth the amplitude. 70 Hz is chosen because it is above the hum due to 50 or 60 Hz power. Figure 13.8(c) shows that the measurement is made by passing the 7 kHz region through a bandpass filter and recovering the depth of amplitude modulation with a demodulator.

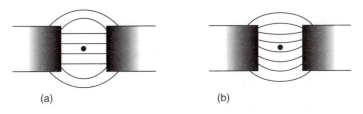

(a)   (b)

Figure 13.10   (a) Flux in gap with no coil current. (b) Distortion of flux needed to create drive force.

A more stringent test for the creation of sidebands is where two high frequencies with a small frequency difference are linearly added and used as the input to the DUT. For example, 19 and 20 kHz will produce a 1 kHz difference or beat frequency if there is non-linearity as shown in Figure 13.8(d). With such a large difference between the input and the beat frequency it is easy to produce a 1 kHz bandpass filter which rejects the inputs. The filter output is a measure of IMD. With a suitable generator, the input frequencies can be swept or stepped with a constant 1 kHz spacing.

More advanced tests exploit not just the beats between the fundamentals, but also those involving the harmonics. In one proposal, shown in Figure 13.8(e) input tones of 8 and 11.95 kHz are used. The fundamentals produce a beat of 3.95 kHz, but the second harmonic of 8 kHz is 16 kHz which produces a beat of 4.05 kHz. This will intermodulate with 3.95 kHz to produce a 100 Hz component which can be measured. Clearly this test will only work in 60 Hz power regions, and the exact frequencies would need modifying in 50 Hz regions.

If a precise spectrum analyser is available, all the above tests can be performed simultaneously. Figure 13.8(f) shows the results of a spectrum analysis of a DUT supplied with two test tones. Clearly it is possible to test with three simultaneous tones or more. Some audio devices, particularly power amplifiers, ADCs and DACs, are relatively benign under steady-state testing with simple signals, but reveal their true colours with a more complex input.

13.6   Subjective testing

Subjective testing can only be carried out by placing the device under test in series with an existing sound-reproduction system. Unless the DUT is itself a loudspeaker, the testing will only be as stringent as the loudspeakers in the system allow. Unfortunately the great majority of loudspeakers do not reach the standard required for meaningful subjective testing of units placed in series and consequently the majority of such tests are of questionable value.

If useful subjective testing is to be carried out, it is necessary to use the most accurate loudspeakers available and to test the loudspeakers themselves before using them as any kind of reference. Whilst simple tests such as on-axis frequency response give an idea of the performance of a loudspeaker, the majority produce so much distortion and modulation noise that the figures are not even published.

Digital audio systems potentially have high signal resolution, but subjective testing of high-performance convertors is very difficult because of loudspeaker limitations. Consequently it is important to find listening tests which will meaningfully assess loudspeakers, especially for linearity and resolution, whilst eliminating other variables as much as possible. Linearity and resolution are essential to allow superimposition of an indefinite number of sounds in a stereo image.

Non-linearity in stereo has the effect of creating intermodulated sound objects which are in a different place in the image from the genuine sounds. Consequently the requirements for stereo are more stringent than for mono. This can be used for speaker testing. One stringent test is to listen to a high-quality stereo recording in which multi-tracking has been used to superimpose a number of takes of a musician or vocalist playing/singing the same material. The use of multi-tracking reduces the effect of intermodulation at the microphone and ADC as these handle only one source at a time. The use of a panpot eliminates any effects due to inadequate directivity in a stereo microphone.

It should be possible to hear how many simultaneous sources are present, i.e. whether the recording is double, triple or quadruple tracked, and it should be possible to concentrate on each source to the exclusion of the others.

It should also be possible to pan each version of a multi-tracked recording to a slightly different place in the stereo image and individually identify each source even when the spacing is very small. Poor loudspeakers smear the width of an image because of diffraction and fail this test.

In another intermodulation test it is necessary to find a good-quality recording in which a vocalist sings solo at some point, and at another point is accompanied by a powerful low-frequency instrument such as a pipe organ or a bass guitar. There should be no change in the imaging or timbre of the vocal whether or not the LF is present.

Another stringent test of linearity is to listen to a recording made on a coincident stereo microphone of a spatially complex source such as a choir. It should be possible to identify the location of each chorister and to concentrate on the voice of each. The music of Tallis is highly suitable. Massed strings are another useful test, with the end of Barber's *Adagio for Strings* being particularly revealing. Coincident stereo recordings should also be able to reproduce depth. It should be possible to resolve two instruments or vocalists one directly behind the other at different distances from the microphone. Loudspeakers which cannot pass these tests are not suitable for subjective quality testing.

When a pair of reference-grade loudspeakers has been found which will demonstrate all the above effects, it will be possible to make meaningful comparisons between devices such as microphones, consoles, analog recorders, ADCs and DACs. Quality variations between the analog outputs of different CD players or DAT machines will be readily apparent. Those which pay the most attention to convertor clock jitter are generally found to be preferable. Very expensive high-end CD players are often disappointing because these units concentrate on one aspect of performance and neglect others.

One myth which has taken a long time to be revealed is the belief that a low-grade loudspeaker should be used in the production process so that an indication of how the mix will sound on mediocre consumer equipment will be obtained. If an average loudspeaker could be obtained this would be possible. Unfortunately the main defect of a poor loudspeaker is that it stamps its own characteristic footprint on the audio. These footprints vary so much that there is no such thing as an average poor loudspeaker and people who make decisions on cheap loudspeakers are taking serious risks. It is a simple fact that an audio production can never be better than the monitor loudspeakers used and the author's extensive collection of defective CDs indicates that good monitoring is rare.

13.7 Digital audio quality

In theory the quality of a digital audio system comprising an ideal ADC followed by an ideal DAC is determined at the ADC. This will be true if the digital signal path is sufficiently well engineered that no numerical errors occur, which is the case with most reasonably maintained equipment. The ADC parameters such as the sampling rate, the wordlength and any noise shaping used put limits on the quality which can be achieved. Conversely, the DAC itself may be transparent, because it only converts data whose quality are already determined back to the analog domain. In other words, the ideal ADC determines the system quality and the ideal DAC does not make things any worse.

In practice both ADCs and DACs can fall short of the ideal, but with modern convertor components and attention to detail the theoretical limits can be approached very closely and at reasonable cost. Shortcomings may be the result of an inadequacy in an individual component such as a convertor chip, or due to incorporating a high-quality component into a poorly though-out system. Poor system design or implementation can destroy the performance of a convertor. Whilst oversampling is a powerful technique for realizing high-quality convertors, its use depends on digital interpolators and decimators whose quality affects the overall conversion quality.[1] Interpolators and decimators with erroneous arithmetic or inadequate filtering performance have been known.

ADCs and DACs have the same transfer function, since they are only distinguished by the direction of operation, and therefore the same terminology can be used to classify the possible shortcomings of both. Figure 13.11 shows the transfer functions resulting from the main types of convertor error.

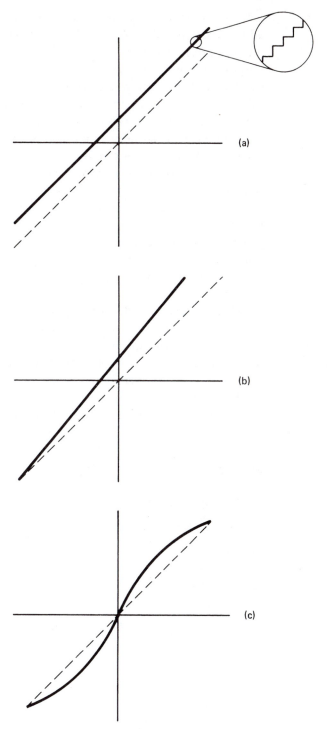

Figure 13.11 Main convertor errors (solid line) compared with perfect transfer line. These graphs hold for ADCs and DACs, and the axes are interchangeable. If one is chosen to be analog, the other will be digital.

Figure 13.11(a) shows *offset error*. A constant appears to have been added to the digital signal. This has no effect on sound quality, unless the offset is gross, when the symptom would be premature clipping. DAC offset is of little consequence, but ADC offset is undesirable since it can cause an audible thump if an edit is made between two signals having different offsets. Offset error is sometimes cancelled by digitally averaging the convertor output and feeding it back to the analog input as a small control voltage. Alternatively, a digital high-pass filter can be used.

Figure 13.11(b) shows *gain error*. The slope of the transfer function is incorrect. Since convertors are referred to one end of the range, gain error causes an offset error. The gain stability is probably the least important factor in a digital audio convertor, since ears, meters and gain controls are logarithmic.

Figure 13.11(c) shows *integral linearity*. This is the deviation of the dithered transfer function from a straight line. It has exactly the same significance and consequences as linearity in analog circuits, since if it is inadequate, distortion will be caused.

Differential non-linearity is the amount by which adjacent quantizing intervals differ in size. This is usually expressed as a fraction of a quantizing interval. In audio applications the differential non-linearity requirement is quite stringent. This is because with properly employed dither, an ideal system can remain linear under low-level signal conditions. When low levels are present, only a few quantizing intervals are in use. If these change in size, clearly waveform distortion will take

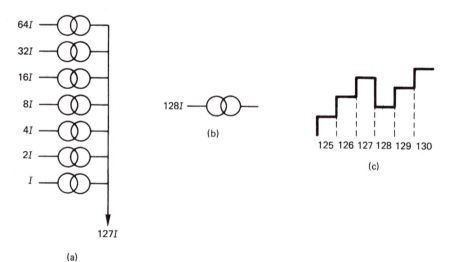

(a)

Figure 13.12 (a) Equivalent circuit of DAC with 127_{10} input. (b) DAC with 128_{10} input. On a major overflow, here from 27_{10} to 128_{10}, one current source ($128I$) must be precisely I greater than the sum of all the lower-order sources. If $128I$ is too small, the result shown in (c) will occur. This is non-monotonicity.

place despite the dither. Enhancing the subjective quality of convertors using noise shaping will only serve to reveal such shortcomings.

Figure 13.12 shows that *monotonicity* is a special case of differential non-linearity. Non-monotonicity means that the output does not increase for an increase in input. Figure 13.12(a) shows that in a DAC with a convertor input code of 01111111 (127 decimal), the seven low-order current sources of the convertor will be on. The next code is 10000000 (128 decimal), shown in Figure 13.12(b), where only the eighth current source is operating. If the current it supplies is in error on the low side, the analog output for 128 may be less than that for 127 as shown in Figure 13.12(c). In an ADC non-monotonicity can result in missing codes. This means that certain binary combinations within the range cannot be generated by any analog voltage. If a device has better than $1/2Q$ linearity it must be monotonic. It is difficult for a one-bit convertor to be non-monotonic.

Absolute accuracy is the difference between actual and ideal output for a given input. For audio it is rather less important than linearity. For example, if all the current sources in a convertor have good thermal tracking, linearity will be maintained, even though the absolute accuracy drifts.

Clocks which are free of jitter are a critical requirement in convertors as was shown in Chapter 4. The effects of clock jitter are proportional to the slewing rate of the audio signal rather than depending on the sampling rate, and as a result oversampling convertors are no more prone to jitter than conventional convertors.[2] Clock jitter is a form of frequency modulation with a small modulation index. Sinusoidal jitter produces sidebands which may be audible. Random jitter raises the noise floor which is more benign but still undesirable. As clock jitter produces artifacts proportional to the audio slew rate, it is quite easy to detect. A spectrum analyser is connected to the convertor output and a low audio frequency signal in input. The test is then repeated with a high audio frequency. If the noise floor changes, there is clock jitter. If the noise floor rises but remains substantially flat, the jitter is random. If there are discrete frequencies in the spectrum, the jitter is periodic. The spacing of the discrete frequencies from the input frequency will reveal the frequencies in the jitter.

Aliasing of audio frequencies is not generally a problem, especially if oversampling is used. However, the nature of aliasing is such that it works in the frequency domain only and translates frequencies to new values without changing amplitudes. Aliasing can occur for any frequency above one half the sampling rate. The frequency to which it aliases will be the difference frequency between the input and the nearest sampling rate multiple. Thus in a non-oversampling convertor, *all* frequencies above half the sampling rate alias into the audio band. This includes radio frequencies which have entered via audio or power wiring or directly. RF can leap-frog an analog anti-aliasing filter capacitively. Thus good RF screening is necessary around ADCs, and the manner of entry of cables to equipment must be such that RF energy on them is directed to earth. Recent legislation regarding the sensitivity of equipment to electromagnetic interference can only be beneficial in this respect.

Oversampling convertors respond to RF on the input in a different manner. Although all frequencies above half the sampling rate are folded into the baseband, only those which fold into the audio band will be audible. Thus an unscreened oversampling convertor will be sensitive to RF energy on the input at frequencies within ±20 kHz of integer multiples of the sampling rate. Fortunately interference from the digital circuitry at exactly the sampling rate will alias to DC and be inaudible.

Convertors are also sensitive to unwanted signals superimposed on the references. In fact the multiplicative nature of a convertor means that reference noise amplitude modulates the audio to create sidebands. Power supply ripple on the reference due to inadequate regulation or decoupling causes sidebands 50, 60, 100 or 120 Hz away from the audio frequencies, yet does not raise the noise floor when the input is quiescent. The multiplicative effect reveals how to test for it. Once more a spectrum analyser is connected to the convertor output. An audio frequency tone is input, and the level is changed. If the noise floor changes with the input signal level, there is reference noise. RF interference on a convertor reference is more insidious, particularly in the case of noise-shaped devices. Noise-shaped convertors operate with signals which must contain a great deal of high-frequency noise just beyond the audio band. RF on the reference amplitude modulates this noise and the sidebands can enter the audio band, raising the noise floor or causing discrete tones depending on the nature of the pickup.

Noise-shaped convertors are particularly sensitive to a signal of half the sampling rate on the reference. When a small DC offset is present on the input, the bit density at the quantizer must change slightly from 50 per cent. This results in idle patterns whose spectrum may contain discrete frequencies. Ordinarily these are designed to occur near half the sampling rate so that they are beyond the audio band. In the presence of half-sampling-rate interference on the reference, these tones may be demodulated into the audio band.

Although the faithful reproduction of the audio band is the goal, the nature of sampling is such that convertor design must respect EMC and RF engineering principles if quality is not to be lost. Clean references, analog inputs, outputs and clocks are all required, despite the potential radiation from digital circuitry within the equipment and uncontrolled electromagnetic interference outside.

Unwanted signals may be induced directly by ground currents, or indirectly by capacitive or magnetic coupling. It is essential practice to separate grounds for analog and digital circuitry, connecting them in one place only. Capacitive coupling uses stray capacitance between the signal source and point where the interference is picked up. Increasing the distance or conductive screening helps. Coupling is proportional to frequency and the impedance of the receiving point. Lowering the impedance at the interfering frequency will reduce the pickup. If this is done with capacitors to ground, it need not reduce the impedance at the frequency of wanted signals.

Magnetic or inductive coupling relies upon a magnetic field due to the source current flow inducing voltages in a loop. Reduction in inductive coupling requires the size of any loops to be minimized. Digital circuitry should always have ground planes in which return currents for the logic signals can flow. At high frequency, return currents flow in the ground plane directly below the signal tracks and this minimizes the area of the transmiting loop. Similarly, ground planes in the analog circuitry minimize the receiving loop whilst having no effect on baseband audio. A further weapon against inductive coupling is to use ground fill between all traces on the circuit board. Ground fill will act like a shorted turn to alternating magnetic fields. Ferrous screening material will also reduce inductive coupling as well as capacitive coupling.

The reference of a convertor should be decoupled to ground as near to the integrated circuit as possible. This does not prevent inductive coupling to the lead frame and the wire to the chip itself. In the future convertors with on-chip references may be developed to overcome this problem.

In summary, spectral analysis of convertors gives a useful insight into design weaknesses. If the noise floor is affected by the signal level, reference noise is a possibility. If the noise floor is affected by signal frequency, clock jitter is likely. Should the noise floor be unaffected by both, the noise may be inherent in the signal or in analog circuit stages.

One interesting technique which has been developed recently for ADC testing is a statistical analysis of the frequency of occurrence of the various code values in data. If, for example, a full-scale sine wave is input to an ADC having a frequency which is asynchronous to the sampling rate, the probability of a particular code occurring in the output of an ideal convertor is a function only of the slew rate of the signal. At the peaks of the sine wave the slew rate is small and the codes there are more frequent. Near the zero crossing the slew rate is high and the probability is lower. Near the zero crossing, the probability of codes being created is nearly equal. However, if one quantizing interval is slightly larger than its neighbours, the signal will take longer to cross it and the probability of that code appearing will rise. Conversely, if the interval is smaller the probability will fall. By collecting a large quantity of data from a test and displaying the statistics it is possible to measure differential non-linearity to phenomenal accuracy.

This technique has been used to show that oversampled noise-shaped convertors are virtually free of differential non-linearity because of the averaging in the decimation process.

In practice signals used are not restricted to high-level sine waves. A low-level sine wave will only exercise a small number of codes near the audiologically sensitive centre of the quantizing range. However, it may be better to use a combination of three sine waves which exercises the whole range. As the test method reveals differences in probability of occurrence of individual codes, it can be used with program material. In this case the exact distribution of code probabilities is not important. Instead it is important that the probability distribution should be smooth. As Figure 13.13 shows, spikes in the distribution indicate an unusually high or low probability for certain codes.

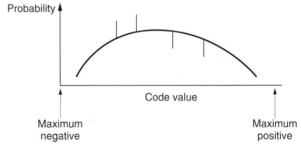

Figure 13.13 An ideal ADC should show a smooth probability curve of code values on real audio signals. The exact curve will vary with signal content. However, a convertor having uneven quantizing intervals will show positive spikes in the probability if the intervals are too wide and negative spikes if they are too narrow.

In an analysis of code probability on a number of commercially available CDs, a disturbing number of those tested had surprising characteristics such as missing codes, particularly in older recordings. Single missing codes can be due to an imperfect ADC, but in some cases there were a large number of missing codes spaced evenly apart. This could only be due to primitive gain controls applied in the digital domain without proper redithering. This may have been required with under-modulated master tapes which would be digitally amplified prior to the cutting process in order to play back at a reasonable level.

Statistical code analysis is quite useful to the professional audio engineer as it can be applied using actual program material at any point in the production and mastering process. Following an ADC it will reveal convertor non-linearities, but used later, it will reveal DSP shortcomings. It is highly likely that a correlation will be found between subjectively perceived resolution and the results of tests of this kind.

13.8   Use of high sampling rates

From time to time there have been proposals to raise the sampling rates used in digital audio to, for example, 96 kHz and even 192 kHz. These are invariably backed with the results of experiments and demonstrations 'proving' that the sampling rate makes a difference. The reality is different because careful study of these experiments show them to be flawed.

The most famous bandwidth myth is the fact that it is possible to hear the difference between a 10 kHz sine wave and a 10 kHz squarewave when the difference between the two starts with the third harmonic at 30 kHz. If we could only hear 20 kHz it wouldn't be audible, but it is. The reason is non-linearity in practical equipment. Even if the signal system, speakers and air were perfectly linear, so we could inject a 10 kHz acoustic squarewave into the ear, we would still hear the difference because the ear itself isn't linear. The ossicles in the ear are a mechanical lever system and have limitations. Consequently hearing a difference between a 10 kHz sine wave and a squarewave doesn't prove anything about the bandwidth of human hearing.

Another classic myth is the experiment shown in Figure 13.14. This takes a 96 kHz source and allows monitoring of the source directly or through a decimation to 48 kHz followed by an interpolation back to 96 kHz. This is supposed to test whether the difference between 48 kHz and 96 kHz is audible. Actually all it proves is that the more stages a signal goes through, the worse it gets. The decimation and interpolation processes will cause degradation of the signal within the 20 kHz band, so it's no wonder that the subjects prefer the 96 kHz path.

What the experiment should have done was to replicate the degradation of the decimate/interpolate path. In other words the elevated noise floor due to two arithmetic roundoffs in series and the ripple and phase response of the filters should also have been present in the 96 kHz path. Tests should have been made to ensure that both paths were identical in all respects up to 20 kHz. Unfortunately they weren't and the conclusions are meaningless because the experiment was not properly designed so that the only difference between the two stimuli was the bandwidth.

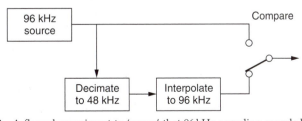

Figure 13.14   A flawed experiment to 'prove' that 96 kHz sampling sounds better.

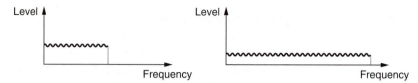

Figure 13.15 A better explanation for the apparent improvement in audio quality using very high sampling rates. If the resolution of the system is inadequate, raising the sampling rate will lower the noise floor.

When a properly designed experiment is performed, in which 96 kHz source material is or is not bandwidth limited to 20 kHz by a psycho-acoustically adequate low-pass filter, it is impossible to hear any difference.[3]

Some ADC manufacturers have demonstrated better sound quality from convertors running at 96 kHz. However, this does not prove that 96 kHz is necessary. Figure 13.15 shows that if an oversampling convertor has suboptimal decimating filters it will suffer from a modulation noise floor which damages resolution. If the sampling rate is doubled, the noise will be spread over twice the bandwidth so the level will be reduced. This is why the high sampling rate convertor sounds better. However, the same sound quality could be obtained by improving the design of the 48 kHz convertor.

13.9   Digital audio interface quality

There are three parameters of interest when conveying audio down a digital interface such as AES/EBU or SPDIF, and these have quite different importance depending on the application. The parameters are:

(a) The jitter tolerance of the serial FM data separator.
(b) The jitter tolerance of the audio samples at the point of conversion back to analog.
(c) The timing accuracy of the serial signal with respect to other signals.

A digital interface is designed to convey discrete numerical values from one place to another. If those samples are correctly received with no numerical change, the interface is perfect. The serial interface carries clocking information, in the form of the transitions of the FM channel code and the sync patterns and this information is designed to enable the data separator to determine the correct data values in the presence of jitter. It was shown in Chapter 8 that the jitter window of the FM code is half a data bit period in the absence of noise. This becomes a quarter of a data bit when the eye opening has reached the minimum allowable in the professional specification as can be seen from Figure 8.2. If jitter is within this limit, which corresponds to about 80 nanoseconds pk–pk, the serial digital interface perfectly reproduces the sample data, irrespective of the intended use of the data. The data separator of an AES/EBU receiver requires a phase-locked loop in order to decode the serial message. This phase-locked loop will have jitter of its own, particularly if it is a digital phase-locked loop where the phase steps are of finite size. Digital phase-locked loops are easier to implement along with other logic in integrated circuits. There is no point in making the jitter of the phase-locked loop vanishingly small as the jitter tolerance of the channel code will absorb it. In fact the digital phase-locked loop is simpler to implement and locks up quicker if it has larger phase steps and therefore more jitter.

This has no effect on the ability of the interface to convey discrete values, and if the data transfer is simply an input to a digital recorder no other parameter is of consequence as the data values will be faithfully recorded. However, it is a further requirement in some applications that a sampling clock for a convertor is derived from a serial interface signal.

It was shown in Chapter 4 that the jitter tolerance of convertor clocks is measured in picoseconds. Thus a phase-locked loop in the FM data separator of a serial receiver chip is quite unable to drive a convertor directly as the jitter it contains will be as much as a thousand times too great. Nevertheless this is exactly how a great many consumer outboard DACs are built, regardless of price. The consequence of this poor engineering is that the serial interface is no longer truly digital. Analog variations in the interface waveform cause variations in the convertor clock jitter and thus variations in the reproduced sound quality. Different types of digital cable 'sound' different and journalists claim that digital optical interfaces are 'sonically superior' to electrical interfaces. The

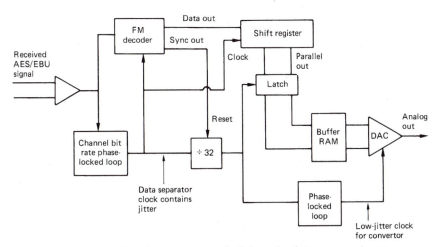

Figure 13.16 In an outboard convertor, the clock from the data separator is not sufficiently free of jitter and additional clock regeneration is necessary to drive the DAC.

digital outputs of some CD players 'sound' better than others and so on. In fact source and cable substitution is an excellent test of outboard convertor quality. A properly engineered outboard convertor will sound the same despite changes in CD player, cable type and length and despite changing from electrical to optical input because it accepts only data from the serial signal and regenerates its own clock. Audible differences simply mean the convertor is of poor design and should be rejected.

Figure 13.16 shows how a convertor should be configured. The serial data separator has its own phase-locked loop which is less jittery than the serial waveform and so recovers the audio data. The serial data are presented to a shift register which is read in parallel to a latch when an entire sample is present by a clock edge from the data separator. The data separator has done its job of correctly returning a sample value to parallel format. A quite separate phase-locked loop with extremely high damping and low jitter is used to regenerate the sampling clock. This may use a crystal oscillator or it may be a number of loops in series to increase the order of the jitter filtering. In the professional channel status, bit 5 of byte 0 indicates whether the source is locked or unlocked. This bit can be used to change the damping factor of the phase-locked loop or to switch from a crystal to a varicap oscillator. When the source is unlocked, perhaps because a recorder is in varispeed, the capture range of the phase-locked loop can be widened and the increased jitter is accepted. When the source is locked, the capture range is reduced and the jitter is rejected.

The third timing criterion is only relevant when more than one signal is involved as it affects the ability of, for example, a mixer to combine two inputs.

In order to decide which criterion is most important, the following may be helpful. A single signal which is involved in a data transfer to a recording medium is concerned only with eye pattern jitter as this affects the data reliability.

A signal which is to be converted to analog is concerned primarily with the jitter at the convertor clock. Signals which are to be mixed are concerned with the eye pattern jitter and the relative timing. If the mix is to be monitored, all three parameters become important.

A better way of ensuring low jitter conversion to analog in digital audio reproducers is to generate a master clock from a crystal adjacent to the convertor, and then to slave the transport to produce data at the same rate. This approach is shown in Figure 13.17. Memory buffering between transport and convertor then ensures that the transport jitter is eliminated. Whilst this can also be done with a remote convertor, it does then require a reference clock to be sent to the transport as in Figure 13.18 so that data can be sent at the correct rate. Unfortunately most consumer CD and DAT players have no reference input and this approach cannot be used. Consumer remote DACs then must regenerate a clock from the

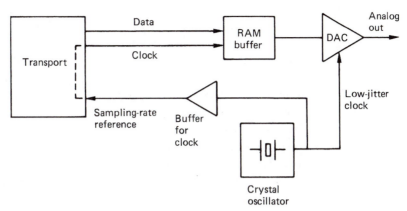

Figure 13.17 Low convertor jitter is easier if the transport is slaved to the crystal oscillator which drives the convertor. This is readily achieved in a single box device.

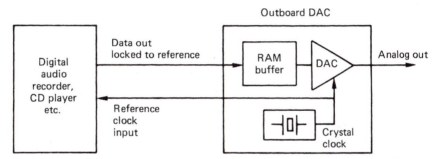

Figure 13.18 If a separate transport is to be slaved to a crystal oscillator in the DAC a reference signal must be sent. Not many consumer transports have external clock inputs.

player and seldom do it accurately enough. In fact it is a myth that outboard convertors are necessary for high quality. For the same production cost, a properly engineered inboard convertor adhering to the quality criteria of Chapter 4 can sound better than a two-box system. The real benefit of an outboard convertor is that in theory it allows several digital sources to be replayed for the cost of one convertor. In practice few consumer devices are available with only a digital output, and the convertors are duplicated in each device.

13.10 Compression in stereo

The human hearing mechanism has an ability to concentrate on one of many simultaneous sound sources based on direction. The brain appears to be able to insert a controllable time delay in the nerve signals from one ear with respect to the other so that when sound arrives from a given direction the nerve signals from both ears are coherent causing the binaural threshold of hearing to be 3–6 dB better than monaural at around 4 kHz. Sounds arriving from other directions are incoherent and are heard less well. This is known as attentional selectivity.

Human hearing can also locate a number of different sound sources simultaneously presented by constantly comparing excitation patterns from the two ears with different delays. Strong correlation will be found where the delay corresponds to the interaural delay for a given source. This delay-varying mechanism will take time and the ear is slow to react to changes in source direction. Oscillating sources can only be tracked up to 2–3 Hz and the ability to locate bursts of noise improves with burst duration up to about 700 ms. Location accuracy is finite.

Stereophonic and surround systems should allow attentional selectivity to function such that the listener can concentrate on specific sound sources in a reproduced image with the same facility as in the original sound.

We live in a reverberant world which is filled with sound reflections. If we could separately distinguish every different reflection in a reverberant room we would hear a confusing cacaphony. In practice we hear very well in reverberant surroundings, far better than microphones can, because of the transform nature of the ear and the way in which the brain processes nerve signals. Because the ear has finite frequency discrimination ability in the form of critical bands, it must also have finite temporal discrimination.

This is good news for the loudspeaker designer because the ear has finite accuracy in frequency, time and spatial domains. This means that a blameless loudspeaker is not just a concept, it could be made real by the application of sufficient rigour.

When two or more versions of a sound arrive at the ear, provided they fall within a time span of about 30 ms, they will not be treated as separate sounds, but will be fused into one sound. Only when the time separation reaches 50–60 ms do the delayed sounds appear as echoes from different directions. As we have evolved to function in reverberant surroundings, most reflections do not impair our ability to locate the source of a sound.

Clearly the first version of a transient sound to reach the ears must be the one which has travelled by the shortest path and this must be the direct sound rather than a reflection. Consequently the ear has evolved to attribute source direction from the time of arrival difference at the two ears of the first version of a transient.

Versions which may arrive from elsewhere simply add to the perceived loudness but do not change the perceived location of the source unless they arrive within the inter-aural delay of about 700 μs when the precedence effect breaks down and the perceived direction can be pulled away from that of the first arriving source by an increase in level. This area is known as the time-intensity trading region. Once the maximum inter-aural delay is exceeded, the hearing mechanism knows that the time difference must be due to reverberation and the trading ceases to change with level.

Unfortunately reflections with delays of the order of 700 µs are exactly what are provided by the legacy rectangular loudspeaker with sharp corners. These reflections are due to acoustic impedance changes and if we could see sound we would double up with mirth at how ineptly the sound is being radiated. Effectively the spatial information in the audio signals is being convolved with the spatial footprint of the speaker. This has the effect of defocusing the image. Now the effect can be measured.

Intensity stereo, the type obtained with coincident mikes or panpots, works purely by amplitude differences at the two loudspeakers. The two signals should be exactly in phase. As both ears hear both speakers the result is that the space between the speakers and the ears turns the intensity differences into time of arrival differences. These give the illusion of virtual sound sources.

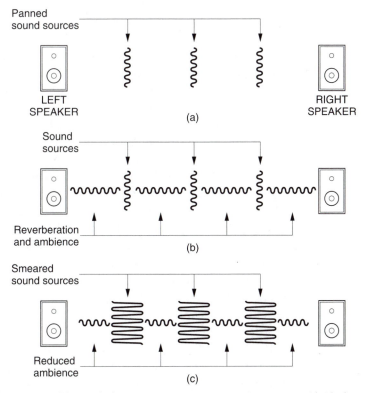

Figure 13.19 At (a) sounds from a pan-pot appear as point sources with ideal speakers. (b) Reverberation and ambience appear between the point sources. (c) Most legacy loudspeakers cause smearing which widens the sound sources and marks the ambience between.

A virtual sound source from a panpot has zero width and on diffraction-free speakers would appear as a virtual point source. Figure 13.19(a) shows how a panpotted dry mix should appear spatially on ideal speakers whereas (b) shows what happens when stereo reverb is added. In fact (b) is also what is obtained with real sources using a coincident pair of mikes. In this case the sources are the real sources and the sound between is reverb/ambience.

Figure 13.19(c) is the result obtained with traditional square box speakers. Note that the point sources have spread so that there are almost no gaps between them, effectively masking the ambience. This represents a lack of spatial fidelity, so we can say that rectangular box loudspeakers cannot accurately reproduce a stereo image, nor can they be used for assessing the amount of reverbertion added to a 'dry' recording. Such speakers cannot meaningfully be used to assess compression codecs.

A compressor works by raising the level of 'noise' in parts of the signal where it is believed to be masked. If this belief is correct, the compression will be inaudible. However, if the codec is tested using a signal path in which there is another masking effect taking place, the results of the test are meaningless. Theoretical analysis and practical measurement that legacy loudspeakers have exactly such a masking process, both temporally and spatially.[4]

If a stereophonic system comprising a variable bit rate codec in series with a pair of speakers is considered to be a communication channel, then it will have a finite information rate in frequency, temporal and spatial domains. If this information rate exceeds the capacity of the human hearing mechanism, it will be deemed transparent. However, in the system mentioned, either the codec or the speakers could be the limiting factor and ordinarily there would be no way to separate the effects.

If a variable bit-rate codec is available, some conclusions can be drawn. Figure 13.20(a) shows what happens as the bit rate is increased with an ideal speaker. The sound quality increases up to the point where the capacity of the ear is reached, after which raising the bit rate appears to have no effect. However, if suboptimal speakers are used, the situation of Figure 13.20(b) arises. Now, as the bit rate is increased, the quality levels off prematurely where the information capacity of the loudspeaker has been reached. As a result simply by varying the bit rate of a coder, it becomes possible to measure the effective bit rate of a pair of loudspeakers.

In subjective compression tests, the configuration of Figure 13.20(c) is used. The listener switches between the uncompressed and compressed versions to see if a difference can be detected. If the speaker of Figure 13.20(b) is used, the experimenter is misled, because it would appear that there is no difference between direct and compressed listening at an artificially low bit rate, whereas in fact the limiting factor is the speaker.

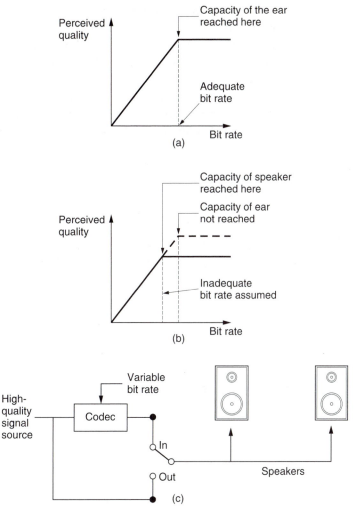

Figure 13.20 (a) If an ideal speaker is used, the quality levels off when the compressor is delivering enough data for the requirement of the ear. However, (b), if an inadequate speaker is used, the quality appears to level off early, giving an incorrect assessment of the bit rate needed. (c) Using compression to test loudspeakers. The better the loudspeaker, the less compression can be used before it becomes audible.

At the point shown in Figure 13.20(b) the masking due to the speaker is equal to the level of artifacts from the coder. At any lower bit rate, compression artifacts will become audible over the footprint of the speaker. The lower the information capacity of the speaker, the lower the bit rate at which the artifacts are audible.

Non-ideal loudspeakers act like bit-rate compressors in that they conceal or mask information in the audio signal. If a real compressor is tested with non-ideal loudspeakers certain deficiencies of the compressor will not be heard and it may erroneously be assumed that the compressor is transparent when in fact it is not. Compression artifacts which are inaudible in mono may be audible in stereo and the spatial compression of non-ideal stereo loudspeakers conceals real spatial compression artifacts.

Precision monitor speakers should be free of reflections in the sub-700 µs trading region so that the imaging actually reveals what is going on spatially. When such speakers are used to assess audio compressors,

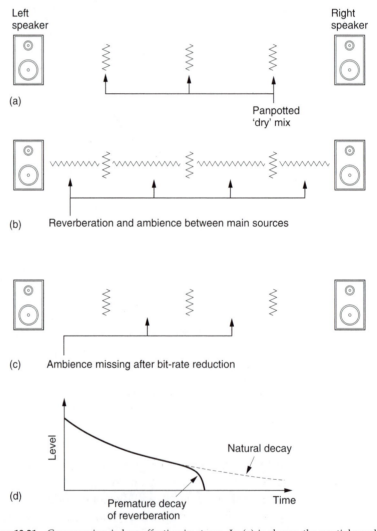

Figure 13.21 Compression is less effective in stereo. In (a) is shown the spatial result of a 'dry' panpotted mix. (b) shows the result after artificial reverberation which can also be obtained in an acoustic recording with coincident mikes. After compression (c) the ambience and reverberation may be reduced or absent. (d) Reverberation may also decay prematurely.

even at high bit rates corresponding to the smallest amount of compression, it is obvious that there is a difference between the original and the compressed result. Figure 13.21 shows graphically what is found. The dominant sound sources are reproduced fairly accurately, but the ambience and reverb between are virtually absent, making the decoded sound much drier than the original.

The effect will be apparent to the same extent with, for example, both MPEG Layer II and Dolby AC-3 coders even though their internal workings are quite different. This is not surprising because both are probably based on the same psychoacoustic masking model. MPEG-3 fares even worse because the bit rate is lower. Transient material has a peculiar effect whereby the ambience will come and go according to the entropy of the dominant source. A percussive note will narrow the sound stage and appear dry but afterwards the reverb level will come back up. All these effects largely disappear when the signals to the speakers are added to make mono, removing the ear's ability to discriminate spatially.

These effects are not subtle and do not require golden ears. The author has successfully demonstrated them to various audiences up to 60 in number in a variety of untreated rooms. Whilst compression may be adequate to deliver post-produced audio to a consumer with mediocre loudspeakers, these results underline that it has no place in a quality production environment. When assessing codecs, loudspeakers having poor diffraction design will conceal artifacts. When mixing for a compressed delivery system, it will be necessary to include the codec in the monitor feeds so that the results can be compensated. Where high-quality stereo is required, either full bit rate PCM or lossless (packing) techniques must be used.

References

1. Lipshitz, S.P. and Vanderkooy, J., Are D/A convertors getting worse? Presented at the 84th Audio Engineering Society Convention (Paris, 1988), Preprint 2586 (D-6)
2. Harris, S., The effects of sampling clock jitter on Nyquist sampling analog to digital convertors and on oversampling delta-sigma ADCs. *J. Audio Eng. Soc.*, **38**, 537–542 (1990)
3. Katz, B., The ultimate listening test. *Audio Media* (April 2000), 116–118
4. Watkinson, J.R., Putting the science back into loudspeakers. Presented at the Multi-channel Audio Summit (London, May 2000)

Index